Stress Analysis
of Fiber-Reinforced
Composite Materials

Stress Analysis of Fiber-Reinforced Composite Materials

Stress Analysis of Fiber-Reinforced Composite Materials

M. W. Hyer

Virginia Polytechnic Institute and State University

Contributions on Fibers, Matrices, Interfaces, and Manufacturing By S. R. White

University of Illinois at Urbana–Champaign

Boston, Massachusetts Burr Ridge, Illinois Dubuque, Iowa
Madison, Wisconsin New York, New York San Francisco, California St. Louis, Missouri

WCB/McGraw-Hill

A Division of The **McGraw·Hill** Companies

Stress Analysis of Fiber-Reinforced Composite Materials

1 2 3 4 5 6 7 8 9 0 DOW/DOW 9 0 9 8 7

ISBN 0-07-016700-1

Editorial director: *Tom Casson*
Senior sponsoring editor: *Debra Riegert*
Marketing manager: *John Wannemacher*
Project manager: *Margaret Rathke*
Production supervisor: *Charlene R. Perez*
Designer: *Michael Warrell*
Compositor: *Techsetters, Inc.*
Typeface: *10/12 Times Roman*
Printer: *R. R. Donnelley & Sons Company*

Library of Congress Cataloging-in-Publication Data

Hyer, M. W.
    Stress analysis of fiber-reinforced composite materials / M.W.
  Hyer : with contributions on fibers, matrices, interfaces, and
  manufacturing by S.R. White
       p.   cm.
    Includes biographical references and index.
    ISBN 0-07-016700-1
    1. Composite materials. 2. Fibrous composites. 3. Laminated
  materials.   I. White, S. R. (Scott R.)   II. Title.
  TA428.9 C6H98   1997   97–2166
  620.1 ' 18—dc21

http://www.mhcollege.com

PREFACE

APPROACH

This book focuses on the mechanics aspects of fiber-reinforced composite materials. By *mechanics* is meant the study of equilibrium, stress, strain, deformation, elastic properties, failure theories, and the linkages between these topics. A significant portion of the book emphasizes the use of mechanics to study the stresses due to applied deformations, loads, and temperature changes. Since interest in fiber-reinforced composite materials stems mainly from their ability to withstand high stress and deformation levels, such an emphasis centers on the important issues.

No prior knowledge of composite materials is assumed. Only the basic concepts introduced in an undergraduate strength-of-materials course are necessary. The book is intended for use at the senior undergraduate or first-year graduate levels in any engineering curriculum designed to explore the behavior and performance of these advanced materials. Mechanical engineers interested in considering composite materials for automobiles or trucks and flywheels for energy storage, civil engineers investigating the application of composite materials to infrastructure, aerospace engineers studying advanced airframe design, and biomedical engineers developing lightweight composite materials for bone replacement and repair will find the book valuable.

A strong feature of the book is the use of a set of examples that is introduced early and then built upon as additional concepts are developed. This set of examples provides continuity to the discussion and allows the reader to evaluate the impact of more complicated issues on the stresses and deformations of fiber-reinforced composite materials as the book progresses. A second strong feature is the reminders of the implications of the various simplifying assumptions used to study the mechanical behavior of fiber-reinforced materials. These reminders are designed so the reader does not misinterpret the theories and results, and is able to evaluate outcomes based on the concepts presented. Many authors do not take the time or space to do this. Another strong feature is the rather substantial coverage of thermal effects in composites, specifically the far-reaching effects of thermally induced deformations and stresses due to residual effects. In addition, with composite materials being envisioned for construction of high-speed civilian airplanes, the coverage of thermal effects is timely. Also, a number of characteristics of fiber-reinforced materials are difficult to include in design and analysis procedures and are often difficult to fully comprehend. Specifically, the elastic couplings inherent in fiber-reinforced materials are often designed around, at some cost in design efficiency, or are assumed to be zero to simplify analysis procedures. This book counters this trend by addressing the topic and by viewing elastic couplings as characteristics that can be used to seek designs not possible with metallic materials.

Because so many subtleties are involved with understanding and effectively using fiber-reinforced composite materials, an in-depth view of a limited number of topics, rather than an overview of many topics, is offered. The reader will be well-versed in the details of important calculations and their impact on results. This book provides enough information so the reader will know what questions to ask and find it easy to proceed to other readings on the subject.

LEARNING AIDS

Many of the chapters include a list of suggested readings. Taken from well-known and readily available archival journals and books, the readings are selected to reinforce the principles presented in this book, expand on the concepts, and provide information on topics not discussed.

The notation used in this book is widely used. The important equations are enclosed in boxes for handy reference. One set of material properties, representing an off-the-shelf intermediate modulus polymer matrix graphite-fiber material, is used throughout. This feature, coupled with the continuing use of a set of examples, provides additional continuity from chapter to chapter. There are assigned exercise sets at the ends of many of the sections which emphasize the fundamentals presented in that section. The exercises, though simple in the early chapters, become more involved as the book progresses. The exercises should be completed as they are assigned, as opposed to doing a large amount of reading and then returning to previous sections to do them.

Computer exercises are also included. A number of steps for studying the response of composite materials are the same from one problem to the next. Programming these steps is recommended, and this activity forms the basis for these computer exercises. In this way it is possible to concentrate on the physics of the results, rather than on the algebra. In fact, the computer programming assignments are such that by the middle of the book the reader will have a computer program that can be used to predict some of the more important responses of composite materials, for example, stresses, strains, and thermal expansion coefficients. More importantly, because the programs are created by the reader, making changes, adapting them to special cases, changing the output, and so on, can be easily done because the reader knows where in the programs to make the changes. In addition, the programs can be used to help complete some of the more complicated assigned exercises. For those interested, several useful programs written in FORTRAN are available on-line from McGraw-Hill.

CONTENTS

Chapter 1 provides a brief overview of the concept of fiber-reinforced materials—why fiber reinforcing can be used to achieve high-performance materials, and how the fiber and the material surrounding it, the matrix, interact. The chapter relies to a large degree on a materials science viewpoint to describe fibers, matrix materials, and fiber sizings. It is important for the mechanician to be aware of the terminology and these basic ideas, particularly to work in an interdisciplinary environment.

Chapter 2 introduces the three-dimensional stress-strain behavior of a composite material that is used as the basis for discussion throughout the text. It is assumed that the fibers and matrix are smeared into a single homogeneous orthotropic material, and the chapter focuses on the response of a small, isolated element of this homogeneous material. The compliance and stiffness matrices are defined and typical material properties for graphite-reinforced and glass-reinforced materials are given. The chapter uses simple examples to emphasize the importance of a three-dimensional stress state. Also discussed is the response of an isolated element of material to a temperature change.

Chapter 3 presents a brief overview of micromechanics. Unit cell models are studied with the aid of finite-element analyses. No finite-element theory is discussed; rather, emphasis is on the stresses within the unit cell as a function of fiber volume fraction. As a contrast to the numerically based finite-element models, the well-known concentric cylinders model, which is based on the theory of elasticity, is briefly introduced. Finally, several rule-of-mixture models are presented. One of the valuable results of Chapter 3 is that simple working expressions are developed to provide estimates of composite elastic and thermal expansion properties from the properties of the fiber and matrix.

As it is one of the most frequently used key assumptions in the analysis of the mechanical behavior of materials, the plane-stress assumption is the sole topic of Chapter 4. The three-dimensional stress-strain behavior of Chapter 2 is simplified to account for the plane-stress assumption, including thermal expansion effects. The consequences of these simplifications are emphasized with numerical examples.

Chapter 4 is coupled with, and leads directly into, Chapter 5, which discusses the plane-stress stress-strain relations in a coordinate system not aligned with the principal material directions—a so-called global, or off-axis, coordinate system. Through simple examples, the response of an element of fiber-reinforced material with its fibers aligned at an angle relative to the coordinate system is described and quantified in detail. Counterpart examples using aluminum to dramatize the unusual response of composite materials are included. The engineering properties of an off-axis element of fiber-reinforced material are defined and the coefficients of mutual influence are introduced.

Chapters 6, 7, and 8 constitute the central theme of the book, namely, the analysis of the response of composite laminates under the assumptions of classical lamination theory. Chapter 6 addresses another key assumption of the analysis of mechanical behavior of materials, the Kirchhoff hypothesis. Its implications are illustrated through a series of examples that are built upon in subsequent chapters. Because the Kirchhoff hypothesis is a kinematic assumption, its impact on the variation of the strains through the thickness of a laminate is first considered. Then, following the previously discussed plane-stress stress-strain relations, the stress variation through the thickness of the laminate is related to the strain variation through the thickness. The strains and stresses in the example problems are considered in detail. From the way the stresses are observed to vary through the thickness of the laminate, the definition of force and moment results seems natural. These quantities are informally defined, and the force and moment resultants for the example problems are computed strictly on physical grounds.

In Chapter 7, the force and moment resultants are formally defined, and as a result, the classical A, B, and D matrices are introduced. The calculation of the elements in each of these three matrices is detailed and simplifications of the matrices for special but important laminates are presented. Considerable discussion is provided to interpret the physical meaning of the various off-diagonal terms in the A and D matrices, and the meaning of the B matrix.

Chapter 8 presents other examples of laminate response based on the assumptions of classical lamination theory. Force-based counterpart examples to the kinematics-based examples of Chapter 6 are presented, and the results contrasted. The emphasis of basic principles such as differentiating between specifying kinematics and specifying forces or moments is felt to be one of the strong features of this book.

Chapters 9 and 10 introduce the topic of failure of fiber-reinforced composite materials. Chapter 9 introduces the maximum stress failure criterion. The various failure modes associated with fiber-reinforced composite materials are introduced, and to illustrate the criterion, and as realistic examples of results, several cases involving both simple and combined loading of laminates are presented. Failure loads and failure modes are predicted.

Chapter 10 introduces the Tsai-Wu failure criterion. This criterion was chosen because it considers interaction of the stress components as a possible cause of failure, in contrast to the maximum stress criterion, which assumes failure is due to only one stress component. The Tsai-Wu criterion is used to predict the failure loads and modes for the same cases considered using the maximum stress criterion in the previous chapter. The results of the two criteria are contrasted.

Chapter 11 discusses the effect of a temperature change on the response of laminates. This is a major chapter in that it revisits classical lamination theory and the failure theories, but with the inclusion of thermal effects. Several of the example failure problems solved earlier without thermal effects are resolved with thermal effects included. The thermal effect considered is the cooling from the consolidation temperature of the laminate, which is a residual thermal effect due to curing of the composite material.

Chapter 12 considers a topic that is often forgotten in books dealing with classical lamination theory—through-thickness strain effects. Through-thickness Poisson's ratios and coefficients of thermal expansion in the thickness direction are defined and illustrated in this brief but important chapter.

In Chapter 13 the mechanics of composite plates are introduced. The plate is assumed to obey all the assumptions of classical lamination theory, and the differential equations governing the plate are derived from equilibrium considerations. The boundary conditions that must be enforced along the edges are presented. For demonstrating some of the important effects found with composite plates, several semi-infinite plate problems that can be solved in closed form are considered. The influence of various boundary conditions, and the coupling of boundary conditions with the B matrix for unsymmetric laminates are discussed. Finally, a finite, square, uniformly loaded, laminated plate is studied by using a series solution. With both the semi-infinite and square plates, stresses are also discussed. The point of the chapter is not a comprehensive view of composite plates; rather, the chapter serves as a bridge between the study of the response of the small, isolated elements of composite materials in the preceding chapters and the study of structural elements—the plate is one of the simplest.

Chapter 14, the appendix, provides an overview of the manufacturing of composites. The fabrication and processing phases are considered and a pictorial essay of the hand lay-up technique is provided so the important steps can be emphasized. The roles of release agents, peel plies, breather plies, and other specialty materials are described. The processing phase, with emphasis on autoclave processing, is briefly considered. Other forms of manufacturing, such as filament winding and pultrusion, are identified.

SUPPLEMENTS

An answer book for all the exercises is available for the instructor. Separate FORTRAN programs are available to compute the following: the engineering properties E_1, v_{12}, E_2, and G_{12} from fiber and matrix properties; the components of the 6×6 compliance and stiffness matrices and the components of the 3×3 transformed reduced compliance and stiffness matrices, for the various stress-strain relations, from the engineering properties; laminate stiffness and thermal expansion properties from layer engineering properties and fiber orientations; and the stresses and strains through the thickness of a laminate due to given deformations, given force and moment resultants, or a given temperature change.

ACKNOWLEDGMENTS

Before closing, I must mention several individuals and organizations that were of enormous help in developing this book. Scott R. White, of the Department of Aeronautical and Astronautical Engineering at the University of Illinois at Urbana-Champaign, contributed the material in Chapter 1 on fibers, matrices, and interfaces, and the material in Chapter 14 on manufacturing. I am indebted to him for providing text and figures for these portions of the book. The contributions are vital to the book. At Virginia Tech Tamara Knott was an invaluable assistant for many of the calculations in Chapter 3. Caroline Scruggs also helped by assembling the calculations in graphical form. Beverly Williams provided considerable assistance with the text. And finally, the savior of many days and near disasters: Paula Davis. Paula worked the formatting and editing of this book into her very busy schedule. Without her this book probably would not exist. In addition, the very stimulating, positive, and always competitive atmosphere of the Department of Engineering Science and Mechanics at Virginia Tech had much to do with the composition of this book. Similarly, the research conducted within the NASA–Virginia Tech Composites Program has provided a foundation for this book. The students in the program, fellow faculty members in the program, and colleagues at the NASA–Langley Research Center all provided important stimuli.

Of immense help in refining the book were the reviewers who provided thousands of helpful comments. Their time and effort are greatly appreciated. The reviewers were: Donald F. Adams, University of Wyoming; Erian A. Armanios, Georgia Institute of Technology; Leif A. Carlsson, Florida Atlantic University; Charles E. Bakis, Pennsylvania State University; Sherrill B. Biggers, Jr., Clemson University; Paul F. Eastman, Brigham Young University; Harold E. Gascoigne, California Polytechnic State University at San Luis Obispo; Yousef Haik, Florida A&M University–Florida State University; Dahsin Liu, Michigan State University; Ozden O. Ochoa, Texas A&M University; Nicholas J. Salamon, Pennsylvania State University; Richard A. Scott, University of Michigan; David L. Sikarskie, Michigan Technological University; C. T. Sun, Purdue University; and Anthony J. Vizzini, University of Maryland.

Finally, the time, patience, and very helpful comments of the two editors involved in this text, John Corrigan and Debra Riegert, are a significant contribution. Editors must balance technical, editorial, production, and marketing issues. Their work is not easy and it is very much appreciated.

M. W. Hyer

TABLE OF CONTENTS

Stress Analysis
of Fiber-Reinforced
Composite Materials

CHAPTER 1

Fiber-Reinforced Composite Materials

1.1
BACKGROUND AND BRIEF OVERVIEW

Studies of strong, stiff, lightweight materials for application to diverse structures—from aircraft, spacecraft, submarines, and surface ships to robot components, prosthetic devices, civil structures, automobiles, trucks, and rail vehicles—focus on using fiber-reinforced materials. But why are fibers getting so much attention? To answer this question, one must know something about material science and, in particular, about the molecular bonds that hold matter together. Even though this book is devoted to the *mechanics* of composite materials, because the fiber form is such a central concern, we will begin this chapter with a presentation of the basic concepts in material science associated with fiber reinforcement. Figure 1.1 illustrates a basic unit of material. At the corners of the unit are atoms or molecules held in place by interatomic bonds. The figure shows that this basic unit of material has directionally dependent properties. To varying degrees, many common materials, including iron, copper, nickel, carbon, and boron, have directionally dependent properties, with the directional dependence being due to the strengths of the interatomic and intermolecular bonds. The bonds are stronger in some directions than in others, and the material unit is very stiff and exhibits considerable strength in the direction of the stronger bond. Unfortunately, the favorable properties found in one direction usually come at the expense of the properties in the other directions. In directions perpendicular to the stiff and strong direction, the material is much softer and weaker. Other properties like electrical conductivity and heat conduction can also be directionally dependent.

When material is processed and fabricated in bulk form (e.g., in manufacturing steel billets, you start with molten steel and pour it into a billet form), the units of material are more or less randomly oriented within the volume of material (see Figure 1.2). As a result of random orientation, the bulk material has the same properties in all directions. Generally, the properties of the bulk material reflect the poorer

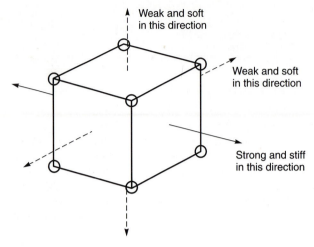

FIGURE 1.1
Basic unit of material

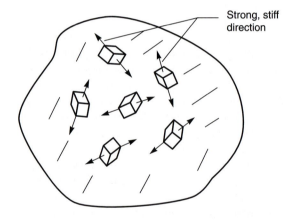

FIGURE 1.2
Several basic units of material oriented randomly in bulk volume of material

properties of the unit in Figure 1.1; the properties of the bulk material are determined more or less by the properties of the weakest link of the unit. Having the same properties in all directions is referred to as having isotropic behavior (the prefix *iso* means equal). Thus, the tensile strength of a specimen cut from a larger piece of steel or aluminum is independent of the direction the tensile specimen is machined from. (In a strict sense, this is not true; for instance, rolling metal into a sheet alters the microstructure of the material and causes it to have different properties in the roll direction than in directions perpendicular to the roll.)

If you can process the material in a manner that permits you to align the strong and stiff directions of all the basic material units, you can preserve some of the high strength and stiffness properties of a single unit, thereby countering isotropic behavior. Processing so that the strong and stiff directions of all the units align results in a long, thin element of material referred to as a whisker (see Figure 1.3). In reality,

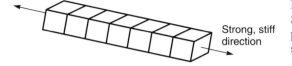

Strong, stiff direction

FIGURE 1.3
Several basic units of material processed so their strong and stiff directions coincide

whiskers are quite small and, compared to bulk material, have very high strength and stiffness in the lengthwise direction. A typical whisker may be $1 - 10 \times 10^{-6}$ m in diameter and 10–100 times as long. With care in processing, the properties of microscopic whiskers can be very close to the ideal properties of a single unit. Unfortunately, attempting to lengthen a whisker by adding more basic units can cause imperfections and impurities. For crystal-like whiskers such as graphite, the imperfection may be a dislocation or the absence of a carbon atom. For polymeric fibers such as Kevlar®, foreign matter is a possibility. These deviations from ideal form significantly influence the strength and stiffness of the whisker and become the weak link in the material. Imperfections also cause an increase in both thermal and electrical resistance, leading to degraded conductive properties. Because there is no way to align them, whiskers used for reinforcement generally have a random orientation within the material, and so the reinforced material still has isotropic properties.

As more basic material units are added to the length of a whisker, it becomes what is called a fiber. Fibers have significant length, so they can be easily aligned in one direction to provide selective reinforcement within another material. A fiber contains many units in its length, and thus it has a greater chance of having an imperfection. As a result, a fiber is weaker than a whisker. The strength properties of fibers are a random variable. Testing 10,000 fibers would result in 10,000 different strength values. Obviously, you can use such raw strength data to form a probability distribution of the strength. The average strength and the scatter (variance) of the strength become important quantities in describing the properties of a fiber. Because of the random nature of fiber strengths, many researchers employ probabilistic methods to study the strengths of composite materials.

Some fibers, especially most forms of graphite fibers, have such small diameters that they are more conveniently handled in groups. A group of fibers is called a fiber tow, or simply a tow, and consists of from hundreds to hundreds of thousands of fibers. A tow is like a rope made of fibers, though generally without the complicated interlocking twist and braid patterns.

1.2
UTILIZING THE STRENGTH OF FIBERS

Once you can produce strong, stiff material in the form of fibers, there immediately comes the challenge of how to make use of the material: The fibers need to be aligned with the load, the load needs to be transferred into the fibers, and the fibers need to remain aligned under the load. Equally important, the fibers need to be in a format that makes them readily available and easy to use. Figure 1.4 illustrates the basic mechanism used to transfer a tensile load, F, into a fiber tow. Essentially, the fiber tow is embedded in, surrounded by, and bonded to another material; see Figure 1.4(a). The material, which is usually softer and weaker, not only surrounds the tow, but it also penetrates the tow and surrounds every fiber in the tow. The embedding material is referred to as the matrix material, or matrix. The matrix transmits the load to the fiber through a shear stress, τ. This can be seen in the section view, Figure 1.4(b), along the length of the fiber. Due to F, a shear stress acts on the outer surface of the fiber. This stress, in turn, causes a tensile stress, σ, within the fiber. Near the ends of the fiber the shear stress on the surface of the fiber is high and the tensile stress within the fiber is low. As indicated in Figure 1.4(c), as the distance from the end of the fiber increases, the shear stress decreases in magnitude and the tensile stress increases. After some length, sometimes referred to as the characteristic distance, the shear stress becomes very small and the tensile stress reaches a maximum value. This tensile condition continues along the length of the fiber. Generally, this characteristic distance is many times smaller than the length of the fiber.

For loading the fiber in compression (see Figure 1.5), the issue of fiber buckling must be addressed. If the shear stress on the end of the fiber in Figure 1.4(c) is

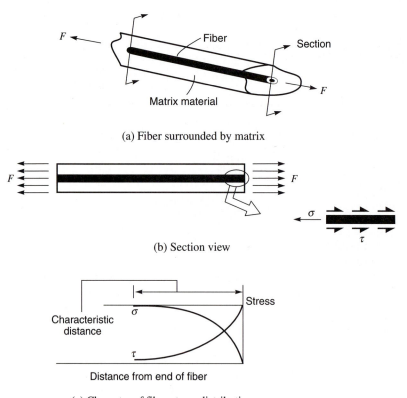

(a) Fiber surrounded by matrix

(b) Section view

(c) Character of fiber stress distribution

FIGURE 1.4
Load transfer to fiber: Tension

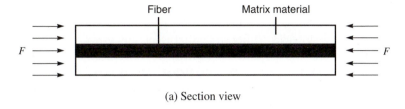

(a) Section view

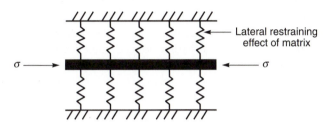

(b) Supporting effect of matrix

FIGURE 1.5
Load transfer to fiber: Compression and the lateral restraint provided by the matrix

reversed, then the stress within the fiber becomes compressive and attains a maximum value at some distance from the end. This is exactly like the tension case except that the fiber responds quite differently to a compressive load; specifically, the fiber tends to buckle. The compressive resistance of some types of fibers is so poor that they will kink and fold, much like a string loaded in compression. Other fibers are quite stiff and act like very thin columns; they fail by what might be considered classic column buckling. To prevent the fiber from kinking, folding, or buckling due to a compressive load, it must be restrained laterally, and the matrix provides this restraint. To use a rough analogy, the fiber and matrix in compression are like a beam-column on an elastic foundation. As you might expect, in the presence of a compressive loading, any slight crookedness or waviness in the fiber can be quite detrimental.

Up to this point, we have focused on the idea of a single fiber or fiber tow, how loads are transmitted into it, and how it is prevented from buckling. The matrix serves both these roles. In addition, the matrix keeps the fibers aligned and in a parallel array. A cross section of a graphite-fiber-reinforced epoxy matrix composite is illustrated in Figure 1.6. The lighter circles are graphite fiber. Evident in Figure 1.6, in the upper left, is a region with no fibers, a so-called resin-rich region. Such regions can occur, and care should be taken to ensure they do not occur frequently. The embedding of strong, stiff fibers in a parallel array in a softer material results in a *fiber-reinforced composite material* with superior properties *in the fiber direction*. Clearly, the material properties perpendicular to the fiber direction are not as good. Recall that in an assemblage of basic units making up the fiber, as in Figure 1.3, the poorer properties of the basic unit are transverse to the lengthwise direction of the fiber. Therefore, to load a composite material perpendicularly to the fiber direction is to load the fiber in the soft and weak diametral direction of the fiber. In addition,

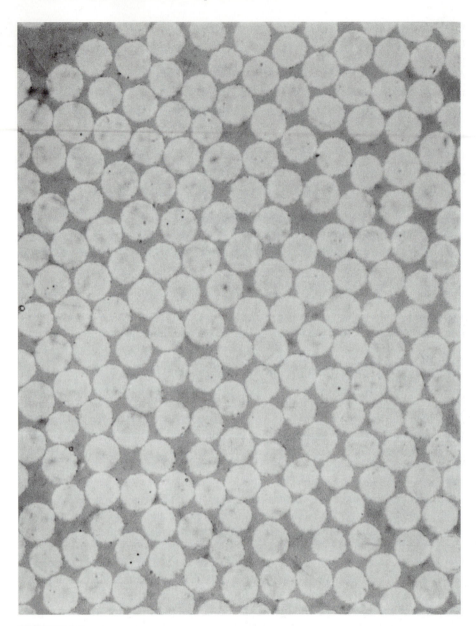

FIGURE 1.6
Cross section of graphite-reinforced material

if a composite material is loaded perpendicularly to the fiber direction, commonly referred to as the transverse direction, not all of the load is transmitted through the fiber. A portion of the load goes around the fiber and is entirely in the matrix material. This can be seen if it is imagined that the cross section of Figure 1.6 was subjected to horizontal tensile forces on the left and right edges of the figure. The fact that the fibers do not touch means some of the load must be transferred through the matrix. The poorer transverse properties of the fiber, coupled with the softer and weaker properties of the matrix, lead to poor properties of the composite in the direction perpendicular to the fibers. In addition, and more importantly, the transverse properties of the composite depend to a large degree on the integrity of the interface bond between the fibers and matrix. If this bond is weak, the transverse properties of the composite material are poor, and a poor interface leads to poor transverse strength. Progressive failure of the interfaces leads to what can be interpreted as low stiffness in the transverse direction. A poor interface results in high resistance to thermal and electrical conduction. Considerable research is directed toward improving the bond at the interface between the fiber and matrix by treating the surface of the fiber before it is combined with the matrix material to form a composite. Thus, as Figure 1.7 summarizes, though the use of fibers leads to large gains in the properties in one direction, the properties in the two perpendicular directions are greatly reduced. In addition, the strength and stiffness properties of fiber-reinforced materials are poor in another important aspect. In Figure 1.8, the three basic components of shear stress are being applied to a small volume of fiber-reinforced material, but in neither case is the inherent strength of the fiber being utilized. In all three cases the strength of the composite depends critically upon the strength of the fiber-matrix interface, either in shear, as in Figure 1.8(a) and (c), or in tension, as in Figure 1.8(b). In addition, the strength of the matrix material is being utilized to a large degree. This lack of good shear properties is as serious as the lack of good transverse properties. Because of their poor transverse and shear properties, and because of the way fiber-reinforced material is supplied, components made of fiber-reinforced composite are usually laminated by using a number of layers of fiber-reinforced material. The number of layers can vary from just a few to several hundred. In a single layer, sometimes referred to as a lamina, all the fibers are oriented in a specific direction. While the majority of the layers in a laminate have their fibers in the direction of the load, some layers have their fibers oriented specifically to counter the poor transverse and shear properties of fiber-reinforced materials. Despite these poor transverse properties, however, the *specific* strength, namely, strength normalized by density, and the *specific* stiffness, stiffness normalized by density, of composite materials are much greater than that of a single homogeneous material. Consequently, the weight

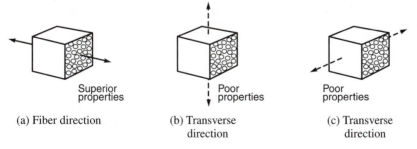

(a) Fiber direction   (b) Transverse   (c) Transverse
                              direction   direction

FIGURE 1.7
Poor transverse properties

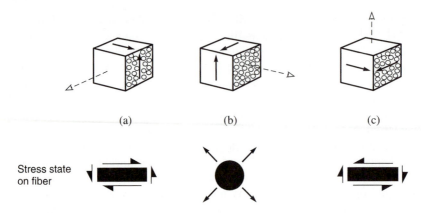

FIGURE 1.8
Poor shear properties

of a structure utilizing fiber reinforcement to meet strength and stiffness requirements is reduced.

1.3
LAMINAE AND LAMINATES

Figure 1.9 shows a through-thickness cross section of an 11-layer flat laminate, with the layers not all having the same fiber orientation. This laminate, which is fabricated from carbon-based fibers in an epoxy matrix and is just over 3 mm thick, consists of five layers with their fibers oriented left and right on the page (the lighter strips), and six layers with their fibers oriented in and out of the page (the darker strips). For the layers with fibers perpendicular to the page, the tows have been sliced perpendicular to their length, while for the layers with fibers in the plane of the page, the tows have been sliced along their length. A closer view of the layers with their fibers oriented out of the page would look like Figure 1.6. This laminate is intended for use in a situation where there is slightly more load out of the page than there is across the page, and it would not be useful for shear loadings in the plane of the laminate.

FIGURE 1.9
Cross section of an 11-layer graphite-reinforced flat laminate

However, instead of having layers with fibers oriented at 90° to one another, if the fiber angles in some of the layers were oriented at 30°, 45°, or 60°, some inplane shear could be tolerated. A laminate subjected to both shear and tension may need fibers oriented at 45° to react the shear load, and at 0° and/or 90° to react the tensile load. The percentage of fibers to use in each orientation depends on the relative magnitudes of the tensile and shear loads.

The issues of fiber orientation, layers, and tension requirements versus shear requirements lead us to questions related to composite materials in a more general sense. How do we determine the fiber orientations for the best performance in a particular application? How many layers are required? How stiff should the fibers be? How strong? How detrimental are the poor transverse and shear strengths? To answer these questions we must develop the tools to help us understand the response of fiber-reinforced composite materials to applied loads. These tools will allow us to answer questions regarding the stress and strain states within a fiber-reinforced material. More importantly, they will allow us to identify the specific advantages of utilizing fiber-reinforced materials. However, these matters must be addressed in the context of a particular application. For instance, one must consider the cost of using composite materials, including the cost of determining the final design. The cost to use composite materials may include education in the subject of composite materials, equipment to process and machine composite materials, and software to analyze composite materials. These are not trivial issues.

In the next chapter, we will begin laying the foundations that will allow us to answer many of these questions. Before proceeding, however, so that the foundations to be presented may be viewed in context, the next sections will present some of the important issues related to the production of fibers and the synthesis of matrix materials. Since most advanced composite materials are based on polymer matrices, we will emphasize these materials. Despite their high cost compared to polymer matrix materials, metals, such as aluminum or titanium, are sometimes used as a matrix, and many of the polymer matrix mechanics issues discussed in later chapters are equally valid for metal matrix composites.

1.4
FIBERS

Boron fibers were used in early composite structures. Currently, three types of fiber reinforcements are in common use in polymer matrix composites, namely, carbon-based or graphite fibers, glass-based fibers, and synthetic polymeric fibers such as Kevlar. The basic building blocks for these three fibers are carbon, silicon, oxygen, and nitrogen, which are characterized by strong covalent interatomic bonds, low density, thermal stability, and relative abundance in nature.

1.4.1 Carbon-Based Fibers

To make carbon-based fibers, you begin with a precursor fiber. Early precursor fibers were made from commonly available rayon. Thornel 40®, from Union Carbide, and

HMG-50® from Hitco are examples of early rayon-derived fibers. The yield of fiber from rayon precursor is relatively low. Currently, polyacrylonitril (PAN) precursor fibers are most commonly used. T300® from Toray and Type A® from the former Hercules are typical PAN-derived fibers. Precursor fibers made from pitch are also in use. Other precursors, such as phenolics, polyimides, and polyvinylalcohols, have been used but to a lesser degree. Ultimate fiber mechanical properties are not significantly affected by the type of precursor. However, the processing techniques are much different among the various precursors. In general, high-modulus carbon-based fibers are produced by carbonizing organic precursor fibers, and then graphitizing them at very high temperatures. Preferential orientation is achieved in the fibers by stretching them at various stages of processing. This stretching results in better alignment of the graphite layer planes, referred to as the basal planes, in the axial direction of the fiber, increasing strength and stiffness in that direction. This was discussed in a cursory fashion in connection with Figure 1.3.

PAN-derived fibers

There are a number of types of PAN precursor fibers. They are all acrylic-based fibers and they contain at least 85 percent of acrylonitrile, but the balance may include secondary polymers or residual spin bath chemicals left over from initial PAN precursor fiber processing. Most of the secondary polymers are trade secrets, and their addition yields small improvements in strength or other specific properties.

PAN precursor, if heated to a high temperature to promote carbonization, will not yield fibers of high strength and stiffness unless a preoxidation, or stabilization, step is used during processing. A thermally stable structure is obtained during this step, so upon further heating the original fiber architecture is retained. Stabilization generally calls for a heat treatment in the range 200°C to 300°C in an oxygen-containing atmosphere, with the result that the polymer backbone of the precursor undergoes a series of chemical reactions that ultimately result in the formation of polynaphthyridine, a substance with the preferred structural form for the formation of graphite. Cross-linking also occurs, induced either by oxidizing agents or by other catalysts. Upon further heating, this precursor structure gives rise to graphitic nuclei whose basal planes of the carbon atoms are oriented parallel to the direction of the polymer chains. Significant shrinkage (e.g., up to 40 percent) occurs during stabilization; this can be reduced by stretching the fibers during the heat treatment or by infiltrating PAN fibers with fine silica aggregates. The silica particles lodge in the interstices between and around the fibers and essentially lock the fiber in place. Using such methods can reduce shrinkage to about 20 percent. Several chemical treatments can be used to speed up the oxidation process. For instance, treating PAN fibers with a diethanolamine or triethanolamine solution prior to heat treatment has been found to substantially reduce the time required for oxidation.

The carbonization step involves heating the stabilized precursor fiber to temperatures up to 1000°C in an inert or mildly oxidative atmosphere. Carbonization can take anywhere from a few minutes to several hours. In one type of process, fibers possessing high modulus and high strength are obtained by first heating PAN fibers in an oxidizing atmosphere until they are permeated with oxygen. The fibers are then heated further, to initiate carbonization, while being held under tension in a nonoxidizing atmosphere. Finally, to increase the ultimate tensile strength, the fibers are heat-treated in an inert atmosphere between 1300°C and 1800°C. Research has shown that if carbonization is done in an oxidizing atmosphere, then the optimal atmosphere should contain between 50 and 170 ppm oxygen. In this study, argon was used as the carrier gas, with the oxygen content varying between 2.8 and 1500 ppm. The fibers were then tested for ultimate tensile strength and modulus. The results of the study are shown in Table 1.1.

Graphitization occurs by heating the carbonized fiber to high temperature (up to 3000°C) in an inert atmosphere. The process can last anywhere from 1 to 20 minutes. Tensioning the fibers during graphitization improves ultimate mechanical properties and reduces any residual shrinkage. Figure 1.10 details an entire process for manufacturing high-modulus graphite fibers. The tension and temperature histories are overlaid on the figure. Precursor fiber begins the process, and tension is applied to the precursor while heating it to an elevated temperature. The tension is reduced before the preoxidation phase and is increased during preoxidation. The fiber is then continuously heated and graphitized under high tension. The graphitized fiber is wound onto a take-up reel and dried. Processing speeds can reach as high as 45 m/hour, but are typically between 6 and 12 m/hour.

Figure 1.10 shows other important steps that can take place after the fibers are dried. First, after graphitization, special coating materials, called sizings, are applied to the fiber surface. Sizings, or fiber surface treatments, are used to provide lubrication and to protect the fibers during subsequent processing and handling. Other chemicals are also applied during the sizing operation to assist in bonding the fibers to the matrix. More will be said about sizing in a subsequent section. Second, the fibers must eventually be combined with the matrix material. If the matrix can be made into a liquid or semiliquid form, then after the fibers are dried and surface treatment has been applied, the fibers can be combined with the matrix. In one type of process (see Figure 1.10) the fibers are unwound from the take-up

TABLE 1.1
Optimal purge gas for PAN carbonization

Atmosphere oxygen content (ppm)	Tensile strength (GPa)	Tensile modulus (GPa)
2.8	1.45	152
18	2.00	152
35	2.07	162
50	2.14	172
110	2.19	214
170	2.17	186
330	1.46	152
430	1.24	165
650	0.82	155
800	0.75	155
1500	0.79	155

Source: Adapted from P.G. Rose; U.S. Patent 3,660,018: May 2, 1972; Rolls Royce Ltd., England.

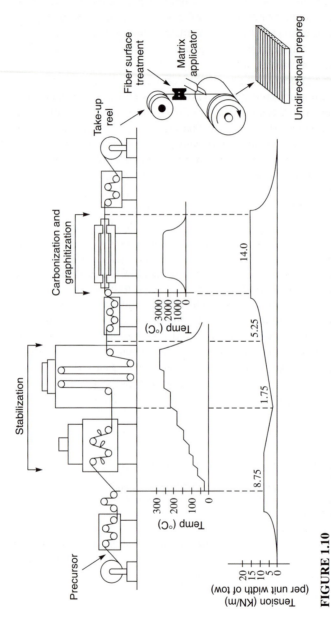

FIGURE 1.10

Continuous production of graphite fiber from PAN precursor showing temperature and tension profiles employed (*Source*: U.S. Patent 3,803,672.)

12

reel, the sizings are applied, the liquid matrix material is applied to a fiber, and the wetted fibers are rewound onto a large drum. The drum is usually covered with a sheet of paper coated to prevent the fibers from sticking to the drum, and to act as a backing paper for holding the fiber-matrix system together. Generally, the winding of the fibers as they are impregnated with the matrix in liquid form and the concurrent winding of the backing paper are a continuous process; the result is layer after layer of impregnated fibers are wound on the drum and separated by sheets of backing paper. This matrix-impregnated material is popularly known as unidirectional "prepreg," and it is generally the form in which material is received from the supplier. Thus, prepreg is short for *preimpregnated*, meaning impregnated before the user receives the material. The preimpregnated material on the drum is generally cut to some standard width, ranging from, say, 3 mm wide to 300 mm or more wide. The narrower forms are often referred to as tapes and are used in machines that automatically fabricate composite structural components.

Pitch-derived fibers

Two particular factors have led to the use of pitch as a precursor: (1) higher yields and (2) faster production rates. However, pitch-derived fibers are more brittle than those derived from PAN, and they have a higher density, leading to lower specific properties. In addition, the steps leading to pitch fibers are slightly different from the steps leading to PAN-derived fibers.

The process of producing specific types of pitch from petroleum products is critically important to the successful production of high-modulus and high-strength carbon-based fibers. The basic process is a distillation of residual oils left over from the thermal or catalytic cracking of crude oil. The residuals from asphalt production, natural asphalt, shale oil, or coal tar can also be used. These heavy oil-based products are introduced into a reactor or series of reactors, where they are heated to temperatures between 350°C and 500°C. Thermal cracking, polymerization, and condensation occur, and the gases and light oils that are released are taken out of the reactor through a condenser. The resulting material is even heavier, has a higher carbon content, and is the basis for pitch fiber precursor.

The preferred raw material for pitch fiber production is liquid crystal, or meso-phase, pitch. The mesophase is a highly anisotropic substance in the form of crystals, called spherulites, mixed in an isotropic pitch medium. The spherulites consist of a collection of relatively long molecules with their long axes normal relative to the boundary of the sphere. Under the influence of heat, the spherulites continue to grow and expand at the expense of the isotropic pitch surrounding them. An interesting feature of mesophase pitch is that it softens above 350°C and it can be mechanically deformed in this state. When the mesophase content reaches about 75 percent, the carbon substance can be subjected to fiber-forming techniques such as melt spinning. The melt spinning of pitch fibers from the raw material can occur after the carbon content reaches the range 91 percent to 96.5 percent and the mean molecular weight is at least 400. The heat treatments of the petroleum products in the reactor vessels must occur for sufficiently long times to ensure that these conditions are satisfied. Additives are sometimes used to increase the molecular weight or to promote better fiber-spinning characteristics. Sulfur, organic compounds containing sulfur, or an organic/inorganic peroxide is added to increase the molecular weight. To improve yields and handling characteristics polymers are sometimes added, for instance, polyethylene, polypropylene, polystyrene, polymethacrylate, and rubber. There are various types of spinning techniques, such as extrusion, centrifugal-type, pressure-extrusion, spraying, or jet-type. When the melt viscosity is high and when long continuous fibers are desired, the extrusion process is used.

After pitch fibers are spun, they are subjected to an oxidizing gas at a temperature below the spinning temperature, or they are subjected to another chemical treatment that renders them infusible. In one method, for example, pitch fibers are treated for seven hours at a temperature of 100°C with air containing ozone; thereafter, the temperature is raised at 1°C/min up to 300°C. This stage of processing is critically important to guarantee that pitch fibers will retain their shape under heat treatment during carbonization and graphitization. However, if they are exposed to oxidization for too long, the fibers become brittle.

The carbonization of pitch fibers occurs at somewhat higher temperatures. The heating rate between 100°C and 500°C is critical to prevent fiber rupture from released volatiles. A typical heating schedule would call for heating from 100° to 500°C at 5°C/hr and from 500° to 1100°C at 10°C/hr. The cooldown from 1100°C is usually controlled to be less than 30°C/hr. If desired, the carbonized pitch fibers can be further heated in an inert atmosphere to produce a graphitic microstructure. Graphitization temperatures are typically between 2500° and 3300°C. Total graphitization times are generally very short, on the order of a few minutes.

Microstructure of graphite fibers

It is important from a mechanics perspective to understand the basic microstructure of graphite fibers after processing. Some of the failure modes of fibers and their composites are dictated by the type of the fiber's microstructure. Although most of the topics presented will apply equally to fibers made from any of the precursors, there are some exceptions. For example, the cross section of a PAN-derived fiber is somewhat circular, while the cross section of a pitch-derived fiber is almost perfectly circular. The effect of these cross sections on the ultimate mechanical properties of fibers is not clearly understood.

Graphite in its pristine form has a crystal structure with planes of carbon atoms arranged in a hexagonal unit cell (see Figure 1.11). The hexagonal unit cells are covalently bonded together at the adjoining carbon atoms, and these planes are the so-called basal planes. The basal planes are stacked upon each other to form a layered microstructure. There is a considerable lack of isotropy to this arrangement. For instance, the extensional modulus of graphite in directions within the basal planes approaches 1000 GPa, while the modulus normal to the planes is only about 30 GPa. In ideal graphite fibers the axial direction corresponds to the planar direction, that is, the long direction indicated in Figure 1.3. However, no graphite fiber exhibits perfect crystallographic alignment along the fiber axis. Various types of defects—like point defects, vacancies, dislocations, and boundaries—all combine to reduce the degree of crystallographic orientation. By stretching the fibers during graphitization and also by increasing the heat-treatment temperature, the degree of orientation along the fiber axis is increased. This leads to increased stiffness in the axial direction.

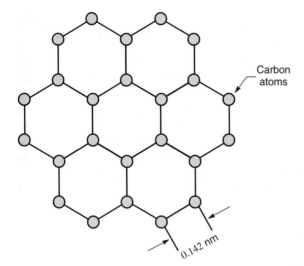

FIGURE 1.11
Graphite crystal structure

Carbon atoms

0.142 nm

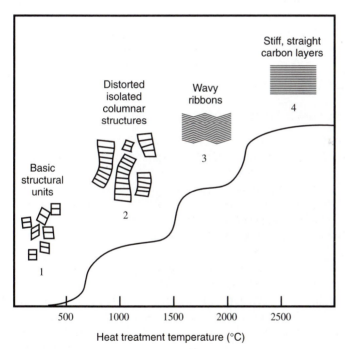

Stiff, straight
carbon layers

Distorted
isolated
columnar
structures

Wavy
ribbons

4

Basic
structural
units

3

2

1

500   1000   1500   2000   2500

Heat treatment temperature (°C)

FIGURE 1.12
Microstructural phases during graphitization (*Source:* Adapted from A. Oberlin, *Carbon*, Copyright 1984, p. 521, with kind permission from Elsevier Science Ltd., The Boulevard, Langford Lane, Kidlington OX5 1GB, Uk.)

Figure 1.12 shows the four distinct phases during processing as the planes arrange themselves into an ordered and layered structure. For temperatures below 800°C (see region 1 in Figure 1.12), the basic structural units, which consist of short lengths (10 Å) of two or three parallel carbon layers, start to pile up and form a disordered columnar structure as impurities and volatiles are released. Between 800° and 1500°C (see region 2 in Figure 1.12), the columnar structure increases in length with a higher degree of orientation of the basic structural unit. Between 1500° and 1900°C (see region 3), the columnar structure disappears as wavy ribbons or wrinkled layers are formed by the joining of adjacent columns. By about 2100°C most of the waviness has disappeared, and inplane defects have been greatly reduced. Above 2100°C (see region 4) stiff, flat carbon layers are observed and three-dimensional crystal growth commences. Thus, graphitization corresponds to the removal of structural defects—first between the layers, which are then flattened—followed by removal of transverse disorientation, that is, increasing the orientation parallel to the fiber axis. Figure 1.13 shows a rendering of a typical microstructure for a high-modulus PAN-derived fiber showing the ribbonlike structure. Notice the folded nature of the carbon planes along the axis of the fiber.

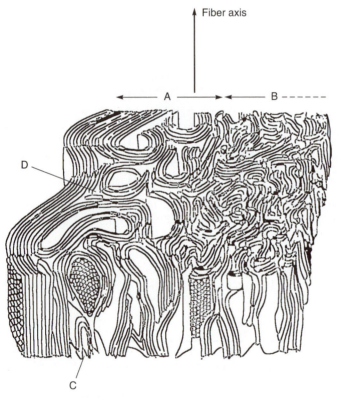

A. A narrow surface region
B. An interior region
C. A "hairpin" defect
D. A wedge disclination

FIGURE 1.13
Rendering of the microstructure of a high-modulus graphite fiber (*Source:* Adapted from S. C. Bennett, D. L. Johnson, and W. Johnson, *Journal of Materials Science* 18 (1983), p. 3337, with kind permission from Chapman and Hall.)

(a) Skin-core

(b) Onion skin

FIGURE 1.14
The four types of graphite
fiber microstructures

(c) Radial

(d) Transversely isotropic

Four basic structural models for graphite fibers have proven to be quite helpful in understanding structure-property relations. The most common is the skin-core structure, Figure 1.14(a), which is typical for PAN-derived graphite fibers. The fibers exhibit a skin that is somewhat more graphitic than the core region. The skin is typically about 1×10^{-6} m thick and is formed by basal planes wrapping around the circumference. The basal planes in the core region are radially aligned or, in some cases, randomly aligned. The relative freedom of the layers near the surface of the fiber to rearrange themselves, unimpeded by the constraints of neighboring layers, leads to the skin-core geometry. The formation of large misoriented crystallites in the skin region is believed to be responsible for fiber failure under applied stress. The circumferential alignment, or onion-skin arrangement, shown in Figure 1.14(b), is observed in benzene-derived carbon-based fibers manufactured at low heat treatment temperatures. Other architectures are the radial alignment, Figure 1.14(c), and the arrangement where the basal planes are randomly oriented within the cross section. This leads to a transversely isotropic fiber, as shown in Figure 1.14(d). The fiber architecture for a pitch-derived fiber has many possibilities, depending upon the type of extrusion conditions used. The stirring of the pitch above the capillary tube during extrusion, the viscosity and temperature of the pitch, and the overall stirring conditions are all important in determining the type of microstructure that develops. Various microstructures can be achieved with the same pitch precursor material, for example, radial (no stirring), random (certain types of nozzle geometries), and circumferential (low viscosity and high temperature).

The mechanical and physical properties of commercial carbon-based fibers are listed in Table 1.2. The quoted mechanical properties in the table are for fibers that have been processed according to accepted and repeatable processing conditions. Obviously, some variation in conditions occurs from batch to batch, which results in a related variation in mechanical properties from fiber to fiber. It is the material supplier's responsibility to minimize this variation so that material properties can be used with confidence in engineering design. Nearly every graphite fiber manufacturer is investigating new and improved methods of processing fibers to yield higher

TABLE 1.2
Properties of carbon-based fibers

Property	PAN			Pitch Type-P[4]	Rayon
	IM[1]	HM[2]	UHM[3]		
Diameter (μm)	8–9	7–10	7–10	10–11	6.5
Density (kg/m^3)	1780–1820	1670–1900	1860	2020	1530–1660
Tensile modulus (GPa)	228–276	331–400	517	345	41–393
Tensile strength (MPa)	2410–2930	2070–2900	1720	1720	620–2200
Elongation (%)	1.0	0.5	0.3–0.4	0.4–0.9	1.5–2.5
Coeff. of thermal expansion ($\times 10^{-6}$/°C)					
Fiber direction	–0.1 to –0.5	–0.5 to –1.2	–1.0	–0.9 to –1.6	—
Perpendicular to fiber direction	7–12	7–12	—	7.8	—
Thermal conductivity (W/m/°C)	20	70–105	140	—	38
Specific heat (J/kg/°K)	950	925	—	—	—

[1]IM = intermediate modulus.
[2]HM = high modulus.
[3]UHM = ultra high modulus.
[4]Mesophase pitch precursor.

levels of strength and stiffness and lower densities. Strengths in excess of 4 GPa and extensional moduli greater than 1000 GPa have been reported for experimental graphite fibers.

1.4.2 Glass-Based Fibers

Glass-based fibers are considered to be somewhat lower performing than graphite fibers, mainly because glass-based fibers have been in existence for a number of years and, in one form or another, appear in playground equipment, recreational items, piping for corrosive chemicals, and many other common applications. Also, the cost of glass-based fibers is considerably lower than the cost of carbon-based fibers.

Silica, SiO_2, forms the basis of nearly all commercial glasses. It exists in the form of a polymer $(SiO_2)_n$. It does not melt, but gradually softens until reaching a temperature of 2000°C, after which it begins to decompose. When silica is heated until fluidlike and then cooled, it forms a random glassy structure. Only prolonged heating above 1200°C will induce crystallization (i.e., a quartz-type structure). Using silica as a glass is perfectly suitable for many industrial applications. However, its drawback is the high processing temperatures needed to form the glass and work it into useful shapes. Other types of glasses were developed to decrease the complexity of processing and increase the commercialization of glass in fiber form. Table 1.3 presents the four predominant glass compositions used to form continuous glass fibers. Type A, a soda-lime glass, was the first used, and it is still retained for a few minor applications. Type E, a borosilicate glass, was developed for better resistance to attack by water and mild chemical concentrations. Relative to Type E, Type C glass

TABLE 1.3

Glass compositions

Constituent	Type A: Common soda-lime glass (%)	Type C: Chemical glass, used for corrosion resistance (%)	Type E: Electrical glass, used in most general purpose composite applications (%)	Type S: Structural glass, has high strength, high modulus, used for high-performance structures (%)
SiO_2	72.0	65.0	55.2	65.0
Al_2O_3	2.5	4.0	14.8	25.0
B_2O_3	0.5	5.0	7.3	—
MgO	0.9	3.0	3.3	10.0
CaO	9.0	14.0	18.7	—
Na_2O	12.5	8.5	0.3	—
K_2O	1.5	—	0.2	—
Fe_2O_3	0.5	0.5	0.3	—
F_2	—	—	0.3	—

Source: Adapted from K.L. Lowenstein, *The Manufacturing Technology of Continuous Glass Fibres*, 3rd Ed., 1993, Table 4.2, p. 32, with kind permission from Elsevier Science, P.O. Box 211, 1000 AE Amsterdam, The Netherlands.

has a much improved durability when exposed to acids and alkalis. The increased strength and stiffness of Type S glass makes it a natural choice for use in high-performance applications, where higher specific strength and specific stiffness are important.

Once the correct glass composition has been achieved, fibers can be produced using a number of techniques. Each method is a variation of the generalized fiber-drawing process, sometimes called attenuation (see Figure 1.15). It consists of five main components:

1. A fiber-drawing furnace consisting of a heat source, tank, and platinum alloy bushing. Raw material is fed into the furnace tank through an inlet, and fibers are drawn out through tiny nozzles in the bushing. By convention, the number of nozzles is usually 200 or a multiple thereof. Special cooling fins are located immediately under the bushing to stabilize the fiber-drawing process. This drawing process can produce material with aligned properties, as discussed in connection with Figure 1.3. However, glass fibers, even though drawn, or stretched, in one direction, do not have properties in the lengthwise direction that are too different from properties across their diameter; in other words, glass fibers tend to be isotropic.
2. A light water spray below the bushing to cool the fibers.
3. A sizing applicator.
4. A gathering shoe to collect the individual fibers and combine them into tows.
5. A collet or winding mandrel to collect the tows. The winding of the collet produces tensioning of the tows, which draws the fibers through the bushing nozzles.

The wound fiber tows at this point are referred to as glass fiber cakes. The sizings applied to the fibers are almost all aqueous based, and thus the cake has a typical water content of about 10 percent. Before the fiber is shipped for end use,

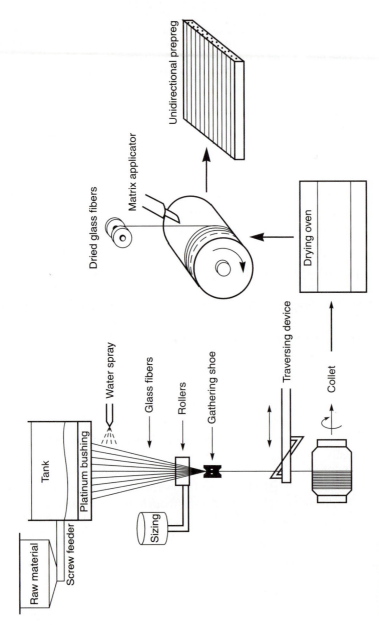

FIGURE 1.15
Glass fiber production

TABLE 1.4
Properties of glass-based fibers

Property	Glass type		
	E	C	S
Diameter (μm)	8–14	—	10
Density (kg/m^3)	2540	2490	2490
Tensile modulus (GPa)	72.4	68.9	85.5
Tensile strength (MPa)	3450	3160	4590
Elongation (%)	1.8–3.2	4.8	5.7
Coeff. of thermal expansion ($\times10^{-6}$/°C)	5.0	7.2	5.6
Thermal conductivity (W/m/°C)	1.3	—	—
Specific heat (J/kg/°K)	840	780	940

this water content has to be substantially reduced. This is accomplished by oven drying. There are two main objectives during oven drying: (1) Water content must be reduced to less than 0.1 percent, and (2) the dry fiber must be subjected to heat treatment to allow conglomerates of sizing particles to flow within the tow to impart certain handling characteristics. A nominal drying schedule calls for a temperature between 115° and 125°C for 4–10 hours. If the fibers are to be preimpregnated, the dried fibers are unwound, the matrix material is applied, and the wetted fibers are rewound on a drum.

Table 1.4 presents the physical and mechanical characteristics of typical glass fibers. Of these, E-glass and S-glass are most often used in structural composites; S-glass is preferred for performance-critical applications.

1.4.3 Polymeric Fibers

A relatively new class of fibers is finding increased use in fiber-reinforced composite materials. Polymeric fibers, using a suitable processing method, can exhibit high strength and stiffnesses. This happens as a result of the alignment of the polymer chains along the axis of the fiber. Several commercial polymeric fibers are now available and many others are being developed.

Kevlar is perhaps the most common polymer fiber. It was developed by the DuPont Co. in 1968 and is an aromatic polyamide called poly(paraphenylene terephthalamide). The aromatic rings make the fiber fairly rigid. Spectra® is an oriented polyethylene fiber developed by AlliedSignal and is produced by solution or gel spinning followed by drawing (20–100 percent) to orient the polymer chains. Its advantages as a reinforcement include good chemical resistance and low density, but the maximum use temperature is relatively low (100°C). Many other polymers can be produced as fibers, but their commercial applications have been very limited. These include aromatic copolyesters, aromatic heterocyclic polymers like poly(benzobisoxazole) (PBO), and a new class, polyimides. Polyimide fibers such as Avimid®, produced by the DuPont Company, will find wide application due to their high maximum use temperatures (> 300°C).

Production of polymeric fibers

The production of synthetic polymeric fibers differs from the production of inorganic fibers because of the one-dimensional nature of the polymer chains. To produce fibers that are strong and stiff, the polymer chains must be extended and oriented along the fiber axis. After orientation, external forces acting upon the fiber are absorbed by the strong interatomic covalent bonds along the polymer backbone, resulting in a strong and stiff fiber.

The commercial processes for producing strong and stiff polymer-based fibers fall into two basic categories: (1) melt or dry jet spinning from a liquid crystalline phase and (2) melt or gel spinning and extension of conventional random-coil polymers. Polymers which exhibit a liquid crystalline phase can be spun in such a way that the already rodlike molecules are uniaxially oriented after exiting the spinneret. For conventional polymers in which the polymer chains are highly coiled and interwoven, orientation is achieved by subjecting the fibers to extremely high elongations after spinning.

Kevlar fibers are manufactured by the extrusion and spinning processes. A solution of the polymer and a solvent is held at a low temperature, between $-50°$ and $-80°C$, before being extruded into a hot-walled cylinder at $200°C$. The solvent then evaporates and the fibers are wound onto a drum. The fibers at this stage have low strength and stiffness. The fibers are subsequently subjected to hot stretching to align the polymer chains along the axis of the fiber. Afterwards, the aligned fibers show significant increases in strength and stiffness.

Microstructure of polymeric fibers

The microstructure of Kevlar has been extensively studied using diffraction techniques and electron microscopy. Figure 1.16 shows a schematic of its structure. The polymer molecules form rigid planar sheets, and the sheets are stacked on top of each other with only weak hydrogen bonding between them. These sheets are folded in the axial direction and are oriented radially, with the fold generally occurring along the hydrogen-bonding line. This type of structure is very similar to the radial orientation in graphite fibers; see Figure 1.14(c). The microstructure of Kevlar influences the mechanical properties in a number of ways. Kevlar fibers have a low longitudinal shear modulus, poor transverse properties, and, in particular, a very low axial compressive strength as a result of the weak bonding between the oriented sheets. This is typical for nearly all polymer fibers, as the polymer chains are extended along the fiber axis, leaving relatively weak bonding transverse to this axis.

Polymeric fibers in general are characterized by their low density, good chemical resistance, and high tensile strength. Table 1.5 shows some typical mechanical and physical properties for commercial polymeric fibers.

1.5
MATRICES

Polymers used as matrix materials are commonly referred to as resins. The matrix resin generally accounts for 30 to 40 percent, by volume, of a composite material. In

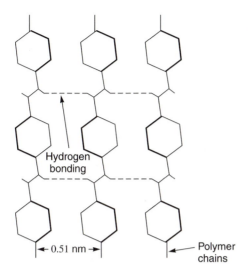

FIGURE 1.16
The structure of Kevlar
(*Source:* Adapted from
M. G. Dobb, D. L. Johnson,
and B. P. Saville, *Phil. Trans. of
Roy. Soc. of London*, Copyright
1980, vol. 294A, p. 483–5,
with kind permission of The
Royal Society, 6 Carlton House
Terrace, London SW1 Y5AG
UK.)

(a) Lamella sheet

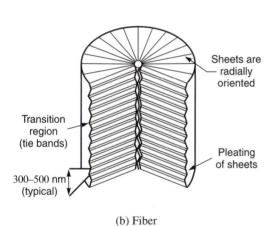

(b) Fiber

addition to maintaining the shape of the composite structure, aligning the reinforce-ments, and acting as a stress transfer medium, the matrix protects the fibers from abrasion and corrosion. More importantly, the limitation of a composite may well be a function of matrix properties. For example, the thermal stability and maximum use temperature of a composite are largely determined by the matrix properties. Additionally, an agressive chemical environment, a moist environment, or exposure to other adverse conditions may well degrade the performance of the matrix before the fibers are degraded. In general, any condition that lowers the so-called glass transition temperature, T_g, of the matrix material is detrimental to the composite. The glass transition temperature defines the transition between the soft rubbery state of a polymer and its more stiff, or glassy, state. The latter state occurs below the glass transition temperature and is the better state for the matrix to transfer load

TABLE 1.5
Properties of polymeric fibers

Property	Fiber type		
	Kevlar-29	Kevlar-49	Spectra 900 (polyethylene)
Diameter (μm)	12	12	38
Density (kg/m^3)	1440	1479	970
Tensile modulus (GPa)	62	131	117
Tensile strength (MPa)	2760	2800–3792	2580
Elongation (%)	3–4	2.2–2.8	4–5
Coeff. of thermal expansion ($\times 10^{-6}$/°C)			
Fiber direction	–2	–2	—
Perpendicular to fiber direction	59	59	—
Thermal conductivity (W/m/°C)	—	0.04–0.5	—
Specific heat (J/kg/°K)	—	1420	—

to the fibers, provide support against fiber buckling, and maintain alignment of the fibers. When operating above the glass transition temperature the matrix becomes soft and does not perform these functions well. Moisture, for example, lowers the glass transition temperature. Thus, with a glass transition temperature of 200°C and the composite operating at a temperature of 175°C in a dry environment, the matrix in the composite would be below its glass transition temperature. Adding moisture to the matrix could lower its glass transition temperature to 150°C. In this situation, the operating environment of 175°C would be above the glass transition temperature of the matrix and the performance of the composite would be degraded due to softening of the matrix.

The processability and processing history of the resin also influence the ultimate performance of the composite. For example, the interfacial bonding between the matrix and the reinforcement is severely degraded if the resin does not penetrate the fiber tows and wet all the fibers. Additionally, flaws such as voids or unreacted resin will contribute to a loss in composite strength.

The two basic classes of resins are thermosets and thermoplastics. The difference between the two arises from the their unique behavior when heated. Thermosets undergo an irreversible chemical change when they are heated, called curing. They chemically cross-link and develop a network structure that sets them in shape. If they are heated after they have been cured, they do not melt. They will retain their shape until they begin to thermally decompose at high temperatures. On the other hand, thermoplastics reversibly melt when heated and solidify when cooled. Once they have been initially melted to form the composite, they can be reshaped by heating above a lower forming temperature. Thus, thermoplastic composites have the unique ability to be repaired once they have been placed into service.

1.5.1 Thermosets

Thermosetting resins are the most common type of matrix system for composite materials. They have become popular for a number of reasons, including low melt viscosity, good fiber impregnation, and fairly low processing temperatures. They are also lower in cost compared to thermoplastic resins.

Epoxy resins are the predominant choice for the advanced composite materials market. They are popular because of their excellent mechanical properties, their retention of mechanical properties when operating in hot and moist environments, and their good chemical resistance. They also possess good dimensional stability, are easily processed, are low in cost, and exhibit good adhesion to a variety of fibers. The most common epoxy resin is based on the reaction of epichlorohydrin and bisphenol A. The character of the epoxide chain affects the processability and the cross-link density. The curing of the epoxide network is accomplished by adding a curing agent that reacts with the epoxide and is ultimately incorporated into the network structure. Thus, the mechanical properties of the epoxy are dependent on the type of curing agent used. Acid anhydrides and multifunctional amines are most commonly used. Aliphatic amines yield fast cure times, whereas aromatic amines are less reactive but result in higher glass transition temperatures. For 121°C curing epoxies, dicyandiamide (dicy) is used as the curing agent. Most 177°C curing epoxies use a curing agent based on the tetraglycidyl derivative of 4, 4'-methylene dianiline and 4, 4'-diaminodiphenyl sulfone. Once the curing agent and epoxide are mixed, the liquid resin is converted to a solid by applying heat. Epoxies cure very slowly and several hours may be required for complete curing. Once fully cured, they are brittle. Much research over the past several years has been devoted to improving fracture toughness, moisture resistance, and thermal stability. These improvements have been the result of the addition of functional thermoplastics or multifunctional epoxides into the epoxide/curing agent mixture before curing.

Another class of thermosetting polymers, addition-polyimide resins, is primarily used in high-temperature applications. Their development has given polymer composites new opportunities for applications requiring moderately high service temperatures (250°C). Bismaleimide resins are addition-polyimides in which the imide monomer is terminated by reactive maleimides. Free radical cross-linking occurs across the terminal double bonds and is thermally initiated. These reactive sites can react with themselves or with other coreactants such as vinyl, allyl, or amine functionalities. Bismaleimides possess better processing characteristics than linear high-molecular-weight polyimides. The main disadvantage is their inherent brittleness. They can be toughened by introducing polysulfone, polyetherimide, or other thermoplastic phases into the resin. Decreasing the cross-link density also has been shown to toughen the resin. Two other types of addition polyimides, ethynyl and norbornene (nadic)-terminated imide oligomers, have been used as matrices for polymer composites. The norbornene-type oligomers led to the development of PMR-15 polyimides, which are formed by the in situ polymerization of reactive monomers to form an oligomer with an average molecular weight of about 1500 g/mol. Thermally activated cross-linking yields a network structure with a glass transition temperature of about 300°C.

Cyanates are another class of thermosetting polymers suitable for high-temperature applications. Cyanate resins are esters of bisphenols that have a cyanate functional end group. Once heat is applied in the presence of a catalyst, cyclotrimerization occurs. Curing occurs by heating, followed by addition polymerization. The unique trimerization feature of cyanates contributes to their high glass transition temperature (250–290°C) and high toughness. The cyanate resins offer performance in hot and wet environments that ranges between bismaleimides and epoxies.

The most widely used class of thermosets in the automotive and construction market is unsaturated polyesters. Their normal preparation consists of reacting saturated dialcohols with a mixture of unsaturated and saturated dibasic acids (or anhydrides). The most common reactants are propylene glycol, maleic anhydride, and phthalic anhydride. Usually a diluent such as styrene, divinyl benzene, or methyl methacrylate is added to reduce the viscosity during impregnation and to increase the degree of cross-linking after cure. The entire mixture is then heated (sometimes a catalyst and an accelerator are added to reduce processing times) to form the polymer network structure. Glass fibers are the most common reinforcement used with polyester matrices such as vinyl esters. Vinyl esters are often used because of their cost, the ease and speed of processing, and their good resistance to wet environments. Table 1.6 lists the representative mechanical and physical properties of thermosetting resins.

1.5.2 Thermoplastics

Thermoplastic resins are not cross-linked. They derive their strength and stiffness from the inherent properties of the monomer units and the degree of entanglement of the polymer chains. There are two types of thermoplastic resins: (1) amorphous and (2) crystalline. Amorphous thermoplastics exhibit a high degree of entanglement of the polymer chains, which act like cross-links. Upon heating, the chains

TABLE 1.6
Typical room temperature properties of thermosetting polymers

Property	Thermosetting polymer				
	Polyester	**Vinyl ester**	**Epoxy**	**Bismaleimide**	**Polyimide**
Density (kg/m^3)	1100–1500	1150	1100–1400	1320	1430–1890
Tensile modulus (GPa)	1.2–4.5	3–4	2–6	3.6	3.1–4.9
Shear modulus (GPa)	0.7–2	—	1.1–2.2	1.8	—
Tensile strength (MPa)	40–90	65–90	35–130	48–78	70–120
Compressive strength (MPa)	90–250	127	100–200	200	—
Elongation (%)	2–5	1–5	1–8.5	1–6.6	1.5–3
Coeff. of thermal expansion ($\times 10^{-6}$/°C)	60–200	53	45–70	49	90
Thermal conductivity (W/m/°C)	0.2	—	0.1–0.2	—	—
Specific heat (J/kg/°K)	—	—	1250–1800	—	—
Glass transition temperature (°C)	50–110	100–150	50–250	250–300	280–320
Water absorption (%) [24h @ 20°C]	0.1–0.3	—	0.1–0.4	—	0.3
Shrinkage on curing (%)	4–12	1–6	1–5	—	—

disentangle and the resin becomes a viscous fluid. The resin can then be formed and subsequently cooled to solidify the part. Crystalline thermoplastics show a high degree of molecular order and alignment. When heated, the crystalline phase melts, and the resin reverts to an amorphous, viscous liquid. In practice, thermoplastics that exhibit crystalline behavior are actually semicrystalline with both amorphous and crystalline phases present.

Because they can be processed much more quickly, thermoplastics offer the potential for reduced manufacturing costs. As no cross-linking reaction occurs, there is no need to maintain elevated temperatures for an extended period of time. Some thermoplastics have much higher glass transition and maximum-use temperatures compared to epoxies and bismaleimides. Table 1.7 gives a comparison of the glass transition temperatures for a wide variety of thermosetting and thermoplastic resins. Because they yield and undergo large deformations before final fracture, thermoplastics have a much higher toughness compared to thermosets. To their detriment, however, they also exhibit time-dependent deformation, or creep, under the influence of sustained loading.

Aromatic polyimides are considered to be one of the first high-temperature polymers. High-molecular-weight polyimides are difficult to process unless they are modified. Processable polyimides are obtained by incorporating flexible linking units, bulky side groups, and asymmetry into the polyimide backbone, or by copoly-

TABLE 1.7
Glass transition and processing temperatures of several polymers

Polymer	Trade name	T_g (°C)	Process temperature (°C)
Poly(ether ether ketone)	Victrex PEEK	143	370
Poly(ether ketone ketone)	PEKK	156	370
Polyarylene ketone	PXM 8505	265	—
Polyphenylene sulfide	PPS, Ryton	85	343
Polyarylene sulfide sulfone	PAS-2	215	329
Polyarylamide	J-2	156	300
Polyamide imide	Torlon C	275	350–400
Polyamide imide	Torlon AIX638/696	243	350
Polyether imide	Ultem	217	343
Polyimide	NR 150 B2	360	400
Polyimide	Avimid K-III	251	350
Polyimide	LARC-TPI	264	350
Polyimide	K-1	210	300–350
Polysulfone	Udel P-1700	190	300
Polyarylsulfone	Radel A400	220	330
Polyester	Xydar SRT-300	350	400
Polyester	Vectra	175	350
Epoxy	3501-6	206	177
Epoxy	MY720	240	180
Epoxy	8551-7	182	177
Bismaleimide	HG9107	258	180–227

Source: T.L. Vigo and B.J. Kinzig, eds., *Composite Applications: The Role of Matrix, Fiber, and Interface*, 1992, p.18, Table 1.6, reprinted by permission of John Wiley and Sons, Inc.

merizing with a more flexible component. Examples of these methods include the incorporation of ether (General Electric's Ultem®), amide (Amoco's Torlon®), and hexafluoroisopropylidene (DuPont's Avimid), or the use of the bulky indane group (Ciba Geigy's 5218®).

The condensation reaction of a variety of bis(n-diamine)s with aromatic dicarboxylic acids yields a class of polymers known as polybenzimidazoles (PBI). The processing cycle is fairly long and complicated due to the liberation of significant amounts of byproducts during condensation. The most attractive PBI for commercial use is poly[2, 2'-(m-phenylene)-5, 5'-bibenzimidazole]. Its glass transition temperature (420°C) is the highest of any commercially available polymer. PBIs are well suited for use at temperatures up to 250°C, and they offer good stability after aging. At higher temperatures oxidative degradation occurs and their strength gradually diminishes with time.

Polyphenylene sulfide (PPS) is a semicrystalline aromatic thermoplastic. It has a melting temperature of 285°C and a glass transition of 85°C and can be prepared from relatively inexpensive monomers. It exhibits excellent chemical and thermal stability. Interestingly, cross-linking and/or chain extension occurs at high temperature in the presence of oxygen. As a result of these changes, toughness, molecular weight, and melt viscosity are all increased. A number of different poly(arylene sulfide)s can be obtained by the addition or substitution of different monomer groups during polymerization. These copolymers generally exhibit an amorphous or less crystalline morphology and a higher glass transition temperature.

Polyarylene ethers represent a family of polymers obtained from nucleophilic or electrophilic reaction. Poly(ether ether ketone) (PEEK) is perhaps the most well known of this class. It is synthesized by the aromatic nucleophilic step or condensation reaction of 4, 4'-difluoro-benzophenone and hydroquinone in a solvent of diphenylsulfone. Other poly(ether ketone)s have been formulated having different ratios of ketone to ether groups. Polyarylene ester resins are excellent engineering thermoplastics, but they have found limited use in high-performance composite applications due to inadequate chemical resistance and adhesion. PEEK thermoplastic is manufactured by Fiberite. It is known as Victrex® in the resin-only form, and it is referred to as neat resin and as APC-2® (an acronym for aromatic polymer composite) when preimpregnated into the fiber. Table 1.8 gives the representative properties of several thermoplastic resins.

1.6
FIBER SURFACE TREATMENTS

Though a composite may be made from a strong fiber and a well-suited matrix, the result may not necessarily be a strong material. The reason is that the strength of the fiber-matrix interface is equally important in determining the mechanical performance of a composite. The surface area of the fiber-matrix interfaces for a single layer of a typical graphite-epoxy composite is about 50 times the total surface area of that layer. Ultimately, the successful development of composite materials is determined by the quality of the fiber-matrix interface. To enhance the qualities of

TABLE 1.8
Typical room temperature properties of thermoplastic polymers

Property	Aromatic polyimides		PPS: Polyphenylene sulfide	PEEK: Poly(ether ether ketone)	PS: Polysulfone	PP: Polypropylene	Nylon: Nylon 6, 6	PC: Polycarbonate
	Polyether imide (Ultem)	Polyamide imide (Torlon)						
Density (kg/m³)	1270	1400	1340	1320	1240	900	1140	1060–1200
Tensile modulus (GPa)	3	5	3.3	—	2.5	1–1.4	1.4–2.8	2.2–2.4
Tensile strength (MPa)	105	95–185	70–75	92–100	70–75	25–38	60–75	45–70
Compressive strength (MPa)	140	276	110	—	—	—	34	86
Elongation (%)	60	12–18	3	150	50–100	300	40–80	50–100
Coeff. of thermal expansion ($\times 10^{-6}$/°C)	62	36	54–100	—	56–100	110	90	70
Thermal conductivity (W/m/°C)	—	—	—	—	—	0.2	0.2	0.2
Glass transition temperature (°C)	217	243–274	85	143	190	−20 to −5	50–60	133
Water absorption (%) [24h @ 20°C]	0.25	0.3	0.2	0.1	0.2	0.03	1.3	0.1

the interface, the surface of the fiber is treated by a number of agents or processes, referred to collectively as interfacial treatments and sizings, that produce chemical change of the surface. Many different interfacial treatments are used in the composites industry. Lubricants and protectants are used immediately after fiber formation to protect the fibers from damage as they pass over guide rollers and winders. Coupling agents are used to increase adhesion between the fiber and matrix. Special coatings are sometimes used to protect the fibers from environmental attack, such as corrosion from salt water. Most fiber sizes are formulated so that several different objectives are met with the same compound. For instance, the same compound that protects the fibers during drawing and winding may later be used as an adhesion promoter between the fiber and matrix. Other interfacial treatments are also used, including plasma treatment, acid etching, irradiation, and oxidation. For a number of reasons the technology of fiber sizing is very complicated. Usually, fiber sizes are aqueous dispersions or solutions. The adhesives used in sizes are particulates in dilute suspensions. Most of the technologies of colloid stabilization and surfactant chemistry are closely guarded trade secrets developed through empirical observation. In fact, many of the sizes used in industry were developed for water-based paints and adhesives. With the multiplicity of objectives for sizing compounds, it is not uncommon to find that improper performance can be traced to problems with sizing, either in its application or its chemistry. For polymer composites perhaps the most important issue is the promotion of adhesion between the fiber and the matrix.

Currently, there are efforts to have some of the sizing migrate a short distance from the surface of the fiber and combine with the matrix. This provides a gradual transition from matrix to fiber and sizing properties in the vicinity of the fiber. This region of transition is sometimes referred to as the interphase region, and its presence is thought to enhance fiber-matrix interaction.

1.6.1 Graphite Fiber Treatment

Graphite fibers are generally fragile and subject to abrasion during handling. To protect the fibers from abrasion, epoxy sizings are applied to the fiber surface. In some cases, a vinyl addition polymer may be incorporated into the epoxy sizing to improve handling characteristics. The sizing compounds are generally in the form of a solution and may contain lubricants and film-formers.

Pyrolytic coatings have been shown to improve the tensile strength and increase the oxidation resistance of graphite fibers. These coatings are applied by decomposing the source gas (hydrocarbons, elemental halides) onto the heated surface of the fibers. Coating uniformity is difficult to control, but can be improved if the pyrolysis is carried out in a vacuum. Improvements in tensile strength are also found when a bromine treatment is given to an untreated fiber. In this technique carbon fibers are immersed in liquid bromine or bromine dissolved in a solvent. The bromine is subsequently removed, but some remains within the fiber.

Boron nitride coatings have also been found to improve the oxidative resistance of graphite fibers. Coatings are applied by mixing boric acid with urea, passing the fibers through the solution, drying the fibers to drive off the water, and firing in a nitrogen atmosphere for about one minute at 1000°C. During firing the urea-boric acid complex is reduced to boron nitride. The coating concentration is about 4 percent after firing and the boron nitride is molecularly bonded to the carbon surface. This type of bonding assures a permanent joining of the boron nitride and the fiber. Boron nitride-coated fibers are extremely stable at high temperatures in oxidizing atmospheres. Metal carbides have also been used instead of boron nitride to protect fibers from oxidation. The carbide impregnates the fiber surface and lodges in the crevices of surface irregularities. When the carbide surface is exposed to an oxidizing atmosphere above 400°C, the carbide is converted to a refractory oxide. This oxide protects the fiber from further oxidation. The refractory oxide acts not only as a chemical barrier but also as a thermal shield, and it improves the overall durability of the fiber.

First attempts at improving fiber-matrix bonding concentrated on chemical mod-ifications to the resin system. The resin's wetting ability has been used as a criterion to judge the suitability of using a particular resin system with graphite fiber re-inforcement. However, some resins having poor wetting ability, such as aromatic polyphenylene resins, possess other desirable characteristics that dictate their use in composite structures. Thus, there has been a need to develop techniques to al-ter the fiber surface to overcome the poor wetting ability and promote increased fiber-matrix adhesion. As a result, several surface treatment methods exist. The most well-developed are acid treatments, oxidation treatments, plasma treatments, carbon coatings, resin coatings, ammonia treatments, and electrolytic treatments.

Acids are used to promote a strong interfacial bond between graphite fibers and resins with poor wetting ability. For instance, fibers are wetted with a sulfonic acid solution, dried to drive off the solvent, heated to allow the acid to react with the surface or with itself, washed to remove any unreacted acid, and then dried to remove the washing solvent. Hypochlorus acid has also been used successfully, and the resulting composites show significantly increased shear strengths.

Heat treating graphite fibers in an ammonia atmosphere at 1000°C prior to im-pregnation with matrix resin increases the shear strength of the composite material. Heating is usually accomplished by passing a current through the fibers as they pass through a controlled atmosphere containing 10 to 100 percent ammonia. The balance can be any nonoxidizing gas like nitrogen, argon, hydrogen, or helium. The exposure time is usually very short, from 1 to 60 seconds. As the concentration of ammonia is increased, the shear strength of the composite material increases; however, there may also be a decrease in fiber tensile strength and tensile strength of the composite. For optimal properties the competing influences of the ammonia content of the atmo-sphere, exposure time, and fiber temperature must be balanced. Resins that bond well with fibers treated with ammonia are those that bond to amine functions during cure. This class of resins includes epoxies, polyimides, polyethylene, and polypropylene.

Composite materials made from low-modulus carbon fibers, typically, have a high shear strength. Compared to high modulus fibers, these types of fibers show superior bonding ability with resins. It is believed that this behavior is a direct result of the presence of an isotropic surface layer of carbon on the low-modulus carbon fibers. The fundamental structural units of these fibers are small and have a lay-ered structure, but are randomly oriented. Therefore, a large portion of the fiber surface will consist of exposed layer edges uniformly distributed over the surface. The exposed edges are believed to be highly reactive and may even bond chemically with epoxy resins. In contrast, high-modulus carbon fibers are more ordered due to graphitization and orientation of the crystallites during manufacture. The surface is anisotropic, consisting of relatively large areas of exposed crystallite basal planes and little exposure of edges, such as the skin-core microstructure in Figure 1.14(a). Crystallite basal planes have low reactivity, and they show poor bonding ability with most resins. One technique for improving interfacial bonding for high-modulus graphite fibers is to deposit a coating of isotropic carbon on the surface. Two methods exist to accomplish this. In the first method, graphite fibers are electrically heated to about 1200°C and then exposed to an atmosphere containing methane and nitrogen. The methane decomposes on the fiber surface, creating a uniform carbon coating. Composites made using these fibers show a twofold increase in shear strength compared to untreated fibers. There is a small loss in tensile strength for these composites. In the second method, the fibers are impregnated with a thermally carbonizable organic precursor such as phenylated polyquinoxaline. The precursor is then pyrolyzed at a high temperature and it carbonizes on the surface of the fibers.

Carbon and graphite fibers can be electrolytically treated to improve their surface characteristics for improved bonding to matrices. Electrolysis is used either (1) to change the reactivity of the fiber surface or (2) to deposit chemical groups on the surface of the fiber that will bond to the matrix. Electrolysis is accomplished, as in Figure 1.17, by pulling the fibers through a series of rollers that are electrically charged in positive/negative pairs. The negatively charged rollers are immersed in the electrolytic solution. After passing through the electrolysis, the fibers are dried before being wound onto a take-up reel. Electrolytic solutions are usually an aqueous-based caustic mixture. If the fibers are used as the cathode and vinyl monomer is added to the electrolysis solution, then the fibers will be covered with the vinyl polymerization product, greatly improving the bond strength between the fiber and some matrix resins.

Oxidation of carbon and graphite fibers begins at the fiber surface in regions that are irregularly shaped. By heating the fiber to 1000°C in an oxidizing atmosphere, a pitted fiber surface is obtained. The increased surface area leads to greater bond strength compared to untreated fibers. The exposure time must be controlled so that weight loss is less than 1 percent. In another type of treatment, fibers are treated by exposure to formates, acetates, and nitrate salts of copper, lead, cobalt, cadmium, and vandium pentoxide. The fibers are subsequently oxidized by exposing them to air or oxygen in the temperature range 200–600°C. This method uniformly roughens the fiber surface.

Significant improvements in interfacial bonding can be realized by plasma treating the fiber surface. Fibers are drawn through a plasma chamber in which is generated a thermal plasma of argon or oxygen, or mixtures of hydrogen and nitrogen or carbon fluoride and oxygen. Argon plasma generally introduces active sites that are able to react subsequently with atmospheric oxygen. Oxygen plasma introduces oxygen both by direct reaction and by active sites. Carbon fluoride/oxygen plasma essentially results in an etching and oxidation of the fiber surface. Nitrogen/oxygen plasma introduces amine-like groups on the fiber surface that are able to participate

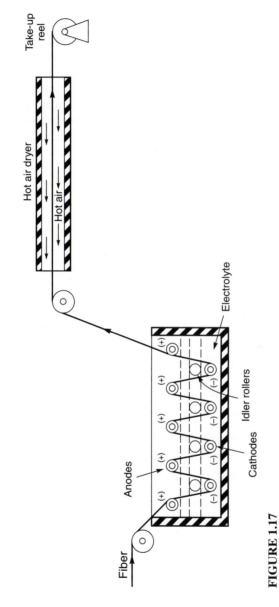

FIGURE 1.17
Electrolytic surface treatment of graphite fibers.
(*Source:* U.S. Patent 3,832,297 Alliant Techsystems, Inc.)

in the cross-linking reaction for epoxies. The temperature within the plasma can reach 8000°C. Carbon fiber is continuously drawn through the plasma, and the residence time within the plasma must be carefully controlled so that the fiber surface temperature does not get too high.

A polymer coating on the surface is sometimes incorporated to enhance interfacial bonding. Thermoplastic polymers like polysulfone or polycarbonate are coated onto the surface of fibers before they are impregnated with a thermosetting matrix. The result is an interphase region between fiber and matrix. The resulting composite material shows a significant increase in shear strength. Elastomers like urethane polymers are also used to provide a compatible interphase region between the fiber surface and the matrix. The elastomer can be applied to the fiber in the form of a sizing, or it can be blended with matrix material. The development of an interphase region provides for a more efficient distribution of stresses and reduces the tendency for cracks to develop at the interface. In addition, compatible elastomer and matrix material combinations, a result of using a common curing agent in both the elastomer and the matrix, will result in a gradual transition of properties within the interphase. These composites have enhanced toughness and high impact strength. Tailoring of interphase properties may provide for significant improvements in composite mechanical properties.

Many other techniques have been used to some degree of success to improve fiber-matrix bond strengths. For instance, whiskerization is a technique in which silicon carbide single crystals are grown on the surface of carbon fibers to roughen the surface and to change the surface chemistry. Typical silicon carbide crystals are 0.01 to 1×10^{-6} m in diameter. They provide a mechanical interlock with the matrix material surrounding the fiber. There is also some evidence that irradiation by neutrons during cure improves fiber-resin bonding. Oxides, organometallics, isocyanates, and metal halides have all been used with some success by suitably coating the carbon fiber surface and then impregnating the fibers to form the composite. With the multiplicity of surface treatment techniques available, much of the development work in characterizing new composite materials revolves around empirical analysis of optimal combinations of surface treatments to produce composites with high tensile and shear strengths, good toughness, and reasonable costs.

The illustrations in Figure 1.18 present a microscopic view of various effects of applying—or not applying—surface treatments to graphite fibers. Figure 1.18(a) shows a failed graphite fiber–reinforced composite that was not treated. Here we see inadequate fiber-matrix bonding; the fiber have very little matrix material attached to them, which is evidence of poor interfacial bonding. Figure 1.18(b) also shows a failed, untreated composite, but this example exhibits much better interfacial bonding. Here the failure has occurred in the matrix itself, some of which remains attached to the fiber. Figure 1.18(c) and (d) provide an alternative point of view and dramatically illustrate the influence of graphite fiber surface treatment on the tendency of a matrix to bond to the fibers. Figure 1.18(c) illustrates the effects of a poor surface treatment; the thermoplastic matrix, because it is forming small spheres as a consequence of the treatment, is making minimal contact with the fiber. By contrast, in Figure 1.18(d), the contact area between the fiber and the matrix material has been maximized as a result of an improved surface treatment.

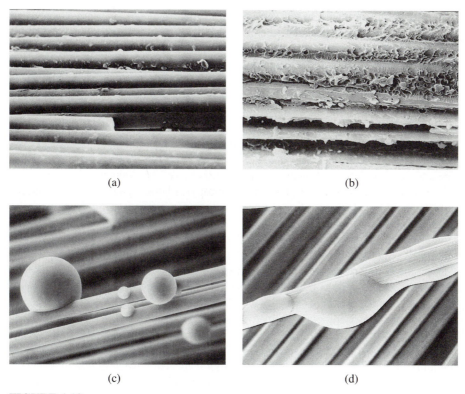

(a)   (b)

(c)   (d)

FIGURE 1.18
Effect of surface treatment on matrix bonding (a) Poor bonding (b) Good bonding
(c) Poor fiber-matrix attraction (d) Good fiber-matrix attraction

1.6.2 Glass Fiber Treatment

A fiber sizing for glass fibers to be used in polymer matrix composites must accomplish several objectives simultaneously. The fiber sizing should: (1) promote good adhesion between the glass fiber and the polymer matrix; (2) promote good cohesion between the fibers that make up the tow; (3) impart certain handling characteristics, like hardness or choppability (for shorter fiber composites); (4) provide adequate protection to the fiber during processing; and (5) impart antistatic properties so that static charges do not build up on the fiber surface.

For glass fibers to act efficiently as a reinforcement, some method of coupling the hydrophilic fibers to the hydrophobic polymer matrix must be used. (A material is said to be *hydrophobic* if it does not absorb water; it is said to be *hydrophilic* if it can absorb water.) Coupling agents are a class of chemicals that are organo-metallic and, in most cases, organo-silicon, possessing dual functionality. Each metal or silicon atom has attached to it one or more groups that can react with the glass surface, and one or more groups that can coreact with the resin during its polymerization.

With a coupling agent, a chemical bridge is formed between the glass surface and the polymer. In actual practice, the function of coupling agents is a little more complex. There is significant evidence to suggest that several layers (about eight monomolecular layers) react with the glass surface. This gives rise to an interphase layer within the glass that possess mechanical properties different from those of the glass fibers and the polymer matrix. There is also evidence to suggest that the flexibility of the interphase is sufficient to permit the breaking and reforming of bonds to the glass surface when the composite is under stress. Table 1.9 shows the effects of coupling agents on the strength of glass fiber-reinforced composite and gives results for a glass-polyester system that was tested for flexural strength under dry and wet conditions. The dry strength is increased by 42 percent using the A 174 silane coupling agent (g-methacryloxypropyltrimethoxy-silane). The strength enhancement by silane coupling is even more dramatic considering the test results for specimens immersed in boiling water for two hours. For this case the flexural strength is increased by 128 percent over the untreated specimens. In addition, the wet strength retention (ratio of the dry strength to wet strength) is improved. To a large degree, the pathways to strength enhancement through the use of coupling agents are empirically driven, as the surface chemistry of adhesion is not fully developed for many composite systems.

Another sizing component, film-formers, are materials used to bind the individual fibers together as a tow. The vast bulk of fiber sizes employ polyvinyl acetate (PVA) as a film-former. PVA is in the form of a suspension of particles in an aqueous medium. Photomicrographs of a tow show that PVA is deposited in globules on and between fibers and that the tow is held together by these globules' forming bridges from one fiber to the next. One drawback to the use of PVA as a film-former is that it is an unwelcome addition to the composite material. PVA remains after the fiber tows are impregnated with resin, leading to a possible reduction in mechanical properties of the composite material. Other film-formers more compatible with the intended matrix system are being developed. Certain polyesters have been shown to be suitable for epoxy or polyester matrices, and acrylic polymers have been introduced in fiber sizes as a film-former for thermoplastic composites.

Plasticizers are added to PVA emulsions (8–20 percent) to increase flexibility and reduce the softening temperature of the fiber tow. By increasing the amount of

TABLE 1.9

Effect of various coupling agents on the flexural strength of otherwise identical glass cloth-polyester laminates

Trade name of silane coupling agent	Flexural strength at 25°C (MPa)		Wet strength retention (%)
	Dry	**After 2-hour boil**	
None	42.3	24.7	58
Volan A (DuPont)	50.8	43.7	86
A 172 (Union Carbide)	50.8	48.0	94
A 174 (Union Carbide)	60.0	56.4	94

Source: Adapted from K.L. Lowenstein, *The Manufacturing Technology of Continuous Glass Fibres*, 3rd Ed., 1993, Table 6.2, p. 258, with kind permission from Elsevier Science, P.O. Box 211, 1000 AE Amsterdam, The Netherlands.

plasticizer added to fiber sizing, the flexibility of the fiber tow can be increased. The most common plasticizers used are phthalates, phosphates, and polyesters. The use of polyesters as plasticizing agents is of particular interest to the glass fiber-reinforced composites industry since they are much more compatible with epoxy and polyester matrices.

Lubricants are added to fiber sizes in concentrations from 0.2 to 2 percent. Most lubricants are cationic surface-active agents, as they will be attrached to the negative charges normally present on the surface of a glass fiber. The cationic group is usually an amine to which a fatty acid or other lubricating group is attached.

Static electricity on fibers is created by friction as they are drawn over rollers or as they slide relative to each other. Static charging increases until the losses from conduction are balanced by the rate at which charges are generated. Conduction along the surface of the fiber is possible if it is moist enough. Thus, one technique used to control static electricity is to carry out the processing of fibers in a humid environment. A relative humidity level of about 70 percent is usually sufficient. If this is not practical, then antistatic agents must be used to conduct electricity along the fiber surface. The problem that glass fiber manufacturers face is that the amount of antistatic material that must be supplied to ensure an adequate conduction path may be so large that other properties of the tow are sacrificed. For example, the impregnation of resin into the tows may be slowed if the concentration of antistatic agents is too large. Chemicals suitable as antistatic agents must be able to ionize to conduct electricity. For these agents to ionize they must be hydrophilic. Both lithium chloride and magnesium chloride have been used successfully as antistatic agents.

The exact formulation for a particular sizing is dictated by a number of factors, such as intended matrix material, cost of components, compatibility among different components, handling characteristics of the fiber tow, and stability of the size in the diluted form. Most of the common fiber sizings are formulated with the concentration ranges listed in Table 1.10.

1.6.3 Polymer Fiber Treatment

Relatively little data have been assembled concerning the surface treatment of polymer fibers. The data that exist can be grouped into two classes: (1) protective coatings and (2) adhesion promoters.

TABLE 1.10
Typical formulation for glass fiber sizing

Component	(%)
Coupling agent	0.3–0.6
Film-former	3.5–15.0
(including plasticizer)	
Lubricants	0.1–0.3
Surfactants	0–0.5
Antistatic agents	0–0.3
Distilled water	83.3–96.1

Kevlar fibers are susceptible to surface damage during processing operations such as weaving. To minimize this damage, they are coated with a polyvinyl alcohol sizing, which serves as a protective layer covering the fiber surface. Conventional coupling agents used in glass and carbon fiber sizings to improve adhesion do not work well with Kevlar fibers. However, the fiber surface of Kevlar shows a good affinity for some epoxy resins. Thus, a light pretreatment with an epoxy resin has been shown to give improved adhesion with other polymer matrices. Spectra fibers can be plasma treated to increase the strength of the interfacial bond with epoxy matrices. Flexural strength has been shown to increase by a factor of three over untreated Spectra-epoxy composites.

1.7
SUMMARY

Our discussion has focused on the basic ingredients of a fiber-reinforced composite, namely, the fiber and the matrix and, to some extent, the interface. Though this chapter relied largely on a materials science and chemistry perspective, it is nevertheless important for mechanicians to be aware of this terminology and these basic ideas, particularly if the mechanician is to work in an interdisciplinary environment. We now turn to some of the basic concepts and principles for predicting the mechanical response of fiber-reinforced composite materials.

1.8
SUGGESTED READINGS

1. Peebles, L. H. *Carbon Fibers: Formation, Structure, and Properties.* Ann Arbor, MI: CRC Press, 1995.
2. Loewenstein, K. L. *The Manufacturing Technology of Continuous Glass Fibres.* 3rd ed. Amsterdam: New York: Elsevier Publishing Co., 1993. This is volume 6 in the Glass Science and Technology Series.
3. Vigo, T., and B. Kinzig, eds. *Composite Applications: The Role of the Matrix, Fiber, and Interface.* New York: VCH Publishers, 1992.
4. Dresselhaus, M. S.; G. Dresselhaus; K. Sugihara; I. L. Spain; and H. A. Goldberg. *Graphite Fibers and Filaments.* Berlin and New York: Springer-Verlag, 1988. This is volume 5 in the Springer Series in Materials Science.
5. Hull, D. *An Introduction to Composite Materials.* Cambridge, UK: Cambridge University Press, 1981. This volume is part of the Cambridge Solid State Science Series.
6. Delmonte, J. *Technology of Carbon and Graphite Fiber Composites.* New York: Van Nostrand Reinhold Co., 1981.
7. Sittig, M., ed. *Carbon and Graphite Fibers: Manufacture and Applications.* Park Ridge, NJ: Noyes Data Corp. and *Chemical Technology Review* no. 162, 1980.
8. Mohr, J. G., and W. P. Rowe. *Fiberglass.* New York: Van Nostrand Reinhold, 1978.
9. Carroll-Porczynski, C. Z. *Manual of Man-Made Fibers: Their Manufacture, Properties, and Identification.* New York: Chemical Publishing Co., 1961.

CHAPTER 2

Linear Elastic Stress-Strain Characteristics of Fiber-Reinforced Material

As the previous chapter shows, the study of the mechanics of fiber-reinforced composites could begin at several points. Because the fiber plays a key role in the performance of the material, one logical starting point would be to study the interaction of the fiber with the matrix. The photograph of Figure 1.6 provides ample motivation for studying the fiber and the matrix as separate constituents and focusing on their interaction. We might address a host of problems: stresses in the fiber, stresses in the matrix around the fiber, adhesive stresses at the interface, breaking of the fiber, cracking of the matrix, the interaction of two or more fibers, the effects on stresses in the matrix of moving two fibers farther apart or closer together, the effect on neighboring fibers of a broken fiber, or local yielding of either the matrix or fiber. This localized look at the interaction of the fibers and the matrix is called micromechanics. It is indeed a logical starting point for a study of the mechanics of fiber-reinforced composites. On the other end of the spectrum, interest centers on the behavior of structures made using fiber-reinforced material. Deflections, maximum allowable loads, vibration frequencies, vibration damping, energy absorption, buckling loads, the effects of geometric discontinuities such as holes and notches, and many other more global responses are all of interest. In between these two extremes, the response of an individual layer or the response of a group of layers can be of interest. How does an individual layer respond when subject to stresses? How much does it deform? How much load can it sustain? Similarly, how does a laminate respond when subject to stresses? How much does it deform? How much load can it sustain? What is the influence of neighboring or adjacent layers on any particular layer? What is the effect on the laminate of changing the fiber orientation of any particular layer? What is the effect of changing the material properties of any one particular layer? The list of questions is almost endless. With the continuous introduction of new fibers having greater strength and stiffness, old questions have to be reanswered and new ones are asked. Moreover, new polymer matrix materials are introduced frequently, and issues need to be reexamined in the context of these new materials.

In this study of the mechanics of composite materials and structures, we will present methodologies that will enable engineers, scientists, and designers to answer some of these questions. The starting point will be an examination of the deformations of an element of material taken from a single layer. The element, though small, is assumed to contain many fibers. Rather than starting with the examination of fiber and matrix interaction, this starting point is used because it allows an orderly and smooth transition to the analysis of composite structures, the final products for any fiber-reinforced material. Also, in the analysis of a complete structure, it is impossible to include the response of every fiber to the surrounding matrix material. Computers are not big enough to allow this. On the other hand, what is responsible for the failure of composite structures is the breaking of the fibers, the breaking of the fiber-matrix interface bond, and the breaking of the matrix. Therefore, micromechanics cannot be overlooked. A glimpse of micromechanics will be provided in the next chapter when the issue of estimating the elastic properties of a composite is addressed.

2.1
STRESSES AND DEFORMATIONS
IN FIBER-REINFORCED MATERIALS

In discussing the mechanics of fiber-reinforced materials, it is convenient to use an orthogonal coordinate system that has one axis aligned with the fiber direction. We will do so here and identify the system as the 1-2-3 coordinate system or the *principal material coordinate system*. Figure 2.1 illustrates an isolated layer and the orientation of the principal material coordinate system. The 1 axis is aligned with the fiber direction, the 2 axis is in the plane of the layer and perpendicular to

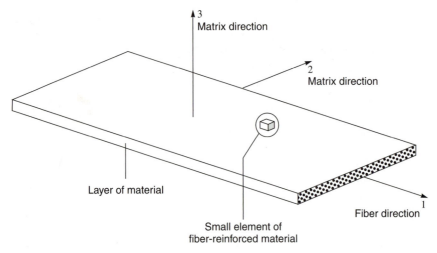

FIGURE 2.1
Principal material coordinate system

the fibers, and the 3 axis is perpendicular to the plane of the layer and thus also perpendicular to the fibers. The 1 direction is the *fiber direction*, while the 2 and 3 directions are the *matrix directions*. As we mentioned in Chapter 1, the direction perpendicular to the fibers is also called the *transverse direction*. The terms *matrix direction* and *transverse direction* are somewhat ambiguous because there are two directions that fit either of these descriptions. We shall use *matrix direction*, and the specific direction (i.e., 2 or 3) will be made clear from the context of the problem. Stresses, strains, and strengths will ultimately be referred to the principal material coordinate system.

The study of the stress-strain response of a single layer is equivalent to deter-mining the relations between the stresses applied to the bounding surfaces of the layer and the deformations of the layer as a whole. The strain of an individual fiber or element of matrix is of no consequence at this level of analysis. The effect of the fiber reinforcement is smeared over the volume of material, and we assume that the two-material fiber-matrix system is replaced by a single homogenous material. This is an important concept because it makes the analysis of a fiber-reinforced composite easier. Equally important is the fact that this single material does not have the same properties in all directions. It is obviously stronger and stiffer in the 1 direction than in the 2 or 3 directions. In addition, just because the 2 and 3 directions are both perpendicular to the fiber direction, the properties in the 2 and 3 directions are not necessarily equal to each other. A material with different properties in three mutually perpendicular directions is called an *orthotropic* material. As a result, a layer is said to be *orthotropic*. The 1-2, 1-3, and 2-3 are three planes, and the material properties are symmetric with respect to each of these planes. As mentioned in Chapter 1, a material with the same properties in all directions is said to be *isotropic*.

Figure 2.2 illustrates a small element of smeared fiber-reinforced material sub-ject to stresses on its six bounding surfaces. As Figure 2.1 shows, this small volume of material has been considered removed from a layer. The normal stress acting on the element face with its outward normal in the 1 direction is denoted as σ_1. The shear stress acting in the 2 direction on that face is denoted as τ_{12}, and the shear stress acting in the 3 direction on that face is denoted as τ_{13}. The normal and shear stresses acting on the other faces are similarly labeled. The extensional strain responses of

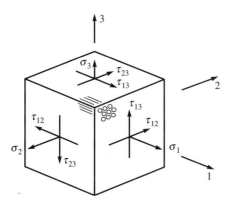

FIGURE 2.2
Stresses acting on a small element of fiber-reinforced material

the element as referenced in the 1-2-3 coordinate system are denoted as ε_1, ε_2, and ε_3, while the engineering shearing strain responses are denoted as γ_{12}, γ_{23}, and γ_{13}. With this notation ε_1 is the stretching of the element in the fiber direction, γ_{12} is the change of right angle in the 1-2 plane, and so on. The stress-strain relation for the small element of material will be constructed by considering the response of the element to each of the six stress components. As only linear elastic response is to be considered, superposition of the responses will be used to determine the response of the element to a complex or combined stress state.

Figure 2.3(a) illustrates the element subjected to only a tensile normal stress in the 1 direction, σ_1. Figure 2.3(b)–(d) illustrate three views of the element indicating how it would be deformed by this tensile stress. The tensile normal stress σ_1 causes extension of the element in the 1 direction and, due to Poisson effects, contraction in the 2 and 3 directions. There is no a priori reason to believe that the contractions in the 2 and 3 directions are the same. In addition, nothing has been said so far to indicate that the element of material actually contracts. It could expand. In reality, this is not the case for a single layer. However, laminates can be made that will expand rather than contract. Laminate Poisson's ratios will be discussed later.

The extensional strain in the 1 direction is related to the tensile normal stress in the 1 direction by the tensile, or extensional, modulus of the equivalent smeared material in the fiber direction E_1. The relation between these quantities is

$$\varepsilon_1 = \frac{\sigma_1}{E_1} \tag{2.1}$$

If a Poisson's ratio relating contraction in the 2 direction to extension in the 1 direction

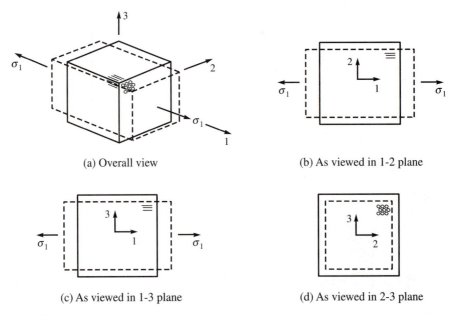

(a) Overall view   (b) As viewed in 1-2 plane

(c) As viewed in 1-3 plane   (d) As viewed in 2-3 plane

FIGURE 2.3
Deformation of an element due to σ_1

is defined to be

$$\nu_{12} = -\frac{\varepsilon_2}{\varepsilon_1} \tag{2.2}$$

then

$$\varepsilon_2 = -\nu_{12}\varepsilon_1 = -\nu_{12}\frac{\sigma_1}{E_1} \tag{2.3}$$

The subscripts, and their order, on Poisson's ratio are important. The first subscript refers to the direction of the applied tensile stress and the resulting extensional strain. The second subscript refers to the direction of the contraction. According to this convention, the contraction in the 3 direction is related to the extension in the 1 direction by ν_{13}, specifically,

$$\nu_{13} = -\frac{\varepsilon_3}{\varepsilon_1} \tag{2.4}$$

We rewrite this equation as

$$\varepsilon_3 = -\nu_{13}\varepsilon_1 = -\nu_{13}\frac{\sigma_1}{E_1} \tag{2.5}$$

If instead of applying a tensile normal stress in the 1 direction, a tensile normal stress is applied in the 2 direction, then the element of smeared material will deform as in Figure 2.4. Because of the softness of the material perpendicular to the fibers, the element will extend more easily in the 2 direction than in the 1 direction. Because the stiff fibers tend to counter any Poisson effect, the contraction in the 1 direction

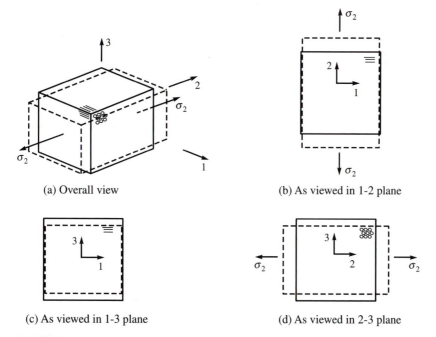

(a) Overall view

(b) As viewed in 1-2 plane

(c) As viewed in 1-3 plane

(d) As viewed in 2-3 plane

FIGURE 2.4
Deformation of an element due to σ_2

will be minimal. In contrast, the contraction in the 3 direction will be large; there is only matrix and the soft diametral direction of the fiber resisting deformation. For the loading of Figure 2.4, the tensile normal stress in the 2 direction, σ_2, and the extensional strain in the 2 direction are related by another extensional modulus through the relation

$$\varepsilon_2 = \frac{\sigma_2}{E_2} \tag{2.6}$$

As might be expected, because the stress is acting perpendicularly to the fibers, E_2 is much smaller than E_1. Using the subscript convention established for Poisson's ratios, the contraction in the 1 direction is related to the extension in the 2 direction by yet another Poisson's ratio, namely,

$$\nu_{21} = -\frac{\varepsilon_1}{\varepsilon_2} \tag{2.7}$$

As a result, due to σ_2,

$$\varepsilon_1 = -\nu_{21}\varepsilon_2 = -\nu_{21}\frac{\sigma_2}{E_2} \tag{2.8}$$

Similarly, the contraction in the 3 direction is related to the extension in the 2 direction by ν_{23}, ν_{23} being defined as

$$\nu_{23} = -\frac{\varepsilon_3}{\varepsilon_2} \tag{2.9}$$

Rearrangement gives

$$\varepsilon_3 = -\nu_{23}\varepsilon_2 = -\nu_{23}\frac{\sigma_2}{E_2} \tag{2.10}$$

It is important to recognize that the definitions being made apply only to the case of a single stress acting on the element of material. The deformation is being examined with σ_1 alone acting on the element of material, then with σ_2 alone acting. *The definitions of extensional moduli and Poisson's ratios are valid only in the context of the element being subjected to a simple tensile or compressive stress.*

Finally, if only a tensile normal stress σ_3 is applied, the strains in the three directions are given by

$$\varepsilon_3 = \frac{\sigma_3}{E_3}$$

$$\varepsilon_2 = -\nu_{32}\varepsilon_3 = -\nu_{32}\frac{\sigma_3}{E_3} \tag{2.11}$$

$$\varepsilon_1 = -\nu_{31}\varepsilon_3 = -\nu_{31}\frac{\sigma_3}{E_3}$$

In the above, E_3 is the extensional modulus in the 3 direction, ν_{32} relates contraction in the 2 direction and extension in the 3 direction, and ν_{31} relates contraction in the 1 direction and extension in the 3 direction. This is for the case of only stress σ_3 being applied. Again, due to the relative stiffness in the fiber and matrix directions, the σ_3 stress will not cause much contraction in the 1 direction. Contraction in the 2 direction will be much larger, as Figure 2.5 shows.

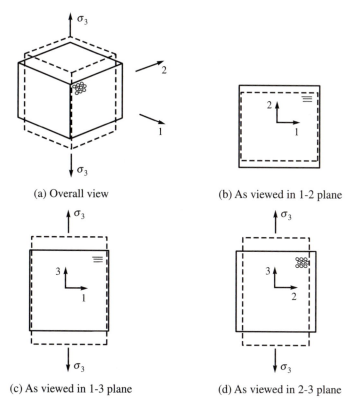

(a) Overall view   (b) As viewed in 1-2 plane

(c) As viewed in 1-3 plane   (d) As viewed in 2-3 plane

FIGURE 2.5
Deformation of an element due to σ_3

If all three tensile stresses are applied simultaneously, the strain in any one direction is a result of the combined effects, namely,

$$
\left\{
\begin{array}{c}
\varepsilon_1 \\
\varepsilon_2 \\
\varepsilon_3
\end{array}
\right\}
=
\left[
\begin{array}{ccc}
\dfrac{1}{E_1} & \dfrac{-\nu_{21}}{E_2} & \dfrac{-\nu_{31}}{E_3} \\[2ex]
\dfrac{-\nu_{12}}{E_1} & \dfrac{1}{E_2} & \dfrac{-\nu_{32}}{E_3} \\[2ex]
\dfrac{-\nu_{13}}{E_1} & \dfrac{-\nu_{23}}{E_2} & \dfrac{1}{E_3}
\end{array}
\right]
\left\{
\begin{array}{c}
\sigma_1 \\
\sigma_2 \\
\sigma_3
\end{array}
\right\}
\tag{2.12}
$$

Note that a given component of extensional strain, say, ε_2, is a result of the combined effects of the three components of normal stress. The normal stresses and the extensional strains are completely coupled; that is, the matrix is full. Poisson's ratio ν_{12} is generally referred to as the major Poisson's ratio.

The effects of the shearing stress are less complicated. For an orthotropic material there is no coupling among the three shear deformations. Figure 2.6 illustrates the

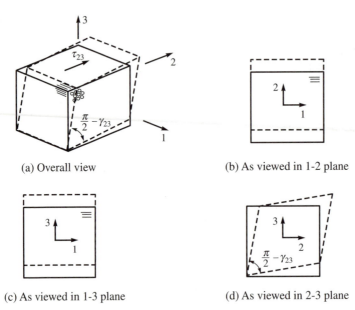

(a) Overall view

(b) As viewed in 1-2 plane

(c) As viewed in 1-3 plane

(d) As viewed in 2-3 plane

FIGURE 2.6
Deformation of an element due to τ_{23}

deformation of the small element of composite subjected to a shear stress τ_{23}, with the shearing stress causing right angles in the 2-3 plane to change. All other angles of the element remain orthogonal. It is important to note the sense of the change in right angles in the various corners of the element due to a positive shear stress, with some right angles decreasing, some increasing. As the engineering shear stress γ denotes the *change* in right angle, the relation between the applied shear stress in the 2-3 plane and the change in right angle in the 2-3 plane is given by

$$\gamma_{23} = \frac{\tau_{23}}{G_{23}} \tag{2.13}$$

The quantity G_{23} is called the shear modulus in the 2-3 plane and γ_{23} is the engineering shear strain in the 2-3 plane. The convention established for shearing in the 2-3 plane can be easily extended to the 1-3 and 1-2 planes. As Figure 2.7 illustrates, a shear stress τ_{13} causes right angles in the 1-3 plane to change but all other angles in the cube remain orthogonal. Similarly, as in Figure 2.8, a shear stress τ_{12} causes only the angles in the 1-2 plane to deform. As a result,

$$\gamma_{13} = \frac{\tau_{13}}{G_{13}} \qquad \gamma_{12} = \frac{\tau_{12}}{G_{12}} \tag{2.14}$$

The quantities G_{13} and G_{12} are the shear moduli in the 1-3 and 1-2 planes, respectively, and in general G_{12}, G_{13}, and G_{23} have different values. It is easy to envision, however, that the values of G_{12} and G_{13} could be approximately the same. The same can be said of E_2 and E_3, and ν_{12} and ν_{13}. Note that the shearing action caused by τ_{23} does nothing to the fibers except roll them over one another. The stiffness and

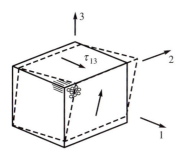

(a) Overall view

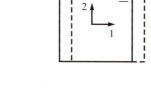

(b) As viewed in 1-2 plane

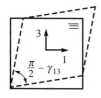

(c) As viewed in 1-3 plane

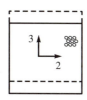

(d) As viewed in 2-3 plane

FIGURE 2.7
Deformation of an element due to τ_{13}

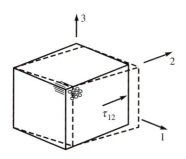

(a) Overall view

(b) As viewed in 1-2 plane

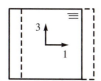

(c) As viewed in 1-3 plane

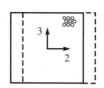

(d) As viewed in 2-3 plane

FIGURE 2.8
Deformation of an element due to τ_{12}

strength of the fiber are not involved in this shearing action. The same can be said of the shearing action caused by τ_{13} and τ_{12}. Though the fibers are not rolled over each other, they are slid along each other, without their superior stiffness and strength being involved. As a result, the magnitudes of G_{12}, G_{13}, and G_{23} can be expected to be about the same as the magnitudes of E_2 and E_3. The various extensional moduli, Poisson's ratios, and shear moduli are collectively referred to as *engineering constants*, or *engineering properties*. The analysis of fiber-reinforced composites depends on knowing the numerical values of these engineering constants.

Before we proceed, it is important to state that implicit in the discussion so far has been the fact that

$$\tau_{21} = \tau_{12} \qquad \tau_{31} = \tau_{13} \qquad \tau_{32} = \tau_{23} \qquad (2.15)$$

For this reason the stresses τ_{21}, τ_{31}, and τ_{32} have not been mentioned explicitly. Because of the above equalities, the definitions of the quantities G_{21}, G_{31}, and G_{32} are superfluous and have not been introduced. They would have to equal G_{12}, G_{13}, and G_{23}, respectively.

Also, in the application of the theory of elasticity to the analysis of composite materials, the definitions of the tensor shear strains, namely,

$$\varepsilon_{12} = \frac{\gamma_{12}}{2} \qquad \varepsilon_{13} = \frac{\gamma_{13}}{2} \qquad \varepsilon_{23} = \frac{\gamma_{23}}{2} \qquad (2.16)$$

are sometimes more convenient to use. If this is done, then the extensional and shear deformations will all be tensor quantities. Despite this, much analysis and nomenclature have been developed for composite materials that are based on engineering shear strain. Here use will be made of both strain measures, though the engineering shear strain will be normally considered because of its direct association with the change in right angles.

Finally, it will be assumed that the elastic properties of the composite in compression in the 1, 2, and 3 directions are the same as the elastic properties in tension.

All the relationships between the stresses and strains take the collective form

$$\begin{Bmatrix} \varepsilon_1 \\ \varepsilon_2 \\ \varepsilon_3 \\ \gamma_{23} \\ \gamma_{13} \\ \gamma_{12} \end{Bmatrix} = \begin{bmatrix} \frac{1}{E_1} & \frac{-\nu_{21}}{E_2} & \frac{-\nu_{31}}{E_3} & 0 & 0 & 0 \\ \frac{-\nu_{12}}{E_1} & \frac{1}{E_2} & \frac{-\nu_{32}}{E_3} & 0 & 0 & 0 \\ \frac{-\nu_{13}}{E_1} & \frac{-\nu_{23}}{E_2} & \frac{1}{E_3} & 0 & 0 & 0 \\ 0 & 0 & 0 & \frac{1}{G_{23}} & 0 & 0 \\ 0 & 0 & 0 & 0 & \frac{1}{G_{13}} & 0 \\ 0 & 0 & 0 & 0 & 0 & \frac{1}{G_{12}} \end{bmatrix} \begin{Bmatrix} \sigma_1 \\ \sigma_2 \\ \sigma_3 \\ \tau_{23} \\ \tau_{13} \\ \tau_{12} \end{Bmatrix} \qquad (2.17)$$

The square six by six matrix of material properties is called the *compliance matrix*, commonly denoted by S. In terms of S, the stress-strain relations are written as

$$
\begin{Bmatrix} \varepsilon_1 \\ \varepsilon_2 \\ \varepsilon_3 \\ \gamma_{23} \\ \gamma_{13} \\ \gamma_{12} \end{Bmatrix} = \begin{bmatrix} S_{11} & S_{12} & S_{13} & 0 & 0 & 0 \\ S_{21} & S_{22} & S_{23} & 0 & 0 & 0 \\ S_{31} & S_{32} & S_{33} & 0 & 0 & 0 \\ 0 & 0 & 0 & S_{44} & 0 & 0 \\ 0 & 0 & 0 & 0 & S_{55} & 0 \\ 0 & 0 & 0 & 0 & 0 & S_{66} \end{bmatrix} \begin{Bmatrix} \sigma_1 \\ \sigma_2 \\ \sigma_3 \\ \tau_{23} \\ \tau_{13} \\ \tau_{12} \end{Bmatrix} \tag{2.18}
$$

With this notation

$$
\begin{aligned}
S_{11} &= \frac{1}{E_1} & S_{12} &= \frac{-\nu_{21}}{E_2} & S_{13} &= \frac{-\nu_{31}}{E_3} \\[2mm]
S_{21} &= -\frac{\nu_{12}}{E_1} & S_{22} &= \frac{1}{E_2} & S_{23} &= -\frac{\nu_{32}}{E_3} \\[2mm]
S_{31} &= -\frac{\nu_{13}}{E_1} & S_{32} &= -\frac{\nu_{23}}{E_2} & S_{33} &= \frac{1}{E_3} \\[2mm]
S_{44} &= \frac{1}{G_{23}} & S_{55} &= \frac{1}{G_{13}} & S_{66} &= \frac{1}{G_{12}}
\end{aligned} \tag{2.19}
$$

The inverse of the compliance matrix is called the *stiffness matrix*, sometimes called the *modulus matrix* or *elasticity matrix*, and is commonly denoted by C. With the inverse defined, the stress-strain relations become

$$
\begin{Bmatrix} \sigma_1 \\ \sigma_2 \\ \sigma_3 \\ \tau_{23} \\ \tau_{13} \\ \tau_{12} \end{Bmatrix} = \begin{bmatrix} C_{11} & C_{12} & C_{13} & 0 & 0 & 0 \\ C_{21} & C_{22} & C_{23} & 0 & 0 & 0 \\ C_{31} & C_{32} & C_{33} & 0 & 0 & 0 \\ 0 & 0 & 0 & C_{44} & 0 & 0 \\ 0 & 0 & 0 & 0 & C_{55} & 0 \\ 0 & 0 & 0 & 0 & 0 & C_{66} \end{bmatrix} \begin{Bmatrix} \varepsilon_1 \\ \varepsilon_2 \\ \varepsilon_3 \\ \gamma_{23} \\ \gamma_{13} \\ \gamma_{12} \end{Bmatrix} \tag{2.20}
$$

Clearly the C_{ij} can be written in terms of the S_{ij}, and ultimately in terms of the engineering constants. For shorthand notation, the relations between stress and strain will be abbreviated by

$$
\{\varepsilon\}_1 = [S]\{\sigma\}_1 \quad \{\sigma\}_1 = [C]\{\varepsilon\}_1 \tag{2.21}
$$

Here

$$
\{\varepsilon\}_1 = \begin{Bmatrix} \varepsilon_1 \\ \varepsilon_2 \\ \varepsilon_3 \\ \gamma_{23} \\ \gamma_{13} \\ \gamma_{12} \end{Bmatrix} \quad \{\sigma\}_1 = \begin{Bmatrix} \sigma_1 \\ \sigma_2 \\ \sigma_3 \\ \tau_{23} \\ \tau_{13} \\ \tau_{12} \end{Bmatrix} \tag{2.22}
$$

The subscript 1 outside the brackets means that the stresses and strains are referred to the 1-2-3 coordinate system.

As seen by equation (2.19), the compliance matrix involves 12 engineering properties: three extensional moduli (E_1, E_2, E_3), six Poisson's ratios (ν_{12}, ν_{21}, ν_{13}, ν_{31}, ν_{23}, ν_{32}), and three shear moduli (G_{23}, G_{13}, G_{12}). As a result, the stiffness matrix also depends on 12 engineering constants. However, the 12 engineering properties are not all independent. This is a very important point. There are actually only nine independent material properties. So-called *reciprocity relationships* can be established among the extensional moduli and the Poisson's ratios. As a result of these reciprocity relationships, the compliance and stiffness matrices are symmetric. To establish these relations, it is convenient to use the Maxwell-Betti Reciprocal Theorem. In the next two sections the theorem is briefly reviewed and used to establish the reciprocity relations.

2.2
MAXWELL-BETTI RECIPROCAL THEOREM

Consider an elastic body acted upon by two sets of loads, $\vec{P}_1$, $\vec{P}_2$, ..., $\vec{P}_M$ and $\vec{p}_1$, $\vec{p}_2$, ..., $\vec{p}_N$, where the overbar arrow denotes a vector quantity. These two sets of loads act at different locations on the body. For this discussion the two sets of loads are to be applied to the body in two specific sequences. Figure 2.9(a) illustrates a body with the two sets of loads, $\vec{P}_m$, $i = 1$, M and $\vec{p}_n$, $n = 1$, N. Assume, as in Figure 2.9(b), load set $\vec{P}_m$ is first applied to the body. The body deforms. These deformations are of no consequence here. Subsequent application of load set $\vec{p}_n$, as in Figure 2.9(c), causes the body to deform further. In particular, application of the load set $\vec{p}_n$ causes displacements at points of application of load set $\vec{P}_m$. Denoting these displacements by $\vec{d}_m$, the work done by load set $\vec{P}_m$ due to these displacements is given by the dot product

$$W_{P/p} = \sum_{m=1}^{M} \vec{P}_m \cdot \vec{d}_m \tag{2.23}$$

The subscript on W denotes the fact that the work is due to loads $\vec{P}_m$ moving through the displacements caused by the application of loads $\vec{p}_n$. Obviously, load set $\vec{P}_m$ does work as it is initially applied to the body and the body deforms. Like the initial deformations caused by $\vec{P}_m$, this work is not involved in this discussion.

Conversely, as in Figure 2.9(d), assume that load set $\vec{p}_n$ is first applied to the body. The body deforms. Neither these deformations nor the work done by $\vec{p}_n$ are of concern. Subsequent application of load set $\vec{P}_m$, as in Figure 2.9(e), causes further deformation. In particular, the body deforms at points of application of the load set $\vec{p}_n$. These displacements are denoted by $\vec{D}_n$; the work done by load set $\vec{p}_n$ due to these displacements is given by

$$W_{p/P} = \sum_{n=1}^{N} \vec{p}_n \cdot \vec{D}_n \tag{2.24}$$

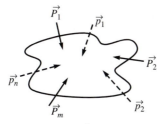

(a) Load sets $\vec{P}_m$ and $\vec{p}_n$

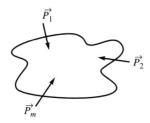

(b) Load set $\vec{P}_m$ initially applied
(ignore deformations)

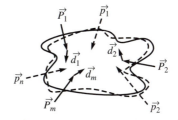

(c) Application of load set $\vec{p}_n$
causes body to deform
(broken line)

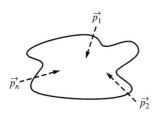

(d) Load set $\vec{p}_n$ initially applied
(ignore deformations)

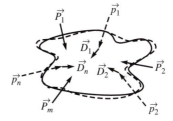

(e) Application of load set $\vec{P}_m$
causes body to deform
(broken line)

FIGURE 2.9
Maxwell-Betti Reciprocal Theorem

The Maxwell-Betti Reciprocal Theorem states that these two quantities of work
are equal; that is,

$$W_{P/p} = W_{p/P} \tag{2.25}$$

or, according to equations (2.23) and (2.24):

$$\sum_{m=1}^{M} \vec{P}_m \cdot \vec{d}_m = \sum_{n=1}^{N} \vec{p}_n \cdot \vec{D}_n \tag{2.26}$$

In other words: *The work done by load set $\vec{P}_m$ due to the displacements caused
by the application of load set $\vec{p}_n$ equals the work done by load set $\vec{p}_n$ due to the
displacements caused by the application of load set $\vec{P}_m$.*

2.3
RELATIONSHIPS AMONG MATERIAL PROPERTIES

To illustrate that there is a relationship among some of the engineering properties introduced, the Maxwell-Betti Reciprocal Theorem will be applied to the same small volume of fiber-reinforced material that we have been working with. At this time, however, the dimensions of the element must be specified. Figure 2.10(a) shows the element of material with the three dimensions, Δ_1, Δ_2, and Δ_3, indicated. Applied to the element are stresses σ_1 and σ_2. These two stresses constitute the two load sets discussed in the statement of the Maxwell-Betti Reciprocal Theorem. To apply the Maxwell-Betti Reciprocal Theorem to this situation, the work generated by σ_2 due to the deformations caused by the application of σ_1 will be computed. Then the work generated by σ_1 due to the deformations caused by the application of σ_2 will be computed. The Maxwell-Betti Reciprocal Theorem states that these two quantities of work are equal. From this equality, a relation among four engineering properties will evolve.

Consider the case with a stress σ_2 initially applied, Figure 2.10(b). Deformations caused by this stress are ignored. As in Figure 2.10(c), when the stress σ_1 is applied to the element, it contracts in the 2 direction due to Poisson's ratio ν_{12}. The application of σ_1 causes σ_2 to do work. Since σ_2 moves in the direction opposite to which it is acting, this work is actually negative. [For simplicity, Figure 2.10(c) shows all the contraction occurring at the upper component of σ_2.] Let us compute this work: By definition

$$\varepsilon_2 = -\nu_{12}\varepsilon_1 \tag{2.27}$$

The actual displacement of σ_2, denoted by $\delta\Delta_2$, is the strain times the element dimension in the 2 direction. That is:

$$\delta\Delta_2 = \varepsilon_2\Delta_2 \tag{2.28}$$

If we substitute from equation (2.27),

$$\delta\Delta_2 = -\nu_{12}\varepsilon_1\Delta_2 \tag{2.29}$$

The strain ε_1 is caused by σ_1 and is given by equation (2.1), namely,

$$\varepsilon_1 = \frac{\sigma_1}{E_1} \tag{2.30}$$

Therefore, the displacement of σ_2 due to the application of σ_1 can be expressed as

$$\delta\Delta_2 = -\frac{\nu_{12}}{E_1}\sigma_1\Delta_2 \tag{2.31}$$

The force due to σ_2 is

$$F_2 = \sigma_2\Delta_1\Delta_3 \tag{2.32}$$

or the stress times the area on which it acts, and thus the work done by σ_2 due to the deformations caused by the subsequent application of σ_1 is

$$W_{2/1} = F_2\delta\Delta_2 = (\sigma_2\Delta_1\Delta_3)\left(-\frac{\nu_{12}}{E_1}\sigma_1\Delta_2\right) = -\frac{\nu_{12}}{E_1}\sigma_1\sigma_2\Delta_1\Delta_2\Delta_3 \tag{2.33}$$

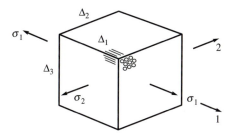

(a) Element with σ_1 and σ_2 applied

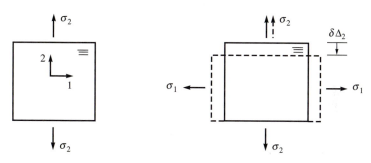

(b) σ_2 initially applied

(c) Deformations caused by application of σ_1

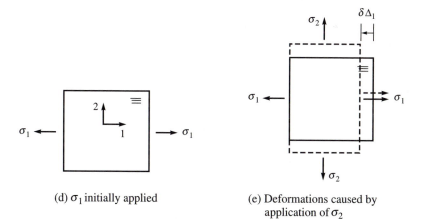

(d) σ_1 initially applied

(e) Deformations caused by application of σ_2

FIGURE 2.10
Application of the Maxwell-Betti Reciprocal Theorem to an element of fiber-reinforced material

Conversely, consider the case with a σ_1 initially applied, Figure 2.10(d). If a stress σ_2 is subsequently applied, then the element contracts in the 1 direction an amount

$$\varepsilon_1 = -\nu_{21}\varepsilon_2 \tag{2.34}$$

Using equation (2.34), the displacement in the 1 direction is given by

$$\delta\Delta_1 = -\varepsilon_1\Delta_1 = -\nu_{21}\varepsilon_2\Delta_1 \tag{2.35}$$

From equation (2.6), the strain ε_2 caused by the stress σ_2 is

$$\varepsilon_2 = \frac{\sigma_2}{E_2} \tag{2.36}$$

so the displacement in the 1 direction becomes

$$\delta\Delta_1 = -\frac{\nu_{21}}{E_2}\sigma_2\Delta_1 \tag{2.37}$$

This displacement in the 1 direction causes the stress σ_1 to do work. The force due to σ_1 is

$$F_1 = \sigma_1\Delta_2\Delta_3 \tag{2.38}$$

so the work of the force is given by

$$W_{1/2} = F_1\delta\Delta_1 = (\sigma_1\Delta_2\Delta_3)\left(\frac{-\nu_{21}}{E_2}\sigma_2\Delta_2\right) = -\frac{\nu_{21}}{E_2}\sigma_1\sigma_2\Delta_1\Delta_2\Delta_3 \tag{2.39}$$

By the Maxwell-Betti Reciprocal Theorem,

$$W_{2/1} = W_{1/2} \tag{2.40}$$

or, using the results of equations (2.33) and (2.39),

$$-\frac{\nu_{12}}{E_1}\sigma_1\sigma_2\Delta_1\Delta_2\Delta_3 = \frac{-\nu_{21}}{E_2}\sigma_1\sigma_2\Delta_1\Delta_2\Delta_3 \tag{2.41}$$

After we simplify, a relation among the material properties involved is immediately obvious, namely:

$$\boxed{\frac{\nu_{12}}{E_1} = \frac{\nu_{21}}{E_2}} \tag{2.42}$$

From this equation it is clear that the two extensional moduli E_1 and E_2 and the two Poisson's ratios ν_{12} and ν_{21} for the material are not completely arbitrary. If one knows any three of the properties, the fourth one can be determined. Similar considerations for work by pairs of stresses σ_1 and σ_3, and σ_2 and σ_3 lead to reciprocity relations for the other extensional moduli and Poisson's ratios, namely:

$$\boxed{\frac{\nu_{13}}{E_1} = \frac{\nu_{31}}{E_3} \qquad \frac{\nu_{23}}{E_2} = \frac{\nu_{32}}{E_3}} \tag{2.43}$$

Because of the three reciprocity relations, only nine independent constants are needed to describe the linear elastic behavior of a fiber-reinforced material. Also, because of the reciprocity relations, from equation (2.19),

$$S_{21} = -\frac{\nu_{12}}{E_1} = -\frac{\nu_{21}}{E_2} = S_{12}$$

$$S_{31} = -\frac{\nu_{13}}{E_1} = -\frac{\nu_{31}}{E_3} = S_{13} \tag{2.44}$$

$$S_{32} = -\frac{\nu_{23}}{E_2} = -\frac{\nu_{32}}{E_3} = S_{23}$$

As a result the compliance matrix, and therefore the stiffness matrix, are symmetric. The symmetry of these two matrices is an important property.

After we incorporate the reciprocity relations, the stress-strain relations in terms of the compliances are

$$
\begin{Bmatrix} \varepsilon_1 \\ \varepsilon_2 \\ \varepsilon_3 \\ \gamma_{23} \\ \gamma_{13} \\ \gamma_{12} \end{Bmatrix} = \begin{bmatrix} S_{11} & S_{12} & S_{13} & 0 & 0 & 0 \\ S_{12} & S_{22} & S_{23} & 0 & 0 & 0 \\ S_{13} & S_{23} & S_{33} & 0 & 0 & 0 \\ 0 & 0 & 0 & S_{44} & 0 & 0 \\ 0 & 0 & 0 & 0 & S_{55} & 0 \\ 0 & 0 & 0 & 0 & 0 & S_{66} \end{bmatrix} \begin{Bmatrix} \sigma_1 \\ \sigma_2 \\ \sigma_3 \\ \tau_{23} \\ \tau_{13} \\ \tau_{12} \end{Bmatrix} \tag{2.45}
$$

In the above equation

$$
\begin{aligned}
S_{11} &= \frac{1}{E_1} & S_{12} &= -\frac{\nu_{12}}{E_1} & S_{13} &= -\frac{\nu_{13}}{E_1} \\
S_{22} &= \frac{1}{E_2} & S_{23} &= -\frac{\nu_{23}}{E_2} & S_{33} &= \frac{1}{E_3} \\
S_{44} &= \frac{1}{G_{23}} & S_{55} &= \frac{1}{G_{13}} & S_{66} &= \frac{1}{G_{12}}
\end{aligned} \tag{2.46}
$$

The inverse relations are

$$
\begin{Bmatrix} \sigma_1 \\ \sigma_2 \\ \sigma_3 \\ \tau_{23} \\ \tau_{13} \\ \tau_{12} \end{Bmatrix} = \begin{bmatrix} C_{11} & C_{12} & C_{13} & 0 & 0 & 0 \\ C_{12} & C_{22} & C_{23} & 0 & 0 & 0 \\ C_{13} & C_{23} & C_{33} & 0 & 0 & 0 \\ 0 & 0 & 0 & C_{44} & 0 & 0 \\ 0 & 0 & 0 & 0 & C_{55} & 0 \\ 0 & 0 & 0 & 0 & 0 & C_{66} \end{bmatrix} \begin{Bmatrix} \varepsilon_1 \\ \varepsilon_2 \\ \varepsilon_3 \\ \gamma_{23} \\ \gamma_{13} \\ \gamma_{12} \end{Bmatrix} \tag{2.47}
$$

In terms of the compliances, the components of the stiffness matrix are given by

$$
\begin{aligned}
C_{11} &= \frac{S_{22} S_{33} - S_{23} S_{23}}{S} & C_{12} &= \frac{S_{13} S_{23} - S_{12} S_{33}}{S} \\
C_{22} &= \frac{S_{33} S_{11} - S_{13} S_{13}}{S} & C_{13} &= \frac{S_{12} S_{23} - S_{13} S_{22}}{S} \\
C_{33} &= \frac{S_{11} S_{22} - S_{12} S_{12}}{S} & C_{23} &= \frac{S_{12} S_{13} - S_{23} S_{11}}{S} \\
C_{44} &= \frac{1}{S_{44}} \quad C_{55} = \frac{1}{S_{55}} & C_{66} &= \frac{1}{S_{66}}
\end{aligned} \tag{2.48}
$$

where

$$S = S_{11} S_{22} S_{33} - S_{11} S_{23} S_{23} - S_{22} S_{13} S_{13} - S_{33} S_{12} S_{12} + 2 S_{12} S_{23} S_{13}$$ (2.49)

If needed, the compliances in equation (2.48) can be written in terms of the engineering properties, and therefore the stiffnesses can be expressed directly in terms of the engineering properties.

Despite the reciprocity relations among some of the material properties, an element of fiber-reinforced material still has different material properties in each of the three principal material directions. As we mentioned previously, such a material is called orthotropic. The zero entries in the upper right and lower left portions of both the compliance and stiffness matrices characterize orthotropic behavior. If a material is orthotropic and the stress-strain relations are written in the principal coordinate material system, then the compliance and stiffness matrices will always have these zero entries. We shall see at a later time that if the material is orthotropic, but the stress-strain relations are written in a coordinate system other than the principal one, then some of the zero entries become nonzero. It is possible to find materials that have nonzero entries in the upper right and lower left portions of their compliance and stiffness matrices for every coordinate system. Such a material is said to be *anisotropic*. We choose here to concentrate on an orthotropic material because it is the building block of most composite materials.

For isotropic materials

$$E_1 = E_2 = E_3 = E \qquad \nu_{23} = \nu_{13} = \nu_{12} = \nu$$

$$G_{23} = G_{13} = G_{12} = G = \frac{E}{2(1 + \nu)}$$ (2.50)

As a result, the compliance matrix is

$$\begin{bmatrix} S_{11} & S_{12} & S_{12} & 0 & 0 & 0 \\ S_{12} & S_{11} & S_{12} & 0 & 0 & 0 \\ S_{12} & S_{12} & S_{11} & 0 & 0 & 0 \\ 0 & 0 & 0 & S_{44} & 0 & 0 \\ 0 & 0 & 0 & 0 & S_{44} & 0 \\ 0 & 0 & 0 & 0 & 0 & S_{44} \end{bmatrix}$$ (2.51a)

where $$S_{11} = \frac{1}{E} \qquad S_{12} = -\frac{\nu}{E} \qquad S_{44} = \frac{1}{G} = \frac{2(1 + \nu)}{E}$$ (2.51b)

The lack of any directional dependence is reflected in the fact that the off-diagonal terms are identical, and the on-diagonal terms for the three components of shear are identical, as are the other three on-diagonal terms. Likewise, the stiffness matrix becomes

$$\begin{bmatrix} C_{11} & C_{12} & C_{12} & 0 & 0 & 0 \\ C_{12} & C_{11} & C_{12} & 0 & 0 & 0 \\ C_{12} & C_{12} & C_{11} & 0 & 0 & 0 \\ 0 & 0 & 0 & C_{44} & 0 & 0 \\ 0 & 0 & 0 & 0 & C_{44} & 0 \\ 0 & 0 & 0 & 0 & 0 & C_{44} \end{bmatrix} \qquad (2.52a)$$

with

$$C_{11} = \frac{E}{(1+v)(1+2v)} \qquad C_{12} = \frac{(1-v)E}{(1+v)(1+2v)} \qquad C_{44} = G = \frac{E}{2(1+v)} \qquad (2.52b)$$

Between orthotropic material behavior and isotropic material behavior lies a third type of material behavior, namely, *transversely isotropic* behavior. For an element of fiber-reinforced material it is often assumed that the material behavior in the 2 direction is identical to the material behavior in the 3 direction. As these directions are both perpendicular to the fiber direction, assuming identical properties in these directions is understandable. For this situation,

$$E_2 = E_3 \qquad v_{12} = v_{13} \qquad G_{12} = G_{13} \qquad (2.53a)$$

and more importantly,

$$G_{23} = \frac{E_2}{2(1+v_{23})} \qquad (2.53b)$$

If equations (2.53a) and (2.53b) are true, then the material is said to be isotropic in the 2-3 plane, or transversely isotropic in the 2-3 plane. Any particular property is independent of direction within that plane. With this characteristic the compliance matrix becomes

$$\begin{bmatrix} S_{11} & S_{12} & S_{12} & 0 & 0 & 0 \\ S_{12} & S_{22} & S_{23} & 0 & 0 & 0 \\ S_{12} & S_{23} & S_{22} & 0 & 0 & 0 \\ 0 & 0 & 0 & S_{44} & 0 & 0 \\ 0 & 0 & 0 & 0 & S_{55} & 0 \\ 0 & 0 & 0 & 0 & 0 & S_{55} \end{bmatrix} \qquad (2.54a)$$

where

$$S_{11} = \frac{1}{E_1} \qquad S_{12} = -\frac{v_{12}}{E_1} \qquad S_{22} = \frac{1}{E_2} \qquad S_{23} = -\frac{v_{23}}{E_2}$$

$$S_{44} = \frac{1}{G_{23}} = \frac{2(1+v_{23})}{E_2} \qquad S_{55} = \frac{1}{G_{12}} \qquad (2.54b)$$

and the stiffness matrix becomes

$$
\begin{bmatrix}
C_{11} & C_{12} & C_{12} & 0 & 0 & 0 \\
C_{12} & C_{22} & C_{23} & 0 & 0 & 0 \\
C_{12} & C_{23} & C_{22} & 0 & 0 & 0 \\
0 & 0 & 0 & C_{44} & 0 & 0 \\
0 & 0 & 0 & 0 & C_{55} & 0 \\
0 & 0 & 0 & 0 & 0 & C_{55}
\end{bmatrix}
\tag{2.55}
$$

For a transversely isotropic material, there are five independent material properties: E_1, E_2, v_{12}, v_{23}, and G_{12}.

2.4
TYPICAL MATERIAL PROPERTIES

Though they will not be developed using micromechanical models until the next chapter, representative numerical values of the engineering properties of two common fiber-reinforced composite materials are given in Table 2.1, namely, an intermediate-modulus graphite-reinforced polymeric material and an S-glass-reinforced polymeric material. For consistency the numerical values in the table will be used throughout this book. The numerical values of aluminum are provided for comparison and future reference. Table 2.1 includes values of coefficients of thermal expansion and coefficients of moisture expansion for the materials. These

TABLE 2.1
Typical engineering properties of several materials

	Graphite-polymer composite[1]	Glass-polymer composite	Aluminum
E_1	155.0 GPa	50.0 GPa	72.4 GPa
E_2	12.10 GPa	15.20 GPa	72.4 GPa
E_3	12.10 GPa	15.20 GPa	72.4 GPa
v_{23}	0.458	0.428	0.300
v_{13}	0.248	0.254	0.300
v_{12}	0.248	0.254	0.300
G_{23}	3.20 GPa	3.28 GPa	—[2]
G_{13}	4.40 GPa	4.70 GPa	—[2]
G_{12}	4.40 GPa	4.70 GPa	—[2]
α_1	$-0.01800 \times 10^{-6}/°C$	$6.34 \times 10^{-6}/°C$	$22.5 \times 10^{-6}/°C$
α_2	$24.3 \times 10^{-6}/°C$	$23.3 \times 10^{-6}/°C$	$22.5 \times 10^{-6}/°C$
α_3	$24.3 \times 10^{-6}/°C$	$23.3 \times 10^{-6}/°C$	$22.5 \times 10^{-6}/°C$
β_1	$146.0 \times 10^{-6}/\%M$	$434 \times 10^{-6}/\%M$	0
β_2	$4770 \times 10^{-6}/\%M$	$6320 \times 10^{-6}/\%M$	0
β_3	$4770 \times 10^{-6}/\%M$	$6320 \times 10^{-6}/\%M$	0

[1] In the chapters to follow it will be assumed that a layer thickness is 150×10^{-6} m, or 0.150 mm.
[2] $G = E/2(1 + v)$.

will be discussed shortly. As can be seen from the table, for the graphite-reinforced polymer the extensional modulus in the fiber direction is about 10 times greater than the extensional modulus perpendicular to the fibers. For glass-reinforced materials, the difference is not as great. For both fiber-reinforced materials, the three shear moduli are similar.

As an example of the magnitude of the components of the compliance and stiffness matrices, for the graphite-reinforced material in Table 2.1, according to equation (2.46),

$$
\begin{array}{ll}
S_{11} = 6.45 \ (\text{TPa})^{-1} & S_{12} = -1.600 \\
S_{22} = 82.6 & S_{13} = -1.600 \\
S_{33} = 82.6 & S_{23} = -37.9 \\
S_{44} = 312 & S_{55} = 227 \\
S_{66} = 227 &
\end{array}
\tag{2.56}
$$

In the above, TPa denotes terraPascals, or 10^{12} Pa. For example,

$$
6.45 \ (\text{TPa})^{-1} = 6.45 \frac{1}{10^{12} \ \text{Pa}} = 6.45 \times 10^{-12} \frac{1}{\text{Pa}}
\tag{2.57}
$$

Using equation (2.48), the stiffnesses for the graphite-reinforced material are

$$
\begin{array}{ll}
C_{11} = 158.0 \ \text{GPa} & C_{12} = 5.64 \\
C_{22} = 15.51 & C_{13} = 5.64 \\
C_{33} = 15.51 & C_{23} = 7.21 \\
C_{44} = 3.20 & C_{55} = 4.40 \\
C_{66} = 4.40 &
\end{array}
\tag{2.58}
$$

Here GPa denotes the more familiar gigaPascal, or 10^9 Pa. With this notation,

$$
158.0 \ \text{GPa} = 158.0 \times 10^9 \ \text{Pa}
\tag{2.59}
$$

These numbers should be contrasted with those for aluminum, which generally is assumed to be isotropic, where

$$
\begin{array}{l}
S_{11} = S_{22} = S_{33} = 13.81 \ (\text{TPa})^{-1} \\
S_{12} = S_{13} = S_{23} = -4.14 \\
S_{44} = S_{55} = S_{66} = 36.0
\end{array}
\tag{2.60}
$$

and

$$
\begin{array}{l}
C_{11} = C_{22} = C_{33} = 97.5 \ \text{GPa} \\
C_{12} = C_{13} = C_{23} = 41.8 \\
C_{44} = C_{55} = C_{66} = 27.8
\end{array}
\tag{2.61}
$$

In matrix form, for the graphite-reinforced material

$$[S] = \begin{bmatrix} 6.45 & -1.600 & -1.600 & 0 & 0 & 0 \\ -1.600 & 82.6 & -37.9 & 0 & 0 & 0 \\ -1.600 & -37.9 & 82.6 & 0 & 0 & 0 \\ 0 & 0 & 0 & 312 & 0 & 0 \\ 0 & 0 & 0 & 0 & 227 & 0 \\ 0 & 0 & 0 & 0 & 0 & 227 \end{bmatrix} (\text{TPa})^{-1} \qquad (2.62)$$

$$[C] = \begin{bmatrix} 158.0 & 5.64 & 5.64 & 0 & 0 & 0 \\ 5.64 & 15.51 & 7.21 & 0 & 0 & 0 \\ 5.64 & 7.21 & 15.51 & 0 & 0 & 0 \\ 0 & 0 & 0 & 3.20 & 0 & 0 \\ 0 & 0 & 0 & 0 & 4.40 & 0 \\ 0 & 0 & 0 & 0 & 0 & 4.40 \end{bmatrix} \text{GPa} \qquad (2.63)$$

The degree of directional dependence, or *degree of orthotropy*, for a material can be evaluated by examining differences among the 11, 22, and 33 terms. For extensional effects in graphite-reinforced material there is a high degree of directional dependence, or orthotropy. Shear effects exhibit little orthotropy, but Poisson coupling is directionally dependent, for example, S_{13} versus S_{23}. As will be discussed, thermal expansion characteristics are highly dependent on direction.

For all calculations cited throughout the text, results will be quoted to three significant digits, unless the number starts with 1, in which case four significant digits will be cited. This will not be the case for layer thickness, however, where only three significant digits will be cited (i.e., 150×10^{-6} m), or in other cases where even numbers are chosen to make example problems convenient (e.g., 125 kN). The engineering properties in Table 2.1 are taken to be exact as shown, with only zeros being dropped from the numbers given in the table. That is, for graphite-reinforced material, G_{12} could be written as $4.400000\ldots000$ GPa, or ν_{12} could be written as $0.24800000\ldots000$.

Exercises for Section 2.4

1. Verify equation (2.48). Hint: Inasmuch as σ_1, σ_2, and σ_3 interact only with ε_1, ε_2, and ε_3, the upper six of equation (2.48) can be verified by considering only the upper left hand 3×3 submatrix of equation (2.45). Verifying the lower three of equation (2.48) is trivial.

2. Use the appropriate definitions to show that the stiffness components, C_{ij}, can be written directly in terms of the engineering properties as follows:

$$C_{11} = \frac{(1 - \nu_{23}\nu_{32})E_1}{1 - \nu}$$

$$C_{12} = \frac{(\nu_{21} + \nu_{31}\nu_{23})E_1}{1 - \nu} = \frac{(\nu_{12} + \nu_{32}\nu_{13})E_2}{1 - \nu}$$

$$C_{13} = \frac{(\nu_{31} + \nu_{21}\nu_{32})E_1}{1 - \nu} = \frac{(\nu_{13} + \nu_{12}\nu_{23})E_3}{1 - \nu}$$

$$C_{22} = \frac{(1 - \nu_{13}\nu_{31})E_2}{1 - \nu}$$

$$C_{23} = \frac{(\nu_{32} + \nu_{12}\nu_{31})E_2}{1 - \nu} = \frac{(\nu_{23} + \nu_{21}\nu_{13})E_3}{1 - \nu}$$

$$C_{33} = \frac{(1 - \nu_{12}\nu_{21})E_3}{1 - \nu}$$

$$C_{44} = G_{23} \qquad C_{55} = G_{13} \qquad C_{66} = G_{12}$$

where $\qquad \nu = \nu_{12}\nu_{21} + \nu_{23}\nu_{32} + \nu_{31}\nu_{13} + 2\nu_{21}\nu_{32}\nu_{13}$

Computer Exercise

Write a user-friendly computer program to prompt you for the engineering constants and, through equations (2.46) and (2.48), or the results of Exercise 2 above, compute and print the values of the compliances and stiffnesses in matrix form. Use the program to compute the values of the compliance and stiffness matrices for the glass-polymer composite. Comment on the degree of orthotropy compared to the graphite-polymer composite and aluminum.

2.5
IMPORTANT INTERPRETATION OF STRESS-STRAIN RELATIONS

The stress-strain relations just established can be interpreted in several ways. First, the relations can be considered simple algebraic relations between 12 quantities, $\sigma_1, \sigma_2, \ldots, \tau_{12}$ and $\varepsilon_1, \varepsilon_2, \ldots, \gamma_{12}$. Given any six quantities, the other six can be determined by the rules of algebra. However, there is a second, more important and more physical interpretation. The 12 stresses and strains should be considered in pairs: $\sigma_1\text{-}\varepsilon_1$; $\sigma_2\text{-}\varepsilon_2$; $\sigma_3\text{-}\varepsilon_3$; $\tau_{23}\text{-}\gamma_{23}$; $\tau_{13}\text{-}\gamma_{13}$; $\tau_{12}\text{-}\gamma_{12}$. Though it is possible to prescribe any six and solve for the other six, physically one can only prescribe either a stress or a strain from each pair, but not both. Though they are two of the six prescribed quantities, it is not physically correct to prescribe both σ_3 and ε_3, for example. Only one or the other can be prescribed. This is obvious when the shear response is considered. Because

$$\tau_{12} = G_{12}\gamma_{12} \tag{2.64}$$

it is impossible to specify both τ_{12} and γ_{12}. To do so violates the shear stress–shear strain relation. Similarly, to specify both σ_1 and ε_1, or both σ_2 and ε_2, or both σ_3 and ε_3 also violates the stress-strain relation, but not in so obvious a fashion. There is one situation, however, where efforts are made to know both the stress and strain. For example, if the extensional modulus in the 1 direction is to be determined from an experiment, then both σ_1 and ε_1 have to be known. If all the stresses are zero except σ_1, and the strain ε_1 is measured, then the extensional modulus can be computed directly from equation (2.1). Similarly for E_2, E_3, and the other engineering properties. However, even in this case of determining material properties, σ_1 is specified from the level of applied load and cross-sectional area, but ε_1 is not really specified also. The strain ε_1 is *measured* and E_1 is *inferred*.

As an example of the above comments regarding pairing of stress and strain components, and to illustrate the use and some of the implications of the stress-strain relations for fiber-reinforced materials, consider the following: A 50-mm cube of graphite-reinforced material, shown in Figure 2.11(a), is subjected to a 125-kN compressive force perpendicular to the fiber direction, specifically in the 2 direction. In one situation, Figure 2.11(b), the cube is free to expand or contract; in the second situation, Figure 2.11(c), the cube is constrained against expansion in the 3 direction; and in the third case, Figure 2.11(d), the cube is constrained against expansion in the 1 direction. Of interest are the changes in the 50-mm dimensions in each of these three cases, and the stresses, and hence forces, required to provide the constraints in the latter two cases.

To begin, we shall assume that the compressive force is uniformly distributed over the 2-direction faces. Further, we shall assume that stress resulting from the force is distributed uniformly throughout the volume of the cube. Then

$$\sigma_2 = \frac{P}{\Delta_1 \Delta_3} = \frac{-125,000 \text{ N}}{(0.050 \text{ m})(0.050 \text{ m})} = -50,000,000 \text{ Pa} = -50.0 \text{ MPa} \quad (2.65)$$

For the first case, Figure 2.11(b), the cube is free from any stress on the other faces. Thus, throughout the volume of the cube,

$$\sigma_2 = -50.0 \text{ MPa} \qquad \sigma_1 = \sigma_3 = \tau_{23} = \tau_{13} = \tau_{12} = 0 \quad (2.66)$$

Because of the previous discussions of pairing of stresses and strains, nothing can be said regarding ε_1, ε_2, ε_3, γ_{23}, γ_{13}, or γ_{12}. These are all unknown and will be

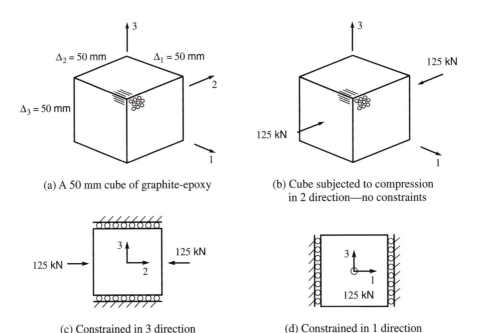

(a) A 50 mm cube of graphite-epoxy

(b) Cube subjected to compression in 2 direction—no constraints

(c) Constrained in 3 direction

(d) Constrained in 1 direction

FIGURE 2.11
Cube of material subjected to compressive stress in 2 direction

determined. It is most convenient to use the compliance form of the stress-strain relation, equation (2.45). Specifically, if we use equation (2.66) in equation (2.45),

$$\begin{Bmatrix} \varepsilon_1 \\ \varepsilon_2 \\ \varepsilon_3 \\ \gamma_{23} \\ \gamma_{13} \\ \gamma_{12} \end{Bmatrix} = \begin{bmatrix} S_{11} & S_{12} & S_{13} & 0 & 0 & 0 \\ S_{12} & S_{22} & S_{23} & 0 & 0 & 0 \\ S_{13} & S_{23} & S_{33} & 0 & 0 & 0 \\ 0 & 0 & 0 & S_{44} & 0 & 0 \\ 0 & 0 & 0 & 0 & S_{55} & 0 \\ 0 & 0 & 0 & 0 & 0 & S_{66} \end{bmatrix} \begin{Bmatrix} 0 \\ \sigma_2 \\ 0 \\ 0 \\ 0 \\ 0 \end{Bmatrix} \qquad (2.67)$$

Expanding,

$$\varepsilon_1 = S_{12}\sigma_2 \qquad \varepsilon_2 = S_{22}\sigma_2 \qquad \varepsilon_3 = S_{23}\sigma_2 \qquad \gamma_{23} = \gamma_{13} = \gamma_{12} = 0 \quad (2.68)$$

As the stresses are uniform throughout the cube, the strains are also uniform throughout the cube. By definition,

$$\varepsilon_1 = \frac{\delta\Delta_1}{\Delta_1} \qquad \varepsilon_2 = \frac{\delta\Delta_2}{\Delta_2} \qquad \varepsilon_3 = \frac{\delta\Delta_3}{\Delta_3} \qquad (2.69)$$

where $\delta\Delta_1$, $\delta\Delta_2$, $\delta\Delta_3$ denote the change in the 50 mm length in the 1, 2, and 3 directions, respectively. Rearranging equation (2.69), and using equation (2.68) and values from equation (2.56),

$$\delta\Delta_1 = \Delta_1\varepsilon_1 = \Delta_1 S_{12}\sigma_2 = (0.050)(-1.600 \times 10^{-12})(-50.0 \times 10^6)$$
$$= 4.00 \times 10^{-6} \text{ m} = 0.00400 \text{ mm}$$

$$\delta\Delta_2 = \Delta_2\varepsilon_2 = \Delta_2 S_{22}\sigma_2 = (0.050)(82.6 \times 10^{-12})(-50.0 \times 10^6)$$
$$= -207 \times 10^{-6} \text{ m} = -0.207 \text{ mm} \qquad (2.70)$$

$$\delta\Delta_3 = \Delta_3\varepsilon_3 = \Delta_3 S_{23}\sigma_2 = (0.050)(-37.9 \times 10^{-12})(-50.0 \times 10^6)$$
$$= 94.6 \times 10^{-6} \text{ m} = 0.0946 \text{ mm}$$

In this situation the fibers provide considerable constraint and prevent the expansion in the 1 direction from being anywhere near as large as the expansion in the 3 direction. Of course, there is compression in the 2 direction and obviously

$$\gamma_{23} = \gamma_{13} = \gamma_{12} = 0 \qquad (2.71)$$

Consider the second case, Figure 2.11(c). For this situation we do not know the normal stress in the 3 direction that is providing the constraint. Because the displacement in the 3 direction is restrained to be zero along two sides of the cube, $\delta\Delta_3 = 0$, and because the strains are assumed to be uniform throughout the cube, $\varepsilon_3 = 0$. Therefore, instead of equation (2.66) we have

$$\sigma_2 = -50.0 \text{ MPa} \qquad \sigma_1 = \varepsilon_3 = \tau_{23} = \tau_{13} = \tau_{12} = 0 \qquad (2.72)$$

and we must solve for ε_1, ε_2, σ_3, γ_{23}, γ_{13}, and γ_{12}, the other half of each of the pairs involved in equation (2.72). Again referring to the compliance form of the stress-strain relations,

$$\begin{Bmatrix} \varepsilon_1 \\ \varepsilon_2 \\ 0 \\ \gamma_{23} \\ \gamma_{13} \\ \gamma_{12} \end{Bmatrix} = \begin{bmatrix} S_{11} & S_{12} & S_{13} & 0 & 0 & 0 \\ S_{12} & S_{22} & S_{23} & 0 & 0 & 0 \\ S_{13} & S_{23} & S_{33} & 0 & 0 & 0 \\ 0 & 0 & 0 & S_{44} & 0 & 0 \\ 0 & 0 & 0 & 0 & S_{55} & 0 \\ 0 & 0 & 0 & 0 & 0 & S_{66} \end{bmatrix} \begin{Bmatrix} 0 \\ \sigma_2 \\ \sigma_3 \\ 0 \\ 0 \\ 0 \end{Bmatrix} \qquad (2.73)$$

or $\qquad \varepsilon_1 = S_{12}\sigma_2 + S_{13}\sigma_3 \qquad \varepsilon_2 = S_{22}\sigma_2 + S_{23}\sigma_3 \qquad 0 = S_{23}\sigma_2 + S_{33}\sigma_3 \quad (2.74)$

For this case we must use the third equation to solve for a relation between σ_2 and σ_3, namely,

$$\sigma_3 = -\frac{S_{23}}{S_{33}}\sigma_2 \qquad (2.75)$$

Solving the first and second equations and substituting for σ_3 results in

$$\varepsilon_1 = S_{12}\sigma_2 + S_{13}\sigma_3 = \left(S_{12} - \frac{S_{13}S_{23}}{S_{33}} \right)\sigma_2$$

$$\varepsilon_2 = S_{22}\sigma_2 + S_{23}\sigma_3 = \left(S_{22} - \frac{S_{23}S_{23}}{S_{33}} \right)\sigma_2 \qquad (2.76)$$

For this case also,

$$\gamma_{23} = \gamma_{13} = \gamma_{12} = 0 \qquad (2.77)$$

We often refer to the combination of compliances in equation (2.76) as apparent or reduced compliances. In the second of equation (2.76), the term $(S_{23}S_{23}/S_{33})$ subtracts from S_{22} and indicates the material behaves in a less compliant manner than determined by S_{22} alone. Certainly it makes physical sense that a constraint of any kind, here a constraint against expansion in the 3 direction, makes the system less compliant. Equation (2.69) relates the change of length to the strain and so, using equation (2.76) and values from equation (2.56),

$$\delta\Delta_1 = \Delta_1\varepsilon_1 = \Delta_1\left(S_{12} - \frac{S_{13}S_{23}}{S_{33}} \right)\sigma_2$$

$$= (0.050)(-2.33 \times 10^{-12})(-50.0 \times 10^6)$$

$$= 5.83 \times 10^{-6} \text{ m} = 0.00583 \text{ mm}$$

$$\delta\Delta_2 = \Delta_2\varepsilon_2 = \Delta_2\left(S_{22} - \frac{S_{23}S_{23}}{S_{33}} \right)\sigma_2 \qquad (2.78)$$

$$= (0.050)(65.3 \times 10^{-12})(-50.0 \times 10^6)$$

$$= -163.3 \times 10^{-6} \text{ m} = -0.1633 \text{ mm}$$

$$\delta\Delta_3 = 0 \text{ (by definition)}$$

The stress to constrain ε_3 to be zero is given by equation (2.75), namely,

$$\sigma_3 = -\frac{S_{23}}{S_{33}}\sigma_2 = -22.9 \text{ MPa} \tag{2.79}$$

If we compare the deformations for the case of Figure 2.11(c), given by equation (2.78), with the deformations for the case of Figure 2.11(b), given by equation (2.70), we can see that with the constraint in the 3 direction the cube expands about 50 percent more in the fiber direction, and compresses about 25 percent less in the 2 direction. The constraint in the 3 direction makes the cube stiffer in the 2 direction and forces the inevitable volume change to be reflected with increased expansion in the fiber direction.

Finally, consider the third case, Figure 2.11(d). For this situation the normal stress in the 3 direction is again zero and the extensional strain in that direction is unknown. The extensional strain in the fiber direction is known to be zero, but the corresponding stress, σ_1, is not known. With this

$$\sigma_2 = -50.0 \text{ MPa} \qquad \varepsilon_1 = \sigma_3 = \tau_{23} = \tau_{13} = \tau_{12} = 0 \tag{2.80}$$

and we must solve for σ_1, ε_2, ε_3, γ_{23}, γ_{13}, and γ_{12}. Equation (2.45) becomes

$$\begin{Bmatrix} 0 \\ \varepsilon_2 \\ \varepsilon_3 \\ \gamma_{23} \\ \gamma_{13} \\ \gamma_{12} \end{Bmatrix} = \begin{bmatrix} S_{11} & S_{12} & S_{13} & 0 & 0 & 0 \\ S_{12} & S_{22} & S_{23} & 0 & 0 & 0 \\ S_{13} & S_{23} & S_{33} & 0 & 0 & 0 \\ 0 & 0 & 0 & S_{44} & 0 & 0 \\ 0 & 0 & 0 & 0 & S_{55} & 0 \\ 0 & 0 & 0 & 0 & 0 & S_{66} \end{bmatrix} \begin{Bmatrix} \sigma_1 \\ \sigma_2 \\ 0 \\ 0 \\ 0 \\ 0 \end{Bmatrix} \tag{2.81}$$

or $\qquad 0 = S_{11}\sigma_1 + S_{12}\sigma_2 \qquad \varepsilon_2 = S_{12}\sigma_1 + S_{22}\sigma_2 \qquad \varepsilon_3 = S_{13}\sigma_1 + S_{23}\sigma_2$ (2.82)

If we use the first equation

$$\sigma_1 = -\frac{S_{12}}{S_{11}}\sigma_2 \tag{2.83}$$

the second and third equations become

$$\varepsilon_2 = S_{12}\sigma_1 + S_{22}\sigma_2 = \left(S_{22} - \frac{S_{12}S_{12}}{S_{11}} \right)\sigma_2$$

$$\varepsilon_3 = S_{13}\sigma_1 + S_{23}\sigma_2 = \left(S_{23} - \frac{S_{12}S_{13}}{S_{11}} \right)\sigma_2 \tag{2.84}$$

Of course $\qquad\qquad \gamma_{23} = \gamma_{13} = \gamma_{12} = 0$ (2.85)

If we use equation (2.69) and numerical values,

$$\delta\Delta_1 = 0 \text{ (by definition)}$$

$$\delta\Delta_2 = \Delta_2\varepsilon_2 = \Delta_2\left(S_{22} - \frac{S_{12}S_{12}}{S_{11}}\right)\sigma_2$$

$$= (0.050)(82.2 \times 10^{-12})(-50.0 \times 10^6)$$

$$= -206 \times 10^{-6} \text{ m} = -0.206 \text{ mm} \tag{2.86}$$

$$\delta\Delta_3 = \Delta_3\varepsilon_3 = \Delta_3\left(S_{23} - \frac{S_{12}S_{13}}{S_{11}}\right)\sigma_2$$

$$= (0.050)(-38.2 \times 10^{-12})(-50.0 \times 10^6)$$

$$= 95.6 \times 10^{-6} \text{ m} = 0.0956 \text{ mm}$$

The stress in the fiber direction is given by equation (2.83) as

$$\sigma_1 = -12.4 \text{ MPa} \tag{2.87}$$

It is important to note that the deformations in the 2 and 3 directions for this case are not very different from the case of no restraint, equation (2.70). This is because the fibers constrain deformation in the 1 direction to a considerable degree, and adding a constraint in the 1 direction has less influence than adding a constraint in the 3 direction. Likewise, the stress necessary to constrain the deformation in the fiber direction is, by equation (2.87), -12.4 MPa, whereas the stress to constrain deformation perpendicular to the fibers was, by equation (2.79), -22.9 MPa, about twice as much.

It is instructive to repeat the same exercise, but using aluminum in place of graphite-reinforced composite. This will provide a feel for the effects of the lack of orthotropy, and a comparison of the magnitude of the deformations of aluminum and of a common fiber-reinforced composite material. Before studying the case of aluminum, however, it is important to comment on the physical aspects of the example just discussed. First, the 50.0 MPa compressive stress in the 2 direction is about 25 percent of the compressive failure stress of graphite-reinforced material in the 2 direction and is the failure strength in the 2 direction in tension. Thus, the 50.0 MPa in the example is a realistic level of stress. (Strengths of composite materials will be the subject of a subsequent chapter.) Second, it is important to note that the magnitudes of the deformations, $\delta\Delta_1$, $\delta\Delta_2$, and $\delta\Delta_3$, are quite small— submillimeter in size. This level of deformation cannot be detected with the eye. Special instrumentation is needed to measure deformations this small and in practice that is what is used. Finally, it is important to become familiar with realistic strain levels associated with composite materials. In the unconstrained case, Figure 2.11(b), the strains ε_1, ε_2, and ε_3 are given in equation (2.68) as

$$\varepsilon_1 = S_{12}\sigma_2$$
$$\varepsilon_2 = S_{22}\sigma_2 \tag{2.88}$$
$$\varepsilon_3 = S_{23}\sigma_2$$

If we use numerical values,

$$\varepsilon_1 = (-1.600 \times 10^{-12})(-50.0 \times 10^6) = 80.0 \times 10^{-6} \text{ m/m}$$
$$= 80.0 \times 10^{-6} \text{ mm/mm} = 80 \ \mu\text{mm/mm}$$
$$\varepsilon_2 = (82.6 \times 10^{-12})(-50.0 \times 10^6) = -4130 \times 10^{-6} \text{ m/m}$$
$$= -4130 \times 10^{-6} \text{ mm/mm} = -4130 \ \mu\text{mm/mm} \tag{2.89}$$
$$\varepsilon_3 = (-37.8 \times 10^{-12})(-50.0 \times 10^6) = -1893 \times 10^{-6} \text{ m/m}$$
$$= -1893 \times 10^{-6} \text{ mm/mm} = -1893 \ \mu \text{ mm/mm}$$

In the above section we noted that the computations lead to extensional strains in m/m, but as that is a dimensionless quantity, mm/mm is just as valid. There is no standard notation for reporting strain. Reporting ε_1 as 0.000080 can be done, or it can be reported as 0.008 percent. Because the displacements that result from the strains are usually small, if units are to be assigned, then mm/mm is more appropriate, although light-years/light-years is valid, but ridiculous. In the above section, the notation for micro has also been introduced; that is,

$$\mu = 10^{-6} \tag{2.90}$$

Herein the 10^{-6} or the μ mm/mm forms will be used. For the unconstrained element of material, then, according to equation (2.89) the 50 MPa stress in the 2 direction produces over 4000 μ mm/mm compressive strain in the 2 direction, and through Poisson effects, about 100 μ mm/mm elongation strain in the fiber direction and about 2000 μ mm/mm elongation strain in the 3 direction. These are realistic strain levels and as more examples are discussed, familiarity with strain levels associated with other applied stress levels will be established.

If instead of subjecting an element of graphite-reinforced material to the 50 MPa compressive stress, an element of aluminum is compressed, the deformations of the unconstrained element are given by the analog of equation (2.70), namely,

$$\delta\Delta_1 = \Delta_1\varepsilon_1 = \Delta_1 S_{12}\sigma_2 = (0.050)(-4.14 \times 10^{-12})(-50.0 \times 10^6)$$
$$= 10.36 \times 10^{-6} \text{ m} = 0.01036 \text{ mm}$$
$$\delta\Delta_2 = \Delta_2\varepsilon_2 = \Delta_2 S_{22}\sigma_2 = (0.050)(13.8 \times 10^{-12})(-50.0 \times 10^6)$$
$$= -34.5 \times 10^{-6} \text{ m} = -0.0345 \text{ mm} \tag{2.91}$$
$$\delta\Delta_3 = \Delta_3\varepsilon_3 = \Delta_3 S_{23}\sigma_2 = (0.050)(-4.14 \times 10^{-12})(-50.0 \times 10^6)$$
$$= 10.36 \times 10^{-6} \text{ m} = 0.01036 \text{ mm}$$

The compliances used in the above calculation are given in equation (2.60), and the use of a 1-2-3 coordinate system is only to provide a reference system. With aluminum, there is no difference in material properties in the 3 coordinate directions. As a result, due to the compressive stress in the 2 direction, the increase in length in the 1 direction of the aluminum cube is identical to the increase in length in the 3 direction. This certainly was not the case for the fiber-reinforced material. By the results of equation (2.70), we see that for the graphite-reinforced material, expansion in the 3 direction is about 25 times the expansion in the 1 direction!

If the aluminum is constrained in the 3 direction, as in the second case, the deformations are given by the analog to equation (2.78), namely,

$$\delta\Delta_1 = \Delta_1\varepsilon_1 = \Delta_1\left(S_{12} - \frac{S_{13}S_{23}}{S_{33}}\right)\sigma_2$$

$$= (0.050)(-5.39 \times 10^{-12})(-50.0 \times 10^6)$$

$$= 13.47 \times 10^{-6} \text{ m} = 0.01347 \text{ mm}$$

$$\delta\Delta_2 = \Delta_2\varepsilon_2 = \Delta_2\left(S_{22} - \frac{S_{23}S_{23}}{S_{33}}\right)\sigma_2 \qquad (2.92)$$

$$= (0.050)(12.57 \times 10^{-12})(-50.0 \times 10^6)$$

$$= -31.4 \times 10^{-6} \text{ m} = -0.0314 \text{ mm}$$

$$\delta\Delta_3 = 0 \text{ (by definition)}$$

The constraining stress is given by equation (2.79) as

$$\sigma_3 = -\frac{S_{23}}{S_{33}}\sigma_2 = -15.00 \text{ MPa} \qquad (2.93)$$

Finally, for a constraint in the 1 direction, the analog to equation (2.86) is used. Because the material is isotropic, we can immediately write, referring to equations (2.92) and (2.93),

$$\delta\Delta_1 = 0 \text{ (by definition)}$$

$$\delta\Delta_2 = -0.0314 \text{ mm}$$

$$\delta\Delta_3 = 0.01347 \text{ mm} \qquad (2.94)$$

$$\sigma_1 = -15.00 \text{ MPa}$$

If we use equations (2.83) and (2.84) as a check,

$$\sigma_1 = -\frac{S_{12}}{S_{11}}\sigma_2 = -15.00 \text{ MPa}$$

$$\delta\Delta_1 = 0$$

$$\delta\Delta_2 = \Delta_2\varepsilon_2 = \Delta_2\left(S_{22} - \frac{S_{12}S_{12}}{S_{11}}\right)\sigma_2$$

$$= (0.050)(12.57 \times 10^{-12})(-50 \times 10^6) = -0.0314 \text{ mm} \qquad (2.95)$$

$$\delta\Delta_3 = \Delta_3\varepsilon_3 = \Delta_3\left(S_{23} - \frac{S_{12}S_{13}}{S_{11}}\right)\sigma_2$$

$$= (0.050)(-5.39 \times 10^{-12})(-50 \times 10^6) = 0.01347 \text{ mm}$$

These examples provide an indication of the strains and deformations that result from applying a force to a fiber-reinforced material. More importantly, the results indicate how important the fibers are in controlling the response of composite materials, even though the fibers may not be loaded directly. The exercises below will give further examples to illustrate the domination of the fiber-direction properties.

Exercises for Section 2.5

1. Compute the strains, in μ mm/mm, for the unconstrained element of aluminum subjected to a 50 MPa compressive stress in the 2 direction, that is, the situation given by Figure 2.11(b). This is the aluminum analog to equation (2.89). Unlike the composite, the isotropy of the aluminum leads to identical strains in the 1 and 3 directions.

2. Consider the cube of graphite-reinforced material of Figure 2.11. Suppose the cube is simultaneously constrained against deformation in both the 1 and 3 directions. (a) What is the change in the 50 mm length in the 2 direction? (b) How does the apparent or reduced compliance in the 2 direction compare to the cases when there was only one constraint, equations (2.78) or (2.86)? That the apparent compliance is different indicates there is an interaction between the two constraints. This had better be the case because the stress-strain relations indicate all extensional strains are related to all normal stress components. The interaction can also be demonstrated by computing the stresses required to enforce the constraints. Therefore: (c) Compute the stress required to constrain the deformations in the 1 direction when there are constraints in both the 1 and 3 directions. Compare this with the stress required to constrain the deformations in the 1 direction when there is no constraint in the 3 direction, equation (2.87).

3. In studying the deformations of a cube of material to applied loads, we have used the compliance matrix, as in equations (2.67), (2.73), and (2.81). Exactly the same answer must result if the formulation had started with the stiffness matrix. Rework the problem of the cube with no constraints and subjected to the compressive load in the 2 direction, Figure 2.11(b), but start with the stiffness formulation; that is:

$$
\begin{Bmatrix} 0 \\ \sigma_2 \\ 0 \\ 0 \\ 0 \\ 0 \end{Bmatrix} = \begin{bmatrix} C_{11} & C_{12} & C_{13} & 0 & 0 & 0 \\ C_{12} & C_{22} & C_{23} & 0 & 0 & 0 \\ C_{13} & C_{23} & C_{33} & 0 & 0 & 0 \\ 0 & 0 & 0 & C_{44} & 0 & 0 \\ 0 & 0 & 0 & 0 & C_{55} & 0 \\ 0 & 0 & 0 & 0 & 0 & C_{66} \end{bmatrix} \begin{Bmatrix} \varepsilon_1 \\ \varepsilon_2 \\ \varepsilon_3 \\ \gamma_{23} \\ \gamma_{13} \\ \gamma_{12} \end{Bmatrix}
$$

where $\sigma_2 = -50.0$ MPa. Show that you obtain the results of equation (2.70).

4. A 50 mm cube of graphite-reinforced material is extended in the 1 direction by 0.50 mm and is constrained against contraction in the 3 direction. There is no constraint in the 2 direction. (a) What is the change in dimension in the 2 direction? (b) Determine the stresses in the 1 and 3 directions required to cause these deformations.

2.6
FREE THERMAL STRAINS

To this point the only deformations we have studied have been those that are a result of an applied load. However, when a fiber-reinforced composite material is heated or cooled, just as with an isotropic material, the material expands or contracts. This is a deformation that takes place independently of any applied load. If a load is applied, then we must contend with the deformations due both to applied load and to thermal expansion effects. Unlike an isotropic material, the thermal expansion of a fiber-reinforced material is different in each of the three principal material directions. Graphite fibers contract along their length when heated (see Table 1.2). Polymers, aluminum, boron, ceramics, and most other matrix materials expand when heated. Therefore, when heated, a composite will expand, contract, or possibly exhibit no change in length in the fiber direction, depending on the relative effects of the fiber and matrix materials. In the other two directions, due to the dominance of the properties of the matrix, the composite will expand. Figure 2.12(a) shows our small element of material with its temperature at some reference temperature. The element is not part of any structure; rather, it is isolated in space and free of stresses on its six faces. The temperature is uniform within the element, and at this reference temperature the element has dimensions Δ_1, Δ_2, and Δ_3. As the temperature of the element is changed, the element changes dimensions slightly in the fiber direction and moderately in the other two directions. The change in length of the element per

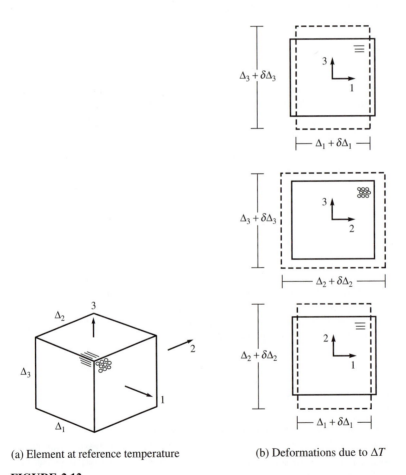

(a) Element at reference temperature   (b) Deformations due to ΔT

FIGURE 2.12
Thermal expansion of an element of fiber-reinforced material

unit original length is defined to be the thermal strain. Because the element has no tractions acting on any of its six faces, the strain is termed *free thermal strain*. The free thermal strains in the three coordinate directions will be denoted as

$$\varepsilon_1^T (T, T_{ref}) \qquad \varepsilon_2^T (T, T_{ref}) \qquad \varepsilon_3^T (T, T_{ref}) \qquad (2.96)$$

The superscript T indicates that the strain is a free thermal strain. T_{ref} is the temperature at the reference state, and T is the temperature at the state of interest, a temperature above or below the reference temperature. The two arguments T_{ref} and T emphasize the fact that the free thermal strain involves *both* of these temperatures. The reference state is, of course, arbitrary and depends on the specific problem being studied. For polymer-based fiber-reinforced composite materials, the reference state is often taken to be the stress-free processing condition. The reference temperature would then be the temperature corresponding to that condition. The physical interpretation of these free thermal strains is indicated in Figure 2.12(b), which shows the three views of the dimensional changes experienced by an element of graphite-reinforced material as the temperature is increased. Since, as mentioned before, graphite contracts when heated, the length of the element in the fiber direction is assumed to decrease. The dimensions in the directions perpendicular to the fibers are assumed to increase. In addition, the increases in the dimensions perpendicular to the fibers are much larger than the decrease in the fiber direction, reflecting the overwhelming influence of polymer expansion perpendicular to the fibers. For glass-reinforced materials, the dimensional changes perpendicular to the fibers would be similar to those changes for graphite-reinforced materials. Parallel to the fibers, however, an element of glass-reinforced material would expand, as opposed to contract, but less than the expansion perpendicular to the fibers. Aluminum, of course, expands the same amount in all three directions.

As Figure 2.12 indicates, the free thermal expansion in the various directions is directly related to the change in length in those directions. Specifically,

$$\varepsilon_1^T (T, T_{ref}) = \frac{\delta \Delta_1}{\Delta_1} \qquad \varepsilon_2^T (T, T_{ref}) = \frac{\delta \Delta_2}{\Delta_2} \qquad \varepsilon_3^T (T, T_{ref}) = \frac{\delta \Delta_3}{\Delta_3} \qquad (2.97)$$

where, as in the past, $\delta \Delta_1$, $\delta \Delta_2$, and $\delta \Delta_3$ are the changes in the lengths of the sides of the element parallel to the 1, 2, and 3 coordinate directions, respectively.

If the thermal expansion is linear with temperature change, then it is only the *difference* between the reference temperature and the temperature at the state of interest that is important, and the free thermal strains can be written as

$$\varepsilon_1^T (T, T_{ref}) = \alpha_1 \Delta T \qquad \varepsilon_2^T (T, T_{ref}) = \alpha_2 \Delta T \qquad \varepsilon_3^T (T, T_{ref}) = \alpha_3 \Delta T \qquad (2.98)$$

where

$$\Delta T = T - T_{ref} \qquad (2.99)$$

The quantities α_1, α_2, and α_3 are referred to as the *coefficients of thermal expansion*, or CTE for short. They have units of 1/°C, or to emphasize they are related to strains, mm/mm/°C. Obviously, ΔT is the temperature difference between the reference temperature pertinent to the problem and the temperature of interest. It is implicit that the thermal strains are defined to be zero at the reference temperature. A positive ΔT is associated with a temperature increase.

Several important points should be made relative to the above formulation. First, strictly speaking, α_1, α_2, and α_3 are referred to as the *linear* coefficients of thermal expansion. Expansional effects are linearly proportional to the temperature change, and as will be seen, the thermally induced stresses in a laminate will double if the temperature change doubles. Second, if expansional effects are not linearly proportional to the temperature change, then the linear coefficient of thermal expansion is meaningless. To study thermally induced deformation effects in these cases it is necessary to have explicit forms of ε_1^T, ε_2^T, and ε_3^T as a function of temperature. These functional forms are usually derived from experiments by the least-squares fitting of simple polynomial functions of temperature to the data. As implied by the notation of equation (2.96), in these cases the reference temperature and the temperature of interest must be part of the functional form for a complete description of the free thermal expansion effects.

A third important point is rather obvious but it should be mentioned explicitly. In the principal material coordinate system, the free thermal strains do not involve any shearing deformations. The strains are strictly dilatational, not distortional. (Dilatation effects are those identified with a change in volume of an element, whereas distortional effects are those identified with a change of shape of an element.)

Finally, for an element of material that is isolated in space, the strains that accompany any temperature change do not result in stresses on any of the six surfaces. This is counter to the stress-strain relations already presented. In the previous formulations the extensional strains ε_1, ε_2, and ε_3 are accompanied by normal stresses σ_1, σ_2, and σ_3. Clearly the stress-strain relations must be modified to account for free thermal strains that do not, in an unrestrained and completely free element, produce stresses.

Typical values for the coefficients of thermal expansion for graphite- and glass-reinforced composites, and aluminum, are presented in Table 2.1. For graphite-reinforced materials, the coefficient in the fiber direction can be slightly negative, while the coefficient perpendicular to the fibers is similar to the expansion of aluminum. Parallel to the fibers the stiffness of the fibers dominates, and the free thermal strains are governed by the contraction properties of the fibers. Perpendicular to the fiber direction the free thermal strains are governed by the large expansion properties of the matrix and the lack of stiffness of the fibers.

In linear thermal expansion, the changes in length of the three dimensions of the element of material in Figure 2.12 are given by

$$\delta\Delta_1 = \Delta_1\alpha_1\Delta T \qquad \delta\Delta_2 = \Delta_2\alpha_2\Delta T \qquad \delta\Delta_3 = \Delta_3\alpha_3\Delta T \qquad (2.100)$$

where equations (2.97) and (2.98) have been employed. For a 50 mm × 50 mm × 50 mm element of graphite-reinforced composite material, if the element is heated 50°C, the changes in dimensions are given by, from Table 2.1 and equation (2.100),

$$\delta\Delta_1 = (50)(-0.018 \times 10^{-6})(50) = -0.0000450 \text{ mm}$$
$$\delta\Delta_2 = (50)(-24.3 \times 10^{-6})(50) = 0.0608 \text{ mm} \qquad (2.101)$$
$$\delta\Delta_3 = (50)(-24.3 \times 10^{-6})(50) = 0.0608 \text{ mm}$$

Thus the dimensions of the heated element are: in the 1 direction

$$\Delta_1 + \delta\Delta_1 = 49.999955 \text{ mm} \qquad (2.102a)$$

in the 2 direction

$$\Delta_2 + \delta\Delta_2 = 50.0608 \text{ mm} \qquad (2.102b)$$

and in the 3 direction

$$\Delta_3 + \delta\Delta_3 = 50.0608 \text{mm} \qquad (2.102c)$$

2.7
STRESS-STRAIN RELATIONS, INCLUDING THE EFFECTS OF FREE THERMAL STRAINS

Before proceeding, a comment should be made regarding the term *free thermal expansion*. We have been considering an element of material wherein the fibers and matrix are smeared into a single equivalent homogeneous material. Free thermal strain refers to the fact that the smeared element is free of any stresses if the temperature is changed. When one considers an unsmeared material and deals with the individual fibers and the surrounding matrix, a temperature change can create significant stresses in the fiber and matrix. Clearly, if graphite contracts in the fiber direction and polymers expand, and the materials are combined, such stresses will certainly exist. However, when the stresses are integrated (i.e., smeared) over the volume of the element, the net result is zero.

To accommodate the fact that in a smeared element of material with no constraints on its bounding surfaces free thermal strains do not cause stresses, the stress-strain relation as it has been presented to this point has to be reinterpreted. The simplest interpretation, and the one that is consistent with the definitions of stress, strain, and free thermal strain, is that the strains in the stress-strain relations are the *mechanical* strains. In the context of free thermal strains this is interpreted to mean that mechanical strains are the *total* strains minus the free thermal strains. The total strains are a measure of the change in dimensions of an element of material, specifically the change in length per unit length. In the context of our element of material, these changes in length per unit length are given by $\delta\Delta_1/\Delta_1$, $\delta\Delta_2/\Delta_2$, $\delta\Delta_3/\Delta_3$. Thus

$$\varepsilon_1 \equiv \frac{\delta\Delta_1}{\Delta_1} \qquad \varepsilon_2 \equiv \frac{\delta\Delta_2}{\Delta_2} \qquad \varepsilon_3 \equiv \frac{\delta\Delta_3}{\Delta_3} \qquad (2.103)$$

where it is implied the strains are the total strains. In the previous section, referring to Figure 2.12, the total strains were equal to the thermal strains (because there were no stresses) and hence the notation of equation (2.97). However, in the presence of stresses on the faces of an element of material, equation (2.97) is not valid.

The concept of mechanical strain is something of an artifact. Unlike the dimensional changes associated with free expansion and those associated with total strains, no specific dimensional changes can be associated with mechanical strains. Nonetheless, the concept is useful for gaining insight into a problem if it is realized that these so-called mechanical strains are the key elements in the stress-strain relations. If we use the concept of mechanical strains to accommodate the free thermal strains, the stress-strain relations can be rewritten as

$$
\begin{Bmatrix}
\varepsilon_1 - \varepsilon_1^T(T, T_{ref}) \\
\varepsilon_2 - \varepsilon_2^T(T, T_{ref}) \\
\varepsilon_3 - \varepsilon_2^T(T, T_{ref}) \\
\gamma_{23} \\
\gamma_{13} \\
\gamma_{12}
\end{Bmatrix}
=
\begin{bmatrix}
S_{11} & S_{12} & S_{13} & 0 & 0 & 0 \\
S_{12} & S_{22} & S_{23} & 0 & 0 & 0 \\
S_{13} & S_{23} & S_{33} & 0 & 0 & 0 \\
0 & 0 & 0 & S_{44} & 0 & 0 \\
0 & 0 & 0 & 0 & S_{55} & 0 \\
0 & 0 & 0 & 0 & 0 & S_{66}
\end{bmatrix}
\begin{Bmatrix}
\sigma_1 \\
\sigma_2 \\
\sigma_3 \\
\tau_{23} \\
\tau_{13} \\
\tau_{12}
\end{Bmatrix}
\tag{2.104}
$$

where the total strains are

$$
\begin{Bmatrix}
\varepsilon_1 \\
\varepsilon_2 \\
\varepsilon_3 \\
\gamma_{23} \\
\gamma_{13} \\
\gamma_{12}
\end{Bmatrix}
\tag{2.105}
$$

and the mechanical strains are

$$
\begin{Bmatrix}
\varepsilon_1^{mech} \\
\varepsilon_2^{mech} \\
\varepsilon_3^{mech} \\
\gamma_{23}^{mech} \\
\gamma_{13}^{mech} \\
\gamma_{12}^{mech}
\end{Bmatrix}
=
\begin{Bmatrix}
\varepsilon_1 - \varepsilon_1^T(T, T_{ref}) \\
\varepsilon_2 - \varepsilon_2^T(T, T_{ref}) \\
\varepsilon_3 - \varepsilon_3^T(T, T_{ref}) \\
\gamma_{23} \\
\gamma_{13} \\
\gamma_{12}
\end{Bmatrix}
\tag{2.106}
$$

In the principal material system, the mechanical shear strains and the total shear strains are identical. If the free thermal strain is linearly dependent on temperature, the stress-strain equations take the form

$$
\begin{Bmatrix}
\varepsilon_1 - \alpha_1 \Delta T \\
\varepsilon_2 - \alpha_2 \Delta T \\
\varepsilon_3 - \alpha_3 \Delta T \\
\gamma_{23} \\
\gamma_{13} \\
\gamma_{12}
\end{Bmatrix}
=
\begin{bmatrix}
S_{11} & S_{12} & S_{13} & 0 & 0 & 0 \\
S_{12} & S_{22} & S_{23} & 0 & 0 & 0 \\
S_{13} & S_{23} & S_{33} & 0 & 0 & 0 \\
0 & 0 & 0 & S_{44} & 0 & 0 \\
0 & 0 & 0 & 0 & S_{55} & 0 \\
0 & 0 & 0 & 0 & 0 & S_{66}
\end{bmatrix}
\begin{Bmatrix}
\sigma_1 \\
\sigma_2 \\
\sigma_3 \\
\tau_{23} \\
\tau_{13} \\
\tau_{12}
\end{Bmatrix}
\tag{2.107}
$$

The inverse relations for this linear thermal expansion case are

$$
\begin{Bmatrix} \sigma_1 \\ \sigma_2 \\ \sigma_3 \\ \tau_{23} \\ \tau_{13} \\ \tau_{12} \end{Bmatrix} = \begin{bmatrix} C_{11} & C_{12} & C_{13} & 0 & 0 & 0 \\ C_{12} & C_{22} & C_{23} & 0 & 0 & 0 \\ C_{13} & C_{23} & C_{33} & 0 & 0 & 0 \\ 0 & 0 & 0 & C_{44} & 0 & 0 \\ 0 & 0 & 0 & 0 & C_{55} & 0 \\ 0 & 0 & 0 & 0 & 0 & C_{66} \end{bmatrix} \begin{Bmatrix} \varepsilon_1 - \alpha_1 \Delta T \\ \varepsilon_2 - \alpha_2 \Delta T \\ \varepsilon_3 - \alpha_3 \Delta T \\ \gamma_{23} \\ \gamma_{13} \\ \gamma_{12} \end{Bmatrix} \tag{2.108}
$$

where the mechanical strains are given by

$$
\begin{Bmatrix} \varepsilon_1^{mech} \\ \varepsilon_2^{mech} \\ \varepsilon_3^{mech} \\ \gamma_{23}^{mech} \\ \gamma_{13}^{mech} \\ \gamma_{12}^{mech} \end{Bmatrix} = \begin{Bmatrix} \varepsilon_1 - \alpha_1 \Delta T \\ \varepsilon_2 - \alpha_2 \Delta T \\ \varepsilon_3 - \alpha_3 \Delta T \\ \gamma_{23} \\ \gamma_{13} \\ \gamma_{12} \end{Bmatrix} \tag{2.109}
$$

These definitions of strains are the most general and include the cases of stresses with no thermal effects, thermal effects but no stresses (i.e., free thermal strains), and thermal effects with stresses. For stresses with no temperature change, $\Delta T = 0$ and equations (2.107) and (2.108) reduce directly to equations (2.45) and (2.47). For a temperature change but no stresses, the free thermal strain case,

$$
\sigma_1 = \sigma_2 = \sigma_3 = \tau_{23} = \tau_{13} = \tau_{12} = 0 \tag{2.110}
$$

and by using either equation (2.107) or (2.108), the total strains are

$$
\varepsilon_1 = \alpha_1 \Delta T \qquad \varepsilon_2 = \alpha_2 \Delta T \qquad \varepsilon_3 = \alpha_3 \Delta T \qquad \gamma_{23} = \gamma_{13} = \gamma_{12} = 0 \tag{2.111}
$$

For this case, if we use equations (2.109) and (2.111), the mechanical strains are given by

$$
\begin{aligned}
\varepsilon_1^{mech} &= \varepsilon_1 - \alpha_1 \Delta T = \alpha_1 \Delta T - \alpha_1 \Delta T = 0 \\
\varepsilon_2^{mech} &= \varepsilon_2 - \alpha_2 \Delta T = \alpha_2 \Delta T - \alpha_2 \Delta T = 0 \\
\varepsilon_3^{mech} &= \varepsilon_3 - \alpha_3 \Delta T = \alpha_3 \Delta T - \alpha_3 \Delta T = 0 \\
\gamma_{23}^{mech} &= \gamma_{23} = 0 \\
\gamma_{13}^{mech} &= \gamma_{13} = 0 \\
\gamma_{12}^{mech} &= \gamma_{12} = 0
\end{aligned} \tag{2.112}
$$

that is, the mechanical strains are all zero. On the other hand, if an element of material is fully restrained against deformation,

$$\delta\Delta_1 = \delta\Delta_2 = \delta\Delta_3 = 0 \tag{2.113}$$

By equation (2.103),

$$\varepsilon_1 = \varepsilon_2 = \varepsilon_3 = 0 \tag{2.114}$$

and we know

$$\gamma_{23} = \gamma_{13} = \gamma_{12} = 0 \tag{2.115}$$

If the temperature is changed an amount ΔT, the stresses caused by the temperature change are

$$
\begin{Bmatrix} \sigma_1 \\ \sigma_2 \\ \sigma_3 \\ \tau_{23} \\ \tau_{13} \\ \tau_{12} \end{Bmatrix}
=
\begin{bmatrix}
C_{11} & C_{12} & C_{13} & 0 & 0 & 0 \\
C_{12} & C_{22} & C_{23} & 0 & 0 & 0 \\
C_{13} & C_{23} & C_{33} & 0 & 0 & 0 \\
0 & 0 & 0 & C_{44} & 0 & 0 \\
0 & 0 & 0 & 0 & C_{55} & 0 \\
0 & 0 & 0 & 0 & 0 & C_{66}
\end{bmatrix}
\begin{Bmatrix} -\alpha_1\Delta T \\ -\alpha_2\Delta T \\ -\alpha_3\Delta T \\ 0 \\ 0 \\ 0 \end{Bmatrix}
\tag{2.116}
$$

In this case, if we use equations (2.109), (2.114), and (2.115), the mechanical strains are given by

$$\varepsilon_1^{mech} = -\alpha_1\Delta T \qquad \varepsilon_2^{mech} = -\alpha_2\Delta T \qquad \varepsilon_3 = -\alpha_3\Delta T$$
$$\gamma_{23}^{mech} = \gamma_{13}^{mech} = \gamma_{12}^{mech} = 0 \tag{2.117}$$

The concept of mechanical strain is an important one. As an example, consider a fully restrained element of graphite-reinforced material. Substituting numerical values for the stiffnesses from equation (2.58), and numerical values for the coefficients of thermal expansion from Table 2.1, we find that if the temperature is raised 50°C from some reference temperature, the changes in the stresses in a fully restrained material are, from equation (2.116),

$$\sigma_1 = -(\alpha_1 C_{11} + \alpha_2 C_{12} + \alpha_3 C_{13})\Delta T = -13.55 \text{ MPa}$$

$$\sigma_2 = -(\alpha_1 C_{12} + \alpha_2 C_{22} + \alpha_3 C_{23})\Delta T = -27.6 \text{ MPa}$$

$$\sigma_3 = -(\alpha_1 C_{13} + \alpha_2 C_{23} + \alpha_3 C_{33})\Delta T = -27.6 \text{ MPa} \tag{2.118}$$

$$\tau_{23} = \tau_{13} = \tau_{12} = 0$$

Though these normal stresses do not seem significant, a lowering of the temperature would cause the sign of the stresses to become positive, and the roughly 30 MPa stress perpendicular to the fiber direction (i.e., σ_2 or σ_3) is a substantial percentage of the stress to cause failure of the material in tension in these two directions. In addition, due to the interaction of the three components of normal stress with the three components of extensional strain, the stress in the fiber direction, σ_1, is compressive. This is the case despite the fact that in the context of free thermal strains, the material contracts in the fiber direction when heated. Contraction in a direction when heated would lead one to believe the thermally induced stress would be tensile in that direction. The compressive stress in the fiber direction is a result of Poisson effects coupling the fiber-direction stress with the other two stresses. The interaction of thermal and Poisson effects is subtle but quite important. The mechanical strains for this fully constrained case are, from equation (2.117)

$$\varepsilon_1^{mech} = 0.900 \times 10^{-6} \text{ mm/mm} = 0.900 \ \mu \text{ mm/mm}$$

$$\varepsilon_2^{mech} = -1215 \times 10^{-6} \text{ mm/mm} = -1215 \ \mu \text{ mm/mm}$$

$$\varepsilon_3^{mech} = -1215 \times 10^{-6} \text{ mm/mm} = -1215 \ \mu \text{ mm/mm} \tag{2.119}$$

$$\gamma_{23}^{mech} = \gamma_{13}^{mech} = \gamma_{12}^{mech} = 0$$

A significant point should be made at this time. In reality the elastic properties of a fiber-reinforced material are dependent on temperature. It is the absolute temperature that is important, not the temperature relative to some reference temperature. Then, in the stress-strain relations the stiffnesses and compliances should be considered as functions of temperature; that is,

$$C_{ij} = C_{ij}(T) \qquad i, j = 1, 6 \tag{2.120}$$

and $$S_{ij} = S_{ij}(T) \qquad i, j = 1, 6 \tag{2.121}$$

The interpretation in this case is that at a particular temperature T, the stresses and strains are related in accordance with equation (2.104), where it has been assumed that the free thermal expansion is a function of T and of the reference temperature, T_{ref}. How the material was heated to the particular temperature or what the material properties are at other temperatures are not important. The relationship is path-independent, and the material properties depend only on the current temperature, and for the free thermal strain, also on the reference temperature. For moderate increases of temperature relative to room temperature, the elastic properties can be assumed to be independent of temperature. As the temperature increase approaches the processing temperature, however, the elastic properties are a function of temperature. At temperatures lower than room temperature, the elastic properties for any particular material should also be checked for temperature dependence.

Exercise for Section 2.7

Consider a 50 mm × 50 mm × 50 mm element of graphite-reinforced material that is heated 50°C above some reference state and is restrained in the 2 direction. (a) What are the changes in dimensions of the heated element in the 1 and 3 directions? (b) What stress σ_2 is required to restrain the element against deformation in the 2 direction? Assume the material properties are independent of temperature. (c) How do the changes in dimensions of the heated element for this partially restrained case compare numerically with the changes in dimensions of the heated element for the case of free thermal strain, equation (2.102)? Why are they different? (d) What are the mechanical strains for this case? Numerically compare the mechanical strains with the fully constrained case, that is, equation (2.117). Hint: To solve the problem use either equation (2.107) or (2.108), assume $\varepsilon_1 \neq 0$ and $\varepsilon_3 \neq 0$, and solve for ε_1, ε_3, and σ_2.

2.8
STRESS-STRAIN RELATIONS, INCLUDING THE EFFECTS OF FREE MOISTURE STRAINS

When exposed to a liquid, polymers absorb a certain amount of that liquid and, in general, expand. The amount of liquid absorbed is not significant, however. Weight gains in excess of 3 or 4% are unusual. As a result, it is not the weight gain that is important; rather, it is the expansion, which is similar to the expansion that accompanies a temperature increase, that is the issue. The level of the expansion is close to being linear with the amount of liquid absorbed. Consequently, in analogy to the coefficient of thermal expansion, defining a coefficient of *moisture* expansion represents a viable model for characterizing the free moisture expansion of a polymer. For polymer matrix composites, the use of a linear expansion model is also applicable. As might be expected, however, the moisture expansion in the fiber direction is small, at least for graphite fibers; the fibers usually do not absorb moisture, and the stiffness of the fibers overcomes any tendency for the polymer to expand in that direction. On the other hand, expansion perpendicular to the fibers is significant. Using the analogy to linear thermal expansion, equation (2.98), the free moisture strains are given by

$$\varepsilon_1^M(M, M_{ref}) = \beta_1 \Delta M \qquad \varepsilon_2^M(M, M_{ref}) = \beta_2 \Delta M$$

$$\varepsilon_3^M(M, M_{ref}) = \beta_3 \Delta M$$

(2.122)

The superscript M identifies that the strain is moisture-induced and the dependence on both M and M_{ref} signifies that the free moisture strain is measured relative to some reference moisture state. However, here it is assumed that it is only the *change* in moisture relative to that state that determines the strain relative to that state. The free moisture strains in the reference state are defined to be zero. The β_i are referred to as the *coefficients of moisture expansion*. Generally, β_1 can be taken to be zero and β_2 and β_3 are taken to be equal. For most fiber-reinforced materials, β_2 ranges from 0.003 to 0.005/% moisture. With a 3 percent moisture change, say, from the dry state, this can result in expansional strains upward of 0.015 perpendicular to the fibers. This is a large amount of strain. Table 2.1 includes values of the moisture expansion coefficients for graphite- and glass-reinforced composite materials.

Using the values of β_1, β_2, and β_3 from Table 2.1, we can determine the dimensional changes of a 50 mm × 50 mm × 50 mm element of graphite-reinforced material which has absorbed 0.5% moisture by using the moisture analogy to equation (2.97), namely,

$$\delta \Delta_1 = \Delta_1 \varepsilon_1^M(M, M_{ref}) = \Delta_1 \beta_1 \Delta M = (50)(146 \times 10^{-6})(0.5)$$

$$\delta \Delta_1 = 0.00365 \text{ mm}$$

(2.123a)

$$\delta \Delta_2 = \Delta_2 \varepsilon_2^M(M, M_{ref}) = \Delta_2 \beta_2 \Delta M = (50)(4770 \times 10^{-6})(0.5)$$

$$\delta \Delta_2 = 0.1193 \text{ mm}$$

(2.123b)

$$\delta\Delta_3 = \Delta_3 \varepsilon_3^M(M, M_{ref}) = \Delta_3 \beta_3 \Delta M = (50)(4770 \times 10^{-6})(0.5)$$
$$\delta\Delta_3 = 0.1193 \text{ mm} \tag{2.123c}$$

The dimensional changes in the 2 and 3 directions are significant, and restraining them leads to large moisture-induced stresses.

The manner and the rate at which polymers absorb moisture is an important topic in the study of the mechanics of fiber-reinforced material. The study parallels the characteristics of heat transfer within a solid. For the moment, however, it is only necessary to assume that the composite has absorbed moisture and that the moisture has produced a strain. Incorporating these free moisture strains into the stress-strain relations, in the same manner as the free thermal strains were accounted for, leads to, in analogy to equations (2.107) and (2.108),

$$\begin{Bmatrix} \varepsilon_1 - \alpha_1 \Delta T - \beta_1 \Delta M \\ \varepsilon_2 - \alpha_2 \Delta T - \beta_2 \Delta M \\ \varepsilon_3 - \alpha_3 \Delta T - \beta_3 \Delta M \\ \gamma_{23} \\ \gamma_{13} \\ \gamma_{12} \end{Bmatrix} = \begin{bmatrix} S_{11} & S_{12} & S_{13} & 0 & 0 & 0 \\ S_{12} & S_{22} & S_{23} & 0 & 0 & 0 \\ S_{13} & S_{23} & S_{33} & 0 & 0 & 0 \\ 0 & 0 & 0 & S_{44} & 0 & 0 \\ 0 & 0 & 0 & 0 & S_{55} & 0 \\ 0 & 0 & 0 & 0 & 0 & S_{66} \end{bmatrix} \begin{Bmatrix} \sigma_1 \\ \sigma_2 \\ \sigma_3 \\ \tau_{23} \\ \tau_{13} \\ \tau_{12} \end{Bmatrix} \tag{2.124}$$

and

$$\begin{Bmatrix} \sigma_1 \\ \sigma_2 \\ \sigma_3 \\ \tau_{23} \\ \tau_{13} \\ \tau_{12} \end{Bmatrix} = \begin{bmatrix} C_{11} & C_{12} & C_{13} & 0 & 0 & 0 \\ C_{12} & C_{22} & C_{23} & 0 & 0 & 0 \\ C_{13} & C_{23} & C_{33} & 0 & 0 & 0 \\ 0 & 0 & 0 & C_{44} & 0 & 0 \\ 0 & 0 & 0 & 0 & C_{55} & 0 \\ 0 & 0 & 0 & 0 & 0 & C_{66} \end{bmatrix} \begin{Bmatrix} \varepsilon_1 - \alpha_1 \Delta T - \beta_1 \Delta M \\ \varepsilon_2 - \alpha_2 \Delta T - \beta_2 \Delta M \\ \varepsilon_3 - \alpha_3 \Delta T - \beta_3 \Delta M \\ \gamma_{23} \\ \gamma_{13} \\ \gamma_{12} \end{Bmatrix} \tag{2.125}$$

The mechanical strains, the key strains in the stress-strain relations, are given by

$$\begin{aligned} \varepsilon_1^{mech} &= \varepsilon_1 - \alpha_1 \Delta T - \beta_1 \Delta M \\ \varepsilon_2^{mech} &= \varepsilon_2 - \alpha_2 \Delta T - \beta_2 \Delta M \\ \varepsilon_3^{mech} &= \varepsilon_3 - \alpha_3 \Delta T - \beta_3 \Delta M \end{aligned} \tag{2.126}$$

where, as has consistently been the case, the dimensional changes are reflected in

$$\delta\Delta_1 = \varepsilon_1 \Delta_1$$
$$\delta\Delta_2 = \varepsilon_2 \Delta_2$$
$$\delta\Delta_3 = \varepsilon_3 \Delta_3$$

not   (2.127)

$$\delta\Delta_1 = (\varepsilon_1 - \alpha_1\Delta T - \beta_1\Delta M)\Delta_1$$
$$\delta\Delta_2 = (\varepsilon_2 - \alpha_2\Delta T - \beta_2\Delta M)\Delta_2$$
$$\delta\Delta_3 = (\varepsilon_3 - \alpha_3\Delta T - \beta_3\Delta M)\Delta_3$$

With both free thermal strains and free moisture strains present in various combinations, some unusual effects and stress states are possible. The combination of moisture and heat is, in general, detrimental to polymer matrix composite materials, and it represents an important class of problems for these materials.

As a numerical example, using the values of β_1, β_2, and β_3 from Table 2.1, we find that a fully restrained element of graphite-reinforced composite that has absorbed 0.5% moisture but has experienced no temperature change requires the following stresses to prevent it from expanding:

$$\sigma_1 = -(C_{11}\beta_1 + C_{12}\beta_2 + C_{13}\beta_3)\Delta M = -38.4 \text{ MPa}$$
$$\sigma_2 = -(C_{12}\beta_1 + C_{22}\beta_2 + C_{23}\beta_3)\Delta M = -54.6 \text{ MPa}$$ (2.128)
$$\sigma_3 = -(C_{13}\beta_1 + C_{23}\beta_2 + C_{33}\beta_3)\Delta M = -54.6 \text{ MPa}$$

The stresses generated by moisture absorption are substantial. Again, since β_1 is small, the stress generated in the fiber direction is due to a coupling of the 1, 2, and 3 directions through Poisson effects. Clearly, one needs to be very aware of Poisson effects in materials, particularly in fiber-reinforced materials. The mechanical strains for this fully restrained isothermal case are

$$\varepsilon_1^{mech} = -\beta_1\Delta M = -73 \times 10^{-6} = -73 \ \mu \text{ mm/mm}$$
$$\varepsilon_2^{mech} = -\beta_2\Delta M = -2380 \times 10^{-6} = -2380 \ \mu \text{ mm/mm}$$ (2.129)
$$\varepsilon_3^{mech} = -\beta_3\Delta M = -2380 \times 10^{-6} = -2380 \ \mu \text{ mm/mm}$$
$$\gamma_{23}^{mech} = \gamma_{13}^{mech} = \gamma_{12}^{mech} = 0$$

Like temperature, moisture influences the elastic properties of a fiber-reinforced material. For graphite-reinforced materials, moisture has little influence on properties in the fiber direction. However, moisture does influence other properties, and it can influence properties in the fiber direction for some polymeric fibers. To account for moisture-induced property changes, the stiffnesses and the compliances can be made dependent on the absolute moisture content. As with temperature-dependent material properties, the interpretation in this case is that at particular moisture content M, the stresses and strains are related in accordance with equation (2.124) or (2.125). How the moisture content reached a particular level or what the material properties are at other moisture levels are not important. The relationship is path-independent, and the material properties depend only on the current moisture content, and the free moisture strains depend only on the reference moisture state and the current moisture state.

Exercises for Section 2.8

1. An element of graphite-reinforced composite is completely dry and is constrained against any dimensional changes in the 1 and 2 directions, but it is free to expand in the 3 direction. The element absorbs 1.5% of its dry weight in the form of moisture. (*a*) What is the strain in the 3 direction as a result of the moisture absorption? Why is this strain different from the value given by $\beta_3 \Delta M$? (*b*) What stresses develop in the element? Assume that moisture absorption does not influence the material properties.

2. The moisture expansion coefficient of a graphite-reinforced composite in the fiber direction is much smaller than in the other two directions. Therefore it might be expected that constraining an element of graphite-reinforced material from expansion in the 2 and 3 directions but allowing it to expand in the 1 direction when 0.5% moisture is absorbed would result in stresses not too different from that of the fully restrained case. That is, even with no restraint in the fiber direction, the stiff fibers may act as a restraint. Or is this wrong thinking? Compute the stresses for this partially constrained case and comment on why the stresses are or are not different from the fully restrained case.

3. An element of graphite-reinforced material is cooled $-75°C$ and then it absorbs 0.5% of its weight in moisture. The element is completely restrained against any dimensional changes during cooling and subsequent moisture absorption. Assume that neither the temperature change nor the moisture change influence material properties. Compute the stresses due to just the temperature decrease, then compute the stresses due to the temperature decrease and the moisture absorption. Can we say that the absorption of moisture relieves thermally induced stresses resulting from a temperature decrease? Basing your answer on the material properties given in Table 2.1, do you believe the statement is true just for this case, or do you think it might be a general statement?

2.9
SUMMARY

This chapter has introduced the fundamental ideas of the three-dimensional stress-strain relations for fiber-reinforced composite materials. These three-dimensional relationships form the basis for two-dimensional relationships to be developed and utilized in the remainder of the book. Three fundamental points to keep in mind are:

1. The three-dimensional relations addressed in the chapter have obscured interactions between the fiber and the matrix by smearing these constituents into an equivalent homogeneous orthotropic material.
2. The engineering properties E_1, E_2, E_3, v_{12}, v_{13}, v_{23}, G_{12}, G_{13}, G_{23} form the basis for the compliance and stiffness matrices. These properties exhibit a strong directional dependence.
3. Only one component in each of the pairs σ_1-ε_1, σ_2-ε_2, ..., τ_{12}-γ_{12} can be specified.

4. The thermal expansion coefficients α_1, α_2, α_3 are also directionally dependent and can cause significant stresses to develop.
5. Expansion due to moisture absorption can be viewed in a manner similar to thermal expansion and can also lead to significant stresses.

Since the engineering properties form the basis of the compliance and stiffness matrices, the next chapter discusses methods for computing the engineering properties of a composite, given that the engineering properties of the fiber and matrix are known.

2.10
SUGGESTED READINGS

The form of the stress-strain relations in the principal material coordinate system for an element of fiber-reinforced material, where it is assumed that the fiber and matrix properties are smeared into a single homogeneous material, can be traced back to the forms for various elastic symmetries for crystals of many materials. Crystal classification (triclinic, monoclinic, etc.) and the effect of symmetry on the form of the elastic constants is discussed in

1. Dowling, N. E. *Mechanical Behavior of Materials.* Englewood Cliffs, NJ: Prentice Hall, 1993.
2. Love, A. E. H. *A Treatise on the Mathematical Theory of Elasticity.* 4th ed. New York: Cambridge University Press, 1944. Reprint of a 1927 Dover Publications edition.

Postulating the existence of a strain energy density leads to a form of the stress-strain relations, and imposition of various symmetry constraints leads to the orthotropic form, including thermal expansion effects. See the following:

3. Fung, Y. C. *Foundations of Solid Mechanics.* Englewood Cliffs, NJ: Prentice Hall, 1969.

Methods for determining engineering properties of composites are documented by the American Society for Testing and Materials (ASTM) in the following:

4. *ASTM Standards and Literature References for Composite Materials*, 2nd ed. West Conshohocken, PA: ASTM, 1990.

Particularly interesting are the literature references, which trace the history of research that has led to the methods discussed in the book. These test methods are also documented in the following:

5. *Annual Book of ASTM Standards.* Vol. 15-03, *High-Modulus Fibers and Composites.* West Conshohocken, PA: ASTM, 1996.

There are numerous ASTM Special Technical Publications (STPs) that address composite materials. These can be consulted for the latest test methods, as well as data on the most recent materials. Chapter 10 in the following reference also documents test methods:

6. Gibson, R. F. *Principles of Composite Material Mechanics.* New York: McGraw-Hill, 1994.

The following paper contains an early discussion of the expansion of polymers and its similiarity with thermal expansion:

7. Hahn, H. T., and R. Y. Kim. "Swelling of Composite Laminates." In *Advanced Composite Materials—Environmental Effects.* ASTM STP 658. Ed. J. R. Vinson. Philadelphia: American Society for Testing and Materials, 1978.

CHAPTER 3

<hr style="height:6px; background:#cccccc; border:none;" />

Prediction of Engineering Properties Using Micromechanics

Thus far, engineering properties, stiffnesses, compliances, and the relations between them have been discussed without concern for what is happening at the fiber level. While many important issues regarding the response of fiber-reinforced composite materials and structures can be addressed without knowing what is happening at this level, it is often helpful—and sometimes even necessary—to view composites from this vantage point. The interactions of the elastic properties of the fiber and matrix produce the elastic properties of the composite material. Similarly, the thermal expansion properties of the fiber and matrix interact to produce the thermal expansion properties of the composite. By knowing how the two constituents interact, you can predict the material properties of a composite material. More importantly, by knowing how the two constituents interact, you can design a composite material to achieve particular overall properties. Furthermore, what ultimately determines the load capacity of the composite material is the stress state at this microstructural level, within the matrix, within the fiber, and at the interface between the matrix and fiber. Computing the stresses at this level can be very useful for understanding some of the underlying mechanisms of failure and for constructing failure theories for composite materials.

In general, studying the response of a fiber-reinforced composite material at the microstructural level is quite involved. As with viewing composites from the structural level, you must resolve many questions before you can address the response at the micromechanical level. How are the fibers to be represented? How is the matrix to be represented? How many fibers have to be included in the model to properly reflect a representative volume? How are these fibers arranged? Can you assume that the microstructure is repeating or periodic? What length of fiber needs to be considered? Does a fiber have the same properties in its axial direction as across its diameter? What are the material properties of the fiber? How are they determined? After all, it is difficult to load a fiber across its diameter to determine the properties perpendicular to the axial direction. What are the material properties of the matrix?

If failure is to be studied, what failure criterion is to be used at that level? From a different viewpoint, what stresses can the fiber withstand? What stresses can the matrix withstand? What stresses can the interface between the fiber and matrix withstand? Is the interphase region, if there is one, important? The need to answer these questions is affected by the type of information being sought by viewing composites from the microstructural level. As this list of questions demonstrates, the issue of micromechanics needs to be viewed carefully, lest the results be misinterpreted. For example, if key information is missing or not known with a high degree of accuracy—say, material properties of the fiber—it may not be worth the effort to study composites at the microstructural level. Nevertheless, because it is important to be versed in the issues, the following sections address issues in micromechanics. After this discussion, we will again focus on the larger scale of a layer of fiber-reinforced material, the so-called macromechanical scale. We hope that the discussion of micromechanics will allow developments in macromechanics to be viewed in proper prospective.

3.1
BACKGROUND

One of the earliest models of composite materials considered a single infinitely long fiber surrounded by matrix. We saw in Figure 1.4 how the matrix surrounding the fiber transfers the load to the fiber. The transfer occurs at the end of the fiber, so one might conclude that an infinitely long fiber is not a good model. However, because the length of most fibers is hundreds of times greater than their diameter, the region of stress transfer into the fiber from the matrix is so small that an infinite-length fiber model can be justified. On the other hand, although the infinitely long approximation may be accurate, the single-fiber model ignores the existence of other fibers near the one being considered. However, the issue in these early models was the overall elastic and thermal expansion properties of the composite, not the details of the stresses or deformations at the microstructural level. As long as the volume of fiber relative to the volume of matrix was represented correctly, the single-fiber model could be justified to some extent.

In other views of the microstructure of composites, the fibers were seen as a parallel array of stiff elastic elements joined together by a softer element, the matrix material. A strength-of-materials approach was used to analyze this assemblage of stiff and soft elements. Although such models quite accurately predicted the extensional modulus of the composite in the fiber direction, they were poor predictors of the extensional modulus perpendicular to the fibers. Modifications were introduced to correct these problems, and these modified models have been used with some success.

The following sections discuss several kinds of micromechanical models. Results from the various models are presented and comparisons among the models are compared. The primary interest with the models is the prediction of composite material properties. However, to provide insight into failure, we will also discuss stresses in the fiber and matrix. The approach will be to introduce some of the more complex models first. In considering fibers and the surrounding matrix we will assume that the fibers are spaced periodically in square-packed or hexagonal-packed arrays. We will assume that the fibers are infinitely long. To obtain results from these models, numerical approaches must be used. We will present results from the finite-element method and, as will be seen, either the square-packed or hexagonal-packed array model provides a basis for addressing a number of issues related to studying fiber-reinforced composite materials. Though numerical approaches are beyond the scope of this book, we present such results because they provide an accurate representation of response of this level. As a contrast to the numerically based square- and hexagonal-packed array models, we will base the second approach to micromechanics on elasticity solutions. This approach can lead to intractible formulations, but with simplifications results can be obtained. We will use one such simplification, known as the concentric cylinders model, to obtain results for particular problems. As an alternative to the first two approaches discussed, the third series of models discussed will be the simplest. This approach considers the fibers as stiff parallel fiber elements joined by softer matrix material elements. We will analyze these models, referred to as the rule-of-mixtures models, using a strength-of-materials approach. The simplicity and limitations of these models will become immediately obvious.

3.2
FINITE-ELEMENT RESULTS: SQUARE- AND HEXAGONAL-PACKED ARRAY MODELS

From Figure 1.6 we can see that in a fiber-reinforced material, the fibers are distributed throughout the matrix in a pattern we could describe as somewhat repeating or periodic. There is randomness involved, but as a first approximation the cross section could be idealized as in Figure 3.1(a) or (b). Figure 3.1(a) illustrates the square-packed array and Figure 3.1(b) illustrates an alternative model, the hexagonal-packed array. The names of the arrays are derived from the shape of the polygons that describe the fiber-packing pattern, and generally the hexagonal-packed array is the preferred model of the two.

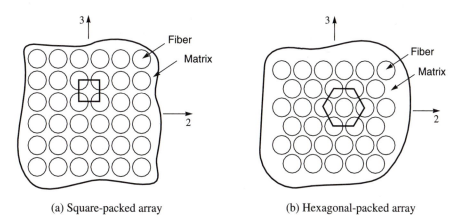

(a) Square-packed array   (b) Hexagonal-packed array

FIGURE 3.1
Cross section idealizations for micromechanics studies

If we assume that either of the two models represents to a reasonable degree of accuracy the microstructure of a fiber-reinforced composite material, how can we determine the stresses at the interface between the fiber and the matrix? How can predictions of the fiber-direction extensional modulus be made? What about thermal expansion effects? The first step to obtaining results is to realize that because of the symmetry, only one, or even only part of one, fiber and the surrounding matrix need be considered. If the cross sections shown in Figure 3.1 are representative, then they go on for quite some extent in both cross-sectional directions. (We assumed that the fibers are of infinite length, so the cross sections of Figure 3.1 also go on indefinitely into and out of the plane of the figure.) If, for example, a load is applied in the fiber direction, out of the plane of the figure, then as each fiber is embedded in a vast array of fibers, each fiber will respond the same as its neighboring fibers and attention can focus on a so-called unit cell. Figure 3.2 illustrates the concept of a unit cell for both the square-packed and the hexagonal-packed arrays. Because each fiber is embedded in a vast array of other fibers, there is a periodicity to the response, and because of this periodicity, we can argue that the straight lines outlining the unit cells in Figure 3.2(a) and (b) remain straight when the composite is subjected to any one of a number of basic loadings, such as a tensile loading in the fiber direction (in the 1 direction out of the plane of the figure), a transverse tensile loading (from left to right in the 2 direction in the figure), a temperature change, and the like. Because these lines remain straight, attention can be directed at the response of just one unit cell, as in Figure 3.3.

By directing attention at a unit cell, we easily see that the cross-sectional area of fiber relative to the total cross-sectional area of the unit cell is a measure of the volume of fiber relative to the total volume of the composite. This fraction is an important parameter in composite materials and is called *fiber volume fraction*. Fiber volume fraction will be denoted V^f and it is a number between 0 and 1, usually 0.5 or greater. We will continue to use the 1-2-3 coordinate system to study the response of a unit cell. Circumferential locations around the fiber-matrix interface will be identified

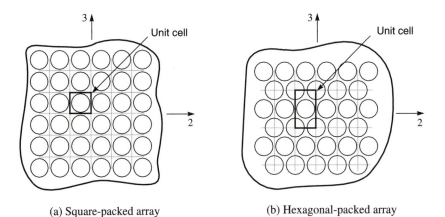

(a) Square-packed array   (b) Hexagonal-packed array

FIGURE 3.2
Concept of unit cells

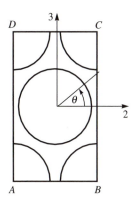

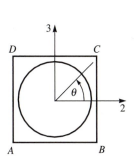

(a) Square-packed array   (b) Hexagonal-packed array

FIGURE 3.3
Details of unit cells

by the angle θ; θ will be measured counterclockwise from the 2 axis. Enforcement of the condition that the boundaries of the unit cell remain straight ensures that the isolated unit cell behaves as if it is part of a larger array of unit cells. The disadvantage of these unit cell models is that to obtain results, it is necessary to rely on a finite-element representation of the unit cell. A discussion of the finite-element method of analysis is not possible here, but suffice it to say that a descretization of the cross section such as that shown in Figure 3.4(a) and (b) is the basis for the method. With the material and geometric symmetry of the 1-2-3 coordinate system, depending on the loading, it may actually be sufficient to consider only a portion of the unit cell. Certain symmetry conditions can be enforced along the horizontal and vertical centerlines of the unit cell so that only one-quarter or one-eighth of a unit cell need be modeled. (We will not discuss here the various alternatives that are possible with the modeling.) With a finite-element representation of a unit cell—whether it be a full, one-fourth, or one-eighth model—it is possible to obtain quite accurate estimates of the response, both of overall response (such as determining E_1) and of stresses (such as at the interface between the fiber and the matrix).

To follow are results obtained from both the square-packed and the hexagonal-packed models using the quarter-cell finite-element representations in Figure 3.4. We present overall elastic and thermal expansion properties of a graphite-reinforced composite, as well as information regarding the stresses in the fiber and matrix. Figure 3.5 illustrates the stresses to be discussed, and though they are treated in the context of the square-packed array, these same stresses are definable in the hexagonal-packed array. The normal and shear stresses acting on the interface between the the fiber and matrix, σ_n and τ_{ns}, have the same value on the matrix side of the interface as on the fiber side. These stress components are responsible for interface failure in the material. The circumferential stress component in the matrix, σ_s^m, does not act directly on the interface and does not have the same value as the circumferential stress component in the fiber, σ_s^f. The circumferential stress component in the matrix can be responsible for failures originating in the matrix. Finally, we will discuss the

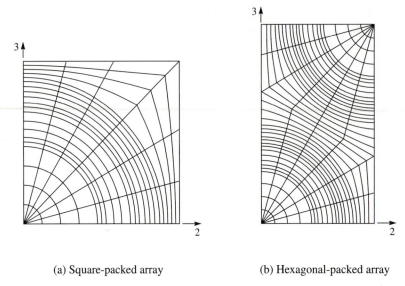

(a) Square-packed array   (b) Hexagonal-packed array

FIGURE 3.4
Finite-element representations of unit cells (quarter models)

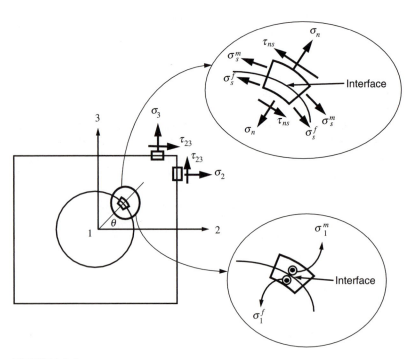

FIGURE 3.5
Stresses of interest within unit cell

stress in the fiber, or 1, direction at the interface. This stress component is not the same in the fiber as it is in the matrix. In fact, σ_1^f can be responsible for fiber failure, while σ_1^m can be responsible for failure of the matrix. These stresses will be studied as a function of angular location, θ, around the fiber-matrix interface. The direction $\theta = 0°$ coincides with the 2 direction, while $\theta = 90°$ coincides with the 3 direction. The stresses will also be studied as a function of the fiber volume fraction of the material. The specific values are 0.2, 0.4, and 0.6; the latter is a realistic number, while the two lower numbers are used to provide insight into trends. We should realize that there is an upper bound to fiber volume fraction. The fibers obviously cannot overlap and the upper bound is achieved when all the fibers just touch each other. In addition, we will also study the stresses at the boundaries of the unit cell (see Figure 3.5). These stresses (σ_2, σ_3, and τ_{23}) are needed to keep the boundaries of the unit cell straight. Because the analysis assumes that the cross section represents the infinite length of the fibers, neither stresses nor strains vary along the fiber direction. Such assumptions categorize the analysis as a generalized plane deformation analysis.

3.2.1 Material Properties of the Fiber and Matrix

The elastic and thermal expansion properties of a graphite fiber are taken to be

$$
\begin{aligned}
E_x^f &= 233 \text{ GPa} & E_r^f &= E_\theta^f = 23.1 \text{ GPa} \\
v_{xr}^f &= v_{x\theta}^f = 0.200 & v_{r\theta}^f &= 0.400 \\
G_{xr}^f &= G_{x\theta}^f = 8.96 \text{ GPa} & G_{r\theta}^f &= 8.27 \text{ GPa} \\
\alpha_x^f &= -0.540 \times 10^{-6}/K & \alpha_r^f &= \alpha_\theta^f = 10.10 \times 10^{-6}/K
\end{aligned}
\tag{3.1}
$$

where the properties are given in the cylindrical coordinate system of the fiber and the superscript f denotes fiber. The above properties correspond to an intermediate-modulus graphite fiber. The negative value of α_x^f reflects the fact that graphite fibers shrink in the axial direction when heated. Obviously the axial direction (x) of the fiber coincides with the 1 direction. With the above properties, the fiber is said to be transversely isotropic in the r-θ plane (i.e., in the cross section of the fiber). This means that the fiber responds the same when subjected to a stress σ_r, for example, as when it is subjected to stress σ_θ. Subjecting the material in the fiber to a stress σ_x results in a different response. Because of the transverse isotropy, when referred to the 1-2-3 coordinate system the fiber properties can be written as

$$
\begin{aligned}
E_1^f &= 233 \text{ GPa} & E_2^f &= E_3^f = 23.1 \text{ GPa} \\
v_{12}^f &= v_{13}^f = 0.200 & v_{23}^f &= 0.400 \\
G_{12}^f &= G_{13}^f = 8.96 \text{ GPa} & G_{23}^f &= 8.27 \text{ GPa} \\
\alpha_1^f &= -0.540 \times 10^{-6}/K & \alpha_2^f &= \alpha_3^f = 10.10 \times 10^{-6}/K
\end{aligned}
\tag{3.2}
$$

For the polymer matrix material the elastic and thermal expansion properties are taken to be

$$E^m = 4.62 \text{ GPa} \qquad \nu^m = 0.360$$
$$\alpha^m = 41.4 \times 10^{-6}/K$$

(3.3)

where the superscript m denotes matrix and the matrix is assumed to be isotropic. All the properties are assumed to be independent of temperature.

3.2.2 Thermal Effects and Determination of Thermal Expansion Coefficients α_1 and α_2

Always of interest are the stresses at the micromechanical level that result from a temperature change. We can study residual thermally induced stresses due to the cooldown from consolidation by examining the effects of a temperature change. We can also assess the micromechanical stresses due to heating or cooling relative to, say, room temperature. Because the material properties are assumed to be independent of temperature and because thermal expansion is linear with temperature, the micromechanical responses due to a temperature change ΔT basically tell the story. We will present the stresses at the interface between the fiber and matrix and at the boundaries of the unit cell that result from a temperature change ΔT. Because the residual stresses due to cooling from the consolidation temperature are an important issue in composite materials, we will assume the temperature change is negative. For convenience and generality, and with no sacrifice in the physical interpretation of the results, we will normalize the stresses with the quantity $E^m \alpha^m |\Delta T|$. This quantity has the units of stress and is the stress in an element of matrix restrained from deforming in one direction. With this normalization the signs of the normalized stresses are the signs that would occur due to cooling from the consolidation temperature. We can easily compute the stresses due to any temperature change by scaling the results by ΔT.

To obtain results representing the effects of a temperature change from the finite-element representation of the unit cell of Figure 3.3, the boundary conditions imposed on the finite-element model are: lines AB, BC, CD, and DA are all free to move and change length but are constrained to remain straight. In reality lines AB, BC, CD, and DA represent surfaces on the sides of the unit cell. Because we assume there are no variations of response in the fiber direction of the unit cell model, the dimension in the fiber direction can be considered unity. Thus, by "free to move," we mean that the net forces acting on the areas represented by the lines AB, BC, CD, and DA are zero. We determine these net forces by the integral of the stresses over the area, with the area being given, for example, by the length of line BC multiplied by 1. More will be said of the integrals of the stresses shortly. In addition to the above stipulations, cross section $ABCD$ is forced to stay planar and perpendicular to the axis of the fiber during the temperature change, but it may move in the fiber direction. The integral over cross section $ABCD$ of the normal stresses σ_1 out of the plane of the figure, namely, the net force in the fiber direction, P, must be zero. Though plane $ABCD$ must remain planar and perpendicular to the fiber axis, it translates in the direction of the fiber axis due to thermal expansion along the fiber direction.

The interface normal and shear stresses between the fiber and the matrix for a graphite-reinforced material are illustrated in Figure 3.6. Results from both the

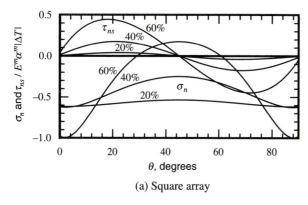

(a) Square array

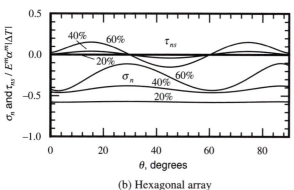

(b) Hexagonal array

FIGURE 3.6
Interface normal and shear stresses in graphite-reinforced material due to a temperature decrease

square- and hexagonal-packed arrays are shown. In these and future figures, because of symmetry considerations, the variations of the interface stresses with circumferential location are plotted only for the range $0° \leq \theta \leq 90°$. Also, in general all three fiber volume fractions considered will be on one figure. As Figure 3.6 shows, both packing models predict a strong influence of the fiber volume fraction on the interface stress. At 0.2 fiber volume fraction, generally referred to as 20 percent fiber volume fraction, the normal stress σ_n is compressive, for the most part independent of circumferential location θ, and has a normalized value of about -0.6. For 20 percent fiber volume fraction the shear stress τ_{ns} is nil. We interpret these results as meaning the fibers that are far enough apart at 20 percent fiber volume fraction that they do not interact (i.e., any particular fiber does not feel the influence of the other fibers in the neighborhood). What occurs at one circumferential location occurs at every other location. As a contrast, at 60 percent fiber volume fraction, the normal stress depends strongly on θ, and there is a shear stress, though it is smaller in magnitude than the normal stress. The square-packed array predicts that the normal stress changes sign with θ, while the hexagonal-packed array predicts the normal stress, while varying with θ, does not change sign and is compressive at all circumferential locations. This strong dependence on θ is to be interpreted as interaction between the fibers, namely, at this volume fraction, the fibers are close enough together that any individual fiber feels the influence of neighboring fibers. Some circumferential locations are closer to neighboring fibers than others and what happens at one circumferential location does not happen at others. For the normal stress the interpretation of the results is as follows: For a negative temperature change and a 20 percent fiber volume fraction, because in the 2-3 plane the matrix material has a larger coefficient of thermal expansion than the fiber, the matrix contracts more and exerts a compressive normal stress on the fiber. Hence, for a negative ΔT the interface normal stresses are compressive, with the value of compressive stress being nearly independent of θ. For 60 percent fiber volume fraction the fibers are closer together, and according to the predictions of the square-packed array, the change in temperature causes tension at the interface in the range $30° \le \theta \le 60°$. The hexagonal-packed array model predicts variations in the compressive stress as θ varies, but not a tensile stress. For the 60 percent fiber volume fraction the square-packed array predicts higher compressive stress levels than the hexagonal-packed array, with the high compressive stress levels occurring where the fibers are in closest proximity to one another, such as at $\theta = 0$ and 90° for the square-packed array and $\theta = 0$ and 60° for the hexagonal array. In both models, the shear stress is predicted to change sign and the square array predicts a higher shear stress.

Another stress of interest for the thermally induced case is the circumferential stress (see Figure 3.7). The circumferential stress in the matrix is, in general, opposite in sign to the circumferential stress in the fiber, though the square-packed array with

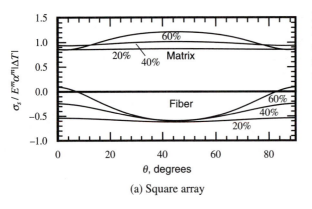

(a) Square array

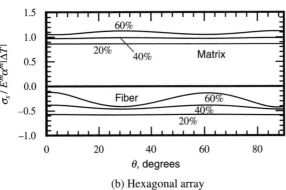

(b) Hexagonal array

FIGURE 3.7
Interface circumferential stresses in graphite-reinforced material due to a temperature decrease

60 percent fiber volume fraction predicts a sign change for the fiber stress. The tensile circumferential stress in the matrix is of particular interest. Being tensile and with a severe enough temperature decrease, the stress in the matrix may exceed failure levels and cause the matrix to crack. As a possible scenario, the crack would be oriented radially from the interface into the matrix (see Figure 3.8). The fiber, on the other hand, is being compressed in the circumferential direction, with no danger of serious fiber failure due to this stress.

Although they are almost completely independent of circumferential location θ, the stresses in the fiber direction (Figure 3.9) are important, particularly when

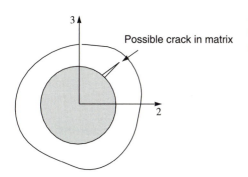

FIGURE 3.8
Possible crack in matrix due to a temperature decrease

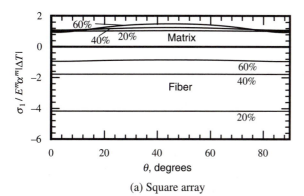

(a) Square array

FIGURE 3.9
Interface fiber-direction stresses in graphite-reinforced material due to a temperature decrease

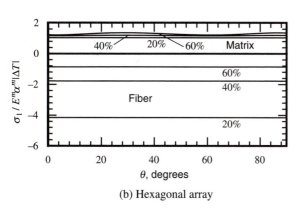

(b) Hexagonal array

interpreted in the context of residual thermal stresses. With the temperature decrease due to cooldown from the consolidation temperature, the stresses in the fiber direction are compressive, while the stresses in the matrix in the fiber direction are tensile. This can be explained as follows: Because the coefficient of thermal expansion of the fiber in the fiber direction, α_1^f, is negative, as the temperature decreases, the fiber tends to become longer. Conversely, the matrix, with its positive coefficient of thermal expansion, tends to contract in the fiber direction when cooled. Because the fiber and matrix are bonded together, the fiber cannot expand as much as it could if it were free, and thus it is in compression in the fiber direction. An extension of the argument leads to the conclusion that the matrix must be in tension. The tensile stress in the matrix could cause matrix cracking to occur. Although it is not shown here, we should mention that in addition to the fact that the fiber-direction stress in the fiber and matrix does not depend strongly on θ, this stress component is tensile and almost uniform within the matrix, and compressive and quite uniform within the fiber. Thus, the fiber-direction stress acting at a point on the cross section is one of only two different values, the value in the matrix or that in the fiber, depending on where the point is.

With the fibers being in close proximity, particularly for the 60 percent volume fraction case, stresses are necessary to keep the unit cell boundaries straight. However, it was also stipulated that there could be no net force on the boundaries of the unit cell. Figure 3.10 shows the distribution of the stresses on the boundaries. The important stresses on the boundaries are the normal stresses, σ_2 and σ_3. The shear stresses on the boundaries, τ_{23}, are negligible. In this and other figures illustrating the boundary stresses, the stresses are plotted as a function of normalized distance along the unit cell boundary. The right boundary is represented by line BC (please refer to Figure 3.3). Due to symmetry, only the upper half of that boundary needs to be considered. Using a normalized distance in the 3 direction, we find that positions on the upper portion of boundary BC for the square-packed array are in the range 0 to 0.5, with 0 coinciding with the 2 axis. Similarly, distances in the 2 direction along the right half of line CD of the square-packed array range from 0 to 0.5, with 0 coinciding with the 3 axis. For the hexagonal-packed array, normalized positions along the upper half of boundary BC range from 0 to $\sqrt{3}/2(= 0.867)$, while normalized positions along the right half of boundary CD range from 0 to 0.5. Note that for the square array the unit cell boundaries are completely in matrix, and for the hexagonal array the unit cell boundaries pass through matrix as well as fiber. The abrupt change in material properties along the boundary lines results in abrupt changes in stress levels.

Considering the square array, we find that the normal stresses, σ_2, along the right boundary, Figure 3.10(a), are compressive near normalized position 0 because the coefficient of thermal expansion of the fiber in the 2-3 plane is less than that of the matrix in that plane. With a temperature decrease, the fiber does not contract as much as the matrix. For line BC to remain straight, a compressive stress is required near 0 to compensate for the lack of fiber contraction. Conversely, because with a temperature decrease the matrix contracts more than the fiber, near position 0.5 tensile stresses are necessary to overcome the greater contraction tendency of the matrix. As the net force along line BC must be zero, the effect of the compressive

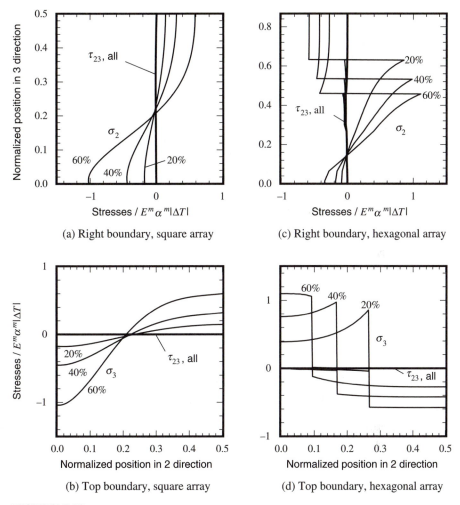

FIGURE 3.10
Stresses on boundaries of unit cell of graphite-reinforced material due to a temperature decrease

stresses near position 0 must cancel the effect of the tensile stresses near position 0.5. As an aid to understanding the sign of the stresses along the boundaries of the unit cell, Figure 3.11 illustrates the deformations of a unit cell due to a temperature decrease and the greater thermal expansion coefficient of the matrix in the 2-3 plane as compared to the thermal expansion coefficient of the fiber in that plane. Figure 3.11(a) shows the undeformed unit cell before the temperature is decreased, while Figure 3.11(b) shows the deformed unit cell, with the boundaries constrained to remain straight. Figure 3.11(c) shows the deformations that would occur to the unit cell if the boundaries were not constrained to remain straight. The stresses along the boundary *BC* shown in Figure 3.10(a) are necessary for the boundary to remain straight. Due to the symmetry of the thermal problem, the physical interpretation of

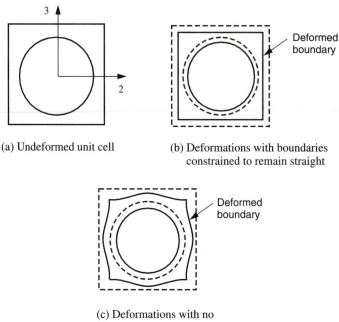

(a) Undeformed unit cell

(b) Deformations with boundaries
constrained to remain straight

(c) Deformations with no
boundary constraints

FIGURE 3.11
Cross-sectional deformations of unit cell of graphite-reinforced material due to a temperature decrease

the stresses along top boundary CD for the square array, Figure 3.10(b), is similar to
the interpretation of the stresses along boundary BC. An analogous interpretation of
the boundary stresses is possible for the case of the hexagonal array, Figure 3.10(c)
and (d).

It is not the purpose of this discussion to dwell on the micromechanical-level
stresses due to a temperature change. However, from the results just discussed, it
is very clear that these stresses can be quite important in determining the integrity
of the composite. If there is too large a mismatch between the thermally induced
deformations of the fiber and matrix, the stresses can be large enough to cause
disbonding of the fiber from the matrix, or cracking in the matrix, or both. Neither
of these results can be tolerated.

A direct result of the analysis of thermally induced stresses is a determination
of the coefficients of thermal expansion of the composite. The change in dimension
across the width of the unit cell due to a temperature change provides a measure
of the coefficient of thermal expansion transverse to the fiber direction, α_2, while
the change in the fiber-direction dimension provides a measure of the coefficient of
thermal expansion along the fiber direction, α_1. With either the square-packed or
hexagonal-packed array, the coefficient of thermal expansion in the 3 direction, α_3,
is equal to α_2. The variations of the coefficients of thermal expansion of the composite with fiber volume fraction are given in Table 3.1, along with other material

TABLE 3.1
Engineering properties of graphite-reinforced composite

Composite property	Square array			Hexagonal array		
	0.2	0.4	0.6	0.2	0.4	0.6
α_1, $\times 10^{-6}$/K	2.69	0.779	0.0884	2.68	0.780	0.0766
$\alpha_2(=\alpha_3)$, $\times 10^{-6}$/K	44.9	35.7	26.4	44.9	35.7	26.6
E_1, GPa	50.2	96.1	141.7	50.4	96.0	141.9
$\nu_{12}(=\nu_{13})$	0.324	0.289	0.259	0.324	0.290	0.257
$E_2(=E_3)$, GPa	6.69	9.03	12.38	6.54	8.43	11.30
ν_{23}	0.507	0.466	0.421	0.518	0.501	0.468
$\nu_{21}(=\nu_{31})$	0.0431	0.0272	0.0224	0.0420	0.0254	0.0205
G_{23}, GPa	1.813	2.17	2.94	2.15[1]	2.81[1]	3.85[1]
G_{12}, GPa[2]	2.23	2.97	4.05	2.23	2.97	4.05
β_1, $\times 10^{-6}$/%M	330	195.0	150.9	329	195.1	147.6
$\beta_2(=\beta_3)$, $\times 10^{-6}$/%M	4350	4570	4700	4350	4580	4750

[1]Assuming transverse isotropy.
[2]Elasticity model, not square or hexagonal array.

properties to be discussed shortly. It is readily apparent from the table that the two packing arrays predict nearly identical values of the two coefficients of thermal expansion. Moreover, with increasing volume fraction, the value of α_1 approaches zero and possibly becomes negative, reflecting the increasing influence of the negative coefficient of thermal expansion of the fiber in the fiber direction. In fact, it is seen that for just the right volume fraction of fiber, there would be no thermal expansion of the composite material in the fiber direction, a very useful property and one unique to graphite fiber-reinforced composites.

3.2.3 Tension in the Fiber Direction and the Determination of Extensional Modulus E_1 and Poisson's Ratios ν_{12} and ν_{13}

Because loading in the fiber direction is the loading mode most favorable to composite materials, it is interesting to examine the stresses at the micromechanical level when the composite is subjected to such a load. With the models being used, this loading condition can be simulated by using the same boundary conditions on lines AB, BC, CD, and DA as were enforced for the thermal response case, namely, that those boundaries remain straight and free of any net force. However, instead of having a zero net force on face $ABCD$, a nonzero value is specified, thus stretching the composite in the fiber direction. The average fiber-direction stress in the composite in the fiber direction, σ_1, can then be determined by dividing this specified load, P, by the area of cross section $ABCD$, expressed as A. In the figures to follow, the stresses are normalized by this average fiber-direction stress, P/A. As might be suspected, the interface normal and shear stresses, as well as the circumferential stresses in the fiber and matrix at the interface, are minor compared to the fiber-direction stress. Because the case of a fiber-direction loading is a fiber-dominated response, and because for the thermal loading the fiber-dominated responses (i.e., α_1) of the square-and hexagonal-packed arrays were in such good agreement, only the results from the square-packed array are shown.

For 20 percent fiber volume fraction, the interface normal stress, shown in Figure 3.12, reflects the fact that for a tensile load in the fiber direction, the magnitude of the Poisson effects in the matrix relative to the fiber causes the matrix to compress the fiber in the radial direction. As the fiber volume fraction increases, this clamping effect of the matrix onto the fiber decreases, with the square array model actually predicting some interface tension near $\theta = 45°$ when the fiber volume fraction is 60 percent. For all volume fractions the interface shear stress is quite small. The circumferential stresses in the matrix, Figure 3.13, also reflect the tendency for the matrix to compress the fiber, with the matrix circumferential stresses causing compression in the fiber as it resists the radial compression effect by trying to push outward on the matrix, something like internal hydrostatic pressure within a void. In turn, the matrix stresses are tensile in the circumferential direction. The most significant stresses for the fiber-direction tensile loading condition are, naturally, the stresses in the fiber direction in the fiber and in the matrix, Figure 3.14. It is abundantly clear that while the magnitude of this component of stress depends on the fiber volume fraction, the magnitude is not a function of circumferential location. As in the thermal case (though not shown), for the fiber-direction loading, the fiber-direction stresses in the fiber and matrix are independent of location within the cross

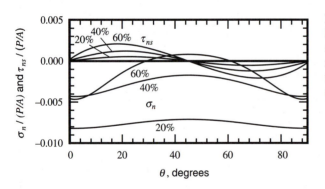

FIGURE 3.12
Interfacial normal and shear stresses in graphite-reinforced material due to applied fiber-direction strain, square array

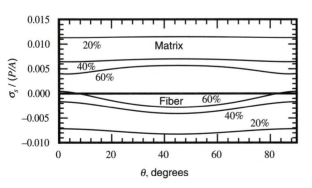

FIGURE 3.13
Interface circumferential stresses in graphite-reinforced material due to applied fiber-direction strain, square array

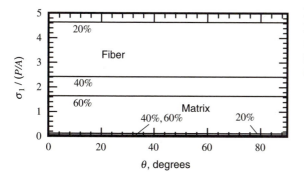

FIGURE 3.14
Interface fiber-direction stresses in graphite-reinforced material due to applied fiber-direction strain, square array

section. We can interpret this as indicating that for this mode of loading, the fibers for this infinitely long unit cell do not interact.

Figure 3.15 illustrates the stresses along the boundaries of the square-packed array unit cell that are required for the boundaries to remain straight. Figure 3.16 shows the deformations of the cross section of the unit cell for the case of the boundaries being constrained to remain straight, Figure 3.16(b), and the boundaries being free to deform, Figure 3.16(c). The cross-sectional deformations that occur if the unit cell boundaries are free to deform result from differences between Poisson's ratio of the matrix, $\nu^m = 0.360$, and Poisson's ratio of the fiber, $\nu^f_{xr} = 0.200$. Because the matrix has the greater Poisson's ratio, it contracts more in the 2-3 plane than the fiber. It is important to note that although the cross-sectional deformations of Figures 3.11 and 3.16 look similar, they each deform the way they do for completely different reasons.

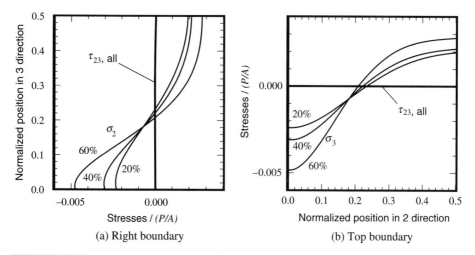

(a) Right boundary   (b) Top boundary

FIGURE 3.15
Stresses on boundaries of unit cell of graphite-reinforced material due to applied fiber-direction strain

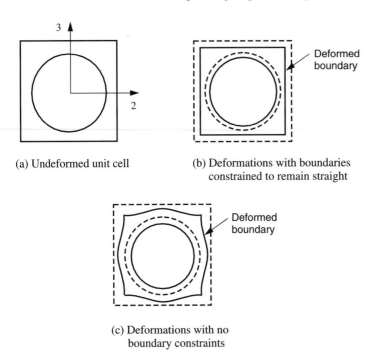

(a) Undeformed unit cell

(b) Deformations with boundaries
constrained to remain straight

(c) Deformations with no
boundary constraints

FIGURE 3.16
Cross-sectional deformations of unit cell of graphite-reinforced material due to applied
fiber-direction strain

We can use the deformations that accompany the fiber-direction tensile loading
to compute the fiber-direction extensional modulus and two of the several Poisson's
ratios for the composite. The strain in the 2 direction divided by the fiber-direction
strain results in one Poisson's ratio, specifically ν_{12}, while the strain in the 3 direction
divided by the fiber-direction strain results in another Poisson's ratio, ν_{13}, which is
identical to ν_{12}. Table 3.1 gives the variations of these composite material properties
with fiber volume fraction. The two packing models predict nearly identical results.

3.2.4 Transverse Tensile Loading and Determination of Extensional Modulus E_2 and Poisson's Ratios ν_{21} and ν_{23}

One of the most interesting and important loadings that can be applied to a fiber-
reinforced material is a tensile load or deformation perpendicular to the fiber direction, here the 2 and 3 directions (i.e., the transverse direction). As there is not a
continuous path through strong and stiff fiber for the load to be transmitted, the load
must either pass through the fiber, across its diameter, through the interface between
the fiber and the matrix, through the matrix, through another interface, through another fiber, and so on, or the load must take a somewhat tortuous path, following
only matrix material. In reality, a portion of the load follows the matrix-only path

and a portion is transmitted through the fiber, with the proportions being determined by the relative stiffness of each constituent and by the fiber volume fraction. With a transverse loading, then, the interaction of fiber and matrix plays a key role in transmitting the stresses through the composite.

To study the effects of a transverse load, a known overall transverse strain is imposed on the unit cell model. This is accomplished by stipulating that lines BC and DA in Figure 3.3 remain straight and move apart a known amount in the 2 direction, and lines AB and CD are free to move in the 3 direction as long as they remain straight and parallel with the 2 direction. All boundary lines can change length. Additionally, it is specified that cross section $ABCD$ remains planar and perpendicular to the axis of the fiber. The integral of the fiber-direction stresses acting on that cross section is again forced to be zero.

Figure 3.17 indicates the interface normal and shear stresses that result from the transverse strain. Because this is a loading transverse to the fibers, the results from both the square- and hexagonal-packed arrays are shown for comparison. In this and subsequent figures the stresses have been normalized by average transverse stress $\bar{\sigma}_2$ acting along the left and right edges of the unit cell. The average stress is the value required to produce the known transverse strain, and it is determined by integrating the stress σ_2 over boundary BC and dividing by the area of boundary BC. The integrals of these stresses were zero in Figures 3.10(a) and 3.15(a),

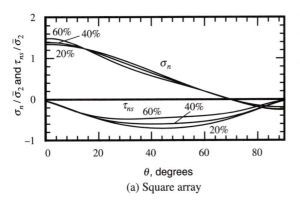

(a) Square array

FIGURE 3.17
Interfacial normal and shear stresses in graphite-reinforced material due to applied transverse tensile strain

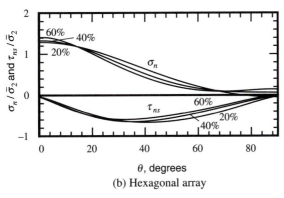

(b) Hexagonal array

cases where the net force in the 2 direction was specified as zero. As expected, the interface normal stress, Figure 3.17, is highest at $\theta = 0°$, the direction of the imposed transverse normal strain. At the $\theta = 0°$ location the matrix wants to pull away from the fiber because it is transmitting the transverse load directly to the fiber there. At $\theta = 90°$ the square-packed array predicts that the normal stress is compressive, while the hexagonal array predicts that the normal stress there is compressive only for the 20 percent fiber volume fraction case. In general, the magnitude of the normalized normal stress is, for the most part, independent of fiber volume fraction. It is important to keep in mind that the average transverse stress $\bar{\sigma}_2$ depends on fiber volume fraction. In absolute terms, then, the interface normal stress does depend on fiber volume fraction. The interface shear stress is of significant magnitude, being highest near $\theta = 45°$. It is of interest to note that in terms of failure, it may be a combination of the interface tension and the interface shear that causes a problem at the interface. With the variations with θ of the two stress components, it may be that interface failure will not occur where the normal stress is highest, nor where the shear stress is highest, but rather at some other circumferential location.

With the transverse loading, the circumferential stress, as Figure 3.18 shows, is an important component. In Figure 3.17 we saw that the interface normal stress is high at $\theta = 0°$. At this location the transverse tensile stress in the composite is in the direction of the interface normal and the high normal tensile stress was interpreted as

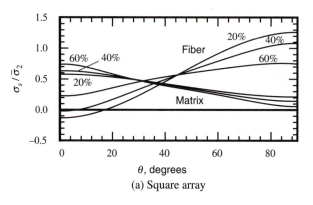

(a) Square array

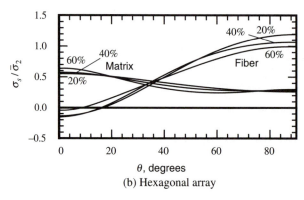

(b) Hexagonal array

FIGURE 3.18
Interface circumferential stresses in graphite-reinforced material due to applied transverse tensile strain

the matrix transmitting the transverse load directly to the fiber. Examination of Figure 3.18 shows that also at $\theta = 90°$ the matrix transmits the transverse load to the fiber. At $\theta = 90°$, for all volume fractions and for both the square- and hexagonal-packed arrays, the circumferential stress in the fiber is much larger than the circumferential stress in the matrix. At $\theta = 90°$ the circumferential direction is aligned with the transverse direction, and it is clear that fiber is taking the majority of the transverse load. The normal and shear stresses from, say, $\theta = 45°$ to $\theta = 90°$ acting in an integrated sense on the fiber-matrix interface, result in the high circumferential stress in the fiber at $\theta = 90°$. Equilibrium considerations require this high fiber circumferential stress. It should be noted that the hexagonal-packed array shows similar trends in the circumferential stress component, both in the fiber and in the matrix.

Figure 3.19 shows the fiber-direction stresses at the interface induced in the matrix and fiber. Both the square- and hexagonal-packed arrays predict that the stress in the fiber is, for the most part, independent of circumferential location θ, while the stress in the matrix depends on circumferential location, with the square array showing some sign reversal of the stress in the matrix near $\theta = 90°$. With a transverse loading, Poisson effects cause the fiber-reinforced material to contract in the fiber direction. As the fiber is stiff in the 1 direction, it resists and is thus in compression. Because the matrix material is relatively soft, the fiber keeps it from contracting as much as it could and the matrix is, in general, in tension. Because

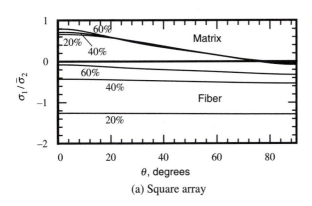

(a) Square array

FIGURE 3.19
Interface fiber-direction stresses in graphite-reinforced material due to applied transverse tensile strain

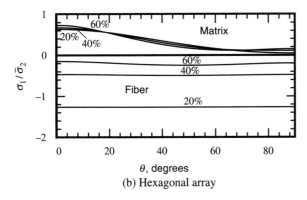

(b) Hexagonal array

there is no overall force in the fiber direction, the net effect of compressive stresses in the fiber and tensile stresses in the matrix must be zero. While there is little variation of the fiber-direction compressive stress in the fiber within the fiber cross section, there is some variation of the fiber-direction tensile stress within the matrix cross section.

Figure 3.20 shows the transverse normal stress σ_2 and the shear stress τ_{23} acting on the right boundary of the square- and hexagonal-packed unit cells. The shear stresses are inconsequential. The transverse normal stress for the square-packed array, Figure 3.20(a), is largest near position 0 and decreases as position 0.5 is approached. Because the modulus of the fiber in the 2-3 plane is so much larger than

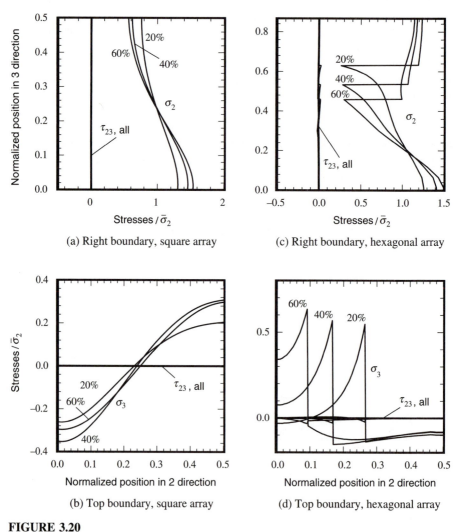

(a) Right boundary, square array

(c) Right boundary, hexagonal array

(b) Top boundary, square array

(d) Top boundary, hexagonal array

FIGURE 3.20

Stresses on boundaries of unit cell of graphite-reinforced material due to applied transverse tensile strain

the modulus of the matrix, a direct result is that the transverse normal stress is high near position 0.

Along the top boundary of the square array, Figure 3.20(b), the normal stress σ_3 dominates. The average stress along this boundary, which is constrained to remain straight, is zero. Near position 0 a compressive normal stress σ_3 is required, while for positions 0.3 and greater a tensile normal stress is required. The primary reason for this behavior is that in the 2 direction the strain in the matrix is much greater than the strain in the fiber, and through Poisson's ratio there is more contraction strain in the 3 direction in the matrix than in the fiber. For line CD to remain straight the stresses in the 3 direction, Figure 3.20(b), are necessary. A similar interpretation can be given to the more abrupt stress distributions of the hexagonal-packed array, Figure 3.20(c) and (d). We can better understand the stresses on the various boundaries of the square-packed unit cell if we study the deformations associated with the transverse tensile loading, Figure 3.21.

The average transverse stress, $\bar{\sigma}_2$, as computed from Figure 3.20(a) or (c), divided by the known applied transverse strain is the transverse extensional modulus for the material, E_2. The compressive fiber-direction strain, ε_1, divided by the known applied transverse strain, ε_2, and the compressive strain in the other transverse direction, ε_3, both induced by the known applied transverse strain, can be used to determine two Poisson's ratios, ν_{21} and ν_{23}, respectively. As Table 3.1 shows, these engineering properties are dependent on the fiber volume fraction. Though the predicted magnitudes of E_2, ν_{23}, and ν_{21} depend somewhat on the packing array used,

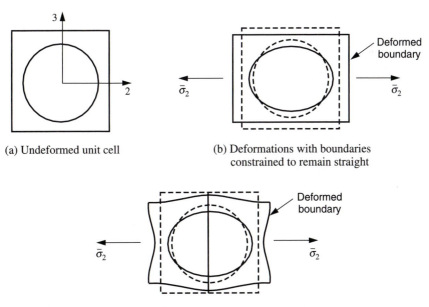

(a) Undeformed unit cell

(b) Deformations with boundaries constrained to remain straight

(c) Deformations with no boundary constraints

FIGURE 3.21
Cross-sectional deformations of unit cell of graphite-reinforced material due to applied transverse tensile strain

the predicted trends with fiber volume fraction do not depend on packing array. It is interesting to note from Table 3.1 that in spite of the approximations to the microstructure used in these finite-element models, the properties E_1, E_2, ν_{12}, and ν_{21} closely satisfy the reciprocity relation, equation (2.42).

3.2.5 Transverse Shear Loading and Determination of Shear Modulus G_{23}

Another interesting and extremely important case to examine at the micromechanical level is called the transverse shear response. *Transverse shear* is used when the unit cell is subjected to a shear stress, or strain, in the 2-3 plane. As with transverse loading, fiber strength and stiffness do not contribute to resisting shear deformation. As a shear response can be decomposed into effective tensile and compressive responses, each oriented at 45° from the direction of shear, the transverse shear loading is somewhat like the transverse tensile loading in that the load must be transmitted partially through the fibers and partially through the matrix; the proportion of load through each constituent depends on the fiber volume fraction and the stiffness of one constituent relative to the other.

To study the transverse shear strain response, the outer boundaries of the unit cell are given a displacement that represents subjecting the unit cell to a prescribed overall shear strain. As the square- and hexagonal-packed arrays do not differ substantially in their predictions for the transverse tensile loading case, only the square-packed array will be used to study transverse shear case. The four corners of the unit cell are given displacements in the 2 and 3 directions, and the four boundaries are constrained to remain straight; see Figure 3.22. The displacements at each of the four boundaries are similar, and the magnitudes are the same but with differing signs. Plane *ABCD* is assumed to remain planar and perpendicular to the fiber direction and the integral of the fiber-direction stresses over the area of that plane is required to be zero. There will be stresses along the four boundaries that are necessary to produce the prescribed overall shear strain.

Figure 3.23 illustrates the interface normal and shear stresses due to the specified overall transverse shear strain as a function of circumferential position around the interface and fiber volume fraction. The stresses have been normalized by the average

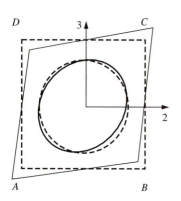

FIGURE 3.22
Transverse shear deformations of a square-packed array unit cell

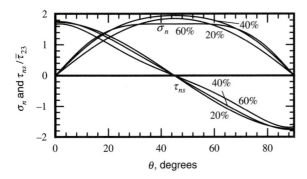

FIGURE 3.23
Interfacial normal and shear stresses in graphite-reinforced material due to applied transverse shear strain, square array

transverse shear stress $\bar{\tau}_{23}$. Although the stresses have been shown only for the range $0 \leq \theta \leq 90°$, note that the normal stress is symmetric about $\theta = 45°$, while the shear stress is antisymmetric about that location. In Figure 3.23 we see that at $\theta = 45°$ the normal stress has a maximum tensile value, and that this direction corresponds to the direction of maximum overall tensile strain for the unit cell, that is, from corner A to corner C in Figure 3.22. The interface normal stress is zero at $\theta = 0°$ and $90°$ but the interface shear stress is maximum at these locations. The two stress components have similar magnitudes. Though it appears that neither the normalized interface normal nor the normalized interface shear stress depend strongly on fiber volume fraction, the average shear stress *increases* with fiber volume fraction. Thus, the absolute level of these interface stresses does depend on fiber volume fraction.

As with the transverse loading, the circumferential stresses in fiber and matrix, Figure 3.24, are important. At $\theta = 135°$ (not shown) the circumferential tensile stress in the fiber is a maximum because at this location the softer matrix transfers to the fiber the tensile stress oriented in the $45°$ direction. That the maximum tensile stress in the fiber occurs at a location $90°$ from the direction of the effective overall tensile strain is just as it was for the transverse tensile loading case, Figure 3.18(a). There the applied tensile strain oriented in the 2 direction resulted in a maximum tensile stress in the fiber at the $90°$ location. With the transverse shear case there is the added effective compressive loading aligned with the $\theta = 135°$ direction. If the fiber is weak across its diameter, then transverse shear failure may result from the fiber's failing in tension at the $135°$ location. If the matrix is weak in tension, then failure

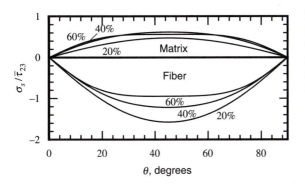

FIGURE 3.24
Interface circumferential stresses in graphite-reinforced material due to applied transverse shear strain, square array

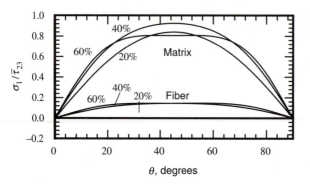

FIGURE 3.25

Interface fiber-direction stresses in graphite-reinforced material due to applied transverse shear strain, square array

may occur in the matrix at 45°. This assumes that failure has not occurred at the interface due to σ_n, or τ_{ns}, or both. The effect of combined tension and compression results in the normalized interface stresses being larger for this transverse shear case than they were for the transverse tension case, though there are analogies between the two loadings. Figure 3.25 shows the normalized fiber-direction stresses in the fiber and matrix at the interface region. These stresses reach extreme values at 45°.

Figure 3.26 shows the stresses along the right and top boundaries of the unit cell. The shear stress, τ_{23}, is not constant along edge BC and a normal stress, σ_2, is required to enforce the deformations of Figure 3.22. The magnitudes of the normal and shear stress are similar, but because the normal stress for negative positions in the 3 direction is negative, the average normal stress is zero. As expected, the character of the shear and normal stress along the top edge is identical to the character of these

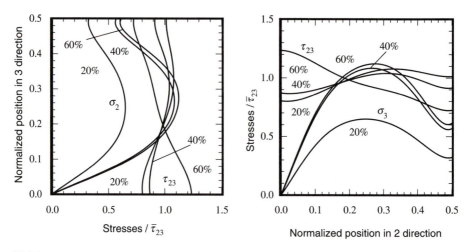

FIGURE 3.26
Stresses on boundaries of unit cell of graphite-reinforced material due to applied transverse shear strain, square array

stress components along the right edge. The deformations of Figure 3.27 can be used to help interpret the character of the stresses along the edges.

The average shear stress along the unit cell boundaries, $\bar{\tau}_{23}$, determined by the integration of the shear stresses of Figure 3.26, and the known shear deformation of Figure 3.22 can be used to determine the transverse shear modulus of the composite, G_{23}. The variation of this material property with fiber volume fraction is given in Table 3.1; the value of G_{23} increases with fiber volume fraction due to the increasing influence of the fiber. Note, however, that the values of G_{23}, ν_{23}, and $E_2(= E_3)$ for the square-packed array do not satisfy equation (2.53b). This is because the square array does not lead to transversely isotropic properties even though the constituents are transversely isotropic. The hexagonal-packed array results in transverse isotropy if the constituents are transversely isotropic. The entries for G_{23} in Table 3.1 for the hexagonal-packed array are computed using equation (2.53b).

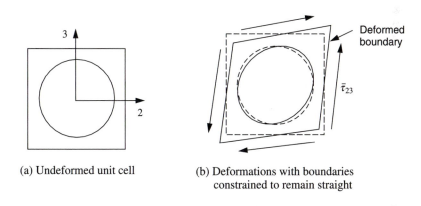

(a) Undeformed unit cell   (b) Deformations with boundaries
                                      constrained to remain straight

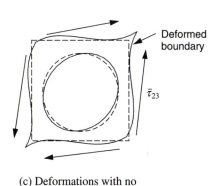

(c) Deformations with no
    boundary constraints

FIGURE 3.27
Cross-sectional deformations of unit cell of graphite-reinforced material due to applied transverse shear strain

3.3
THEORY OF ELASTICITY RESULTS: CONCENTRIC CYLINDER MODELS

When fiber-reinforced materials were first used, numerical methods were not as readily available as they have become. Therefore, some of the early approaches to studying the response of composite materials at the micromechanics level were based on classical elasticity solutions. Solutions to elasticity problems can be quite difficult to determine, and without simplifying assumptions, obtaining solutions is sometimes impossible. As many of the early elasticity solutions were derived for the purpose of determining composite properties from the properties of the constituents, as opposed to studying the details of the stresses at, say, the fiber-matrix interface, some of the simplifying assumptions were not too limiting. One of the key simplifying assumptions was that the volume of fibers and matrix in a composite could be filled with an assemblage of cylindrical fibers and surrounding matrix material, with the fibers being of various sizes to the degree that the fiber-matrix combination of cylinders completely filled the volume of the composite. This notion is shown in Figure 3.28 and is called the composite cylinders model or composite cylinders

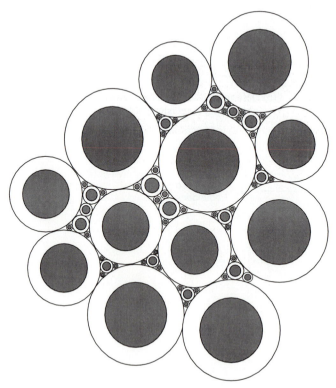

FIGURE 3.28
Philosophy of the concentric cylinders model

assemblage. For each fiber-matrix combination, the ratio of the diameter of the fiber to the diameter of the surrounding matrix is the same; this ratio represents the volume fraction of fiber in the composite. For a representative fiber-matrix combination with fiber radius b and matrix outer radius c, the fiber volume fraction is given by

$$V^F = \frac{\pi b^2}{\pi c^2} = \frac{b^2}{c^2} \tag{3.4}$$

The elasticity approaches are concentrated on an isolated fiber-matrix combination from this assemblage. Such a combination is shown in Figure 3.29, with the cylindrical x-θ-r and the composite 1-2-3 principal material coordinate systems indicated.

3.3.1 Fiber-Direction Tension

To study the response of the composite to tension in the fiber direction, we assume that the response of the concentric cylinders is axisymmetric. In addition, if attention is concentrated away from the ends of the fibers, the stresses, and hence the strains, are assumed to be independent of the axial coordinate, namely the x, or 1, direction. For such conditions, of the three equilibrium equations in cylindrical coordinates, only one is important, and that equation reduces to

$$\frac{d\sigma_r}{dr} + \frac{\sigma_r - \sigma_\theta}{r} = 0 \tag{3.5}$$

As there are no shear stresses for the axisymmetric axial loading case, the stress-strain relations reduce to

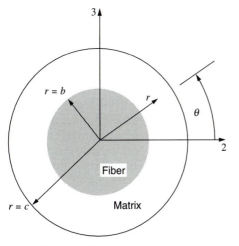

The x and the 1 axes are
perpendicular to the page.

FIGURE 3.29
Isolated fiber-matrix combination of concentric cylinders model

$$\sigma_x = C_{11}\varepsilon_x + C_{12}\varepsilon_\theta + C_{13}\varepsilon_r$$

$$\sigma_\theta = C_{12}\varepsilon_x + C_{22}\varepsilon_\theta + C_{23}\varepsilon_r \tag{3.6}$$

$$\sigma_r = C_{13}\varepsilon_x + C_{23}\varepsilon_\theta + C_{33}\varepsilon_r$$

These equations were introduced in equation (2.47) in the context of the smeared properties of the composite. However, they are equally valid individually for the fiber and for the matrix. Here we will assume that the fiber is transversely isotropic in the r-θ plane and the matrix is isotropic, that is, equation (2.55). Thus, the stress-strain relations for the two constituents simplify, for the fiber, to

$$\sigma_x^f = C_{11}^f\varepsilon_x + C_{12}^f\varepsilon_\theta + C_{12}^f\varepsilon_r$$

$$\sigma_\theta^f = C_{12}^f\varepsilon_x + C_{22}^f\varepsilon_\theta + C_{23}^f\varepsilon_r \tag{3.7}$$

$$\sigma_r^f = C_{12}^f\varepsilon_x + C_{23}^f\varepsilon_\theta + C_{22}^f\varepsilon_r$$

and for the matrix, to

$$\sigma_x^m = C_{11}^m\varepsilon_x + C_{12}^m\varepsilon_\theta + C_{12}^m\varepsilon_f$$

$$\sigma_\theta^m = C_{12}^m\varepsilon_x + C_{11}^m\varepsilon_\theta + C_{12}^m\varepsilon_f \tag{3.8}$$

$$\sigma_r^m = C_{12}^m\varepsilon_x + C_{12}^m\varepsilon_\theta + C_{11}^m\varepsilon_f$$

where the superscripts f and m denote, respectively, fiber and matrix. Using the engineering properties of the fiber and the matrix, equations (3.2) and (3.3), and the definitions of the S_{ij} and the C_{ij}, equations (2.46) and (2.48), C_{11}^f, C_{12}^f, C_{22}^f, C_{23}^f, C_{11}^m, and C_{12}^m can be computed.

The strains in the above equations are related to the displacements by the strain-displacement relations simplified by the assumptions of axisymmetry and independence of strain of the x coordinate. There is no shear strain response, only normal strains given by

$$\varepsilon_x = \frac{\partial u}{\partial x}$$

$$\varepsilon_\theta = \frac{w}{r} \tag{3.9}$$

$$\varepsilon_r = \frac{dw}{dr}$$

In equation (3.9) u is the axial displacement and w is the radial displacement. Because of the assumption of axisymmetric response, the circumferential displacement, v, is assumed to be zero. Note that the partial derivative of u with respect to x is used and that the ordinary derivative of w with respect to r is used because we can argue that w cannot be a function of x, or else the fiber would not be straight when loaded, and at this point in the development, u can be a function of both x and r.

Substituting the strain-displacement relations into the stress-strain relations and these, in turn, into equation (3.5), which is the third equilibrium equation, leads to an equation for the radial displacement, namely,

$$\frac{d^2w}{dr^2} + \frac{1}{r}\frac{dw}{dr} - \frac{w}{r^2} = 0 \tag{3.10}$$

which has the solution

$$w(r) = Ar + \frac{B}{r} \tag{3.11}$$

The quantities A and B are constants of integration that must be solved for by applying boundary and other conditions. The above solution is valid for both the fiber and the matrix. For a given fiber-matrix combination, then, the radial displacement is given by

$$w^f(r) = A^f r + \frac{B^f}{r} \qquad 0 \le r \le b \tag{3.12a}$$

$$w^m(r) = A^m r + \frac{B^m}{r} \qquad b \le r \le c \tag{3.12b}$$

where the range of r for each portion of the solution is given. As the strains do not vary with the axial coordinate, the solution for the axial displacement u is given by

$$u^f(x, r) = \varepsilon_1^f x \qquad 0 \le r \le b \tag{3.13a}$$

$$u^m(x, r) = \varepsilon_1^m x \qquad b \le r \le c \tag{3.13b}$$

where ε_1^f and ε_1^m are constants.

According to equation (3.12a), if B^f is not zero, the radial displacement at the center of the fiber, $r = 0$, is predicted to be infinite. This is physically impossible, so the condition

$$B^f = 0 \tag{3.14}$$

is stipulated. As a result, the strains within the fiber are given by equation (3.9) as

$$\varepsilon_x^f = \frac{\partial u^f}{\partial x} = \varepsilon_1^f$$

$$\varepsilon_\theta^f = \frac{w^f}{r} = A^f \tag{3.15}$$

$$\varepsilon_r^f = \frac{dw^f}{dr} = A^f$$

while those in the matrix are given by

$$\varepsilon_x^m = \frac{\partial u^m}{\partial x} = \varepsilon_1^m$$

$$\varepsilon_\theta^m = \frac{w^m}{r} = A^m + \frac{B^m}{r^2} \tag{3.16}$$

$$\varepsilon_r^m = \frac{dw^m}{dr} = A^m - \frac{B^m}{r^2}$$

With the strains defined, the stresses in the fiber and matrix, respectively, can be written as

$$\sigma_x^f = C_{11}^f \varepsilon_1^f + 2C_{12}^f A^f$$
$$\sigma_\theta^f = C_{12}^f \varepsilon_1^f + \left(C_{22}^f + C_{23}^f\right) A^f \qquad (3.17a)$$
$$\sigma_r^f = C_{12}^f \varepsilon_1^f + \left(C_{23}^f + C_{22}^f\right) A^f$$

$$\sigma_x^m = C_{11}^m \varepsilon_1^m + 2C_{12}^m A^m$$
$$\sigma_\theta^m = C_{12}^m \varepsilon_1^m + \left(C_{11}^m + C_{12}^m\right) A^m + \left(C_{11}^m - C_{12}^m\right) \frac{B^m}{r^2} \qquad (3.17b)$$
$$\sigma_r^m = C_{12}^m \varepsilon_1^m + \left(C_{11}^m + C_{12}^m\right) A^m + \left(C_{12}^m - C_{11}^m\right) \frac{B^m}{r^2}$$

When subjected to any loading, in particular an axial load P, the displacements at the interface between the fiber and matrix are continuous; that is,

$$w^f(b) = w^m(b)$$
$$u^f(b) = u^m(b) \qquad (3.18)$$

By substituting expressions for the displacements, equations (3.12) and (3.13), this condition leads to

$$A^f b = A^m b + \frac{B^m}{b} \qquad (3.19)$$
$$\varepsilon_1^f x = \varepsilon_1^m x$$

The second equation leads to the conclusion that the axial strain is the same in the fiber as in the matrix, namely,

$$\varepsilon_1^f = \varepsilon_1^m = \varepsilon_1 \qquad (3.20)$$

This strain is indeed the strain in the composite in the fiber direction, hence the notation ε_1, as we have been using all along.

As discussed in conjunction with the finite-element results, the radial stress σ_r must be the same on the fiber side of the interface as on the matrix side, or

$$\sigma_r^f(b) = \sigma_r^m(b) \qquad (3.21)$$

In terms of the unknown constants, substituting for the stresses from equations (3.17a) and (3.17b), equation (3.21) becomes

$$C_{12}^f \varepsilon_1 + \left(C_{23}^f + C_{22}^f\right) A^f = C_{12}^m \varepsilon_1 + \left(C_{11}^m + C_{12}^m\right) A^m + \left(C_{12}^m - C_{11}^m\right) \frac{B^m}{b^2} \qquad (3.22)$$

At the outer radius of the matrix, if it is assumed that the radial stress must vanish;

$$\sigma_r^m(c) = 0 \qquad (3.23)$$

Using equation (3.17b),

$$C_{12}^m \varepsilon_1 + \left(C_{11}^m + C_{12}^m\right) A^m + \left(C_{12}^m - C_{11}^m\right) \frac{B^m}{c^2} = 0 \qquad (3.24)$$

As a final condition of the problem, the applied axial load P is actually the integral of the axial stresses over the cross-sectional area of the fiber-matrix combination, namely,

$$2\pi \left\{ \int_0^b \sigma_x^f r\, dr + \int_b^c \sigma_x^m r\, dr \right\} = P \tag{3.25}$$

or $\left\{ C_{11}^f \varepsilon_1 + 2C_{12}^f A^f \right\} \pi b^2 + \left\{ C_{11}^m \varepsilon_1 + 2C_{12}^m A^m \right\} \pi \left(c^2 - b^2 \right) = P$ (3.26)

If we use equations (3.19), (3.22), and (3.24), the constants A^f, A^m, and B^m can be solved for in terms of ε_1. As the early elasticity-based micromechanics analyses were focused on determining overall properties of the composite, in the present case an estimate for E_1 can be obtained by substituting the expressions for A^f and A^m into equation (3.26), resulting in an equation of the form

$$\sigma_1 = E_1 \varepsilon_1 \tag{3.27}$$

where σ_1 is average stress in the axial, or 1, direction. This is the stress in the composite in the principal material 1 direction and is given by

$$\sigma_1 = P/\pi c^2 \tag{3.28}$$

the axial force divided by the cross-sectional area of the fiber-matrix combination. The expression for E_1 in equation (3.27) is complicated but it can be written in the form

$$E_1 = E_1^f \left(1 + \gamma \right) V^f + E^m \left(1 + \delta \right) \left(1 - V^f \right) \tag{3.29}$$

The quantities γ and δ are functions of the extensional moduli and Poisson's ratios of the fiber and matrix and the fiber volume fraction. They are given by

$$\gamma = \frac{2v_{21}^f E^m \left(1 - v^f - 2v_{12}^f v_{21}^f \right) V^f \left(v_{12}^f - v^m \right)}{E_2^f \left(1 + v^m \right) \left(1 + V^F \left(1 - 2v^m \right) \right) + E^m \left(1 - v^f - 2v_{12}^f v_{21}^f \right) \left(1 - V^f \right)} \tag{3.30a}$$

and $$\delta = \frac{2E_2^f v^m V^f \left(v^m - v_{12}^f \right)}{E_2^f \left(1 + v^m \right) \left(1 + V^f \left(1 - 2v^m \right) \right) + E^m \left(1 - v^f - 2v_{12}^f v_{21}^f \right) \left(1 - V^f \right)} \tag{3.30b}$$

In the above definitions, as the fiber has been assumed to be transversely isotropic, use has been made of the fact that

$$E_3^f = E_2^f$$
$$v_{13}^f = v_{12}^f \tag{3.31}$$
$$v_{23}^f = v_{32}^f = v^f$$

For typical properties of the fiber and matrix, γ and δ are much less than one. Hence, to a very good first approximation, E_1 can be written as

$$E_1 = E_1^f V^f + E^m \left(1 - V^f \right) \tag{3.32}$$

When the major Poisson's ratio of the fiber and Poisson's ratio of the matrix are equal, $v_{12}^f = v^m$, both δ and γ are identically zero and equation (3.32) is exact.

A comparison between the predictions of the finite-element approach and equation (3.29) is given in Figure 3.30. In the figure, we use equation (3.29) to plot the composite extensional modulus, normalized by the modulus of the fiber in the axial direction, as a function of fiber volume fraction. The finite-element results from Table 3.1 are indicated at the volume fractions studied in that approach. As can be seen, the comparison is excellent. If equation (3.32) had been used instead of equation (3.29), it would have been impossible to detect the difference in Figure 3.30. Thus equation (3.32) can be considered an accurate equation for determining E_1.

Even though the presense of other fibers encourages us to say that the axisymmetric assumption is invalid, it is interesting to compute the stresses predicted by the fiber-matrix cylinder model and compare them with the stresses computed by the finite-element model. The finite-element model, of course, was not restricted by the assumption of axisymmetry and the comparisons are shown in Figures 3.31–3.33. In these figures the various stress components at the fiber-matrix interface are plotted as a function of circumferential location around the interface. The finite-element calculations are taken from Figures 3.12–3.14. The finite-element calculations for the hexagonal array, though not included in Figures 3.12–3.14, are added to Figures 3.31–3.33.

Examination of Figure 3.31 reveals that for a 20 percent fiber volume fraction, the axisymmetric assumption of the concentric cylinders model is a good approximation. None of the major stress components vary greatly with θ and the concentric cylinders model provides a good indication of the average level of stress. The interfacial shear stress τ_{ns} varies but its value is quite small compared to the interfacial normal stress σ_n. With a 20 percent fiber volume fraction it would appear that the fibers are far enough apart that they act independently of each other when subjected to an axial

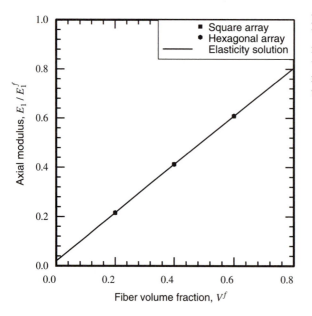

FIGURE 3.30
Variation of composite modulus E_1 with fiber volume fraction for graphite-reinforced material, concentric cylinders model

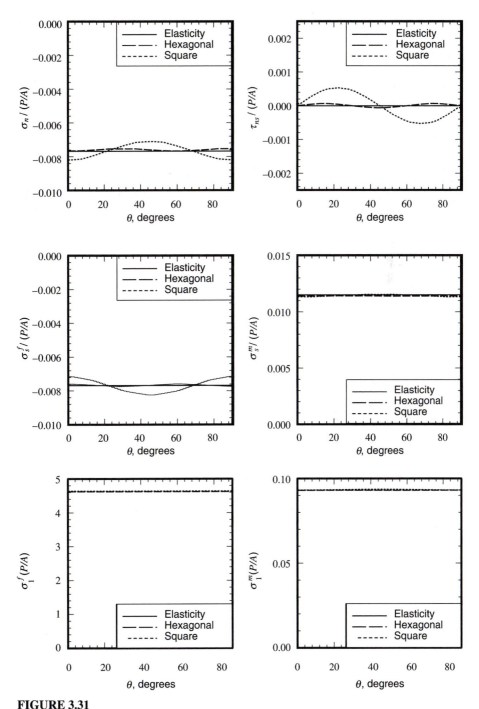

FIGURE 3.31

Comparison of stresses between concentric cylinders model elasticity solution and finite-element calculations for graphite-reinforced material, fiber-direction loading and 20 percent fiber volume fraction

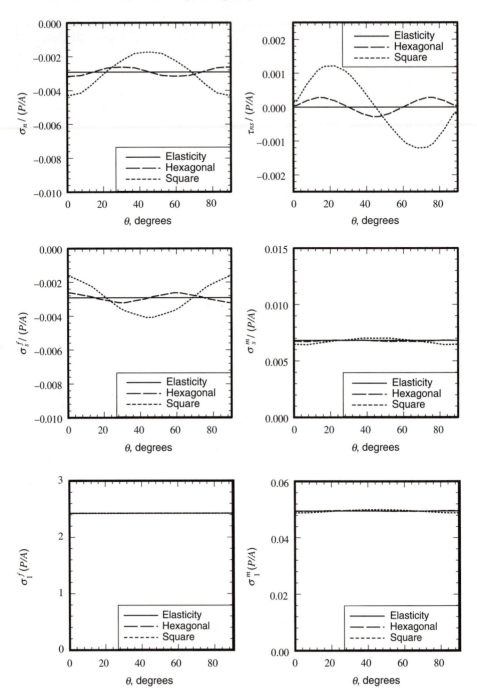

FIGURE 3.32
Comparison of stresses between concentric cylinders model elasticity solution and finite-element calculations for graphite-reinforced material, fiber-direction loading and 40 percent fiber volume fraction

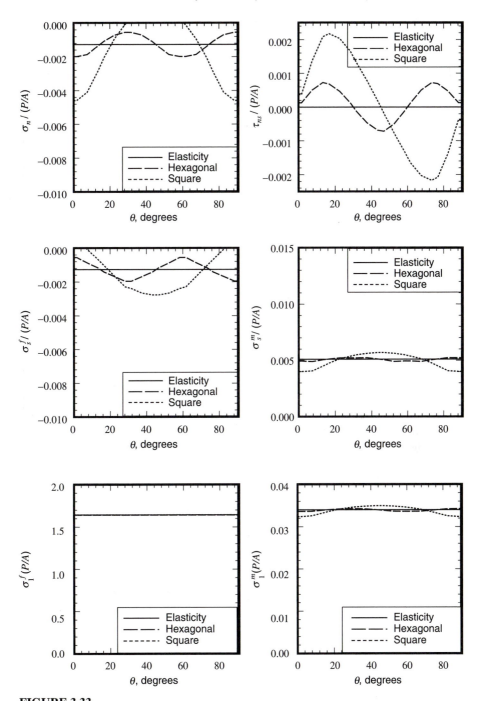

FIGURE 3.33
Comparison of stresses between concentric cylinders model elasticity solution and finite-element calculations for graphite-reinforced material, fiber-direction loading and 60 percent fiber volume fraction

load. When the fiber volume fraction is 40 percent, Figure 3.32, the assumption of axisymmetry appears to be less valid, though for the stress components σ_s^m, σ_1^f, and σ_1^m the assumption appears quite valid. For the other stress components, the comparison of the elasticity solution with the hexagonal-packed array is still quite reasonable; the interfacial shear stress τ_{ns} exhibits the poorest comparison, although it is not a major stress component. For a 60 percent fiber volume fraction comparison between the elasticity solution and either the square- or hexagonal-packed array finite-element calculations indicate that for the stress components σ_n, τ_{ns}, and σ_s^f, the axisymmetric assumption is not particularly good. However, for σ_s^m, σ_1^f, and σ_1^m the assumption of axisymmetry provides a good estimate of the stresses. Despite the lack of perfect correlation between the stress predictions of the elasticity solution and the finite-element calculations, the prediction of E_1 is obviously not influenced by these errors. The prediction of E_1, a property that can be considered an averaged quantity, really depends on the stresses being accurately predicted in an *average* sense. The axisymmetric elasticity model does predict the average value of the stresses very well; the variations with θ are not accurately predicted for the higher volume fractions.

3.3.2 Axial Shear

The elasticity approach can also be used to estimate values of the shear modulus G_{12}; this modulus is often referred to as the *axial* shear modulus. The approach to this problem is based on a slightly different view of the fiber-matrix combination, namely, that of Figure 3.34. In this figure the fiber-matrix combination is viewed by looking along the 3 axis toward the 1-2 plane and the boundaries of the fiber-matrix combination are deformed by shearing in the 1-2 plane. The boundaries of the portion of the fiber-matrix combination shown have the following displacements

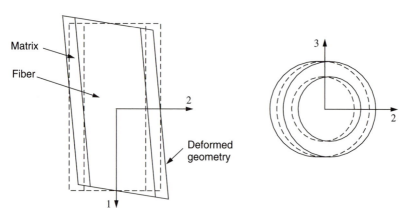

FIGURE 3.34
Axial shear deformation applied to concentric cylinders model

relative to the undeformed state:

$$u_1 = \frac{\gamma_{12}^o}{2} x_2$$

$$u_2 = \frac{\gamma_{12}^o}{2} x_1 \tag{3.33}$$

$$u_3 = 0$$

where x_1, x_2, and x_3 are used to denote the coordinates in the 1-2-3 principal material directions, and u_1, u_2, and u_3 are the displacements in those directions. The quantity γ_{12}^o is the shear strain imposed on the boundary. Though equation (3.33) defines a fairly specific deformation on the boundary, in the interior of the fiber-matrix combination the displacements are assumed to be expressible as

$$u_1 = \phi(x_2, x_3) - \frac{\gamma_{12}^o}{2} x_2$$

$$u_2 = \frac{\gamma_{12}^o}{2} x_1 \tag{3.34}$$

$$u_3 = 0$$

The function $\phi(x_2, x_3)$ defines the shear strains in the interior of the fiber-matrix combination, with the particular form of equation (3.34) leading to relatively simple expressions for the shear strains, namely,

$$\gamma_{12} = \frac{\partial \phi}{\partial x_2}$$

$$\gamma_{13} = \frac{\partial \phi}{\partial x_3} \tag{3.35}$$

It is important to note that there is a function ϕ for the fiber region and a different function ϕ for the matrix region.

As the other strains are zero, the only stresses present in the fiber-matrix combination due to the displacements given in equation (3.34) are

$$\tau_{13} = G\gamma_{13}$$

$$\tau_{12} = G\gamma_{12} \tag{3.36}$$

or if we use equation (3.35),

$$\tau_{13} = G \frac{\partial \phi}{\partial x_3}$$

$$\tau_{12} = G \frac{\partial \phi}{\partial x_2} \tag{3.37}$$

where it has been assumed that the material is either isotropic or transversely isotropic, namely,

$$G_{12} = G_{13} = G \tag{3.38}$$

Because all the stresses except τ_{12} and τ_{13} are zero, the equilibrium conditions reduce to

$$\frac{\partial \tau_{12}}{\partial x_2} + \frac{\partial \tau_{13}}{\partial x_3} = 0 \tag{3.39}$$

Using equation (3.37) in equation (3.39) results in

$$\frac{\partial^2\phi}{\partial x_2^2} + \frac{\partial^2\phi}{\partial x_3^2} = 0 \tag{3.40}$$

and at this point it is convenient to express the problem in cylindrical coordinates; the principal material coordinate system (1-2-3) and the cylindrical coordinate system $(x\text{-}r\text{-}\theta)$ are shown in Figure 3.35. Relations between the coordinates in the two systems are given by

$$\begin{aligned} x_2 &= r\cos\theta \\ x_3 &= r\sin\theta \end{aligned} \tag{3.41a}$$

and the inverse

$$\begin{aligned} r &= \left(x_2^2 + x_3^2\right)^{1/2} \\ \theta &= \tan^{-1}\left(\frac{x_3}{x_2}\right) \end{aligned} \tag{3.41b}$$

The nonzero stresses in the cylindrical coordinate system are related to τ_{12} and τ_{13} by

$$\begin{aligned} \tau_{xr} &= \cos\theta\,\tau_{12} + \sin\theta\,\tau_{13} \\ \tau_{x\theta} &= \cos\theta\,\tau_{13} - \sin\theta\,\tau_{12} \end{aligned} \tag{3.42a}$$

and the inverse relations

$$\begin{aligned} \tau_{12} &= \cos\theta\,\tau_{xr} - \sin\theta\,\tau_{x\theta} \\ \tau_{13} &= \cos\theta\,\tau_{x\theta} + \sin\theta\,\tau_{xr} \end{aligned} \tag{3.42b}$$

In terms of ϕ, the two shear stresses in the cylindrical coordinate system are

$$\begin{aligned} \tau_{xr} &= G\frac{\partial\phi}{\partial r} \\ \tau_{x\theta} &= G\frac{\partial\phi}{r\partial\theta} \end{aligned} \tag{3.43}$$

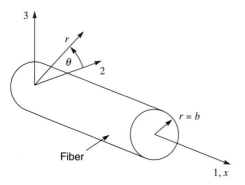

FIGURE 3.35
The x-y-z and 1-2-3 coordinate systems for the fiber-matrix combination

Writing equation (3.40) in the cylindrical coordinate system results in

$$\frac{\partial^2 \phi}{\partial r^2} + \frac{1}{r}\frac{\partial \phi}{\partial r} + \frac{1}{r^2}\frac{\partial^2 \phi}{\partial \theta^2} = 0 \qquad (3.44)$$

Equations (3.40) and (3.44) are Laplace's equation for ϕ in the two different coordinate systems.

The separation of variables technique can be used to solve equation (3.44); the solution is given by

$$\phi(r, \theta) = a_0 + \sum_{n=1}^{\infty}\left(A_n r^n + B_n r^{-n}\right)\left(C_n \sin(n\theta) + D_n \cos(n\theta)\right) \qquad (3.45)$$

For the problem here, only terms to $n = 1$ are necessary. Additionally, $a_o = 0$ and the $\sin(n\theta)$ terms are not needed. Finally, for the stresses to remain bounded at the center of the fiber, B_n for the fiber must be zero. As with past practice, using superscripts f to denote the fiber and m to denote the matrix, the functions ϕ for these two materials in the fiber-matrix combination are

$$\phi^f(r, \theta) = a_1^f r \cos\theta$$

$$\phi^m(r, \theta) = \left(a_1^m r + \frac{b_1^m}{r}\right)\cos\theta \qquad (3.46)$$

where

$$a_1 = A_1 D_1 \quad \text{and} \quad b_1 = B_1 D_1 \qquad (3.47)$$

As a result,

$$\tau_{xr}^f = G^f a_1^f \cos\theta$$

$$\tau_{x\theta}^f = -G^f a_1^f \sin\theta \qquad (3.48)$$

and

$$\tau_{xr}^m = G^m\left(a_1^m - \frac{b_1^m}{r^2}\right)\cos\theta$$

$$\tau_{x\theta}^m = -G^m\left(a_1^m + \frac{b_1^m}{r^2}\right)\sin\theta \qquad (3.49)$$

Continuity of the three components of displacement at the interface between the fiber and matrix reduces to enforcement of

$$\phi^f(b, \theta) = \phi^m(b, \theta) \qquad (3.50)$$

or

$$a_1^f b \cos\theta = \left(a_1^m b + \frac{b_1^m}{b}\right)\cos\theta \qquad (3.51)$$

where, recall, b is the radius of the fiber. Continuity of the stresses at the fiber-matrix interface reduces to a single condition, namely,

$$\tau_{xr}^f(b, \theta) = \tau_{xr}^m(b, \theta) \qquad (3.52)$$

which from equations (3.48) and (3.49) can be written as

$$G^f a_1^f \cos\theta = G^m\left(a_1^m - \frac{b_1^m}{b^2}\right)\cos\theta \qquad (3.53)$$

Using equation (3.34) to enforce the conditions of equation (3.33) at the boundary of the fiber-matrix combination (i.e., at $r = c$ and $\theta = 0$), provides a final condition that can be written as

$$\phi^m = \gamma_{12}^o c \tag{3.54}$$

Substitution from equation (3.46) yields

$$a_1^m c + \frac{b_1^m}{c} = \gamma_{12}^o c \tag{3.55}$$

and solving equations (3.51), (3.53), and (3.55) leads to the solution for a_1^f, a_1^m, and b_1^m, namely,

$$a_1^f = \left\{ \frac{2G^m}{\left(\dfrac{G^m - G^f}{b^2}\right) + \left(\dfrac{G^m - G^f}{c^2}\right)} \right\} \frac{\gamma_{12}^o}{b^2}$$

$$a_1^m = \left\{ \frac{\left(G^m + G^f\right)}{\left(\dfrac{G^m - G^f}{c^2}\right) + \left(\dfrac{G^m + G^f}{b^2}\right)} \right\} \frac{\gamma_{12}^o}{b^2} \tag{3.56}$$

$$b_1^m = \left\{ \frac{\left(G^m - G^f\right)}{\left(\dfrac{G^m - G^f}{c^2}\right)} + \left(\dfrac{G^m + G^f}{b^2}\right) \right\} \gamma_{12}^o$$

Accordingly, using equation (3.49) in the transformation relations, equation (3.42b), leads to

$$\tau_{12}^m = G^m \left\{ \left(a_1^m - \frac{b_1^m}{r^2}\right) \cos\theta + \left(a_1^m + \frac{b_1^m}{r^2}\right) \sin\theta \right\} \tag{3.57}$$

Evaluating this at $r = c$ and $\theta = 0$ and substituting for a_1^m and b_1^m from above leads to the following relation between τ_{12} and γ_{12}^o:

$$\tau_{12}^m = G^m \left\{ \frac{\left(G^m + G^f\right) - V^f \left(G^m - G^f\right)}{\left(G^m + G^f\right) + V^f \left(G^m - G^f\right)} \right\} \gamma_{12}^o \tag{3.58}$$

This is an important expression because it provides an estimate of the composite axial shear modulus G_{12} through the relation

$$\tau_{12}^m = G_{12}\gamma_{12}^o \tag{3.59}$$

specifically $$G_{12} = G^m \left\{ \frac{\left(G^m + G^f\right) - V^f \left(G^m - G^f\right)}{\left(G^m + G^f\right) + V^f \left(G^m - G^f\right)} \right\} \tag{3.60}$$

The dependence of G_{12} on fiber volume fraction for the graphite-reinforced material is illustrated in Figure 3.36. As the fiber volume fraction increases toward 65 percent, the shear stiffness of the fiber produces about a factor of three increase

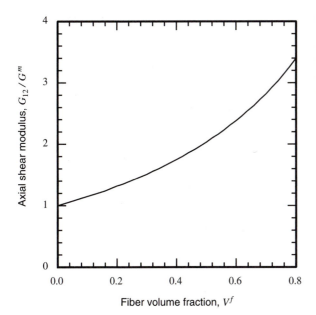

FIGURE 3.36
Variation of composite axial
shear modulus G_{12} with
fiber volume fraction for
graphite-reinforced material,
concentric cylinders model

in the shear stiffness of the composite relative to the matrix, a trend seen with the
other composite elastic properties.

Exercises for Section 3.3

1. Compute the values of γ and δ in equations (3.30a) and (3.30b) using the material properties of the fiber and matrix for a graphite-reinforced composite and show that these
   quantities are small compared to unity. Show this for the range $0.15 \leq V^f \leq 0.75$. The
   material properties are given in equations (3.2) and (3.3).

2. For graphite-reinforced material, compute ν_{12} using the elasticity solution and compare
   the results with the finite-element results. Draw a figure similar to Figure 3.30. Hint: For
   the axial loading,

$$\nu_{12} = -\frac{\dfrac{\Delta c}{c}}{\varepsilon_1} = -\frac{\dfrac{w^m(r=c)}{c}}{\varepsilon_1}$$

where Δc is the change in the radius of the fiber-matrix combination of Figure 3.29 and
ε_1 is the axial strain of the composite caused by the fiber-direction loading. The results
should compare quite well with the finite-element calculations.

3.4
STRENGTH-OF-MATERIALS RESULTS

An important result that can be obtained from the concept of the unit cell and the use
of finite elements, or the use of elasticity solutions, is an estimate of the overall elastic

and thermal expansion properties of the composite. With these models it is possible to evaluate how the overall properties are influenced by fiber volume fraction, fiber properties, matrix properties, and the assumptions of how the fibers are packed (i.e., square- or hexagonal-packed arrays). While the use of finite-element representations of unit cells provides detailed information about the stresses in the fibers, in the matrix, and at the interface between the fiber and matrix, often it is only the elastic or thermal expansion property estimates that are of interest. This was the case in early micromechanics studies, whereby strength-of-materials approaches were used to provide insight into the elastic properties. These approaches can be considered to be at the opposite end of the spectrum from the finite-element or elasticity approaches. The strength-of-materials approaches do not concern themselves with the details of the stresses at the fiber-matrix interface, the packing arrangements, or the many other characteristics that can be considered with unit cell finite-element models. However, reasonable estimates of some of the elastic and thermal expansion properties of composite materials can be obtained with these approaches. They shall be studied next to provide a contrast to the unit cell finite-element methods. In addition, the strength-of-materials approaches result in rather simple algebraic expressions for the elastic and thermal expansion properties of the composite as a function of fiber and matrix properties. These algebraic expressions can be conveniently used for parametric studies or for embedding within other analyses.

3.4.1 Model for E_1 and ν_{12}

The strength-of-materials models, sometimes called *rule-of-mixtures models*, also rely on what could be termed a unit cell. The unit cell used in the strength-of-materials models is quite different from the unit cells of the previous sections, and the particular unit cell considered depends on the composite property being studied. To study E_1 and ν_{12} for the composite, consider a section cut from a single layer of fiber-reinforced material. The section consists of side-by-side alternating regions of fiber and matrix, the fibers arranged in parallel arrays, as in Figure 3.37, the widths of each of the regions of fiber and matrix denoted by W^f and W^m, respectively. The figure shows the 1 and 2 principal material directions. The thickness of the layer is not important at the moment, and can be taken as unity. In fact, in these rule-of-mixtures models the cross-sectional shape of the fibers is not important. They can be considered circular, square, elliptical, or any other shape. For simplicity, assume they are square. As it will turn out, only the cross-sectional areas of the fiber and matrix will be important. Figure 3.38(a) shows details of a "unit cell" cut from a single layer, the length of the cell denoted by L and the cross-sectional areas of the fiber and matrix denoted as A^f and A^m. Assume as in Figure 3.38(b) that the unit cell is subjected to a stress σ_1 such that it stretches in the 1 direction and, due to Poisson effects, contracts in the 2 direction. Because the fiber and matrix are bonded together, they both stretch the same amount in the 1 direction, namely, ΔL; the strain in the 1 direction in both the fiber and matrix is given by

$$\varepsilon_1^f = \varepsilon_1^m = \frac{\Delta L}{L} \tag{3.61}$$

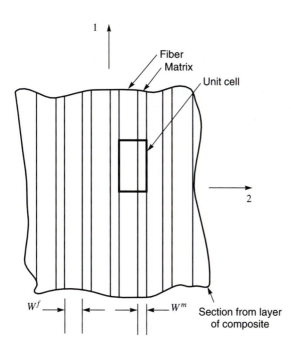

FIGURE 3.37
Section cut from a fiber-
reinforced composite and a
unit cell

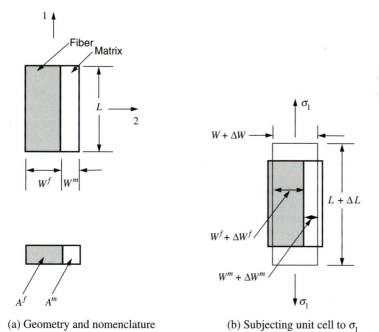

(a) Geometry and nomenclature   (b) Subjecting unit cell to σ_1

FIGURE 3.38
Details of unit cell and rule-of-mixtures model for composite extensional modulus E_1
and major Poisson's ratio ν_{12}

Because the fiber and matrix have different Poisson's ratios, they will not contract the same amount in the widthwise, or 2, direction. The combined contraction of the fiber and matrix results in the overall contraction of the composite in the 2 direction. That will be addressed shortly.

Treating the two constituents as if they were each in a one-dimensional state of stress, we find the stresses in the fiber and matrix are

$$\sigma_1^f = E_1^f \varepsilon_1^f = E_1^f \frac{\Delta L}{L}$$

$$\sigma_1^m = E^m \varepsilon_1^m = E^m \frac{\Delta L}{L}$$

(3.62)

and accordingly the forces in the 1 direction in the fiber and matrix are given by

$$F_1^f = \sigma_1^f A^f = E_1^f \frac{\Delta L}{L} A^f$$

$$F_1^m = \sigma_1^m A^m = E^m \frac{\Delta L}{L} A^m$$

(3.63)

The total force in the 1 direction divided by the total cross-sectional area of the unit cell, A, where $A = A^f + A^m$, is defined as the composite stress σ_1, namely,

$$\sigma_1 = \frac{F_1^f + F_1^m}{A} = \left(E_1^f \frac{A^f}{A} + E^m \frac{A^m}{A} \right) \frac{\Delta L}{L}$$

(3.64)

But $\frac{\Delta L}{L}$ is the composite strain ε_1, so equation (3.64) gives

$$\sigma_1 = E_1 \varepsilon_1$$

(3.65)

where

$$E_1 = E_1^f V^f + E^m V^m$$

(3.66)

with the quantities V^f and V^m being the *area* fractions of fiber and matrix, respectively. By the geometry of the unit cell, however, V^f and V^m represent volume fractions. Because

$$V^m = 1 - V^f$$

(3.67)

equation (3.66) can be written as

$$E_1 = E_1^f V^f + E^m (1 - V^f)$$

(3.68)

Note that this equation is identical to equation (3.32), the approximation to the prediction of E_1 from the elasticity model; the exact prediction is given by equation (3.29). Thus, the strength-of-materials and elasticity solution predictions are closely related. Equation (3.68) is referred to as the rule-of-mixtures equation for E_1.

As a summary of all the results obtained, Figure 3.39 shows the composite fiber direction modulus, E_1, as a function of fiber volume fraction for a graphite-reinforced material as predicted by the rule of mixtures, the two finite-element arrays, and the exact expression from the theory of elasticity, which use the material properties from equations (3.2) and (3.3). The relation is obviously linear and clearly the simple rule-of-mixtures model is quite accurate; the differences among the various approaches are indistinguishable.

The overall contraction of the unit cell in the 2 direction can be used to compute the major Poisson's ratio of the composite, specifically ν_{12}. For the situation in Figure

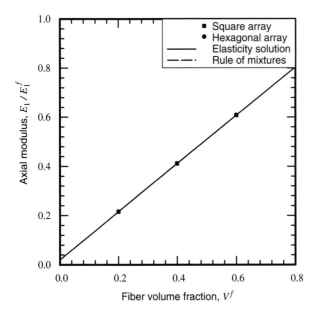

FIGURE 3.39
Rule-of-mixtures prediction for variation of composite modulus E_1 with fiber volume fraction for graphite-reinforced material

3.38(b), because the unit cell is being subjected to a simple uniaxial stress state, ν_{12} is defined as minus the ratio of the contraction strain in the 2 direction divided by the elongation strain in the 1 direction, namely,

$$\nu_{12} = -\frac{\dfrac{\Delta W}{W}}{\dfrac{\Delta L}{L}} \tag{3.69}$$

From Figure 3.38
$$W = W^f + W^m \tag{3.70}$$

and
$$\Delta W = \Delta W^f + \Delta W^m \tag{3.71}$$

Using the definition of Poisson's ratio for each constituent and the fact that each constituent is assumed to be in a state of uniaxial stress, we find the contraction of each constituent is

$$\frac{\Delta W^f}{W^f} = -\nu_{12}^f \frac{\Delta L}{L}$$
$$\frac{\Delta W^m}{W^m} = -\nu^m \frac{\Delta L}{L} \tag{3.72}$$

or
$$\Delta W^f = -\nu_{12}^f W^f \frac{\Delta L}{L}$$
$$\Delta W^m = -\nu^m W^m \frac{\Delta L}{L} \tag{3.73}$$

Substituting these relations into equation (3.71) and dividing both sides by W results in

$$\frac{\Delta W}{W} = -\left(v_{12}^f \frac{W^f}{W} + v^m \frac{W^m}{W}\right) \frac{\Delta L}{L} \tag{3.74}$$

The fiber and matrix volume fractions can be identified, and as a result, equation (3.74) becomes

$$\frac{\Delta W}{W} = -\left(v_{12}^f V^f + v^m V^m\right) \frac{\Delta L}{L} \tag{3.75}$$

From the definition of Poisson's ratio in equation (3.69),

$$v_{12} = v_{12}^f V^f + v^m V^m \tag{3.76}$$

or using equation (3.67), we find

$$v_{12} = v_{12}^f V^f + v^m (1 - V^f) \tag{3.77}$$

This is the rule-of-mixtures expression for the major Poisson's ratio v_{12}. It is very similar to the rule-of-mixtures expression for the modulus E_1 in that it is linear in all of the variables.

Figure 3.40 shows the variation of v_{12} with fiber volume fraction for a graphite-reinforced composite, and a comparison of equation (3.77) with the finite-element results. Again, the accuracy of the simple rule-of-mixtures results for v_{12} is quite obvious and it is similar to the accuracy of the results for E_1. If interest centers only on knowing E_1 and v_{12}, then there is really no reason for using the finite-element approach, as the rule-of-mixtures equations are very accurate. As noted before, the derivations and results are independent of the geometry of the fiber or matrix cross section, an interesting finding in itself.

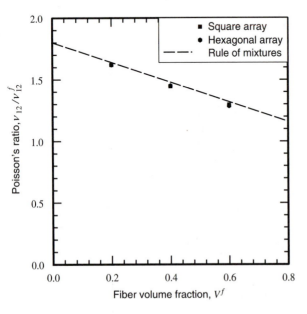

FIGURE 3.40
Rule-of-mixtures prediction for variation of composite major Poisson's ratio v_{12} with fiber volume fraction for graphite-reinforced material

3.4.2 Models for E_2

We can approach one of the most basic considerations for the determination of E_2 by studying the unit cell of Figure 3.37 when it is subjected to a transverse stress, σ_2, as in Figure 3.41. Isolating the fiber and matrix elements, we can argue by equilibrium that each element is subjected to the same transverse stress, σ_2. If this is the case, then the transverse strain in the fiber and matrix are, respectively,

$$\varepsilon_2^f = \frac{\Delta W^f}{W^f}$$
$$\varepsilon_2^m = \frac{\Delta W^m}{W^m}$$

(3.78)

Considering again a one-dimensional stress state, we find the stress and strain in the fiber and matrix are related by

$$\sigma_2^f = \sigma_2 = E_2^f \varepsilon_2^f = E_2^f \frac{\Delta W^f}{W^f}$$
$$\sigma_2^m = \sigma_2 = E^m \varepsilon_2^m = E^m \frac{\Delta W^m}{W^m}$$

(3.79)

These equations can be rearranged and written as

$$\Delta W^f = \frac{W^f}{E_2^f}\sigma_2$$
$$\Delta W^m = \frac{W^m}{E^m}\sigma_2$$

(3.80)

The overall change in the transverse dimension of the unit cell is

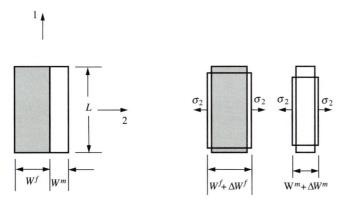

(a) Geometry and nomenclature   (b) Subjecting unit cell to a σ_2

FIGURE 3.41
Rule-of-mixtures model for composite extensional modulus E_2

$$\Delta W = \Delta W^f + \Delta W^m \tag{3.81}$$

so the definition of overall transverse strain is

$$\varepsilon_2 = \frac{\Delta W}{W} = \frac{\Delta W^f + \Delta W^m}{W} \tag{3.82}$$

Using equation (3.80) in equation (3.82), we find

$$\varepsilon_2 = \frac{\left(\dfrac{W^f}{E_2^f} + \dfrac{W^m}{E^m}\right)\sigma_2}{W} \tag{3.83}$$

Using the definitions of the fiber and matrix volume fractions and the geometry of the unit cell yields

$$\varepsilon_2 = \left(\frac{V^f}{E_2^f} + \frac{V^m}{E^m}\right)\sigma_2 \tag{3.84}$$

Comparing the above equation with the form

$$\varepsilon_2 = \frac{\sigma_2}{E_2} \tag{3.85}$$

where E_2 is the composite transverse modulus, we see that

$$\frac{1}{E_2} = \frac{V^f}{E_2^f} + \frac{V^m}{E^m} \tag{3.86}$$

or

$$\frac{1}{E_2} = \frac{V^f}{E_2^f} + \frac{(1 - V^f)}{E^m} \tag{3.87}$$

This equation is the rule-of-mixtures relation for the transverse modulus as a function of the transverse moduli of the fiber and matrix, and the fiber volume fraction. Like rule-of-mixtures relations for E_1 and ν_{12}, this relation is a simple linear relation, in this case among the inverse moduli and volume fractions of the two constituents. Figure 3.42 illustrates the comparison between the above rule-of-mixtures expression and the results from the finite-element unit cell model. The rule-of-mixtures expression for E_2 is not in the good agreement with the finite-element results seen with E_1 and ν_{12}. Perhaps this could be expected from the onset, as the free-body diagrams of the fiber and matrix elements in Figure 3.41 are something of an over-simplification of the interaction between the fiber and the matrix when the composite is subjected to a transverse stress. For one thing, the one-dimensional state of stress may not be accurate because, like the situation depicted in Figure 3.38, the fiber and matrix elements are bonded together and hence change length together in the 1 direction. Considering the one-dimensional stress state, we find the diagram of Figure 3.41 indicates that due to the different Poisson's ratios, the element of fiber is allowed to contract in the 1 direction differently than the element of matrix. Thus, a modification of the model would be to have the length change in the 1 direction of the fiber element and the matrix element be the same, an approach to be taken shortly. However, another difficulty with the simplification of Figure 3.41 is the diagram assumes that both the fiber and matrix are subjected to transverse stress σ_2. The

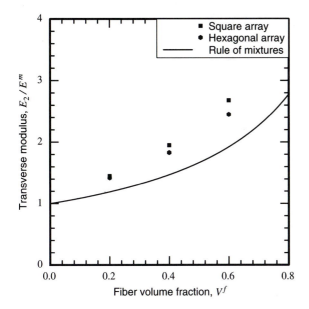

FIGURE 3.42
Rule-of-mixtures prediction
for variation of composite
modulus E_2 with fiber volume
fraction for graphite-reinforced
material

finite-element unit cell models demonstrated that a portion of the transverse stress is transmitted through the fiber, and a portion is transmitted around the fiber, through the matrix material. To use a specific nomenclature, the transverse stress σ_2 in the fiber and matrix is *partitioned* differently than is implied by Figure 3.41. To correct the rule-of-mixtures model for E_2, a so-called *stress-partitioning factor* is often introduced into equation (3.87) to account for the error in the assumption that both the fiber and the matrix are subjected to the full value σ_2. The stress-partitioning factor accounts for a more proper division of the stress in each of the two constituents. To incorporate a partitioning factor, consider equation (3.83) rewritten in slightly different form, specifically,

$$\varepsilon_2 = \frac{\left(\dfrac{W^f}{E_2^f} + \dfrac{W^m}{E^m}\right)\sigma_2}{W^f + W^m} \tag{3.88}$$

Dividing the numerator and denominator by W and again using the geometry of the unit cell as it relates to the fiber and matrix volume fractions, we find equation (3.88) becomes

$$\varepsilon_2 = \frac{\left(\dfrac{V^f}{E_2^f} + \dfrac{V^m}{E^m}\right)\sigma_2}{V^f + V^m} \tag{3.89}$$

Now consider that instead of V^m being the volume fraction of matrix that is subjected to stress level σ_2, assume the volume fraction is less than that, namely, ηV^m, where η will be referred to as a partitioning factor and $0 < \eta < 1$. The volume fraction of fiber that is subjected to stress level σ_2 is still V^f. As a result of this new nomenclature, the total effective volume of fiber and matrix is now

$$V^f + \eta V^m \tag{3.90}$$

Equation (3.89) now takes the form

$$\varepsilon_2 = \frac{\left(\dfrac{V^f}{E_2^f} + \dfrac{\eta V^m}{E^m}\right)\sigma_2}{V^f + \eta V^m} \tag{3.91}$$

and by analogy with equation (3.85), the composite modulus E_2 is given as

$$\frac{1}{E_2} = \frac{\dfrac{V^f}{E_2^f} + \dfrac{\eta V^m}{E^m}}{V^f + \eta V^m} \tag{3.92}$$

or

$$\frac{1}{E_2} = \frac{\dfrac{V^f}{E_2^f} + \dfrac{\eta(1 - V^f)}{E^m}}{V^f + \eta(1 - V^f)} \tag{3.93}$$

This expression for E_2 is referred to as the modified rule-of-mixtures model for E_2. The stress-partitioning factor η generally must be determined empirically. If it can be determined for a specific material by measuring E_2 at a particular volume fraction, and if the elastic properties of the fiber and matrix are known, then the value of η can be determined and used for parameter studies involving other fiber volume fractions. Note that when $\eta = 1$ in equation (3.93), the original rule-of-mixtures relation, equation (3.87), is recovered.

Figure 3.43 illustrates the predictions of the modified rule-of-mixtures model for E_2 as a function of fiber volume fraction, where three values of the stress partitioning factor η are used for purposes of the example. The finite-element results for both square- and hexagonal-packed arrays are used for comparison. A value of $\eta = 0.4$ gives a good comparison with the square-packed array, while a value of $\eta = 0.5$ gives a good comparison with the hexagonal-packed array.

As mentioned previously, the rule-of-mixtures model for E_2 violates intuition regarding the response of the fiber and matrix elements in that the fiber element is allowed to change length independently of the change in length of the matrix element. Figure 3.41 can be modified to account for the elements changing length by the same amount when subjected to a transverse stress σ_2. This will imply, of course, that each element is subjected to a stress in the 1 direction, σ_1. The free-body diagrams of the elements with the two stress components are shown in Figure 3.44, and the deformed transverse widths of the fiber and matrix elements are, respectively,

$$W^f + \Delta W^f \quad \text{and} \quad W^m + \Delta W^m \tag{3.94}$$

while the deformed length of *both* elements is

$$L + \Delta L \tag{3.95}$$

Considering each element to be in a state of stress such that

$$\tau_{12} = \sigma_3 = \tau_{31} = \tau_{32} = 0 \tag{3.96}$$

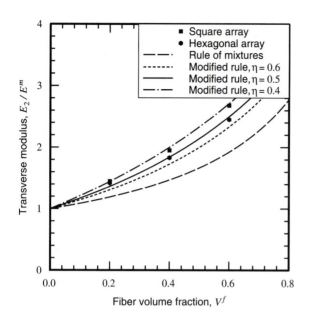

FIGURE 3.43
Modified rule-of-mixtures prediction for variation of composite modulus E_2 with fiber volume fraction for graphite-reinforced material

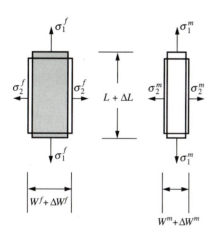

FIGURE 3.44
Alternative rule-of-mixtures model for composite extensional modulus E_2

we find the stress-strain relations for the fiber element are

$$\sigma_1^f = Q_{11}^f \varepsilon_1^f + Q_{12}^f \varepsilon_2^f$$
$$\sigma_2^f = Q_{12}^f \varepsilon_1^f + Q_{22}^f \varepsilon_2^f \tag{3.97a}$$

while for the matrix element,

$$\sigma_1^m = Q_{11}^m \varepsilon_1^m + Q_{12}^m \varepsilon_2^m$$
$$\sigma_2^m = Q_{12}^m \varepsilon_1^m + Q_{22}^m \varepsilon_2^m \tag{3.97b}$$

In the above

$$Q_{11}^f = \frac{E_1^f}{1 - v_{12}^f v_{21}^f} \qquad Q_{22}^f = \frac{E_2^f}{1 - v_{12}^f v_{21}^f}$$

$$Q_{12}^f = \frac{v_{12}^f E_2^f}{1 - v_{12}^f v_{21}^f} = \frac{v_{21}^f E_1^f}{1 - v_{12}^f v_{21}^f}$$

(3.98a)

and

$$Q_{11}^m = \frac{E^m}{1 - (v^m)^2} = Q_{22}^m$$

$$Q_{12}^m = \frac{v^m E^m}{1 - (v^m)^2}$$

(3.98b)

Using the assumptions given by equation (3.96) for both the fiber and the matrix, equations (3.97a) and (3.97b) are derivable from either equation (2.45) or equation (2.47). This will be done in a later chapter in the context of a single layer of material. Due to equilibrium considerations

$$\sigma_2^f = \sigma_2^m = \sigma_2$$

(3.99)

and more importantly

$$\int_{A^m + A^f} \sigma_1 \, dA = 0$$

(3.100)

This latter equation results from the fact there should be no net force in the 1 direction when the composite is subjected to a transverse stress. If we assume that the stress in the 1 direction is constant within the fiber element, and also within the matrix element, then the condition given in equation (3.100) can be written as

$$\sigma_1^f A^f + \sigma_1^m A^m = 0$$

(3.101)

Also, because of the geometry of the deformation,

$$\varepsilon_1^f = \varepsilon_1^m = \frac{\Delta L}{L} \qquad \varepsilon_2^f = \frac{\Delta W^f}{W^f} \qquad \varepsilon_2^m = \frac{\Delta W^m}{W^m}$$

(3.102)

Using equations (3.99) and (3.102) in the stress-strain relations equations (3.97a) and (3.97b), as well as using equation (3.101), results in equations that can be used to find ΔW^f and ΔW^m, and ultimately E_2. With some rearrangement these equations are

$$Q_{22}^f \frac{\Delta W^f}{W^f} + Q_{12}^f \frac{\Delta L}{L} = \sigma_2$$

$$Q_{22}^m \frac{\Delta W^m}{W^m} + Q_{12}^m \frac{\Delta L}{L} = \sigma_2$$

(3.103)

$$\left(Q_{11}^f \frac{\Delta L}{L} + Q_{12}^f \frac{\Delta W^f}{W^f} \right) A^f + \left(Q_{11}^m \frac{\Delta L}{L} + Q_{12}^m \frac{\Delta W^m}{W^m} \right) A^m = 0$$

Solving for ΔW^f and ΔW^m and using the basic definition of ε_2, namely equation

(3.82), results in a relation of the form

$$\varepsilon_2 = \left(\frac{\eta^f V^f}{E_2^f} + \frac{\eta^m V^m}{E^m} \right) \sigma_2 \tag{3.104}$$

Dividing by σ_2 yields the expression for E_2, namely,

$$\frac{1}{E_2} = \frac{\eta^f V^f}{E_2^f} + \frac{\eta^m V^m}{E^m} \tag{3.105}$$

or

$$\frac{1}{E_2} = \frac{\eta^f V^f}{E_2^f} + \frac{\eta^m (1 - V^f)}{E^m} \tag{3.106}$$

To arrive at equation (3.106), we've made use of the fact that for the geometry of Figure 3.44, A^f and A^m are directly related to the volume fractions V^f and V^m, respectively. Also, the fiber and matrix partitioning factors are given by

$$\eta^f = \left\{ \frac{E_1^f V^f + [(1 - \nu_{12}^f \nu_{21}^f) E^m + \nu^m \nu_{21}^f E_1^f](1 - V^f)}{E_1^f V^f + E^m (1 - V^f)} \right\}$$
$$\eta^m = \left\{ \frac{[(1 - \nu^{m^2}) E_1^f - (1 - \nu^m \nu_{12}^f) E^m] V^f + E^m V^m}{E_1^f V^f + E^m (1 - V^f)} \right\} \tag{3.107}$$

Equation (3.106) is an alternative version of the rule of mixtures. Figure 3.45 shows comparison of results from the alternative rule-of-mixtures model with the results from the finite-element and rule-of-mixtures models. Relative to the rule-of-mixtures model, there is some improvement in the comparison with the finite-element models. Assuming the empirically derived stress-partitioning factor η is known, we find the

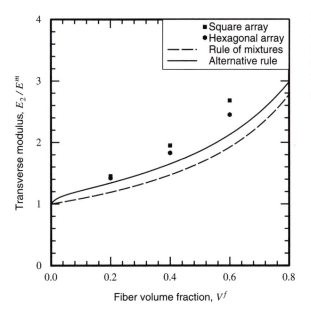

FIGURE 3.45
Alternative rule-of-mixtures prediction for variation of composite modulus E_2 with fiber volume fraction for graphite-reinforced material

modified rule-of-mixtures model of equation (3.93) appears to be most accurate for the incorporation of a simple formula into a parameter study. However, if an empirically derived stress-partitioning factor is not available, the alternative rule-of-mixtures model can be used for improved accuracy relative to the rule-of-mixtures model, and there is a physical basis for the model.

3.4.3 Model for G_{12}

The rule-of-mixtures model for the axial, or fiber-direction, shear modulus G_{12} is similar to the rule-of-mixtures model for E_2. The fiber and matrix elements are each considered to be subjected to shear stress τ_{12}, as in Figure 3.46. By equilibrium considerations, the shear stress on the fiber element has to be the same as the shear stress on the matrix element, and thus the shear strains in the elements of fiber and matrix in Figure 3.46 are given by

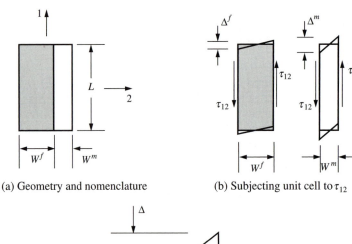

(a) Geometry and nomenclature   (b) Subjecting unit cell to τ_{12}

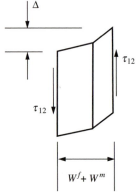

(c) Overall deformation of unit cell

FIGURE 3.46
Rule-of-mixtures model for composite axial shear modulus G_{12}

$$\gamma_{12}^f = \frac{\tau_{12}}{G_{12}^f} \tag{3.108a}$$

$$\gamma_{12}^m = \frac{\tau_{12}}{G^m} \tag{3.108b}$$

where the shear modulus in the fiber and matrix are given by, respectively, G_{12}^f and G^m. By the geometry of the deformation and the definition of shear strain,

$$\Delta^f = \gamma_{12}^f W^f \tag{3.109a}$$

$$\Delta^m = \gamma_{12}^m W^m \tag{3.109b}$$

By considering the fiber and matrix elements joined together, as in Figure 3.46(c), the total deformation of the unit cell is

$$\Delta = \Delta^f + \Delta^m \tag{3.110}$$

The average shear strain for the unit cell is then

$$\gamma_{12} = \frac{\Delta}{W^f + W^m} \tag{3.111}$$

Substituting equation (3.110) into equation (3.111), we find

$$\gamma_{12} = \frac{\Delta^f + \Delta^m}{W^f + W^f} \tag{3.112}$$

or, if we use equation (3.109),

$$\gamma_{12} = \frac{\gamma_{12}^f W^f + \gamma_{12}^m W^m}{W} \tag{3.113}$$

Substituting the stress-strain relations, equation (3.108), and recognizing the definition of the volume fractions, we find equation (3.113) becomes

$$\gamma_{12} = \left(\frac{V^f}{G_{12}^f} + \frac{V^m}{G^m} \right) \tau_{12} \tag{3.114}$$

By analogy,

$$\gamma_{12} = \frac{\tau_{12}}{G_{12}} \tag{3.115}$$

G_{12} being the axial shear modulus of the composite, and from equation (3.114),

$$\frac{1}{G_{12}} = \frac{V^f}{G_{12}^f} + \frac{V^m}{G^m} \tag{3.116}$$

or

$$\frac{1}{G_{12}} = \frac{V^f}{G_{12}^f} + \frac{1 - V^f}{G^m} \tag{3.117}$$

This is the rule-of-mixtures expression for G_{12}. Figure 3.47 illustrates the relationship between G_{12} and fiber volume fraction for the rule-of-mixtures model; the concentric cylinder model elasticity solution prediction, equation (3.60), is shown for comparison. The comparison is not good, as the rule-of-mixtures model generally underpredicts the shear modulus.

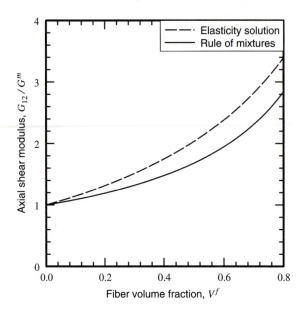

FIGURE 3.47
Rule-of-mixtures prediction for variation of composite axial shear modulus G_{12} with fiber volume fraction for graphite-reinforced material

As with the transverse modulus, E_2, we can modify the partitioning assumption for the shear stress in the fiber and matrix and develop a modified rule-of-mixtures model for G_{12}, resulting in

$$\frac{1}{G_{12}} = \frac{\dfrac{V^f}{G^f_{12}} + \dfrac{\eta'(1 - V^f)}{G^m}}{V^f + \eta'(1 - V^f)} \tag{3.118}$$

where η' is the partitioning factor for the shear stresses. Figure 3.48 shows the variation of G_{12} as predicted by the modified rule-of-mixtures model with $\eta' = 0.6$, along with the elasticity solution predictions; the value of $\eta' = 0.6$ leads to good correlation with the elasticity solution.

3.4.4 Model for α_1 and α_2

The rule-of-mixtures model for the coefficients of thermal expansion in the fiber and transverse directions α_1 and α_2 is similar to the alternative rule-of-mixtures model for the transverse modulus E_2 in that interaction between the fiber and matrix elements must be accounted for. Referring again to Figure 3.44 and considering spatially uniform thermal expansion effects, we find the stress-strain relations for the fiber element are, by analogy with equation (3.97a),

$$\sigma_1^f = Q_{11}^f(\varepsilon_1^f - \alpha_1^f \Delta T) + Q_{12}^f(\varepsilon_2^f - \alpha_2^f \Delta T)$$
$$\sigma_2^f = Q_{12}^f(\varepsilon_1^f - \alpha_1^f \Delta T) + Q_{22}^f(\varepsilon_2^f - \alpha_2^f \Delta T) \tag{3.119a}$$

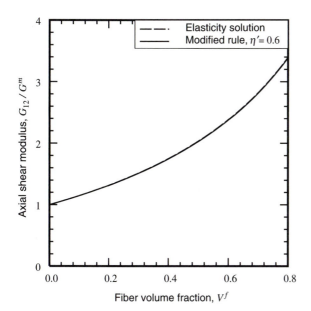

FIGURE 3.48
Modified rule-of-mixtures prediction for variation of composite axial shear modulus G_{12} with fiber volume fraction for graphite-reinforced material

while the stress-strain relations for the matrix element are

$$\sigma_1^m = Q_{11}^m (\varepsilon_1^m - \alpha^m \Delta T) + Q_{12}^m (\varepsilon_2^m - \alpha^m \Delta T)$$
$$\sigma_2^m = Q_{12}^m (\varepsilon_1^m - \alpha^m \Delta T) + Q_{22}^m (\varepsilon_2^m - \alpha^m \Delta T)$$
$$\tag{3.119b}$$

Equations (3.119a) and (3.119b) are derivable from equation (2.107) or (2.108). The relations of equation (3.98) remain valid, and the coefficients of thermal expansion of the fiber in the 1 and 2 directions and the coefficient of thermal expansion in the matrix are given by α_1^f, α_2^f, and α^m, respectively. The temperature change relative to some reference level is ΔT. Because fiber and matrix elements are bonded together and therefore must change length the same amount in the 1 direction,

$$\varepsilon_1^f = \varepsilon_1^m = \frac{\Delta L}{L} \tag{3.120}$$

As with past derivations

$$\varepsilon_2^f = \frac{\Delta W^f}{W^f} \quad \text{and} \quad \varepsilon_2^m = \frac{\Delta W^m}{W^m} \tag{3.121}$$

By equilibrium,

$$\sigma_2^f = \sigma_2^m = \sigma_2 \tag{3.122}$$

and for the case of imposing a temperature change with no force applied in the 2 direction

$$\sigma_2 = 0 \tag{3.123}$$

resulting in

$$\sigma_2^f = \sigma_2^m = 0 \tag{3.124}$$

Likewise, for no applied force in the 1 direction, as in equation (3.100),

$$\int_{A^m + A^f} \sigma_1 \, dA = 0 \tag{3.125}$$

or, again assuming the stress σ_1 is constant within each constituent, we find that

$$\sigma_1^f A^f + \sigma_1^m A^m = 0 \tag{3.126}$$

Thus, from equations (3.124) and (3.126) and the stress-strain relations (3.119a) and (3.119b),

$$Q_{12}^f \frac{\Delta L}{L} + Q_{22}^f \frac{\Delta W^f}{W^f} = \bar{\sigma}_2^f \Delta T$$

$$Q_{12}^m \frac{\Delta L}{L} + Q_{22}^m \frac{\Delta W^m}{W^m} = \bar{\sigma}_2^m \Delta T \tag{3.127}$$

$$(Q_{11}^f A^f + Q_{11}^m A^m) \frac{\Delta L}{L} + Q_{12}^f A^f \frac{\Delta W^f}{W^f} + Q_{12}^m A^m \frac{\Delta W^m}{W^m} = \bar{\sigma}_1^e \Delta T$$

where

$$\bar{\sigma}_2^f = Q_{12}^f \alpha_1^f + Q_{22}^f \alpha_2^f$$

$$\bar{\sigma}_2^m = Q_{12}^m \alpha^m + Q_{22}^m \alpha^m \tag{3.128}$$

$$\bar{\sigma}_1^e = (Q_{11}^f \alpha_1^f + Q_{12}^f \alpha_2^f) A^f + (Q_{11}^m \alpha^m + Q_{12}^m \alpha^m) A^m$$

Solving this set of equations leads to expressions for $\Delta W^f / W^f$, $\Delta W^m / W^m$, and $\Delta L / L$. By definition, for no applied loads, the strain in the 1 direction per unit temperature change is the coefficient of thermal expansion in the 1 direction; that is,

$$\alpha_1 = \left(\frac{\Delta L}{L} \right) \Big/ \Delta T \tag{3.129}$$

Substituting the solution for $\Delta L / L$ from equation (3.127) into equation (3.129) leads to a relation between α_1 and volume fraction, namely,

$$\alpha_1 = \frac{\alpha_1^f E_1^f V^f + \alpha^m E^m V^m}{E_1^f V^f + E^m V^m} \tag{3.130}$$

or

$$\alpha_1 = \frac{\left(\alpha_1^f E_1^f - \alpha^m E^m \right) V^f + \alpha^m E^m}{\left(E_1^f - E^m \right) V^f + E^m} \tag{3.131}$$

The total change in width of the unit cell due to a temperature change is

$$\Delta W = \Delta W^f + \Delta W^m \tag{3.132}$$

Dividing by the total width of the unit cell, and using a slight rearrangement of terms, we find

$$\frac{\Delta W}{W} = \frac{\Delta W^f}{W^f} \frac{W^f}{W} + \frac{\Delta W^m}{W^m} \frac{W^m}{W} \tag{3.133}$$

Recognizing the definitions of volume fraction, we find this equation becomes

$$\frac{\Delta W}{W} = \frac{\Delta W^f}{W^f} V^f + \frac{\Delta W^m}{W^m} V^m \tag{3.134}$$

Again by definition, the coefficient of thermal expansion in the 2 direction is, for the case of no applied loads, the strain in the 2 direction per unit temperature change; that is,

$$\alpha_2 = \left(\frac{\Delta W}{W}\right) \bigg/ \Delta T \tag{3.135}$$

or
$$\alpha_2 = \left(\frac{\Delta W^f}{W^f} V^f + \frac{\Delta W^m}{W^m} V^m\right) \bigg/ \Delta T \tag{3.136}$$

Substituting the solutions for $\Delta W^f / W^f$ and $\Delta W^m / W^m$ from solving equation (3.127) into equation (3.136) provides a relation for α_2 as a function of volume fraction, specifically,

$$\alpha_2 = \left[\alpha_2^f - \left(\frac{E^m}{E_1}\right) v_1^f \left(\alpha^m - \alpha_1^f\right) V^m\right] V^f$$
$$+ \left[\alpha^m + \left(\frac{E_1^f}{E_1}\right) v^m \left(\alpha^m - \alpha_1^f\right) V^f\right] V^m \tag{3.137}$$

where the rule-of-mixtures composite modulus E_1 has been used to simplify the expression for α_2. Note that a simple rule-of-mixtures model for α_2 would lead to

$$\alpha_2 = \alpha_2^f V^f + \alpha^m V^m \tag{3.138}$$

This simple model would not enforce the fact that the change in length in the 1 direction of the fiber and matrix elements must be the same. The additional terms in equation (3.137) relative to equation (3.138) reflect the effect of this geometric constraint. Equation (3.138) can be written as

$$\alpha_2 = \alpha^m + \left(\alpha_2^f - \alpha^m\right) V^f \tag{3.139}$$

while equation (3.137) can be written as

$$\alpha_2 = \alpha^m + \left(\alpha_2^f - \alpha^m\right) V^f$$
$$+ \left(\frac{E_1^f v^m - E^m v_1^f}{E_1}\right) \left(\alpha^m - \alpha_1^f\right) \left(1 - V^f\right) V^f \tag{3.140}$$

Both equation (3.139) and equation (3.140) express α_2 as a function of fiber volume fraction. It should be noted that a simple rule-of-mixtures expression for α_1 does not make sense, as it is clear the matrix and fiber must expand the same amount in the 1 direction when the temperature is changed. The simple rule-of-mixtures model in the fiber direction, as has been seen, ignores this fact.

   The relation between the fiber volume fraction and the coefficient of thermal expansion α_1, equation (3.131), is shown in Figure 3.49. The results given by equation (3.131) agree well with the finite-element predictions and hence the equation can be considered useful for determining α_1. Note that for fiber volume fractions greater than 0.6, α_1 is predicted to be negative, reflecting the dominance of the negative value of the fiber expansion in the 1 direction, α_1^f.

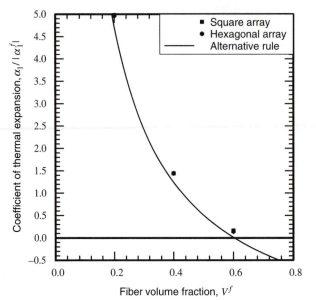

FIGURE 3.49
Rule-of-mixtures prediction for variation of composite coefficient of thermal expansion α_1 with fiber volume fraction for graphite-reinforced material

The predictions for the coefficient of thermal expansion α_2 given by equation (3.140) and the simpler rule-of-mixtures equation (3.139) are shown in Figure 3.50. The more complex relation, equation (3.140), which reflects interaction between the fiber and matrix, more closely matches the predictions of the square and hexagonal arrays, indicating that the constraint effects represented by the more complicated alternative equation are important.

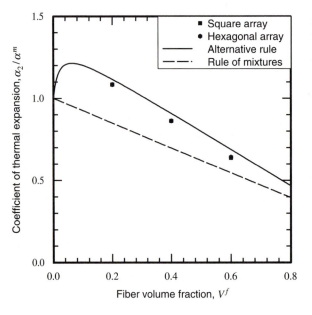

FIGURE 3.50
Rule-of-mixtures prediction for variation of composite coefficient of thermal expansion α_2 with fiber volume fraction for graphite-reinforced material

Exercises for Section 3.4

1. Starting with either equation (2.107) or equation (2.108), develop the expressions given by equation (3.119a) and its isotropic counterpart, equation (3.119b).

2. Derive equation (3.138), the rule-of-mixtures equation for the coefficient of thermal expansion α_2.

3. Derive the rule-of-mixtures expression for the composite extensional modulus E_1 assuming the existence of an interphase region. The starting point for the derivation would be the model shown below. For simplicity, assume the interphase, like the matrix, is isotropic with modulus E^i. With an interphase region there is a volume fraction associated with the interphase (i.e., V^i). For this situation,

$$V^f + V^m + V^i = 1$$

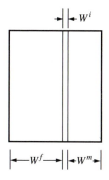

3.5
SUMMARY

This completes the discussion of micromechanics, an involved, important, and interesting view of fiber-reinforced composite materials, a view that can require a great deal of specialization. Several approaches have been presented for studying micromechanics, the approaches depending to a large degree on the information being sought. Most important for the remainder of this book is the fact that this short study of micromechanics has provided estimates of the elastic and thermal expansion properties that are needed to study the response of composite laminates. To that end, accounting for estimates from the finite-element models, the rule-of-mixture models, modified rule-of-mixture models, and the alternative models, the material properties for graphite-reinforced composites will be taken to be those given in Table 2.1. These properties can be considered to correspond to 65 percent fiber volume fraction. Properties for glass-reinforced polymer are taken from similar calculations for that material system. The important point to be aware of is that the material properties used throughout the book are consistent, starting at the micromechanics level with fiber and matrix properties, and proceeding to the laminate properties to be studied in coming chapters.

3.6
SUGGESTED READINGS

To learn more about finite-element technique, the following books can be consulted:

1. Reddy, J. N. *An Introduction to the Finite Element Method.* 2nd ed. New York: McGraw-Hill, 1994.
2. Cook, R. D. *Concepts and Application of Finite Element Analysis.* New York: John Wiley & Sons, 1990.
3. Bathe, K. J. *Finite Element Procedures in Engineering Analysis.* Englewood Cliffs, NJ: Prentice Hall, 1982.

Some of the first papers discussing the use of numerical methods to study micromechanics problems are:

4. Adams, D. F., and D. R. Doner. "Longitudinal Shear Loading of a Unidirectional Composite." *Journal of Composite Materials* 1 (1967), pp. 4–17.
5. Adams, D. F., and D. R. Doner. "Transverse Normal Loading of a Unidirectional Composite." *Journal of Composite Materials* 1 (1967), pp. 152–64.

A recent numerically based paper is:

6. Bigelow, C. A. "The Effects of Uneven Fiber Spacing on Thermal Residual Stresses in a Unidirectional SCS-6/Ti-15-3 Laminate." *Journal of Composites Technology and Research* 14, no. 1 (1992), pp. 211–20.

A paper that compares finite-element and strength-of-materials models is:

7. Caruso, J. J., and C. C. Chamis. "An Assessment of Simplified Composite Micromechanics Using Three-Dimensional Finite Element Analysis." *Journal of Composites Technology and Research* 8 (1986), pp. 77–83.

For general reference to the theory of elasticity, consult:

8. Boresi, A. P., and P. P. Lynn. *Elasticity in Engineering Mechanics.* Englewood Cliffs, NJ: Prentice Hall, 1974.
9. Reismann, H., and P. S. Pawlik. *Elasticity: Theory and Applications.* New York: John Wiley & Sons, 1980.
10. Sokolnikoff, I. S. *Mathematical Theory of Elasticity.* 2nd ed. New York: McGraw-Hill, 1956.
11. Timoshenko, S. P., and J. N. Goodier. *Theory of Elasticity.* 3rd ed. New York: McGraw-Hill, 1970.

The definitive papers on the concentric cylinder elasticity model are:

12. Hashin, Z., and B. W. Rosen. "The Elastic Moduli of Fiber-Reinforced Materials." *Transactions of the ASME, Journal of Applied Mechanics* 31 (1964), pp. 223–32, errata vol. 32 (1965), p. 219.
13. Hashin, Z. "Analysis of Properties of Fiber Composites with Anisotropic Constituents." *Transactions of the ASME, Journal of Applied Mechanics* 46 (1979), pp. 543–50.

Details of the axial shear model for G_{12} are given in:

14. Knott, T. W., and C. T. Herakovich. "Effect of Fiber Orthotrophy on Effective Composite Properties." *Journal of Composite Materials* 25, no. 6 (1991), pp. 732–59.

Another approach to using cylinder models is provided in:

15. Whitney, J. M., and M. B. Riley. "Elastic Properties of Fiber Reinforced Composite Materials." *AIAA Journal* 4, no. 9 (1966), pp. 1537–1542.

Use of the cylinders model and consideration of interaction between fibers can be found in:

16. Mal, A. K., and A. K. Chatterjee. "The Elastic Moduli of Fiber Reinforced Composites." *Transactions of the ASME, Journal of Applied Mechanics* 44, no. 1 (1977), pp. 61–67. (See discussion by J. J. McCoy in vol. 44, no. 2.)

A more complicated elasticity-based micromechanical model can be found in:

17. Jayaraman, K., Z. Gao, and K. L. Reifsnider. "The Interphase in Unidirectional Fiber-Reinforced Epoxies: Effect of Local Stress Fields." *Journal of Composites Technology and Research* 16, no. 1 (1994), pp. 21–31.

A review of a generalization of the earlier elasticity approaches is discussed in:

18. Aboudi, J. "Micromechanical Analysis of Composites by the Method of Cells." *Applied Mechanics Reviews* 42, no. 7 (1989), pp. 193–221.

The use of simplified models for micromechanics analysis is discussed in:

19. Chamis, C. C. "Simplified Composite Micromechanics Equations for Hygral, Thermal, and Mechanical Properties." *SAMPE Quarterly* 15, no. 3 (1984), pp. 14–23.

The concept of using a factor in the strength-of-materials models to account for difficult-to-model interactions between the fiber and matrix was introduced in:

20. Halpin, J. C. "Effect of Environmental Factors on Composite Materials." AFML-TR-67-423, 1969.

The concepts are further discussed in:

21. Halpin, J. C. *Primer on Composite Materials*. 2nd ed. Lancaster, PA: Technomic Publishing Co., Inc., 1992.
22. Tsai, S. W., and H. T. Hahn. *Introduction to Composite Materials*. Lancaster, PA: Technomic Publishing Co., Inc., 1980.
23. Hahn, H. T. "Simplified Formulas for Elastic Modulus of Unidirectional Fiber Composites." *Composite Technology Review* 2, no. 3 (1980), pp. 5–7.

The following paper was the first to address free thermal expansion coefficients, though no. 19, by Chamis, considers thermal effects also:

24. Schapery, R. A. "Thermal Expansion Coefficients of Composite Materials Based on Energy Principles." *Journal of Composite Materials* 2 (1968), pp. 280–404.

Several good overall books on micromechanics are:

25. Christensen, R. M. *Mechanics of Composite Materials*. New York: John Wiley & Sons, 1979.
26. Chou, T.-W. *Microstructural Design of Fiber Composites*. New York: Cambridge University Press, 1992.
27. Aboudi, J. *Mechanics of Composite Materials: A Unified Micromechanical Approach*. New York: Elsevier, 1991.

CHAPTER 4

The Plane-Stress Assumption

Historically, one of the most important assumptions regarding the study of the mechanics of fiber-reinforced materials is that the properties of the fibers and the properties of the matrix can be smeared into an equivalent homogeneous material with orthotropic material properties. In Chapter 2 this assumption helped us develop the stress-strain relations and learn something about the response of fiber-reinforced material that would allow us to move to structural-level response in a tractable manner. Without this assumption, we would still have to deal with the response of the individual fibers embedded in matrix material, as was done in the study of micromechanics in the last chapter. If this assumption had not been made in the development of the mechanics of fiber-reinforced materials, very little progress would have been made in understanding their response. An equally important assumption in the development of the mechanics of fiber-reinforced materials is the *plane-stress assumption*, which is based on the manner in which fiber-reinforced composite materials are used in many structures. Specifically, fiber-reinforced materials are utilized in beams, plates, cylinders, and other structural shapes which have at least one characteristic geometric dimension an order of magnitude less than the other two dimensions. In these applications, three of the six components of stress are generally much smaller than the other three. With a plate, for example, the stresses in the plane of the plate are much larger than the stresses perpendicular to that plane. In all calculations, then, the stress components perpendicular to the plane of the structure can be set to zero, greatly simplifying the solution of many problems. In the context of fiber-reinforced plates, for example (see Figure 2.2), the stress components σ_3, τ_{23}, and τ_{13} are set to zero with the assumption that the 1-2 plane of the principal material coordinate system is in the plane of the plate. Stress components σ_1, σ_2, and τ_{12} are considered to be much larger in magnitude than components σ_3, τ_{23}, and τ_{13}. In fact, σ_1 should be the largest of all the stress components if the fibers are being utilized effectively. We use the term *plane stress* because σ_1, σ_2, and τ_{12} lie in a plane, and stresses σ_3,

τ_{23}, and τ_{13} are perpendicular to this plane and are zero. The small element of Figure 2.2 appears in Figure 4.1 under the assumption of plane stress.

The plane-stress assumption can lead to inaccuracies, some serious and some not so serious. The most serious inaccuracy occurs in the analysis of a laminate near its edge. Laminates tend to come apart in the thickness direction, or *delaminate*, at their edges, much like common plywood. An understanding of this phenomenon, illustrated in Figure 4.2, requires that all six components of stress be included in the analysis. It is exactly the stresses that are set to zero in the plane-stress assumption (i.e., σ_3, τ_{23}, and τ_{13}) that are responsible for delamination, so an analysis that ignores these stresses cannot possibly be correct for a delamination study. Delaminations can also occur away from a free edge, with the layers separating in blister fashion. These are generally caused by the presence of imperfections between the layers. The out-of-plane stress components σ_3, τ_{23}, and τ_{13} are also important in locations where structures or components of structures are joined together; Figures 4.3 and 4.4 illustrate some examples. Figure 4.3 shows a bonded joint consisting of two laminates subjected to tensile load P. For the load to be transferred from one laminate to the other, significant out-of-plane stresses, particularly shear, must develop in the laminates around the interface, as well as at the interface itself. As another example, in many situations stiffeners are used to increase the load capacity of plates, as in Figure 4.4. For the plate-stiffener combination to be effective, the plate must transfer

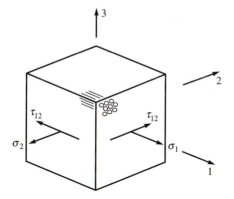

FIGURE 4.1
Stresses acting on a small element of fiber-reinforced material in a state of plane stress

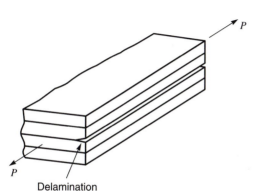

FIGURE 4.2
Example of region of high out-of-plane stresses: Delamination at a free edge

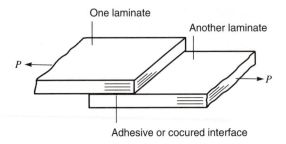

One laminate

Another laminate

P

P

Adhesive or cocured interface

FIGURE 4.3
Example of region of high out-of-plane stresses: Bonded joint

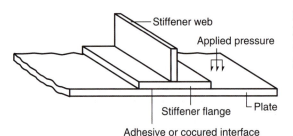

Stiffener web

Applied pressure

Stiffener flange   Plate

Adhesive or cocured interface

FIGURE 4.4
Example of region of high out-of-plane stresses: Stiffened plate

some of the pressure load to the stiffener. Thickness direction stresses must develop in the plate and stiffener flange if load is to be transferred through the interface. In general, all three components of out-of-plane stress, σ_3, τ_{23}, and τ_{13}, develop in this situation. Away from the stiffener the plate may be in a state of plane stress, so not only is there a region of the plate characterized by a fully three-dimensional stress state, there is also a transition region. In this transition region the conditions go from truly plane stress to a fully three-dimensional stress state, making the analysis of such a problem difficult and challenging.

Figure 4.5 illustrates another area where through-the-thickness stresses are important. Often it is necessary, or desirable, to change the thickness of a laminate by gradually terminating some of the layers. Away from the terminated layer region each portion of the laminate could well be in a state of plane stress due to the applied inplane load P. However, the thicker region is in a different state of plane stress than the thinner region. To make the transition between the two stress states, three-dimensional effects occur.

The illustrations in Figures 4.2–4.5 are prime examples of situations encountered in real composite structures. However, the plane-stress assumption is accurate in so many situations that one would be remiss in not taking advantage of its simplifications. The static, dynamic, and thermally induced deflections, and the stresses that result from these, vibration frequencies, buckling loads, and many other responses of

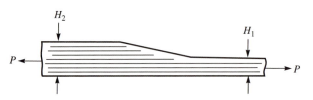

H_2

H_1

P

P

FIGURE 4.5
Example of region of high out-of-plane stresses: Region of terminated layers

composite structures can be accurately predicted using the plane-stress assumption. What is important to remember when applying the plane-stress assumption is that it assumes that three stresses are *small* relative to the other three stresses and they have therefore been set to zero. They do not necessarily have to be exactly zero, and in fact in many cases they are not exactly zero. With the aid of the three-dimensional equilibrium equations of the theory of elasticity, calculations based on the plane-stress assumption can be used to predict the stress components that have been equated to zero. When these results are compared with predictions of the out-of-plane components based on rigorous analyses wherein the out-of-plane components are not assumed to be zero at the outset, we find that in many cases the comparisons are excellent. Thus, a plane-stress, or, using alternative terminology, a two-dimensional analysis, is useful. Two of the major pitfalls associated with using the plane-stress assumption are:

1. The stress components equated to zero are often forgotten and no attempt is made to estimate their magnitude.
2. It is often erroneously assumed that because the stress component σ_3 is zero and therefore ignorable, the associated strain ε_3 is also zero and ignorable.

Regarding the former point, while certain stress components may indeed be small, the material may be very weak in resisting these stresses. As was stated earlier, a fiber-reinforced material is poor in resisting all stresses except stresses in the fiber direction. Thus, several stress components may be small and so the problem conforms to the plane-stress assumption. However, the out-of-plane stresses may be large enough to cause failure of the material and therefore they should not be completely ignored. Often they are. Regarding the second point, the stresses in the 1-2 plane of the principal material coordinate system can cause a significant strain response in the 3 direction. The assumption that ε_3 is zero just because σ_3 is negligible is wrong and, as we shall see shortly, defies the the stress-strain relations that govern material behavior. It is important to keep these two points in mind as we focus our discussion in the following chapters on the plane-stress condition.

4.1
STRESS-STRAIN RELATIONS FOR PLANE STRESS

To see why the plane-stress assumption is important, it is only necessary to see how it simplifies the stress-strain relations. Specifically, for the plane-stress assumption σ_3, τ_{23}, and τ_{13} are set to zero in equations (2.45) and (2.47). Looking at equation (2.45) first, we find

$$
\begin{Bmatrix} \varepsilon_1 \\ \varepsilon_2 \\ \varepsilon_3 \\ \gamma_{23} \\ \gamma_{13} \\ \gamma_{12} \end{Bmatrix} = \begin{bmatrix} S_{11} & S_{12} & S_{13} & 0 & 0 & 0 \\ S_{12} & S_{22} & S_{23} & 0 & 0 & 0 \\ S_{13} & S_{23} & S_{33} & 0 & 0 & 0 \\ 0 & 0 & 0 & S_{44} & 0 & 0 \\ 0 & 0 & 0 & 0 & S_{55} & 0 \\ 0 & 0 & 0 & 0 & 0 & S_{66} \end{bmatrix} \begin{Bmatrix} \sigma_1 \\ \sigma_2 \\ 0 \\ 0 \\ 0 \\ \tau_{12} \end{Bmatrix} \qquad (4.1)
$$

From this relation it is obvious that

$$\gamma_{23} = 0 \qquad \gamma_{13} = 0 \tag{4.2}$$

so with the plane-stress assumption there can be no shear strains whatsoever in the 2-3 and 1-3 planes. That is an important ramification of the assumption. Also,

$$\boxed{\varepsilon_3 = S_{13}\sigma_1 + S_{23}\sigma_2} \tag{4.3}$$

This equation indicates explicitly that for a state of plane stress there is an extensional strain in the 3 direction. To assume that strain ε_3 is zero is *absolutely wrong*. That it is not zero is a direct result of Poisson's ratios ν_{13} and ν_{23} acting through S_{13} and S_{23}, respectively, coupling with the nonzero stress components σ_1 and σ_2. The above equation for ε_3 forms the basis for determining the thickness change of laminates subjected to inplane loads, and for computing through-thickness, or out-of-plane, Poisson's ratios of a laminate.

Despite the fact that ε_3 is not zero, the plane-stress assumption leads to a relation involving only ε_1, ε_2, γ_{12} and σ_1, σ_2, τ_{12}. By eliminating the third, fourth, and fifth equations of equation 4.1, we find

$$\left\{ \begin{array}{c} \varepsilon_1 \\ \varepsilon_2 \\ \gamma_{12} \end{array} \right\} = \left[\begin{array}{ccc} S_{11} & S_{12} & 0 \\ S_{12} & S_{22} & 0 \\ 0 & 0 & S_{66} \end{array} \right] \left\{ \begin{array}{c} \sigma_1 \\ \sigma_2 \\ \tau_{12} \end{array} \right\} \tag{4.4}$$

The definitions of the compliances have not changed from the time they were first introduced, namely,

$$\begin{array}{cc} S_{11} = \dfrac{1}{E_1} & S_{12} = \dfrac{-\nu_{12}}{E_1} = \dfrac{-\nu_{21}}{E_2} \\[3mm] S_{22} = \dfrac{1}{E_2} & S_{66} = \dfrac{1}{G_{12}} \end{array} \tag{4.5}$$

The 3×3 matrix of compliances is called the *reduced compliance matrix*. In matrix notation the lower right hand element of a 3×3 matrix is usually given the subscript 33, though in the analysis of composites it has become conventional to retain the subscript convention from the three-dimensional formulation and maintain the subscript of the lower corner element as 66. For an isotropic material, equation (4.5) reduces to

$$S_{11} = S_{22} = \frac{1}{E} \qquad S_{12} = -\frac{\nu}{E} \qquad S_{66} = \frac{1}{G} = \frac{2(1+\nu)}{E} \tag{4.6}$$

If the plane-stress assumption is used to simplify the inverse form of the stress-strain relation, equation (2.47), the result is

$$
\begin{Bmatrix} \sigma_1 \\ \sigma_2 \\ 0 \\ 0 \\ 0 \\ \tau_{12} \end{Bmatrix} = \begin{bmatrix} C_{11} & C_{12} & C_{13} & 0 & 0 & 0 \\ C_{12} & C_{22} & C_{23} & 0 & 0 & 0 \\ C_{13} & C_{23} & C_{33} & 0 & 0 & 0 \\ 0 & 0 & 0 & C_{44} & 0 & 0 \\ 0 & 0 & 0 & 0 & C_{55} & 0 \\ 0 & 0 & 0 & 0 & 0 & C_{66} \end{bmatrix} \begin{Bmatrix} \varepsilon_1 \\ \varepsilon_2 \\ \varepsilon_3 \\ \gamma_{23} \\ \gamma_{13} \\ \gamma_{12} \end{Bmatrix} \tag{4.7}
$$

With the above, one also concludes that

$$
\gamma_{23} = 0 \qquad \gamma_{13} = 0 \tag{4.8}
$$

In analogy to equation (4.3), the third equation of equation (4.7) yields

$$
0 = C_{13}\varepsilon_1 + C_{23}\varepsilon_2 + C_{33}\varepsilon_3 \tag{4.9}
$$

Rearranged, it becomes

$$
\boxed{\varepsilon_3 = -\frac{C_{13}}{C_{33}}\varepsilon_1 - \frac{C_{23}}{C_{33}}\varepsilon_2} \tag{4.10}
$$

This relationship also indicates that in this state of plane stress ε_3 exists and equation (4.10) indicates it can be computed by knowing ε_1 and ε_2.

The three-dimensional form equation (4.7) cannot be reduced directly to obtain a relation involving only σ_1, σ_2, τ_{12}, and ε_1, ε_2, γ_{12} by simply eliminating equations, as was done with equation (4.1) to obtain equation (4.4). However, equation (4.10) can be used as follows: From equation (4.7), the expressions for σ_1 and σ_2 are

$$
\begin{aligned}
\sigma_1 &= C_{11}\varepsilon_1 + C_{12}\varepsilon_2 + C_{13}\varepsilon_3 \\
\sigma_2 &= C_{12}\varepsilon_1 + C_{22}\varepsilon_2 + C_{23}\varepsilon_3
\end{aligned} \tag{4.11}
$$

Substituting for ε_3 using equation (4.10) leads to

$$
\begin{aligned}
\sigma_1 &= C_{11}\varepsilon_1 + C_{12}\varepsilon_2 + C_{13}\left(-\frac{C_{13}}{C_{33}}\varepsilon_1 - \frac{C_{23}}{C_{33}}\varepsilon_2\right) \\
\sigma_2 &= C_{12}\varepsilon_1 + C_{22}\varepsilon_2 + C_{23}\left(-\frac{C_{13}}{C_{33}}\varepsilon_1 - \frac{C_{23}}{C_{33}}\varepsilon_2\right)
\end{aligned} \tag{4.12}
$$

or

$$
\begin{aligned}
\sigma_1 &= \left(C_{11} - \frac{C_{13}^2}{C_{33}}\right)\varepsilon_1 + \left(C_{12} - \frac{C_{13}C_{23}}{C_{33}}\right)\varepsilon_2 \\
\sigma_2 &= \left(C_{12} - \frac{C_{13}C_{23}}{C_{33}}\right)\varepsilon_1 + \left(C_{22} - \frac{C_{23}^2}{C_{33}}\right)\varepsilon_2
\end{aligned} \tag{4.13}
$$

Including the shear stress–shear strain relation, the relation between stresses and strains for the state of plane stress is written as

$$\left\{\begin{array}{c} \sigma_1 \\ \sigma_2 \\ \tau_{12} \end{array}\right\} = \left[\begin{array}{ccc} Q_{11} & Q_{12} & 0 \\ Q_{12} & Q_{22} & 0 \\ 0 & 0 & Q_{66} \end{array}\right] \left\{\begin{array}{c} \varepsilon_1 \\ \varepsilon_2 \\ \gamma_{12} \end{array}\right\} \tag{4.14}$$

The Q_{ij} are called the *reduced stiffnesses* and from equations (4.13) and (4.7)

$$Q_{11} = C_{11} - \frac{C_{13}^2}{C_{33}}$$

$$Q_{12} = C_{12} - \frac{C_{13}C_{23}}{C_{33}} \tag{4.15}$$

$$Q_{22} = C_{22} - \frac{C_{23}^2}{C_{33}}$$

$$Q_{66} = C_{66}$$

The term *reduced* is used in relations given by equations (4.4) and (4.14) because they are the result of reducing the problem from a fully three-dimensional to a two-dimensional, or plane-stress, problem. However, the numerical values of the stiffnesses Q_{11}, Q_{12}, and Q_{22} are actually less than the numerical values of their respective counterparts for a fully three-dimensional problem, namely, C_{11}, C_{12}, and C_{33}, and so the stiffnesses are reduced in that sense also.

It is very important to note that there is not really a numerically reduced compliance matrix. The elements in the plane-stress compliance matrix, equation (4.5), are simply a subset of the elements from the three-dimensional compliance matrix, equation (4.1), and their numerical values are the identical. On the other hand, the elements of the reduced stiffness matrix, equation (4.15), involve a combination of elements from the three-dimensional stiffness matrix. *It is absolutely wrong to write*

$$\left\{\begin{array}{c} \sigma_1 \\ \sigma_2 \\ \tau_{12} \end{array}\right\} = \left[\begin{array}{ccc} C_{11} & C_{12} & 0 \\ C_{12} & C_{22} & 0 \\ 0 & 0 & C_{66} \end{array}\right] \left\{\begin{array}{c} \varepsilon_1 \\ \varepsilon_2 \\ \gamma_{12} \end{array}\right\}$$

and claim this represents the reduced stiffness matrix. It simply is not so.

By inverting equation (4.4) and comparing it to equation (4.14), it is clear that

$$Q_{11} = \frac{S_{22}}{S_{11}S_{22} - S_{12}^2} \qquad Q_{12} = -\frac{S_{12}}{S_{11}S_{22} - S_{12}^2}$$

$$Q_{22} = \frac{S_{11}}{S_{11}S_{22} - S_{12}^2} \qquad Q_{66} = \frac{1}{S_{66}} \tag{4.16}$$

This provides a relationship between elements of the reduced compliance matrix and elements of the reduced stiffness matrix. A much more convenient form, and one that should be used in lieu of equation (4.16), can be obtained by simply writing the compliance components in equation (4.16) in terms of the appropriate engineering constants, namely,

$$
\begin{aligned}
Q_{11} &= \frac{E_1}{1 - \nu_{12}\nu_{21}} \\[2mm]
Q_{12} &= \frac{\nu_{12}E_2}{1 - \nu_{12}\nu_{21}} = \frac{\nu_{21}E_1}{1 - \nu_{12}\nu_{21}} \\[2mm]
Q_{22} &= \frac{E_2}{1 - \nu_{12}\nu_{21}} \\[2mm]
Q_{66} &= G_{12}
\end{aligned}
\tag{4.17}
$$

This form will be used exclusively from now on. For an isotropic material the reduced stiffnesses become

$$
Q_{11} = Q_{22} = \frac{E}{1 - \nu^2} \qquad Q_{12} = \frac{\nu E}{1 - \nu^2} \qquad Q_{66} = G = \frac{E}{2(1 + \nu)} \tag{4.18}
$$

4.2
IMPORTANT INTERPRETATION OF STRESS-STRAIN RELATIONS REVISITED

When discussing general stress states in Chapter 2, we strongly emphasized that only one of the quantities in each of the six stress-strain pairs σ_1-ε_1, σ_2-ε_2, σ_3-ε_3, τ_{23}-γ_{23}, τ_{12}-γ_{13}, and τ_{12}-γ_{12} could be specified. With the condition of plane stress, this restriction also holds. For the state of plane stress we assume that σ_3, τ_{23}, and τ_{13} are zero. We can say nothing a priori regarding ε_3, γ_{23}, and γ_{13}. However, by using the stress-strain relations, we found, equation (4.2), that γ_{23} and γ_{13} are indeed zero. This is a consequence of the plane-stress condition, not a stipulation. The strain ε_3 is given by equation (4.3), another consequence of the plane-stress condition. Of the three remaining stress-strain pairs, σ_1-ε_1, σ_2-ε_2, and τ_{12}-γ_{12}, only one quantity in each of these pairs can be specified. The other quantity must be determined, as usual, by using the stress-strain relations, either equation (4.4) or (4.14), and the details of the specific problem being solved.

4.3
NUMERICAL RESULTS FOR THE PLANE-STRESS CONDITION

Some of the numerical examples discussed in Section 2.5 are problems which satisfy plane-stress conditions. Specifically, in Figure 2.11(b) the cube of material is subjected to only one stress, as indicated by equation (2.66). Clearly the conditions

$$
\sigma_3 = \tau_{23} = \tau_{13} = 0 \tag{4.19}
$$

are satisfied, so the cube is in a state of plane stress. Using the plane-stress stress-strain relation equation (4.4), we find the strains ε_1, ε_2, and γ_{12} can be determined directly as

$$\varepsilon_1 = S_{12}\sigma_2 \qquad \varepsilon_2 = S_{22}\sigma_2 \qquad \gamma_{12} = 0 \qquad (4.20)$$

The strain ε_3 must now be determined from the condition of equation (4.3), now not a direct part of the stress-strain relations, but rather an auxilary equation. From equation (4.3)

$$\varepsilon_3 = S_{23}\sigma_2 \qquad (4.21)$$

By substituting numerical values for S_{12}, S_{22}, and S_{23} from equation (2.56), and using the definitions of equation (2.69), we find the dimensional changes of the cube, as in equation (2.70).

We can also find the dimensional change of the cube of Figure 2.11(b) from the plane-stress stress-strain relations using equation (4.14), though this approach is not as direct. In particular, for the stress state of Figure 2.11(b), equation (4.14) simplifies to

$$0 = Q_{11}\varepsilon_1 + Q_{12}\varepsilon_2$$
$$\sigma_2 = -50.0 \text{ MPa} = Q_{12}\varepsilon_1 + Q_{22}\varepsilon_2 \qquad (4.22)$$
$$0 = Q_{66}\gamma_{12}$$

From the first and third of these,

$$\varepsilon_1 = -\frac{Q_{12}}{Q_{11}}\varepsilon_2 \quad \text{and} \quad \gamma_{12} = 0 \qquad (4.23)$$

and substituting into the second yields

$$\varepsilon_2 = \left(\frac{Q_{11}}{Q_{11}Q_{22} - Q_{12}^2}\right)\sigma_2 \qquad (4.24)$$

If equation (4.17) is used,

$$\varepsilon_2 = \frac{1}{E_2}\sigma_2 = S_{22}\sigma_2 \qquad (4.25)$$

which is the second equation of equation (2.68). It seems that we have gone in circles—we have not! What we have shown is that the plane-stress stress-strain relations yield results identical to the results obtained by using the general stress-strain relations if the problem is one of plane stress. This is an important point. If the problem is one of plane stress, then using the simpler forms, equations (4.4) and (4.14), rather than the more complicated forms, equations (2.45) and (2.47), gives the correct answer. The auxilary conditions, either equation (4.3) or (4.10), can be used to obtain information about the strain ε_3 that is not a direct part of the plane-stress stress-strain relation. In that context, it is important to keep in mind that the out-of-plane engineering properties are still useful. For example, consider a layer of graphite-reinforced material 100 mm long, 50 mm wide, and 0.150 mm thick. As shown in Figure 4.6, this layer is subjected to a 3750 N inplane force in the fiber direction. The through-thickness strain in the layer can be calculated from equation

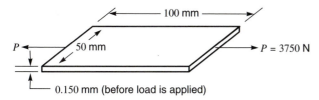

FIGURE 4.6
Layer subjected to inplane forces

(4.3) and the numerical values in equation (2.56) as

$$\varepsilon_3 = S_{13}\sigma_1 + S_{23}\sigma_2 = \left(-1.600 \times 10^{-12}\right)\left(\frac{3750}{(0.050) \times (0.000150)}\right) \qquad (4.26)$$

$$\varepsilon_3 = -800 \times 10^{-6} \text{ m/m} = -800 \times 10^{-6} \text{ mm/mm} = -800 \, \mu \text{ mm/mm}$$

Exercises for Section 4.3

1. Derive equation (4.17), starting with equation (4.16) and substituting the definitions from equation (4.5).

2. Compute the numerical values of the elements of the reduced stiffness matrix for graphite-reinforced material.

3. Suppose in Figure 4.6 the fibers are perpendicular to the 100 mm direction. What would be the through-thickness strain for this situation?

Computer Exercise

Write a computer program to read in and print out the engineering constants E_1, E_2, v_{12}, and G_{12}. Then have the program compute and print the values of the elements of the reduced stiffness matrix, equation (4.17). Use your answers to Exercise 2, above, to check your program. This program, though trivial, is the first step in what will be a series of steps leading to a program that computes the stresses and deformations in a laminate. As we move through this book, more and more will be added to this small program.

4.4
PLANE-STRESS STRESS-STRAIN RELATIONS AND THE EFFECTS OF FREE THERMAL AND FREE MOISTURE STRAINS

If a problem conforms to the plane-stress assumption, namely, $\sigma_3 = \tau_{23} = \tau_{13} = 0$, but free thermal moisture strains are important, then, starting with equation (2.124) and following the same steps that led to equation (4.4), we can conclude that

$$\gamma_{23} = 0 \qquad \gamma_{13} = 0 \qquad (4.27)$$

which is identical to the case of no thermal or moisture expansion effects, as in equation (4.2). This is a direct consequence of the conditions $\tau_{23} = \tau_{13} = 0$. The conclusion regarding ε_3 is not exactly the same as the case with no free thermal or moisture strain effects, and in fact the conclusion is much more far reaching. Specifically, using the condition that $\sigma_3 = 0$ in equation (2.124), we conclude that

$$\varepsilon_3 = \alpha_3 \Delta T + \beta_3 \Delta M + S_{13}\sigma_1 + S_{23}\sigma_2 \tag{4.28}$$

Equation (4.28) will be the basis for determining the through-thickness, or out-of-plane, thermal or moisture expansion effects of a laminate. In this case the through-thickness strain does not depend solely on the through-thickness expansion coefficients α_3 or β_3. The through-thickness strain involves the compliances S_{13} and S_{23}, which in turn involve the Poisson's ratios ν_{13} and ν_{23}, as well as the inplane extensional moduli E_1 and E_2. This coupling of the through-thickness strain with the inplane and out-of-plane elastic properties leads to important consequences. More will be said of this later.

Continuing with the development, we observe that the plane-stress stress-strain relations, including free thermal and moisture strain effects, become

$$\left\{ \begin{array}{c} \varepsilon_1 - \alpha_1 \Delta T - \beta_1 \Delta M \\ \varepsilon_2 - \alpha_2 \Delta T - \beta_2 \Delta M \\ \gamma_{12} \end{array} \right\} = \left[\begin{array}{ccc} S_{11} & S_{12} & 0 \\ S_{12} & S_{22} & 0 \\ 0 & 0 & S_{66} \end{array} \right] \left\{ \begin{array}{c} \sigma_1 \\ \sigma_2 \\ \tau_{12} \end{array} \right\} \tag{4.29}$$

and

$$\left\{ \begin{array}{c} \sigma_1 \\ \sigma_2 \\ \tau_{12} \end{array} \right\} = \left[\begin{array}{ccc} Q_{11} & Q_{12} & 0 \\ Q_{12} & Q_{22} & 0 \\ 0 & 0 & Q_{66} \end{array} \right] \left\{ \begin{array}{c} \varepsilon_1 - \alpha_1 \Delta T - \beta_1 \Delta M \\ \varepsilon_2 - \alpha_2 \Delta T - \beta_2 \Delta M \\ \gamma_{12} \end{array} \right\} \tag{4.30}$$

where the mechanical strains are given by

$$\left\{ \begin{array}{c} \varepsilon_1^{mech} \\ \varepsilon_2^{mech} \\ \gamma_{12}^{mech} \end{array} \right\} = \left\{ \begin{array}{c} \varepsilon_1 - \alpha_1 \Delta T - \beta_1 \Delta M \\ \varepsilon_2 - \alpha_2 \Delta T - \beta_2 \Delta M \\ \gamma_{12} \end{array} \right\} \tag{4.31}$$

As in the three-dimensional stress-strain relations, equations (2.124) and (2.125), $\varepsilon_1, \varepsilon_2, \gamma_{12}$ above are the total strains. The dimensional changes are therefore given by

$$\delta \Delta_1 = \varepsilon_1 \Delta_1 \delta \Delta_2 = \varepsilon_2 \Delta_2 \tag{4.32}$$

not

$$\delta \Delta_1 = (\varepsilon_1 - \alpha_1 \Delta T - \beta_1 \Delta M) \Delta_1 \quad \text{and} \quad \delta \Delta_2 = (\varepsilon_2 - \alpha_2 \Delta T - \beta_2 \Delta M) \Delta_2$$

Equally important, ε_3 in equation (4.28) is the total strain. This is quite obvious when one considers that the right-hand side of that equation explicitly includes free thermal strain effects, free moisture strain effects, and stress-related effects.

To continue with our examples of thermal deformations, consider a 50 mm × 50 mm × 50 mm element of material that is completely constrained in the 1 and 2 directions. To satisfy the plane-stress condition, the element cannot be constrained in the 3 direction. The stresses in the two constrained directions and the deformation in the 3 direction due to a temperature increase of 50°C are to be determined. Moisture absorption is not an issue ($\Delta M = 0$). For this case, because of the stated conditions,

$$\varepsilon_1 = \varepsilon_2 = \gamma_{12} = 0 \tag{4.33}$$

To determine the restraining stresses, equation (4.30) is used to give

$$\sigma_1 = -(Q_{11}\alpha_1 + Q_{12}\alpha_2)\,\Delta T$$

$$\sigma_2 = -(Q_{12}\alpha_1 + Q_{22}\alpha_2)\,\Delta T \tag{4.34}$$

$$\tau_{12} = Q_{66} \times 0$$

To determine the deformation in the 3 direction, equation (4.28) is used to yield

$$\varepsilon_3 = \alpha_3 \Delta T + S_{13}\sigma_1 + S_{23}\sigma_2$$

$$\varepsilon_3 = (\alpha_3 - S_{13}(Q_{11}\alpha_1 + Q_{12}\alpha_2) - S_{23}(Q_{12}\alpha_1 + Q_{22}\alpha_2))\,\Delta T \tag{4.35}$$

$$\varepsilon_3 = (\alpha_3 - (S_{13}Q_{11} + S_{23}Q_{12})\alpha_1 - (S_{13}Q_{12} + S_{23}Q_{22})\alpha_2)\,\Delta T$$

where the various terms in equation (4.35) are retained to show the interaction between two- and three-dimensional elastic properties, and the three coefficients of thermal expansion. Numerically, equations (4.34) and (4.35) give

$$\sigma_1 = -3.52\,\text{MPa}$$

$$\sigma_2 = -14.77\,\text{MPa}$$

$$\tau_{12} = 0 \tag{4.36}$$

$$\varepsilon_3 = 1780 \times 10^{-6}$$

These numbers should be contrasted with the same 50 mm cube constrained in all three directions, equation (2.118) in Chapter 2, where comparison of the two sets of numbers indicates the effect of relaxing the constraint in the 3 direction.

4.5
SUGGESTED READINGS

Analyses of the situations depicted in Figures 4.2–4.5 are discussed in the following. For delamination:

1. Herakovich, C.T. "Free Edge Effects in Laminated Composites." In *Handbook of Composites*: vol. 2, *Structures and Design*, C. T. Herakovich and Y. M. Tarnopol´skii. New York: Elsevier Science Publishing Co., 1989.
2. Armanios, E. A., ed. *Interlaminar Fracture of Composites*. Aedermannsdorf, Switzerland: Trans Tech. Publications, 1989.

For bonded joints, this paper is an often-cited reference:

3. Renton, W. J., and J. R. Vinson. "Analysis of Adhesively Bonded Joints between Panels of Composite Materials." *Transactions of the ASME, Journal of Applied Mechanics* 44, no. 1 (1977), pp. 101–6.

Inclusion of other important effects is discussed in:

4. Reddy, J. N., and S. Roy. "Nonlinear Analysis of Adhesively Bonded Joints." *International Journal of Nonlinear Mechanics* 23, no. 2 (1988), pp. 97–112.

For stiffeners, see:

5. Kassapoglou, C. "Calculation of Stresses at Skin-Stiffener Interfaces of Composite Stiffened Panels under Shear Loads." *International Journal of Solids and Structures* 30, no. 11 (1993), pp. 1491–1501.
6. Cohen, D., and M. W. Hyer. "Influence of Geometric Nonlinearities on Skin-Stiffener Interface Stresses." *AIAA Journal* 30, no. 4 (1992), pp. 1055–62.
7. Hyer, M. W., D. C. Loup, and J. H. Starnes, Jr. "Stiffener/Skin Interactions in Pressure-Loaded Composite Panels." *AIAA Journal* 28, no. 3 (1990), pp. 532–37.
8. Hyer, M. W., and D. Cohen. "Calculation of Stresses in Stiffened Composite Panels." *AIAA Journal* 26, no. 7 (1988), pp. 852–57.

For tapered laminates and regions of terminating plys, see:

9. Harrison, P. N., and E. R. Johnson. "A Mixed Variational Formulation for Interlaminar Stresses in Thickness-Tapered Composite Laminates." *International Journal of Solids and Structures* 33, no. 16 (1996), pp. 2377–99.
10. Fish, J. C., and S. W. Lee. "Delamination of Tapered Composite Structures." *Engineering Fracture Mechanics* 34, no. 1 (1989), pp. 43–54.
11. Wu, C. M. L., and J. P. H. Webber. "Analysis of Tapered (in Steps) Laminated Plates Under Uniform Inplane Load." *Composite Structures* 5 (1981), pp. 87–100.

CHAPTER 5

![bar]

Plane-Stress Stress-Strain Relations in a Global Coordinate System

One of the most important characteristics of structures made of fiber-reinforced materials, and one which dictates the manner in which they are analyzed, is the use of multiple fiber orientations. Generally structural laminates are made of multiple layers of fiber-reinforced material, and each layer has its own specific fiber orientation. To this point we have studied the response of fiber-reinforced materials in the principal material system, whether it is fully three-dimensional stress-strain behavior, as in equations (2.45) and (2.47), or plane-stress behavior, as in equations (4.4) and (4.14). If we are to accommodate multiple layers of fiber-reinforced materials, each with its own fiber orientation, then we will be confronted with using multiple 1-2-3 coordinate systems, each with its own orientation with respect to some global or structural coordinate system. If we are dealing with an x-y-z Cartesian coordinate system to describe the geometry of the structure, then the orientation of each principal material system must be defined with respect to the x-y-z system. If we are dealing with an x-θ-r cylindrical coordinate system to describe the structure, then the orientation of each principal material system must be defined with respect to the x-θ-r system, and so forth, for a spherical coordinate system. This leads to a large number of coordinates systems and orientations for describing the response of the fiber-reinforced structure. As an alternative approach, we can refer the response of each layer of material to the same global system. We accomplish this by transforming the stress-strain relations from the 1-2-3 coordinate system into the global coordinate system. This will be our approach here, in particular; it will be done for a state of plane stress using the standard transformation relations for stresses and strains learned in introductory strength-of-materials courses. Transformation can also be done for a general state of stress. However, transformation here will be limited to the plane-stress state because it will be useful for the development of classical lamination theory, which begins in the next chapter. Equally important, though, is the fact that the transformation of the description of the stress-strain response of fiber-reinforced material from the principal material coordinate system to a global coordinate system results in concepts so different from what one encounters with isotropic materials that it is best to start with a simpler plane-stress state and progress to the more complicated general stress state. When the concepts for the plane-stress stress state response described in a coordinate system other than the principal material coordinate system are fully understood, progression to a three-dimensional stress state is easier.

5.1
TRANSFORMATION RELATIONS

Consider Figure 5.1(a), the familiar picture of an isolated element in the principal material coordinate system. Figure 5.1(b) shows a similar element but one that is isolated in an x-y-z global coordinate system. The fibers are oriented at an angle θ with respect to the $+x$ axis of the global system. The fibers are parallel to the x-y plane and the 3 and z axes coincide. The fibers assumed their orientation by a simple rotation of the principal material system about the 3 axis. The orientation angle θ will be considered positive when the fibers rotate counterclockwise *from* the $+x$ axis *toward* the $+y$ axis. Often the fibers not being aligned with the edges of the element are referred to as an *off-axis* condition, generally meaning the fibers are not aligned with the analysis coordinate system (i.e., off the $+x$ axis). Though we will use the notation for a rectangular Cartesian coordinate system as the global system (i.e., x-y-z), the global coordinate system can be considered to be any orthogonal coordinate system. The use of a Cartesian system is for convenience only and the development is actually valid for any orthogonal coordinate system. The stress-strain relations are a description of the relations between stress and strain at a point within the material. The functional form of these relations does not depend on whether the point is a point in a rectangular Cartesian coordinate system, in a cylindrical coordinate system, or in a spherical, elliptical, or parabolic coordinate system.

The stresses on the small volume of element are now identified in accordance with the x-y-z notation. The six components of stress and strain are denoted as

$$\left\{ \begin{array}{c} \sigma_x \\ \sigma_y \\ \sigma_z \\ \tau_{yz} \\ \tau_{xz} \\ \tau_{xy} \end{array} \right\} \left\{ \begin{array}{c} \varepsilon_x \\ \varepsilon_y \\ \varepsilon_z \\ \gamma_{yz} \\ \gamma_{xz} \\ \gamma_{xy} \end{array} \right\} \tag{5.1}$$

and the six components of stress are illustrated in Figure 5.2. Although we seem interested in describing the stress-strain relation in another coordinate system now that we have developed it and fully understand it in the 1-2-3 system, we should emphasize that Figure 5.2 should be interpreted quite literally. We should interpret the figure as asking, "What is the relation between the stresses and deformations denoted in equation (5.1) for a small volume of material whose fibers are oriented at some angle relative to the boundaries of the element rather than parallel to them?" This is the real issue! Loads will not be always applied parallel to the fibers; our intuition

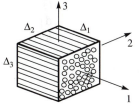

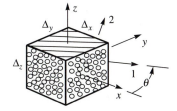

(a) Element in 1-2-3 system   (b) Element in x-y-z system

FIGURE 5.1
Elements of fiber-reinforced material in 1-2-3 and x-y-z coordinate systems

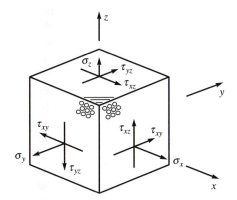

FIGURE 5.2
Stress components in x-y-z coordinate system

indicates that unusual deformations are likely to occur. The skewed orientation of the fibers must certainly cause unusual distortions of the originally cubic volume element. What are these deformations? How do they depend on fiber-orientation? Are they detrimental? Are they beneficial? Fortunately these and other questions can be answered by transforming the stress-strain relations from the 1-2-3 system to the x-y-z system. If we consider the special case shown in Figure 5.1(b), where the two coordinate systems are related to each other through a simple rotation θ about the z axis, then the stresses in the 1-2-3 system are related to the stresses in the x-y-z system by

$$\sigma_1 = \cos^2\theta\,\sigma_x + \sin^2\theta\,\sigma_y + 2\sin\theta\cos\theta\,\tau_{xy}$$

$$\sigma_2 = \sin^2\theta\,\sigma_x + \cos^2\theta\,\sigma_y - 2\sin\theta\cos\theta\,\tau_{xy}$$

$$\sigma_3 = \sigma_z$$

$$\tau_{23} = \cos\theta\,\tau_{yz} - \sin\theta\,\tau_{xz}$$ (5.2)

$$\tau_{13} = \sin\theta\,\tau_{yz} + \cos\theta\,\tau_{xz}$$

$$\tau_{12} = -\sin\theta\cos\theta\,\sigma_x + \sin\theta\cos\theta\,\sigma_y + \left(\cos^2\theta - \sin^2\theta\right)\tau_{xy}$$

For a state of plane stress, σ_3, τ_{23}, and τ_{13} are zero, and upon rearranging, the third, fourth, and fifth components of equation (5.2) give

$$\sigma_z = 0$$

$$\cos\theta\,\tau_{yz} - \sin\theta\,\tau_{xz} = 0 \tag{5.3}$$

$$\sin\theta\,\tau_{yz} + \cos\theta\,\tau_{xz} = 0$$

Because

$$\sin^2\theta + \cos^2\theta = 1 \tag{5.4}$$

the only solution to the last two equations of equation (5.3) is

$$\tau_{yz} = \tau_{xz} = 0 \tag{5.5}$$

leading to the conclusion that for a plane-stress state in the 1-2-3 principal material coordinate system, the out-of-plane stress components in the x-y-z global coordinate system are also zero. This may have been intuitive but here we have shown it directly.

The first, second, and sixth component of equation (5.2) may look more familiar in the form

$$\sigma_1 = \left(\frac{\sigma_x + \sigma_y}{2}\right) + \left(\frac{\sigma_x - \sigma_y}{2}\right)\cos 2\theta + \tau_{xy}\sin 2\theta$$

$$\sigma_2 = \left(\frac{\sigma_x + \sigma_y}{2}\right) - \left(\frac{\sigma_x - \sigma_y}{2}\right)\cos 2\theta - \tau_{xy}\sin 2\theta \tag{5.6}$$

$$\tau_{12} = -\left(\frac{\sigma_x - \sigma_y}{2}\right)\sin 2\theta + \tau_{xy}\cos 2\theta$$

This form is derivable from equation (5.2) by using trigonometric identities and is the form usually found in introductory strength-of-materials courses. The form of equation (5.2) will be most often used as it can be put in matrix form as

$$\begin{Bmatrix} \sigma_1 \\ \sigma_2 \\ \tau_{12} \end{Bmatrix} = \begin{bmatrix} \cos^2\theta & \sin^2\theta & 2\sin\theta\cos\theta \\ \sin^2\theta & \cos^2\theta & -2\sin\theta\cos\theta \\ -\sin\theta\cos\theta & \sin\theta\cos\theta & \cos^2\theta - \sin^2\theta \end{bmatrix} \begin{Bmatrix} \sigma_x \\ \sigma_y \\ \tau_{xy} \end{Bmatrix} \tag{5.7}$$

This transformation matrix of trigonometric functions will be used frequently in the plane-stress analysis of fiber-reinforced composite materials and it will be denoted by $[T]$, $[T]$ being written as

$$[T] = \begin{bmatrix} m^2 & n^2 & 2mn \\ n^2 & m^2 & -2mn \\ -mn & mn & m^2 - n^2 \end{bmatrix} \tag{5.8}$$

where $m = \cos\theta$, $n = \sin\theta$. With the above notation equation (5.7) can be written

$$\begin{Bmatrix} \sigma_1 \\ \sigma_2 \\ \tau_{12} \end{Bmatrix} = [T] \begin{Bmatrix} \sigma_x \\ \sigma_y \\ \tau_{xy} \end{Bmatrix} \tag{5.9}$$

or

$$\begin{Bmatrix} \sigma_1 \\ \sigma_2 \\ \tau_{12} \end{Bmatrix} = \begin{bmatrix} m^2 & n^2 & 2mn \\ n^2 & m^2 & -2mn \\ -mn & mn & m^2 - n^2 \end{bmatrix} \begin{Bmatrix} \sigma_x \\ \sigma_y \\ \tau_{xy} \end{Bmatrix} \qquad (5.10)$$

The inverse of equation (5.10) is

$$\begin{Bmatrix} \sigma_x \\ \sigma_y \\ \tau_{xy} \end{Bmatrix} = \begin{bmatrix} m^2 & n^2 & -2mn \\ n^2 & m^2 & 2mn \\ mn & -mn & m^2 - n^2 \end{bmatrix} \begin{Bmatrix} \sigma_1 \\ \sigma_2 \\ \tau_{12} \end{Bmatrix} \qquad (5.11)$$

which implies

$$[T]^{-1} = \begin{bmatrix} m^2 & n^2 & -2mn \\ n^2 & m^2 & 2mn \\ mn & -mn & m^2 - n^2 \end{bmatrix} \qquad (5.12)$$

In a similar manner, the strains transform according to the specialized relations equation (5.2) as

$$\begin{aligned}
\varepsilon_1 &= \cos^2 \theta \varepsilon_x + \sin^2 \theta \varepsilon_y + 2 \sin \theta \cos \theta \varepsilon_{xy} \\
\varepsilon_2 &= \sin^2 \theta \varepsilon_x + \cos^2 \theta \varepsilon_y - 2 \sin \theta \cos \theta \varepsilon_{xy} \\
\varepsilon_3 &= \varepsilon_z \\
\varepsilon_{23} &= \cos \theta \varepsilon_{yz} - \sin \theta \varepsilon_{xz} \\
\varepsilon_{13} &= \sin \theta \varepsilon_{yz} + \cos \theta \varepsilon_{xz} \\
\varepsilon_{12} &= -\sin \theta \cos \theta \varepsilon_x + \sin \theta \cos \theta \varepsilon_y + (\cos^2 \theta - \sin^2 \theta)\varepsilon_{xy}
\end{aligned} \qquad (5.13)$$

Note very well that the tensor shear strains, ε, not the engineering shear strains are being used in the above. These two measures of strain are different by a factor of two; that is,

$$\varepsilon_{23} = \frac{1}{2}\gamma_{23}$$

$$\varepsilon_{13} = \frac{1}{2}\gamma_{13} \qquad (5.14)$$

$$\varepsilon_{12} = \frac{1}{2}\gamma_{12}$$

If engineering shear strain is used instead, then the transformation relations become

$$\varepsilon_1 = \cos^2\theta\varepsilon_x + \sin^2\theta\varepsilon_y + 2\sin\theta\cos\theta\frac{1}{2}\gamma_{xy}$$

$$\varepsilon_2 = \sin^2\theta\varepsilon_x + \cos^2\theta\varepsilon_y - 2\sin\theta\cos\theta\frac{1}{2}\gamma_{xy}$$

$$\varepsilon_3 = \varepsilon_z \tag{5.15}$$

$$\gamma_{23} = \cos\theta\gamma_{yz} - \sin\theta\gamma_{xz}$$

$$\gamma_{13} = \sin\theta\gamma_{yz} + \cos\theta\gamma_{xz}$$

$$\frac{1}{2}\gamma_{12} = -\sin\theta\cos\theta\varepsilon_x + \sin\theta\cos\theta\varepsilon_y + (\cos^2\theta - \sin^2\theta)\frac{1}{2}\gamma_{xy}$$

As a result of the plane-stress assumption, specifically by equation (4.2),

$$\gamma_{23} = 0 \qquad \gamma_{13} = 0 \tag{5.16}$$

and by analogy to equation (5.3), it is concluded from the fourth and fifth equation of equation (5.15) that

$$\gamma_{yz} = 0 \qquad \gamma_{xz} = 0 \tag{5.17}$$

Also, due to the third equation of equation (5.15) and equation (4.3),

$$\boxed{\varepsilon_z = S_{13}\sigma_1 + S_{23}\sigma_2} \tag{5.18}$$

More importantly, if equation (5.7) is used to transform the stresses, then

$$\boxed{\begin{aligned}\varepsilon_z = {}&(S_{13}\cos^2\theta + S_{23}\sin^2\theta)\sigma_x + (S_{13}\sin^2\theta + S_{23}\cos^2\theta)\sigma_y \\ &+ 2(S_{13} - S_{23})\sin\theta\cos\theta\tau_{xy}\end{aligned}} \tag{5.19}$$

This equation is very important because it indicates that a shear stress in the x-y plane, τ_{xy}, produces an extensional strain, ε_z, perpendicular to that plane! For an isotropic material, $S_{13} = S_{23}$, and this simply will not happen. Shear stresses do not cause extensional strains in isotropic materials! This generation of extensional strains by shear stresses is an important characteristic of fiber-reinforced composite materials.

Returning to equation (5.15) to focus on the strains involved in the plane-stress assumption, we can write the first, second, and sixth equations as

$$\left\{\begin{array}{c}\varepsilon_1 \\ \varepsilon_2 \\ \dfrac{1}{2}\gamma_{12}\end{array}\right\} = [T]\left\{\begin{array}{c}\varepsilon_x \\ \varepsilon_y \\ \dfrac{1}{2}\gamma_{xy}\end{array}\right\} \tag{5.20}$$

or

$$
\begin{Bmatrix} \varepsilon_1 \\ \varepsilon_2 \\ \dfrac{1}{2}\gamma_{12} \end{Bmatrix} = \begin{bmatrix} m^2 & n^2 & 2mn \\ n^2 & m^2 & -2mn \\ -mn & mn & m^2 - n^2 \end{bmatrix} \begin{Bmatrix} \varepsilon_x \\ \varepsilon_y \\ \dfrac{1}{2}\gamma_{xy} \end{Bmatrix} \tag{5.21}
$$

It is very important to note that the transformation retains the factor of $\frac{1}{2}$ with the engineering shear strain.

5.2
TRANSFORMED REDUCED COMPLIANCE

Continuing with the transformation of the stress-strain relations for plane stress to the x-y-z global coordinate system, the stress-strain relations in the 1-2-3 principal material coordinate system, equation (4.4), can be written in a slightly modified form to account for the use of the tensor shear strain rather than the engineering shear strain as

$$
\begin{Bmatrix} \varepsilon_1 \\ \varepsilon_2 \\ \dfrac{1}{2}\gamma_{12} \end{Bmatrix} = \begin{bmatrix} S_{11} & S_{12} & 0 \\ S_{12} & S_{22} & 0 \\ 0 & 0 & \dfrac{1}{2}S_{66} \end{bmatrix} \begin{Bmatrix} \sigma_1 \\ \sigma_2 \\ \tau_{12} \end{Bmatrix} \tag{5.22}
$$

Using the transformations given by equations (5.9) and (5.20) in equation (5.22) leads to

$$
[T] \begin{Bmatrix} \varepsilon_x \\ \varepsilon_y \\ \dfrac{1}{2}\gamma_{xy} \end{Bmatrix} = \begin{bmatrix} S_{11} & S_{12} & 0 \\ S_{12} & S_{22} & 0 \\ 0 & 0 & \dfrac{1}{2}S_{66} \end{bmatrix} [T] \begin{Bmatrix} \sigma_x \\ \sigma_y \\ \tau_{xy} \end{Bmatrix} \tag{5.23}
$$

and multiplying both sides of equation (5.23) by $[T]^{-1}$ results in

$$
\begin{Bmatrix} \varepsilon_x \\ \varepsilon_y \\ \dfrac{1}{2}\gamma_{xy} \end{Bmatrix} = [T]^{-1} \begin{bmatrix} S_{11} & S_{12} & 0 \\ S_{12} & S_{22} & 0 \\ 0 & 0 & \dfrac{1}{2}S_{66} \end{bmatrix} [T] \begin{Bmatrix} \sigma_x \\ \sigma_y \\ \tau_{xy} \end{Bmatrix} \tag{5.24}
$$

Substituting for $[T]$ and $[T]^{-1}$ from equations (5.8) and (5.12), we find that multiplying these three matrices together, and multiplying the third equation through by a factor of 2, yields

$$\left\{ \begin{array}{c} \varepsilon_x \\ \varepsilon_y \\ \gamma_{xy} \end{array} \right\} = \left[\begin{array}{ccc} \bar{S}_{11} & \bar{S}_{12} & \bar{S}_{16} \\ \bar{S}_{12} & \bar{S}_{22} & \bar{S}_{26} \\ \bar{S}_{16} & \bar{S}_{26} & \bar{S}_{66} \end{array} \right] \left\{ \begin{array}{c} \sigma_x \\ \sigma_y \\ \tau_{xy} \end{array} \right\} \tag{5.25}$$

The $\bar{S}_{ij}$ are called the *transformed reduced compliances*. Note that the factor of $1/2$ has been removed and the engineering shear strain reintroduced. Equation (5.25) is a fundamental equation for studying the plane-stress response of fiber-reinforced composite materials. The transformed reduced compliances are defined by

$$\begin{aligned} \bar{S}_{11} &= S_{11}m^4 + (2S_{12} + S_{66})n^2m^2 + S_{22}n^4 \\ \bar{S}_{12} &= (S_{11} + S_{22} - S_{66})n^2m^2 + S_{12}(n^4 + m^4) \\ \bar{S}_{16} &= (2S_{11} - 2S_{12} - S_{66})nm^3 - (2S_{22} - 2S_{12} - S_{66})n^3m \\ \bar{S}_{22} &= S_{11}n^4 + (2S_{12} + S_{66})n^2m^2 + S_{22}m^4 \\ \bar{S}_{26} &= (2S_{11} - 2S_{12} - S_{66})n^3m - (2S_{22} - 2S_{12} - S_{66})nm^3 \\ \bar{S}_{66} &= 2(2S_{11} + 2S_{22} - 4S_{12} - S_{66})n^2m^2 + S_{66}(n^4 + m^4) \end{aligned} \tag{5.26}$$

Equation (5.25) and the definitions equation (5.26) relate the strains of an element of fiber-reinforced material as measured in the x-y-z global coordinate system to the applied stresses measured in that coordinate system. We can look upon these equations as strictly the end result of simple steps in linear algebra, that is, transformations, inversions, and so forth. Alternatively, we can view them as what they actually are, namely, relations that describe what we shall see to be the complex response of an element of fiber-reinforced material in a state of plane stress that is subjected to stresses not aligned with the fibers, nor perpendicular to the fibers. The most profound results of equation (5.25) are that a normal stress σ_x will cause a shearing deformation γ_{xy} through the $\bar{S}_{16}$ term, and similarly a normal stress σ_y will cause a shearing deformation through the $\bar{S}_{26}$ term. Equally important, because of the existence of these same $\bar{S}_{16}$ and $\bar{S}_{26}$ terms at other locations in the compliance matrix, a shear stress τ_{xy} will cause strains ε_x and ε_y. Such responses are totally different from those in metals. In metals, normal stresses do not cause shear strains, and shear stresses do not cause extensional strains. This coupling found in fiber-reinforced composites is termed *shear-extension coupling*. Shear-extension coupling is an important characteristic and is responsible for interesting and important responses of fiber-reinforced composite materials. Recall, equation (5.19) provided another example of shear-extension coupling. Through a series of examples we will examine the response of an element of fiber-reinforced material to simple stress states (i.e., σ_x only, and then τ_{xy} only) and compare the responses with the response of a similar element of metal. After we work through these specific examples, the implications and meaning of shear-extension coupling will be clear.

Figure 5.3 shows the variations with θ of the various elements of the transformed reduced compliance matrix for a graphite-reinforced material. Note that $\bar{S}_{12}$ and all

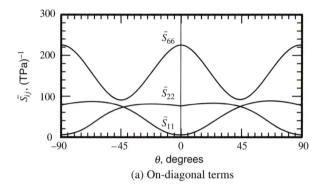

FIGURE 5.3
Variation of transformed reduced compliance with fiber angle θ for graphite-reinforced composite

(a) On-diagonal terms

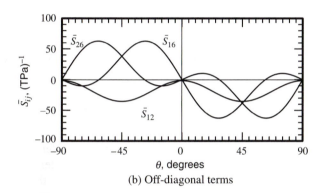

(b) Off-diagonal terms

on-diagonal terms are even functions of θ, while the off-diagonal terms $\bar{S}_{16}$ and $\bar{S}_{26}$ are odd functions of θ. The importance of this will be illustrated shortly. Also note the rapid variation of some of the compliances as θ increases or decreases from $0°$. At $\theta = 30°$, $\bar{S}_{11}$ has increased by a factor of 8 relative to its value at $\theta = 0°$ and $\bar{S}_{16}$ has changed from 0 at $\theta = 0°$ to nearly its maximum magnitude.

Before proceeding with the examples, we should discuss two special cases of equation (5.26). For the first case, consider the situation when the fibers are aligned with the x axis, namely, $\theta = 0°$. With $\theta = 0°$, $m = 1$ and $n = 0$ and equation (5.26) reduces to

$$\bar{S}_{11}(0°) = S_{11} \qquad \bar{S}_{22}(0°) = S_{22}$$

$$\bar{S}_{12}(0°) = S_{12} \qquad \bar{S}_{26}(0°) = 0 \qquad\qquad (5.27)$$

$$\bar{S}_{16}(0°) = 0 \qquad \bar{S}_{66}(0°) = S_{66}$$

where the argument of $0°$ is used as a reminder that the $\bar{S}_{ij}$'s are functions of θ. The results of equation (5.27) simply state that at $\theta = 0°$ the transformed reduced compliance degenerates to the reduced compliance, that is, the compliance in the principal material coordinate system. In the principal material system there is no $\bar{S}_{16}$ or $\bar{S}_{26}$. The quantities S_{11}, S_{12}, S_{22}, and S_{66} are often referred to as the *on-axis compliances*. The barred quantities of equation (5.26) are frequently called the *off-axis compliances*.

For the second special case, consider isotropic materials. The compliances of equation (5.26) reduce to

$$\bar{S}_{11} = \frac{1}{E}$$

$$\bar{S}_{12} = -\frac{v}{E}$$

$$\bar{S}_{16} = 0$$

$$\bar{S}_{22} = \frac{1}{E}$$ (5.28)

$$\bar{S}_{26} = 0$$

$$\bar{S}_{66} = \frac{1}{G} = \frac{2(1+v)}{E}$$

which can be demonstrated by using the definitions of the compliances for an isotropic material, equation (4.6), in equation (5.26). For example:

$$\bar{S}_{11} = \frac{1}{E}m^4 + \left(-\frac{2v}{E} + \frac{2(1+v)}{E}\right)n^2m^2 + \frac{1}{E}n^4$$ (5.29a)

$$\bar{S}_{11} = \frac{1}{E}\left(m^4 + 2m^2n^2 + n^4\right) = \frac{1}{E}\left(m^2 + n^2\right)^2$$ (5.29b)

But

$$m^2 + n^2 = 1$$ (5.30)

so

$$\bar{S}_{11} = \frac{1}{E}$$ (5.31)

The proof that

$$\bar{S}_{22} = \frac{1}{E}$$ (5.32)

is identical. From $\bar{S}_{12}$, we see

$$\bar{S}_{12} = \left(\frac{1}{E} + \frac{1}{E} - \frac{2(1+v)}{E}\right)n^2m^2 - \frac{v}{E}\left(n^4 + m^4\right)$$ (5.33a)

$$\bar{S}_{12} = -\frac{v}{E}\left(2n^2m^2 + n^4 + m^4\right) = -\frac{v}{E}\left(n^2 + m^2\right)^2$$ (5.33b)

so by equation (5.30)

$$\bar{S}_{12} = -\frac{v}{E}$$ (5.34)

From $\bar{S}_{16}$, we see

$$\bar{S}_{16} = \left(\frac{2}{E} - \frac{2v}{E} - \frac{2(1+v)}{E}\right)nm^3 - \left(\frac{2}{E} - \frac{2v}{E} - \frac{2(1+v)}{E}\right)n^3m$$ (5.35a)

$$\bar{S}_{16} = (0)nm^3 - (0)n^3m = 0$$ (5.35b)

Similarly it can be shown

$$\bar{S}_{26} = 0$$ (5.36)

Finally,

$$\bar{S}_{66} = 2\left(\frac{2}{E} + \frac{2}{E} + \frac{4v}{E} - \frac{2(1+v)}{E}\right)n^2m^2 + \frac{2(1+v)}{E}\left(n^4 + m^4\right) \qquad (5.37a)$$

$$\bar{S}_{66} = \frac{2(1+v)}{E}\left(2n^2m^2 + n^4 + m^4\right) = \frac{2(1+v)}{E}\left(n^2 + m^2\right)^2 \qquad (5.37b)$$

Again, by equation (5.30),

$$\bar{S}_{66} = \frac{2(1+v)}{E} \qquad (5.38)$$

Thus, independent of the direction of the coordinate system, for an isotropic material equation (5.28) is true.

We now turn to a series of examples that illustrate the shear-extension coupling predicted by the stress-strain relations of equation (5.25), specifically the deformations caused by a tensile normal stress. Consider a thin element of aluminum subjected to a tensile stress $\sigma_x = 155$ MPa. As in Figure 5.4(a), the aluminum

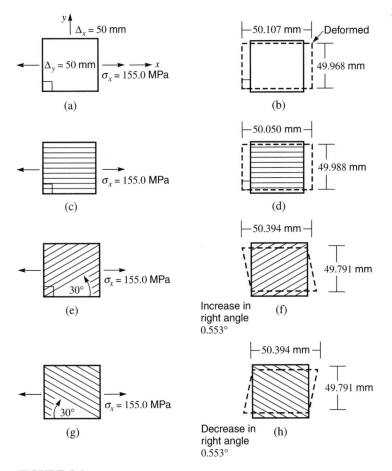

FIGURE 5.4
Response of aluminum and graphite-reinforced composite to tensile normal stress σ_x

element has dimensions 50 mm × 50 mm. Thickness is not important at the moment but consider the element to be thin. As σ_y and τ_{xy} are zero, the stress-strain relations of equation (5.25) reduce to

$$\varepsilon_x = \bar{S}_{11}\sigma_x$$
$$\varepsilon_y = \bar{S}_{12}\sigma_x \tag{5.39}$$
$$\gamma_{xy} = \bar{S}_{16}\sigma_x$$

or, by equation (5.28),

$$\varepsilon_x = \frac{1}{E}\sigma_x$$
$$\varepsilon_y = -\frac{\nu}{E}\sigma_x \tag{5.40}$$
$$\gamma_{xy} = 0 \times \sigma_x$$

Referring to Table 2.1, we note that the strains in the aluminum are

$$\varepsilon_x = \frac{1}{72.4 \times 10^9}155 \times 10^6 = 2140 \ \mu \ \text{mm/mm}$$
$$\varepsilon_y = \frac{-0.3}{72.4 \times 10^9}155 \times 10^6 = -642 \ \mu \ \text{mm/mm} \tag{5.41}$$
$$\gamma_{xy} = 0 \times 250 \times 10^6 = 0$$

The dimensional changes of the square element of aluminum are

$$\delta\Delta_x = \varepsilon_x\Delta_x = (2140 \times 10^{-6})(50) = 0.1070 \ \text{mm}$$
$$\delta\Delta_y = \varepsilon_y\Delta_y = (-642 \times 10^{-6})(50) = -0.0321 \ \text{mm} \tag{5.42}$$

so the deformed dimensions of the aluminum element, in Figure 5.4(b), are

$$\Delta_x + \delta\Delta_x = 50.107 \ \text{mm}$$
$$\Delta_y + \delta\Delta_y = 49.968 \ \text{mm} \tag{5.43}$$

This behavior is well known; the material stretches in the direction of the applied stress and contracts perpendicular to that direction (both in the y direction and the z direction, though the latter is not shown), and all right corner angles remaining right.

Turning to Figure 5.4(c), we now consider a similar-sized square element of graphite-reinforced material with the fibers aligned with the x direction and also subjected to the 155 MPa stress in the x direction. The stress-strain relations of equation (5.25) reduce to

$$\varepsilon_x = \bar{S}_{11}(0°)\sigma_x$$
$$\varepsilon_y = \bar{S}_{12}(0°)\sigma_x \tag{5.44}$$
$$\gamma_{xy} = \bar{S}_{16}(0°)\sigma_x$$

or, if we use the compliances of equation (5.27),

$$\varepsilon_x = S_{11}\sigma_x$$
$$\varepsilon_y = S_{12}\sigma_x \tag{5.45}$$
$$\gamma_{xy} = 0 \times \sigma_x$$

From equation (4.5) and Table 2.1, or, alternatively, directly from equation (2.56),

$$S_{11} = 6.45 \text{ (TPa)}^{-1} \qquad S_{12} = -1.600$$
$$S_{22} = 82.6 \qquad S_{66} = 227 \tag{5.46}$$

resulting in

$$\varepsilon_x = (6.45 \times 10^{-12})(155 \times 10^6) = 1000 \ \mu \text{ mm/mm}$$
$$\varepsilon_y = (-1.600 \times 10^{-12})(155 \times 10^6) = -248 \ \mu \text{ mm/mm} \tag{5.47}$$
$$\gamma_{xy} = 227 \times 10^6 \times 0 = 0$$

The dimensional changes of the graphite-reinforced element are

$$\delta\Delta_x = \varepsilon_x \Delta_x = (1000 \times 10^{-6})(50) = 0.0500 \text{ mm}$$
$$\delta\Delta_y = \varepsilon_y \Delta_y = (-248 \times 10^{-6})(50) = -0.0124 \text{ mm} \tag{5.48}$$

so the deformed dimensions are

$$\Delta_x + \delta\Delta_x = 50.0500 \text{ mm} \qquad \Delta_y + \delta\Delta_y = 49.988 \text{ mm} \tag{5.49}$$

Figure 5.4(d) shows the deformed shape of the graphite-reinforced element, and the deformation is similar to that of the aluminum; the element stretches in the x direction and contracts in the y direction (and in the z direction), and all right corner angles remain right. For the same applied stress level of 155 MPa, the elongation of the fiber-reinforced material in the fiber direction, $\delta\Delta_x$ in equation (5.48), is about one-half the elongation of the aluminum, $\delta\Delta_x$ in equation (5.42); the difference is due to the difference between E_1 for the graphite-reinforced material and E for aluminum. Please note that a tensile stress level of 155 MPa in the fiber direction of graphite-reinforced material, which results in the 1000 μ mm/mm elongation strain, is considerably below the ultimate capacity of that material.

Now consider a square element of graphite-reinforced material with the fibers oriented at $\theta = 30°$ with respect to the x axis and also subjected to a stress, as shown in Figure 5.4(e), $\sigma_x = 155$ MPa. The strains are determined by equation (5.25) as

$$\varepsilon_x = \bar{S}_{11}(30°)\sigma_x$$
$$\varepsilon_y = \bar{S}_{12}(30°)\sigma_x \tag{5.50}$$
$$\gamma_{xy} = \bar{S}_{16}(30°)\sigma_x$$

With $\theta = 30°$, $m = \sqrt{3}/2$ and $n = 1/2$, and equations (5.26) and (5.46) give

$$\bar{S}_{11}(30°) = 50.8 \text{ (TPa)}^{-1} \qquad \bar{S}_{22}(30°) = 88.9$$
$$\bar{S}_{12}(30°) = -26.9 \qquad \bar{S}_{26}(30°) = -3.77 \tag{5.51}$$
$$\bar{S}_{16}(30°) = -62.2 \qquad \bar{S}_{66}(30°) = 126.0$$

As a result, from equation (5.50),

$$\varepsilon_x = (50.8 \times 10^{-12})(155 \times 10^6) = 7880 \ \mu \text{ mm/mm}$$
$$\varepsilon_y = (-26.9 \times 10^{-12})(155 \times 10^6) = -4170 \ \mu \text{ mm/mm} \tag{5.52}$$
$$\gamma_{xy} = (-62.2 \times 10^{-12})(155 \times 10^6) = -9640 \ \mu \text{ rad}(-0.553°)$$

where the notation μ rad has been introduced. The shear strain represents an angle change, in radians, and the prefix μ has been added to represent the factor 10^{-6}, that is,

$$1\mu \text{ rad} = 10^{-6} \text{ rad} = 57.3 \times 10^{-6} \text{ degrees} \tag{5.53}$$

This is the counterpart of μ mm/mm for the strains ε_1 and ε_2. It is interesting to note that at $\theta = 30°$, the angle change of the shear-extension coupling effect of $\bar{S}_{16}$ results in larger deformations than the Poisson contraction effects of $\bar{S}_{12}$ (i.e., -9640 μ rad versus -4170 μ mm/mm). The relative magnitude of these two effects at all values of θ can be determined from the character of $\bar{S}_{12}$ and $\bar{S}_{16}$ as a function of θ, as in Figure 5.3(b). With the above numbers, dimensional changes of the graphite-reinforced element become

$$\delta\Delta_x = \varepsilon_x\Delta_x = (0.00787)(50) = 0.394 \text{ mm}$$
$$\delta\Delta_y = \varepsilon_y\Delta_y = (-0.00417)(50) = -0.209 \text{ mm} \tag{5.54}$$

Unlike the previous two cases, however, the original right corner angles do not remain right. The change in right angle is given by the value of γ_{xy} in equation (5.52), namely, -9640 μ rad, or $-0.553°$. The deformed dimensions of the element are

$$\Delta_x + \delta\Delta_x = 50.394 \text{ mm} \qquad \Delta_y + \delta\Delta_y = 49.791 \text{ mm} \tag{5.55}$$

and Figure 5.4(f) illustrates the deformed shape of the element. It is important to properly interpret the sign of γ_{xy}. *A positive γ_{xy} means that the right angle between two line segments emanating from the origin, one line segment starting from the origin and extending in the $+x$ direction, the other line segment starting from the origin and extending in the $+y$ direction, decreases.* Because γ_{xy} in the above example is negative, the angle in the lower left hand corner of the element *increases*.

As a final example of the effects of tension normal stress in the x direction, consider an element of graphite-reinforced composite with the fibers oriented at $\theta = -30°$ relative to the $+x$ axis, as in Figure 5.4(g). This example illustrates one of the important characteristics of the $\bar{S}_{ij}$ as regards their dependence on θ. In this situation the stress-strain relations of equation (5.25) become

$$\varepsilon_x = \bar{S}_{11}(-30°)\sigma_x$$
$$\varepsilon_y = \bar{S}_{12}(-30°)\sigma_x \tag{5.56}$$
$$\gamma_{xy} = \bar{S}_{16}(-30°)\sigma_x$$

As noted earlier, inspection of the definitions of the off-axis compliance in equation (5.26) reveals that $\bar{S}_{16}$ and $\bar{S}_{26}$ are odd functions of n, and hence of θ, while the remaining $\bar{S}_{ij}$ are even functions of θ. Therefore

$$\bar{S}_{11}(-30°) = +\bar{S}_{11}(+30°) = 50.8 \text{ (TPa)}^{-1}$$
$$\bar{S}_{12}(-30°) = +\bar{S}_{12}(+30°) = -26.9$$
$$\bar{S}_{16}(-30°) = -\bar{S}_{16}(+30°) = 62.2$$
$$\bar{S}_{22}(-30°) = +\bar{S}_{22}(+30°) = 88.9 \tag{5.57}$$
$$\bar{S}_{26}(-30°) = -\bar{S}_{22}(-30°) = 3.77$$
$$\bar{S}_{66}(-30°) = +\bar{S}_{66}(+30°) = 126.0$$

and substituting into equation (5.56) results in

$$\varepsilon_x = (50.8 \times 10^{-12})(155 \times 10^6) = 7880 \ \mu \ \text{mm/mm}$$

$$\varepsilon_y = (-26.9 \times 10^{-12})(155 \times 10^6) = -4170 \ \mu \ \text{mm/mm} \qquad (5.58)$$

$$\gamma_{xy} = (62.2 \times 10^{-12})(155 \times 10^6) = 9640 \ \mu \ \text{rad} \ (0.553°)$$

With these numbers, dimensional changes of the $-30°$ graphite-reinforced element become

$$\delta\Delta_x = \varepsilon_x \Delta_x = (0.00787)(50) = 0.394 \ \text{mm}$$

$$\delta\Delta_y = \varepsilon_y \Delta_y = (-0.00417)(50) = -0.209 \ \text{mm} \qquad (5.59)$$

Like the $+30°$ case, the original right corner angles do not remain right and the change in right angle is given by the value of γ_{xy} in equation (5.58), namely, $9640 \ \mu$ rad, or $0.553°$. The deformed dimensions of the element are the same as the $+30°$ case, namely,

$$\Delta_x + \delta\Delta_x = 50.394 \ \text{mm} \qquad \Delta_y + \delta\Delta_y = 49.791 \ \text{mm} \qquad (5.60)$$

Figure 5.4(h) illustrates the deformed shape of the element. It is important to note that the change in the right corner angle for the $-30°$ case is opposite the change for the $+30°$ case. This ability to change the sign of the deformation by changing the fiber angle is a very important characteristic of fiber-reinforced composite materials. Here the sign change of $\bar{S}_{16}$ was responsible for the sign of the change in right angle. Because it is also an odd function of θ, $\bar{S}_{26}$ changes sign with θ and in other situations it can be responsible for controlling the change in sign of a deformation. In more complicated loadings, specifically with stress components σ_x and σ_y both present, both $\bar{S}_{16}$ and $\bar{S}_{26}$ control the sign of the deformation. The potential for using this characteristic is enormous.

It is important to note that simply rotating the fiber angles by $30°$ increases the strain in the direction of the applied stress by a factor of 8. Equation (5.47) indicates that for $\theta = 0°$, $\varepsilon_x = 1000 \ \mu$ mm/mm, while equation (5.52) shows that for $\theta = 30°$, $\varepsilon_x = 7880 \ \mu$ mm/mm. The loss of stiffness when the fibers rotate away from the loading direction is quite significant.

As mentioned, $\bar{S}_{16}$ and $\bar{S}_{26}$ serve "double duty" in that they couple normal stresses to shear deformation, and they couple the shear stress to extensional deformations. Another series of examples will illustrate this latter coupling and further illustrate the influence of the sign dependence of $\bar{S}_{16}$ and $\bar{S}_{26}$ on θ. The series will again start with an element of aluminum and progress through an element of graphite-reinforced material. This progression, though adding nothing to what we already know about the behavior of aluminum, is taken specifically to show the contrasts, and in some cases the similarities, in the response of fiber-reinforced composites and isotropic materials. Consider, as shown in Figure 5.5(a), a 50×50 mm square of aluminum loaded by a 4.40 MPa shear stress τ_{xy}. Of interest are the deformations caused by the application of this shear stress. Because σ_x and σ_y are zero, the stress-strain

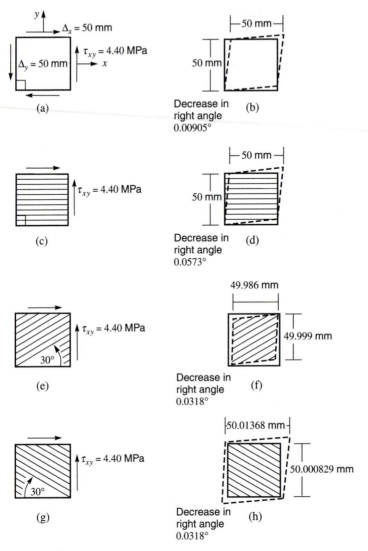

FIGURE 5.5
Response of aluminum and graphite-reinforced composite to a positive shear stress τ_{xy}

relations of equation (5.25) reduce to

$$\varepsilon_x = \bar{S}_{16}\tau_{xy}$$

$$\varepsilon_y = \bar{S}_{26}\tau_{xy} \qquad (5.61)$$

$$\gamma_{xy} = \bar{S}_{66}\tau_{xy}$$

For aluminum, $\bar{S}_{16}$ and $\bar{S}_{26}$ were shown to be zero, in equation (5.28), and as a result

$$\varepsilon_x = 0 \times \tau_{xy}$$

$$\varepsilon_y = 0 \times \tau_{xy}$$

$$\gamma_{xy} = S_{66}\tau_{xy} = \frac{1}{G}\tau_{xy} \qquad (5.62)$$

Using the value of shear modulus for aluminum from Table 2.1 yields

$$\varepsilon_x = 0 \times 4.40 \times 10^6 = 0$$

$$\varepsilon_y = 0 \times 4.40 \times 10^6 = 0$$

$$\gamma_{xy} = \frac{1}{27.8 \times 10^9} 4.40 \times 10^6 = 158.0 \ \mu \ \text{rad} \ (0.00905°) \qquad (5.63)$$

confirming our experience with aluminum that only a shear deformation results; the angle in the lower left corner decreases by 158.0 μ rad, or 0.00905°. The lengths of the sides of the deformed element are still exactly 50 mm; Figure 5.5(b) shows the deformed shape.

Applying a shear stress τ_{xy} to an element of graphite-reinforced composite with the fibers aligned with the x axis, as in Figure 5.5(c), leads to

$$\varepsilon_x = \bar{S}_{16}(0°)\tau_{xy}$$

$$\varepsilon_y = \bar{S}_{26}(0°)\tau_{xy} \qquad (5.64)$$

$$\gamma_{xy} = \bar{S}_{66}(0°)\tau_{xy}$$

Using the compliances of equation (5.27) and the numerical values from equation (5.46) leads to

$$\varepsilon_x = 0 \times \tau_{xy} = 0 \times 4.40 \times 10^6 = 0$$

$$\varepsilon_y = 0 \times \tau_{xy} = 0 \times 4.40 \times 10^6 = 0 \qquad (5.65)$$

$$\gamma_{xy} = S_{66}\tau_{xy} = (227 \times 10^{-12})(4.40 \times 10^6) = 1000 \ \mu \ \text{rad} \ (0.0573°)$$

Again, as in Figure 5.5(d), the only deformation is the shear strain; the 4.40 MPa shear stress τ_{xy} causes a much larger shear strain in the graphite-reinforced material with the fibers aligned with the x axis than in the aluminum. This is because the value of G_{12} for a graphite-reinforced composite is much less than the value of G for aluminum.

Attention now turns to the case of Figure 5.5(e), applying the 4.40 MPa shear stress τ_{xy} to an element of graphite-reinforced material with its fibers oriented at 30° relative to the $+x$ axis. This situation results in an unexpected and unusual response. As with the past cases, the stress-strain relations of equation (5.25) result in

$$\varepsilon_x = \bar{S}_{16}(30°)\tau_{xy}$$

$$\varepsilon_y = \bar{S}_{26}(30°)\tau_{xy} \qquad (5.66)$$

$$\gamma_{xy} = \bar{S}_{66}(30°)\tau_{xy}$$

and using the appropriate numerical values for the off-axis compliances from equation (5.51) yields

$$\varepsilon_x = (-62.2 \times 10^{-12})(4.40 \times 10^6) = -274 \; \mu \text{ mm/mm}$$

$$\varepsilon_y = (-3.77 \times 10^{-12})(4.40 \times 10^6) = -16.58 \; \mu \text{ mm/mm} \qquad (5.67)$$

$$\gamma_{xy} = (126.0 \times 10^{-12})(4.40 \times 10^6) = 555 \; \mu \text{ rad } (0.0318°)$$

The above numbers indicate that due to the shear stress, both the x and y dimensions decrease! This behavior is unlike anything that happens with an isotropic material and is totally unexpected. This coupling of shear stress and extensional deformation again provides unlimited potential for using fiber-reinforced composite materials to achieve results not possible or even conceivable with metals. The dimensional changes associated with the above strains are

$$\delta \Delta_x = \varepsilon_x \Delta_x = (-0.000274)(50) = -0.01368 \text{ mm}$$
$$\delta \Delta_y = \varepsilon_y \Delta_y = (-0.00001658)(50) = -0.000829 \text{ mm} \qquad (5.68)$$

The shear strain is positive so the right corner angle in the lower-left-hand corner of the element decreases by 555 μ rad, or 0.0318°. Figure 5.5(f) illustrates the deformed shape of the element; the lengths of the sides being given by

$$\Delta_x + \delta \Delta_x = 49.986 \text{ mm} \qquad \Delta_y + \delta \Delta_y = 49.999 \text{ mm} \qquad (5.69)$$

Finally, consider the element of graphite-reinforced composite with the fibers oriented at $-30°$ relative to the $+x$ axis, as in Figure 5.5(g). With an applied stress of $\tau_{xy} = 4.4$ MPa, the stress-strain relations of equation (5.25) become

$$\varepsilon_x = \bar{S}_{16}(-30°)\tau_{xy}$$

$$\varepsilon_y = \bar{S}_{26}(-30°)\tau_{xy} \qquad (5.70)$$

$$\gamma_{xy} = \bar{S}_{66}(-30°)\tau_{xy}$$

If we use numerical values from equation (5.57), equation (5.70) becomes

$$\varepsilon_x = (62.2 \times 10^{-12})(4.40 \times 10^6) = 274 \; \mu \text{ mm/mm}$$

$$\varepsilon_y = (3.77 \times 10^{-12})(4.40 \times 10^6) = 16.58 \; \mu \text{ mm/mm} \qquad (5.71)$$

$$\gamma_{xy} = (126.0 \times 10^{-12})(4.40 \times 10^6) = 555 \; \mu \text{ rad } (0.0318°)$$

These numbers indicate that with the fibers at $\theta = -30°$, the sides of the element increase in length! This is exactly opposite the case with $\theta = +30°$. However, the right angle in the lower left corner decreases the same as for the $\theta = +30°$ orientation. The simple switching of the fiber angle has a significant influence on the response. Figure 5.5(h) illustrates the deformed element; the dimensional changes are given by

$$\delta \Delta_x = \varepsilon_x \Delta_x = (0.000274)(50) = 0.01368 \text{ mm}$$
$$\delta \Delta_y = \varepsilon_y \Delta_y = (0.00001658)(50) = 0.000829 \text{ mm} \qquad (5.72)$$

and the new dimensions being

$$\Delta_x + \delta \Delta_x = 50.01368 \text{ mm} \qquad \Delta_y + \delta \Delta_y = 50.000829 \text{ mm} \qquad (5.73)$$

The examples of Figures 5.4 and 5.5 illustrate one of the most important characteristics of the response of fiber-reinforced materials, namely, the coupling effects when the fibers are oriented at some angle relative to the direction of the applied load.

Here because the stress levels considered were small, the deformations were small. For higher stress levels, larger deformations result. The important point is, couplings are present in fiber-reinforced materials, and they can be used to advantage. With some experience, intuition allows one to qualitatively predict these coupling effects for simple cases. With all three components of stress applied, however, relying on intuition can be dangerous and the stress-strain relations of equation (5.25) should be used. Using the stress-strain relations of equation (5.25) is recommended in all cases, even if, in your mind, it is only being used to confirm intuition. Many times your intuition can be fooled, but use of equation (5.25) always leads to the correct answer.

It is important to keep in mind the discussion of Section 2.5 regarding specification of either the stress components or the strain components. In the context of plane stress, either σ_1 or ε_1, and either σ_2 or ε_2, and either τ_{12} or γ_{12} can be stipulated, but not both the stress and the strain component from any one of the pairs. In the context of the fibers being oriented at some angle with respect to the x axis, either σ_x or ε_x, and either σ_y or ε_y, and either γ_{xy} or τ_{xy} can be specified. In the above series of examples, the stress component from each of the pairs was stipulated, two of the three stress components being zero in all the cases. In all cases the strains in each of the stress-strain pairs were being sought. By contrast, consider again the 50 mm × 50 mm square of graphite-reinforced composite, loaded in tension and with its fibers oriented at $-30°$ with respect to the $+x$ axis. Assume that, instead of being completely free to deform, as in the last examples, the off-axis element is constrained from any deformation in the y direction, as in Figure 5.6(a). Of interest here are the deformations that result from this loading. For this situation σ_x is known to be 155 MPa, ε_x is unknown, σ_y is unknown, ε_y is known to be zero, τ_{xy} is known to be zero, and γ_{xy} is unknown. The unknowns involve both stresses and strains, and the knowns involve both stresses and strains. For this particular situation the stress-strain relations of equation (5.25) become

$$\varepsilon_x = \bar{S}_{11}\sigma_x + \bar{S}_{12}\sigma_y + \bar{S}_{16} \times 0$$

$$0 = \bar{S}_{12}\sigma_x + \bar{S}_{22}\sigma_y + \bar{S}_{26} \times 0 \tag{5.74}$$

$$\gamma_{xy} = \bar{S}_{16}\sigma_x + \bar{S}_{26}\sigma_y + \bar{S}_{66} \times 0$$

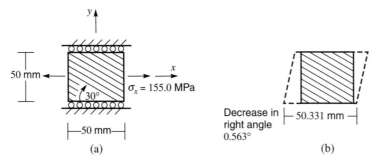

(a)

Decrease in right angle 0.563°

50.331 mm

(b)

FIGURE 5.6
Response of a partially constrained off-axis element of graphite-reinforced material to a normal stress σ_x

where it is understood that the $\bar{S}_{ij}$ are evaluated at $\theta = -30°$. From the second equation

$$\sigma_y = -\frac{\bar{S}_{12}}{\bar{S}_{22}}\sigma_x \qquad (5.75)$$

and using this in the first and third equations results in

$$\varepsilon_x = \left(\bar{S}_{11} - \frac{\bar{S}_{12}^2}{\bar{S}_{22}}\right)\sigma_x$$

$$\gamma_{xy} = \left(\bar{S}_{16} - \frac{\bar{S}_{12}\bar{S}_{26}}{\bar{S}_{22}}\right)\sigma_x \qquad (5.76)$$

The terms in parenthesis are identified as before as reduced compliances. Using numerical values for graphite-reinforced composite, from equation (5.57), we find

$$\varepsilon_x = 6610 \ \mu \ \text{mm/mm} \qquad \gamma_{xy} = 9820 \ \mu \ \text{rad} \ (0.563°) \qquad \sigma_y = 46.9 \ \text{MPa} \quad (5.77)$$

According to these calculations, the applied stress in the x direction causes the x dimension to increase by

$$\delta\Delta_x = \varepsilon_x\Delta_x = (0.00661)(50) = 0.331 \ \text{mm} \qquad (5.78)$$

and the lower left hand right corner angle to decrease by 9820 μ rad, or 0.563°. The stress in the y direction required to maintain the state of zero deformation in that direction is a tensile value of 46.9 MPa. The deformed shape of the element is shown in Figure 5.6(b), and compared to the case of Figure 5.4(g), the addition of the restraint in the y direction decreases the change in length and increases slightly the change in right angle.

Though the deformations in the x and y directions are very important, it is equally important to remember that accompanying these dimensional changes are changes in the z direction. Equation (5.19), or its more fundamental form, equation (5.18), is the expression for the through-the-thickness strain. If we consider the situation in Figure 5.5(e) as an example, equation (5.19) becomes

$$\varepsilon_z = 2(S_{13} - S_{23})\sin\theta\cos\theta\tau_{xy} \qquad (5.79)$$

Using numerical values of S_{13} and S_{23} from equation (2.56) yields

$$\varepsilon_z = 2(-1.600 + 37.9) \times 10^{-12}\sin(30°)\cos(30°) \times 4.40 \times 10^6$$

$$\varepsilon_z = 138.3 \ \mu \ \text{mm/mm} \qquad (5.80)$$

The element of material becomes thicker due to the application of the shear stress. The change in thickness of the element of material, Δh, is given by

$$\Delta h = \varepsilon_z h \qquad (5.81)$$

Exercises for Section 5.2

1. Verify equation (5.26). This can be done by going through the steps discussed for arriving at the equation. The factors of $1/2$ and 2 in the derivation of that equation can be confusing and lead to errors. Be sure you understand the origins of equation (5.26).

2. Suppose a 50 mm × 50 mm square of graphite-reinforced material with its fibers oriented at $+30°$ with respect to the $+x$ axis is somehow restrained from any shear deformation but is free to deform in extension in the x and y directions. The square is compressed by a stress of 25 MPa in the y direction. (*a*) What is the deformation of the element in the x and y directions? (*b*) What shear stress is required to maintain this zero shear deformation condition?

3. As we mentioned earlier, it is often easy to forget that ε_z is not zero for the plane-stress condition. Consider a 50 mm × 50 mm element of graphite-reinforced material subjected to a tensile stress of 25 MPa in the x direction and assume the element thickness h is 0.150 mm. Plot the change in thickness of this element, Δh, as a function of fiber orientation angle θ and note the orientation of maximum and minimum thickness change. Use the range $-\pi/2 \le \theta \le +\pi/2$ and comment on the oddness or evenness of Δh versus θ.

4. Repeat Exercise 3 but consider the thickness change due to an applied shear stress of 25 MPa.

Computer Exercise

It would be useful to write a short computer program to compute the elements of the $\bar{S}$ matrix as given by equation (5.26). Have the program read in and print out the values of E_1, E_2, v_{12}, and G_{12}; also print the values of S_{11}, S_{12}, S_{22}, and S_{66}. Then have the program compute and print the values of the $\bar{S}_{ij}$ as a function of θ, $-\pi/2 \le \theta \le \pi/2$. For glass-reinforced composite, plot these six quantities as a function of θ. Do the $\bar{S}_{ij}$ for glass-reinforced material vary as rapidly with θ as they do for graphite-reinforced materials? What is the reason for this?

5.3
TRANSFORMED REDUCED STIFFNESSES

The inverse of the stress-strain relations of equation (5.25) can be derived by slightly rewriting the stress-strain relation of equation (4.14) to account for the factor of 1/2 in the shear strain as

$$\left\{ \begin{array}{c} \sigma_1 \\ \sigma_2 \\ \tau_{12} \end{array} \right\} = \left[\begin{array}{ccc} Q_{11} & Q_{12} & 0 \\ Q_{12} & Q_{22} & 0 \\ 0 & 0 & 2Q_{66} \end{array} \right] \left\{ \begin{array}{c} \varepsilon_1 \\ \varepsilon_2 \\ \frac{1}{2}\gamma_{12} \end{array} \right\} \tag{5.82}$$

substituting the transformations given by equations (5.9) and (5.20) into equation (5.82), premultiplying both sides of the resulting equation by $[T]^{-1}$, and multiplying the three matrices together. This is all in analogy to equation (5.24), and the result is

$$\left\{ \begin{array}{c} \sigma_x \\ \sigma_y \\ \tau_{xy} \end{array} \right\} = \left[\begin{array}{ccc} \bar{Q}_{11} & \bar{Q}_{12} & \bar{Q}_{16} \\ \bar{Q}_{12} & \bar{Q}_{22} & \bar{Q}_{26} \\ \bar{Q}_{16} & \bar{Q}_{26} & \bar{Q}_{66} \end{array} \right] \left\{ \begin{array}{c} \varepsilon_x \\ \varepsilon_y \\ \gamma_{xy} \end{array} \right\} \tag{5.83}$$

The factors of 1/2 and 2 have been eliminated and the relation written in terms of the engineering strain. The $\bar{Q}_{ij}$ are called the *transformed reduced stiffnesses*, and sometimes the *off-axis reduced stiffnesses*, and they are defined by

$$\bar{Q}_{11} = Q_{11}m^4 + 2(Q_{12} + 2Q_{66})n^2m^2 + Q_{22}n^4$$

$$\bar{Q}_{12} = (Q_{11} + Q_{22} - 4Q_{66})n^2m^2 + Q_{12}(n^4 + m^4)$$

$$\bar{Q}_{16} = (Q_{11} - Q_{12} - 2Q_{66})nm^3 + (Q_{12} - Q_{22} + 2Q_{66})n^3m$$

$$\bar{Q}_{22} = Q_{11}n^4 + 2(Q_{12} + 2Q_{66})n^2m^2 + Q_{22}m^4$$

$$\bar{Q}_{26} = (Q_{11} - Q_{12} - 2Q_{66})n^3m + (Q_{12} - Q_{22} + 2Q_{66})nm^3$$

$$\bar{Q}_{66} = (Q_{11} + Q_{22} - 2Q_{12} - 2Q_{66})n^2m^2 + Q_{66}(n^4 + m^4)$$

(5.84)

Figure 5.7 illustrates the variations of the various components of the transformed reduced stiffness matrix with θ. Like the transformed reduced compliances, the $\bar{S}_{ij}$, the transformed reduced stiffnesses vary significantly with θ. For example, compared to its value at $\theta = 0°$, the value of $\bar{Q}_{11}$ at $\theta = 30°$ decreases 50 percent, while the value of $\bar{Q}_{66}$ increases by about a factor of 8. As with the $\bar{S}_{ij}$, $\bar{Q}_{12}$ and the on-diagonal terms are all even functions of θ, while the off-diagonal terms $\bar{Q}_{16}$ and $\bar{Q}_{26}$ are odd functions of θ. Obviously the $\bar{S}_{ij}$ of equation (5.26) are related to the $\bar{Q}_{ij}$ of equation

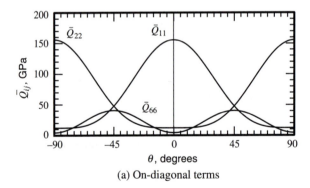

(a) On-diagonal terms

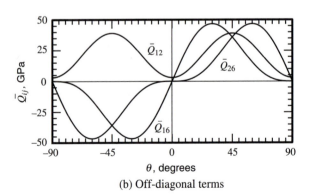

(b) Off-diagonal terms

FIGURE 5.7

Variation of transformed reduced stiffness with fiber angle θ for graphite-reinforced composite

(5.84) by

$$[\bar{Q}] = [\bar{S}]^{-1} \tag{5.85}$$

However, the form of equation (5.84) is more convenient computationally.

In analogy to equation (5.27),

$$\bar{Q}_{11}(0°) = Q_{11} \qquad \bar{Q}_{22}(0°) = Q_{22}$$
$$\bar{Q}_{12}(0°) = Q_{12} \qquad \bar{Q}_{26}(0°) = 0 \tag{5.86}$$
$$\bar{Q}_{16}(0°) = 0 \qquad \bar{Q}_{66}(0°) = Q_{66}$$

where the Q_{11}, Q_{12}, Q_{22}, and Q_{66} are often referred to as the *on-axis reduced stiffnesses* and are given by equation (4.17). For an isotropic material, in analogy to equation (5.28), from equation (4.18),

$$\bar{Q}_{11} = \frac{E}{1 - v^2}$$

$$\bar{Q}_{12} = \frac{vE}{1 - v^2}$$

$$\bar{Q}_{16} = 0$$

$$\bar{Q}_{22} = \frac{E}{1 - v^2} \tag{5.87}$$

$$\bar{Q}_{26} = 0$$

$$\bar{Q}_{66} = G = \frac{E}{2(1 + v)}$$

The transformed reduced stiffness of equation (5.84) and the stress-strain relations that use them, equation (5.83), are very important equations in the analysis of fiber-reinforced composite materials. The transformed stiffnesses will be used more frequently than the transformed reduced compliances. Like the transformed compliances, the transformed stiffnesses relate the strains as defined in the x-y-z global coordinate system to the stresses defined in that system, and the existence of the $\bar{Q}_{16}$ and $\bar{Q}_{26}$ terms, like the existence of the $\bar{S}_{16}$ and $\bar{S}_{26}$ terms, represent shear-extension coupling effects. Whereas the reduced compliances represent the deformations that result from a prescribed stress, the stiffnesses represent the stresses that must be applied to produce a prescribed deformation. Although one can use either the transformed reduced stiffnesses or the transformed reduced compliances to solve a given problem, the physical interpretation of the two quantities is so different that in a given problem it is generally more convenient to use one rather than the other. For example, consider a 50 mm × 50 mm square of material that has been stretched in the x direction by 0.050 mm. To determine the stresses required to achieve this deformation, it is convenient to use the stress-strain relations of equation (5.83) directly, as opposed to the stress-strain relations of equation (5.25).

As a parallel to the series of examples presented in the discussion of the reduced compliance matrix, a similar series of examples will next be discussed to illustrate the physical implications of the terms in the $\bar{Q}$ matrix. Shear-extension coupling and sign sensitivity of the $\bar{Q}_{16}$ and $\bar{Q}_{26}$ terms will again be evident. The particular examples can be considered the complement of the examples presented previously. They are termed *complementary* examples because the strain variable in each stress-strain pair is specified, whereas before, the stress variable of the pair was specified. These examples follow.

A 50 mm × 50 mm element of aluminum, in Figure 5.8(a), is stretched 0.050 mm in the x direction. The y dimension does not change and the right corner angles remain right. Interest focuses on the stresses required to effect this deformation. Assume the element is in a state of plane stress. Note the complementary nature of this problem relative to the problem discussed in Figure 5.4(a). In the present situation the strains of the stress-strain pairs are known and it is the stresses that are unknown and to be solved for. In the situation of Figure 5.4(a) the stresses of the stress-strain pairs were known and it was the strains that were unknown and to be solved for. Specifically, in Figure 5.4(a) σ_x was the only nonzero stress, whereas in Figure 5.7(a) ε_x is the only

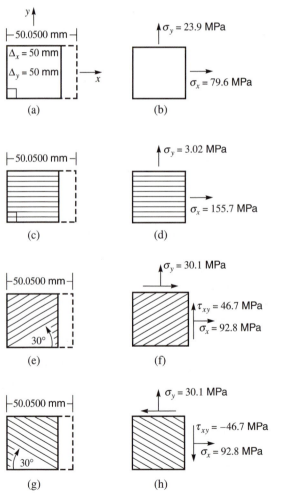

FIGURE 5.8
Stress required in aluminum and graphite-reinforced material to produce extensional strain

nonzero strain. For the deformation described in Figure 5.8(a)

$$\varepsilon_x = \frac{\delta\Delta_x}{\Delta_x} = \frac{0.050}{50} = 1000 \times 10^{-6}$$

$$\varepsilon_y = \frac{\delta\Delta_y}{\Delta_y} = \frac{0}{50} = 0$$

(5.88a)

and because the right corner angles remain right,

$$\gamma_{xy} = 0 \qquad (5.88b)$$

To compute the stresses required to produce this deformation, the stress-strain relations of equation (5.83) can be used, resulting in

$$\sigma_x = \bar{Q}_{11}\varepsilon_x$$

$$\sigma_y = \bar{Q}_{12}\varepsilon_x \qquad (5.89)$$

$$\tau_{xy} = \bar{Q}_{16}\varepsilon_x$$

Because aluminum is isotropic, the $\bar{Q}_{ij}$ of equation (5.87) indicate equation (5.89) becomes

$$\sigma_x = \frac{E}{1-v^2}\varepsilon_x$$

$$\sigma_y = \frac{vE}{1-v^2}\varepsilon_x \qquad (5.90)$$

$$\tau_{xy} = 0 \times \varepsilon_x$$

Substituting in numerical values for aluminum from Table 2.1 yields

$$\sigma_x = \frac{72.4 \times 10^9}{1-(0.3)^2}1000 \times 10^{-6} = 79.6 \text{ MPa}$$

$$\sigma_y = \frac{(0.3)(72.4 \times 10^9)}{1-(0.3)^2}1000 \times 10^{-6} = 23.9 \text{ MPa}$$

(5.91)

$$\tau_{xy} = 0$$

Figure 5.8(b) illustrates this stress state, and the results correlate with our past experience in that a tensile stress in the y direction is required to overcome the tendency of the material to contract in the y direction. Because $\varepsilon_y = 0$, a tensile stress is required to enforce this. Also, the specimen changes thickness in the z direction.

Consider next, as in Figure 5.8(c), an element of graphite-reinforced composite in a state of plane stress with its fibers aligned with the x axis and stretched in the x direction by 0.050 mm, with no deformation in the y direction. Equation (5.88) just applied to aluminum defines the strain state. For this case the stress-strain relations, equation (5.83), reduce to

$$\sigma_x = \bar{Q}_{11}(0°)\varepsilon_x$$

$$\sigma_y = \bar{Q}_{12}(0°)\varepsilon_x \qquad (5.92)$$

$$\tau_{xy} = \bar{Q}_{16}(0°)\varepsilon_x$$

and by equation (5.86) to

$$\sigma_x = Q_{11}\varepsilon_x$$
$$\sigma_y = Q_{12}\varepsilon_x \tag{5.93}$$
$$\tau_{xy} = 0 \times \varepsilon_x = 0$$

For the graphite-reinforced material, we find using equation (4.17) and the numerical values of the engineering properties from Table 2.1, or alternatively, equations (2.58) and (4.15), or equations (2.56) and (4.16), that

$$Q_{11} = 155.7 \text{ GPa} \qquad Q_{22} = 12.16$$
$$Q_{12} = 3.02 \qquad Q_{66} = 4.40 \tag{5.94}$$

Using the applicable stiffness from equation (5.94), the stresses given by equation (5.93) are

$$\sigma_x = (155.7 \times 10^9)(1000 \times 10^{-6}) = 155.7 \text{ MPa}$$
$$\sigma_y = (3.02 \times 10^9)(1000 \times 10^{-6}) = 3.02 \text{ MPa} \tag{5.95}$$
$$\tau_{xy} = 0$$

Because of the relative values of E for aluminum and E_1 for the graphite-reinforced material, stretching the graphite-reinforced material by 0.050 mm takes about twice as much stress as stretching the aluminum the same amount. However, restraining the deformation in the y direction in the graphite-reinforced material, 3.02 MPa, equation (5.95), takes far less than restraining the deformation in the aluminum, 23.9 MPa, equation (5.91), both in terms of absolute stress level, and in terms of stress level relative to σ_x. The smaller value of σ_y required for the graphite-reinforced material is a direct result of the small value of v_{21}.

Consider now the situation in Figure 5.8(e), a 50 mm $\times$ 50 mm element of graphite-reinforced material with the fibers oriented at $+30°$ with respect to the $+x$ axis and stretched 0.050 mm in the x direction. The strains are again given by equation (5.88), and the stresses required to produce these strains are

$$\sigma_x = \bar{Q}_{11}(30°)\varepsilon_x$$
$$\sigma_y = \bar{Q}_{12}(30°)\varepsilon_x \tag{5.96}$$
$$\tau_{xy} = \bar{Q}_{16}(30°)\varepsilon_x$$

With $\theta = 30°$, $m = \sqrt{3}/2$ and $n = 1/2$, and equations (5.94) and (5.84) give

$$\bar{Q}_{11}(30°) = 92.8 \text{ GPa} \qquad \bar{Q}_{22}(30°) = 21.0$$
$$\bar{Q}_{12}(30°) = 30.1 \qquad \bar{Q}_{26}(30°) = 15.5 \tag{5.97}$$
$$\bar{Q}_{16}(30°) = 46.7 \qquad \bar{Q}_{66}(30°) = 31.5$$

Using these numerical values, we find that the stresses required to produce the prescribed deformations are

$$\sigma_x = (92.8 \times 10^9)(1000 \times 10^{-6}) = 92.8 \text{ MPa}$$
$$\sigma_y = (30.1 \times 10^9)(1000 \times 10^{-6}) = 30.1 \text{ MPa} \tag{5.98}$$
$$\tau_{xy} = (46.7 \times 10^9)(1000 \times 10^{-6}) = 46.7 \text{ MPa}$$

Remarkably, a shear stress must be applied to the element of material, in addition to the other two stresses, in order to have the element simply elongate in the x direction. Figure 5.8(f) illustrates the stresses for this example. If this shear stress was not applied, then shearing deformations would result and the deformation would not be just a simple elongation in the x direction. Equally remarkable is the fact that the shear stress required to stop the right corner angles from changing is larger than the stress σ_y required to restrain the material against deformation in the y direction (i.e., resistance to Poisson effects).

As a final example of the stresses required to effect a simple elongation in the x direction, consider an element of graphite-reinforced material with the fibers oriented in the $-30°$ direction, as in Figure 5.8(g). For this situation

$$\sigma_x = \bar{Q}_{11}(-30°)\varepsilon_x$$
$$\sigma_y = \bar{Q}_{12}(-30°)\varepsilon_x \qquad (5.99)$$
$$\tau_{xy} = \bar{Q}_{16}(-30°)\varepsilon_x$$

and from equation (5.97) and the evenness and oddness properties of the $\bar{Q}_{ij}$,

$$\bar{Q}_{11}(-30°) = +\bar{Q}_{11}(+30°) = 92.8 \text{ GPa}$$
$$\bar{Q}_{12}(-30°) = +\bar{Q}_{12}(+30°) = 30.1$$
$$\bar{Q}_{16}(-30°) = -\bar{Q}_{16}(+30°) = -46.7$$
$$\bar{Q}_{22}(-30°) = +\bar{Q}_{22}(+30°) = 21.0 \qquad (5.100)$$
$$\bar{Q}_{26}(-30°) = -\bar{Q}_{26}(+30°) = -15.5$$
$$\bar{Q}_{66}(-30°) = +\bar{Q}_{66}(+30°) = 31.5$$

Substituting these into equation (5.99) yields the stresses required, namely,

$$\sigma_x = (92.8 \times 10^9)(1000 \times 10^{-6}) = 92.8 \text{ MPa}$$
$$\sigma_y = (30.1 \times 10^9)(1000 \times 10^{-6}) = 30.1 \text{ MPa} \qquad (5.101)$$
$$\tau_{xy} = (-46.7 \times 10^9)(1000 \times 10^{-6}) = -46.7 \text{ MPa}$$

Again, shear stresses must be applied to the element, in addition to the other two stresses, to obtain the simple state of elongation in the x direction. However, reversing the fiber orientation is responsible for the change in the sign of the shear stress required to effect the deformation. The stress to restrain the deformation in the y direction is insensitive to the sign of the off-axis fiber orientation, as is the stress in the x direction. Figure 5.8(h) illustrates the stresses of equation (5.101).

This just-completed series of examples illustrates the physical interpretation of the reduced stiffness matrix, and it also illustrates, in a different fashion, the existence, level, and character of shear-extension coupling, a coupling inherent in fiber-reinforced composite materials. As expected, the $\bar{Q}_{16}$ and $\bar{Q}_{26}$ serve double duty in regard to shear-extension coupling. We will illustrate this type of coupling from another viewpoint shortly. First, however, the thickness change in one of the previous examples will be discussed.

Because the stress component σ_z has been specified to be zero in a state of plane stress, ε_z cannot be specified, whether we specify ε_x, ε_y, and γ_{xy}, as in these recent examples, or whether we specify σ_x, σ_y, and τ_{xy}, as in the previous examples. Nevertheless, ε_z is not zero and it can be calculated. Consider the example of Figure 5.8(e), where the element of fiber-reinforced material with its fibers oriented at $+30°$ relative to the $+x$ axis is stretched in the x direction 0.050 mm. No deformation is allowed in the y direction or in shear. The resulting stresses were given by equation (5.98) and illustrated in Figure 5.8(f). Equation (5.19) provides us with the strain in the z direction. Specifically, using $\theta = +30°$, the stresses from equation (5.98), and the material properties for S_{13} and S_{23} from equation (2.56), we find

$$\varepsilon_z = [(-1.600 \times 10^{-12}) \cos^2(30°) + (-37.9 \times 10^{-12}) \sin^2(30°)](92.8 \times 10^6)$$

$$+ [(-1.600 \times 10^{-12}) \sin^2(30°) + (-37.9 \times 10^{-12}) \cos^2(30°)](30.1 \times 10^6)$$

$$+ 2[(-1.600 \times 10^{-12}) - (-37.9 \times 10^{-12})] \sin(30°) \cos(30°)(46.7 \times 10^6)$$

$$\varepsilon_z = -389 \times 10^{-6}$$

$$(5.102)$$

Thus, the element of material becomes thinner as a result of the prescribed deformation in the x-y plane.

Finally, to round out the discussion of the transformed reduced stiffness, consider the situation in Figure 5.9(a), which shows a pure shearing deformation being prescribed for the 50 mm × 50 mm square element of aluminum, with the right corner angle in the lower left hand corner decreasing by 1000 μ rad, or 0.0573°. With a pure shearing deformation the lengths of the sides do not change. The prescribed strain state is thus given by

$$\varepsilon_x = 0$$
$$\varepsilon_y = 0 \qquad (5.103)$$
$$\gamma_{xy} = 1000 \times 10^{-6}$$

and it is assumed the element is in a state of plane stress; that is, σ_z, τ_{xz}, and τ_{yz} are zero. With this, the stress-strain relations of equation (5.83) reduce to

$$\sigma_x = \bar{Q}_{16}\gamma_{xy}$$
$$\sigma_y = \bar{Q}_{26}\gamma_{xy} \qquad (5.104)$$
$$\tau_{xy} = \bar{Q}_{66}\gamma_{xy}$$

For aluminum, from equation (5.87),

$$\sigma_x = 0 \times \gamma_{xy} = 0$$
$$\sigma_y = 0 \times \gamma_{xy} = 0 \qquad (5.105)$$
$$\tau_{xy} = G\gamma_{xy}$$

and from Table 2.1,

$$\sigma_x = 0$$
$$\sigma_y = 0 \qquad (5.106)$$
$$\tau_{xy} = (27.8 \times 10^9)(1000 \times 10^{-6}) = 27.8 \text{ MPa}$$

As our experience tells us, for isotropic materials only a shear stress is required to produce a prescribed shear strain, as in Figure 5.9(b).

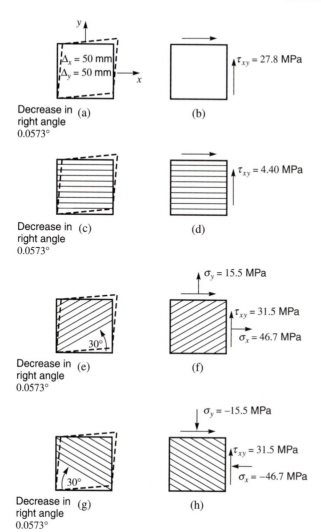

FIGURE 5.9
Stress required in aluminum
and graphite-reinforced
material to produce a
positive shear strain

Prescribing the same pure shear strain state on an element of graphite-reinforced material with its fiber aligned with the x axis, Figure 5.9(c), results in, from equation (5.83),

$$\sigma_x = \bar{Q}_{16}(0°)\gamma_{xy}$$
$$\sigma_y = \bar{Q}_{26}(0°)\gamma_{xy} \qquad (5.107)$$
$$\tau_{xy} = \bar{Q}_{66}(0°)\gamma_{xy}$$

From equations (5.86) and (5.94),

$$\sigma_x = 0 \times \gamma_{xy} = 0$$
$$\sigma_y = 0 \times \gamma_{xy} = 0 \qquad (5.108)$$
$$\tau_{xy} = Q_{66}\gamma_{xy} = (4.40 \times 10^9)(1000 \times 10^{-6}) = 4.40 \text{ MPa}$$

Due to the difference in values between G for aluminum and G_{12}, the level of shear stress required to produce $\gamma_{xy} = 1000\ \mu$ rad in the graphite-reinforced material with its fibers aligned with the x axis is much less than the shear stress required to produce that same shear strain in the aluminum, as Figure 5.9(d) illustrates.

With the fibers oriented at $+30°$ relative to the $+x$ axis, Figure 5.9(e), the stresses required to maintain the pure shear deformation are given by

$$\sigma_x = \bar{Q}_{16}(30°)\gamma_{xy}$$
$$\sigma_y = \bar{Q}_{26}(30°)\gamma_{xy} \tag{5.109}$$
$$\tau_{xy} = \bar{Q}_{66}(30°)\gamma_{xy}$$

From equation (5.97),

$$\sigma_x = (46.7 \times 10^9)(1000 \times 10^{-6}) = 46.7\ \text{MPa}$$
$$\sigma_y = (15.5 \times 10^9)(1000 \times 10^{-6}) = 15.5\ \text{MPa} \tag{5.110}$$
$$\tau_{xy} = (31.5 \times 10^9)(1000 \times 10^{-6}) = 31.5\ \text{MPa}$$

As might be expected by now, equation (5.110) leads to the conclusion that to produce a pure shear deformation with the fibers off-axis at $+30°$ requires not only a shear stress, but also tensile stresses σ_x and σ_y, as in Figure 5.9(f). Without these extensional stresses, the state of pure shear deformation could not exist.

As the final example in this series, consider the case with the fibers oriented at $-30°$, Figure 5.9(g). From equation (5.83)

$$\sigma_x = \bar{Q}_{16}(-30°)\gamma_{xy}$$
$$\sigma_y = \bar{Q}_{26}(-30°)\gamma_{xy} \tag{5.111}$$
$$\tau_{xy} = \bar{Q}_{66}(-30°)\gamma_{xy}$$

and using equation (5.100) leads to

$$\sigma_x = (-46.7 \times 10^9)(1000 \times 10^{-6}) = -46.7\ \text{MPa}$$
$$\sigma_y = (-15.5 \times 10^9)(1000 \times 10^{-6}) = -15.5\ \text{MPa} \tag{5.112}$$
$$\tau_{xy} = (31.5 \times 10^9)(1000 \times 10^{-6}) = 31.5\ \text{MPa}$$

The values of the stresses computed for this case indicate that to effect the prescribed shear strain, when the fibers are at $\theta = -30°$ the element must be *compressed* in the x and y directions. Recall, the results of equation (5.110) indicate that when $\theta = +30°$, *tension* must be applied in the x and y directions! This is strange and unpredictable behavior, to say the least. However, this is the nature of fiber-reinforced composite materials. It emphasizes the fact that it is very dangerous to second-guess the answers. Rely on the equations, use them correctly, and you will not have to guess at the results. Furthermore, as you study composites in greater depth, you will learn that it is in your best interest to not guess. You will be wrong too many times!

This completes the series of examples designed to illustrate the important similarities and differences between the responses of an isotropic material, in this case aluminum, and a fiber-reinforced material, in this case a graphite-reinforced material. The series of examples provides a physical interpretation to the elements of the transformed reduced compliance and stiffness matrices. In summary, the elements of the compliance matrix represent the deformations that result from a simple applied stress, and the elements of the stiffness matrix represent the stresses that must be applied to effect a simple state of deformation.

Exercises for Section 5.3

1. Verify equation (5.84). This can be done by going through the steps discussed for arriving at the equations. The factors of $1/2$ and 2 can be confusing and lead to errors, so be sure you understand the origin of equation (5.84).

2. A 50 mm $\times$ 50 mm square of graphite-reinforced material with its fibers oriented at $+30°$ with respect to the $+x$ axis is compressed 0.050 mm in both the x and y directions. No shear deformations are allowed. What is the state of stress σ_x, σ_y, and τ_{xy} required to produce this bidirectional compression?

3. Consider the elements of material in Figure 5.9. (*a*) What is the thickness change, in mm, of the element of material subjected to the shearing deformation in Figure 5.9(e)? (*b*) Is the thickness change of the case in Figure 5.9(g) the same? Assume the element of material is originally 0.150 mm thick.

Computer Exercise

Modify and add to the computer program written in the exercises for Section 4.3 to have it compute and print the six elements of the $\bar{Q}$ matrix given by equation (5.84). Use the program to compute and print the values of the $\bar{Q}_{ij}$ as a function of θ, $-\pi/2 \leq \theta \leq +\pi/2$. For glass-reinforced material, plot these six quantities as a function of θ. Do the $\bar{Q}_{ij}$ vary as rapidly with θ for glass-reinforced material as for graphite-reinforced material?

5.4
ENGINEERING PROPERTIES IN A GLOBAL COORDINATE SYSTEM

In Chapter 2 we began our studies of the elastic response of fiber-reinforced composites by introducing the engineering properties in the 1-2-3 principal material coordinate system (E_1, E_2, etc.). These engineering properties are, of course, indirectly involved in the transformed reduced stiffnesses and compliances we have just studied. Engineering properties can also be defined in the x-y-z global coordinate system. These are often of much more use than, say, the reduced compliances or stiffnesses. The extensional moduli, Poisson's ratios, and the shear modulus may mean considerably more to many designers and engineers, because the physical interpretation of these quantities is well established and understood. The engineering properties in the x-y-z system are derivable directly from their definitions, just as they were in Chapter 2 for the 1-2-3 system. In this section we shall derive the engineering properties that are important when considering a state of plane stress. They are related to the transformed reduced compliances and hence can ultimately be written in terms of the engineering properties in the 1-2-3 system.

Consider, as in Figure 5.10(a), an off-axis element of fiber-reinforced composite material in the x-y-z system with its fiber oriented at some angle θ with respect to the x axis. The element is subjected to a normal tensile stress σ_x and all other stresses are zero, much like the situation in Figure 5.4(e), though here the angle θ and the magnitude of the tensile stress are arbitrary. In response to this applied stress, the element stretches in the x direction, contracts in the y direction, and because the fibers are not aligned with the x-axis, the right corner angles do not remain right. The extensional strain in the x direction is related to the stress in the x direction by the extensional modulus in the x direction, E_x; the relation between these two quantities is, by definition,

$$\varepsilon_x = \frac{\sigma_x}{E_x} \tag{5.113}$$

But according to the strain-stress relations of equation (5.25), for this situation, as we have used so often,

$$\varepsilon_x = \bar{S}_{11}\sigma_x \tag{5.114}$$

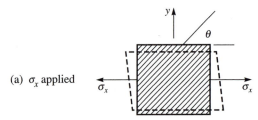

(a) σ_x applied

FIGURE 5.10
Off-axis element with simple stress states for definition of engineering properties

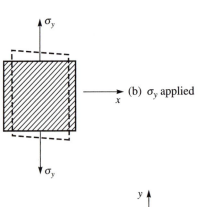

(b) σ_y applied

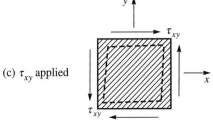

(c) τ_{xy} applied

so by the similiarity of these two equations,

$$\frac{1}{E_x} = \bar{S}_{11} \tag{5.115}$$

we can thus define the extensional modulus in the x direction to be

$$E_x \equiv \frac{1}{\bar{S}_{11}} \tag{5.116}$$

Using the definition of the $\bar{S}_{ij}$, equation (5.26), and the definitions of the S_{ij}, we can write E_x in terms of the principal material coordinate system engineering properties as

$$E_x = \frac{E_1}{m^4 + \left(\dfrac{E_1}{G_{12}} - 2v_{12}\right) n^2 m^2 + \dfrac{E_1}{E_2} n^4} \tag{5.117}$$

It is clear that the extensional modulus in the x direction involves the inplane shear modulus and Poisson's ratio, as well as the extensional modulus, in the 1-2-3 system.

   In addition to stretching in the x direction, the element contracts in the y direction when subjected to a tensile stress in the x direction; see Figure 5.10(a). By definition, the relation between the contraction strain in the y direction and the extensional strain in the x direction, due to simply a tensile stress in the x direction, is a Poisson's ratio. Retaining the convention established for the subscripting of the various Poisson's ratios, we define

$$v_{xy} = -\frac{\varepsilon_y}{\varepsilon_x} \tag{5.118}$$

where the first subscript refers to the direction of the applied stress and second subscript refers to the direction of contraction. Referring again to the strain-stress relations of equation (5.25), for this situation, we find that

$$\varepsilon_y = \bar{S}_{12}\sigma_x \tag{5.119}$$

Using equations (5.114) and (5.119) in the definition of v_{xy}, we obtain

$$v_{xy}' = -\frac{\bar{S}_{12}}{\bar{S}_{11}} \tag{5.120}$$

and using the definitions of the $\bar{S}_{ij}$ and the S_{ij},

$$v_{xy} = \frac{v_{12}(n^4 + m^4) - \left(1 + \dfrac{E_1}{E_2} - \dfrac{E_1}{G_{12}}\right) n^2 m^2}{m^4 + \left(\dfrac{E_1}{G_{12}} - 2v_{12}\right) n^2 m^2 + \dfrac{E_1}{E_2} n^2} \tag{5.121}$$

This Poisson's ratio, like E_x, is a function of many of the properties defined in the 1-2-3 system.

   If instead of subjecting the element of fiber-reinforced material to a stress σ_x, it is subjected to a tensile stress in the y direction, the conditions illustrated in Figure 5.10(b) result. The element stretches in the y direction, contracts in the x direction, and the right angle does not remain right. We can determine the extensional modulus in the y direction by considering that the extensional strain in the y direction, for this simple stress state, is given by

$$\varepsilon_y = \frac{\sigma_y}{E_y} \tag{5.122}$$

and using the fact that

$$\varepsilon_y = \bar{S}_{22}\sigma_y \tag{5.123}$$

we define the extensional modulus in the y direction as

$$E_y \equiv \frac{1}{\bar{S}_{22}} \tag{5.124}$$

or

$$E_y = \frac{E_2}{m^4 + \left(\dfrac{E_2}{G_{12}} - 2\nu_{21}\right)n^2m^2 + \dfrac{E_2}{E_1}n^4} \tag{5.125}$$

Poisson's ratio, due to the stress in the y direction, is

$$\nu_{yx} = -\frac{\varepsilon_x}{\varepsilon_y} \tag{5.126}$$

where for this situation,

$$\varepsilon_x = \bar{S}_{12}\sigma_y \tag{5.127}$$

resulting in

$$\nu_{yx} = -\frac{\bar{S}_{12}}{\bar{S}_{22}} \tag{5.128}$$

This Poisson's ratio can now be written in terms of the engineering properties in the 1-2-3 system as

$$\nu_{yx} = \frac{\nu_{21}(n^4 + m^4) - \left(1 + \dfrac{E_2}{E_1} - \dfrac{E_2}{G_{12}}\right)n^2m^2}{m^4 + \left(\dfrac{E_2}{G_{12}} - 2\nu_{21}\right)n^2m^2 + \dfrac{E_2}{E_1}n^2} \tag{5.129}$$

Finally, if the element is subjected to a shear stress τ_{xy}, it will deform as in Figure 5.10(c). The change in the right corner angle in the x-y plane is denoted by γ_{xy}, and by definition it is related to the applied shear stress by the shear modulus in the x-y plane, namely,

$$\gamma_{xy} = \frac{\tau_{xy}}{G_{xy}} \tag{5.130}$$

The shear strain is also given by

$$\gamma_{xy} = \bar{S}_{66}\tau_{xy} \tag{5.131}$$

so the shear modulus is

$$G_{xy} = \frac{1}{\bar{S}_{66}} \tag{5.132}$$

In terms of engineering properties in the 1-2-3 system,

$$G_{xy} = \frac{G_{12}}{n^4 + m^4 + 2\left(2\dfrac{G_{12}}{E_1}(1 + 2\nu_{12}) + 2\dfrac{G_{12}}{E_2} - 1\right)n^2m^2} \tag{5.133}$$

As expected, the shear modulus in the x-y plane involves engineering properties in the 1-2-3 system, in particular, the fiber-direction modulus, E_1. The large value of E_1 is expected to have an effect as the fiber angle increases from $\theta = 0°$; the effect becomes greatest when $\theta = 45°$ or when the fibers stiffen the diagonal direction of the element.

In summary, the engineering properties are defined as

$$
E_x = \frac{E_1}{m^4 + \left(\dfrac{E_1}{G_{12}} - 2\nu_{12}\right)n^2m^2 + \dfrac{E_1}{E_2}n^4}
$$

$$
\nu_{xy} = \frac{\nu_{12}(n^4 + m^4) - \left(1 + \dfrac{E_1}{E_2} - \dfrac{E_1}{G_{12}}\right)n^2m^2}{m^4 + \left(\dfrac{E_1}{G_{12}} - 2\nu_{12}\right)n^2m^2 + \dfrac{E_1}{E_2}n^2}
$$

$$
E_y = \frac{E_2}{m^4 + \left(\dfrac{E_2}{G_{12}} - 2\nu_{21}\right)n^2m^2 + \dfrac{E_2}{E_1}n^4}
$$ (5.134)

$$
\nu_{yx} = \frac{\nu_{21}(n^4 + m^4) - \left(1 + \dfrac{E_2}{E_1} - \dfrac{E_2}{G_{12}}\right)n^2m^2}{m^4 + \left(\dfrac{E_2}{G_{12}} - 2\nu_{21}\right)n^2m^2 + \dfrac{E_2}{E_1}n^2}
$$

$$
G_{xy} = \frac{G_{12}}{n^4 + m^4 + 2\left(2\dfrac{G_{12}}{E_1}(1 + 2\nu_{12}) + 2\dfrac{G_{12}}{E_2} - 1\right)n^2m^2}
$$

The variations with θ of the off-axis engineering properties for a graphite-reinforced material are illustrated in Figure 5.11. Note how rapidly E_x decreases with off-axis angle. At $\theta = \pm 10° E_x$ is but 50 percent of its value at $\theta = 0°$. This can be interpreted to mean that if the fibers are not oriented exactly as intended, then the value of E_x, and perhaps the performance of the composite, could be considerably less than expected. In analogous fashion, the modulus E_y is quite small but then increases rapidly as θ approaches $\pm 90°$. Note also the maximum value of G_{xy} occurs at $\theta = \pm 45°$ and is greater by roughly a factor of two than its value at $\theta = 0$ and $90°$.

As the series of examples in the previous section indicated, and as Figure 5.10(a) illustrates, when subjecting an off-axis element of fiber-reinforced material to a simple tensile loading, a shear response results. Likewise, when subjecting an off-axis element of fiber-reinforced material to a simple shear loading, extensional strains result. Neither of these responses fits the classical definitions of material properties (i.e., extensional modulus, Poisson's ratio, or shear modulus). For isotropic materials there is not a definition of a material property that relates shear strain to normal stress, or extension strain to shear stress. Yet these responses occur for composite materials. The next section formally defines material properties associated with these coupled responses.

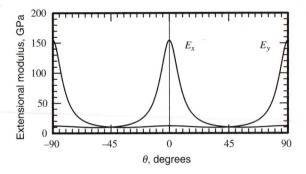

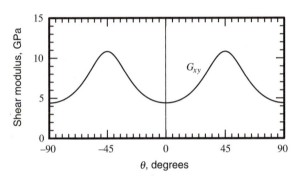

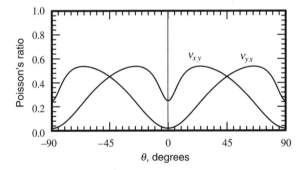

FIGURE 5.11
Variation of engineering properties with fiber angle θ for graphite-reinforced composite

5.5
COEFFICIENTS OF MUTUAL INFLUENCE

For fiber-reinforced composites it is useful to define several other material properties that can be used to categorize response. The properties are defined in the same context as the engineering properties and thus it is appropriate to introduce them at this time. These properties have as their basis the fact that an element of fiber-reinforced material with its fiber oriented at some arbitrary angle exhibits a shear strain when subjected to a normal stress, and it also exhibits extensional strain when subjected to a shear stress. Poisson's ratio is defined as the ratio of extensional strains, given that the element is subjected to only a single normal stress. By analogy, the *coefficient of mutual influence of the second kind* is defined as the ratio of a shear strain to an extensional strain, given that the element is subjected to only a single normal stress. The *coefficient of mutual influence of the first kind* is defined as the ratio of an extensional strain to a shear strain, given that the element is subjected to only a single shear stress. These coefficients of mutual influence can be thought of as a generalization of Poisson's ratios, as they are defined as ratios of strains.

Formally, one coefficient of mutual influence of the second kind is defined as

$$\eta_{xy,x} \equiv \frac{\gamma_{xy}}{\varepsilon_x}$$

(5.135)

when $\sigma_x \neq 0$ and all other stresses are 0

Another coefficient of mutual influence of the second kind is defined as

$$\eta_{xy,y} \equiv \frac{\gamma_{xy}}{\varepsilon_y}$$

(5.136)

when $\sigma_y \neq 0$ and all other stresses are 0

These coefficients relate the shear strains caused by fiber orientation effects and a normal stress to the extensional strain that is a direct result of this normal stress. This normal stress is the only stress present. In terms of the transformed reduced compliances, the coefficients of mutual influence of the second kind are given by

$$\eta_{xy,x} = \frac{\bar{S}_{16}}{\bar{S}_{11}} \qquad \eta_{xy,y} = \frac{\bar{S}_{26}}{\bar{S}_{22}}$$

(5.137)

Figure 5.12 shows the variations with θ of the coefficients of mutual influence of the second kind for graphite-reinforced material. The largest interaction between shearing and extensional effects for this particular material occurs near $\pm 8°$ for

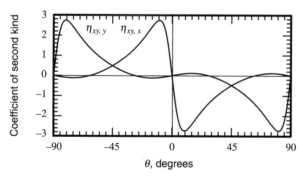

FIGURE 5.12
Variation of coefficients of mutual influence with fiber angle θ for graphite-reinforced composite

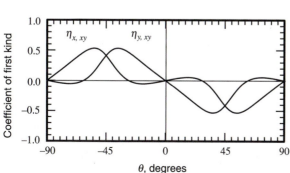

$\eta_{xy,x}$ and near $\pm 82°$ for $\eta_{xy,y}$. One might expect the largest interaction to occur when $\theta = 45°$, but this is not the case.

The coefficients of mutual influence of the first kind are defined as

$$\eta_{x,xy} \equiv \frac{\varepsilon_x}{\gamma_{xy}} \qquad \eta_{y,xy} \equiv \frac{\varepsilon_y}{\gamma_{xy}} \tag{5.138}$$

when $\tau_{xy} \neq 0$ and all other stresses are 0

These coefficients relate the extensional strains caused by fiber orientation effects and a shear stress to the shear strain that is a direct result of this shear stress. The shear stress is the only stress present. In terms of transformed reduced compliances,

$$\eta_{x,xy} = \frac{\bar{S}_{16}}{\bar{S}_{66}} \quad \text{and} \quad \eta_{y,xy} = \frac{\bar{S}_{26}}{\bar{S}_{66}} \tag{5.139}$$

Figure 5.12 illustrates the variation of these two coefficients for graphite-reinforced composite. These two coefficients having similar ranges of values. The coefficient $\eta_{y,xy}$ has the greatest magnitude near $\theta = \pm 35°$, while $\eta_{x,xy}$ has the greatest magnitude near $\theta = \pm 55°$.

Exercises for Section 5.5

1. Plot the variation of the engineering properties as a function of fiber angle θ, $-\pi/2 \leq \theta \leq +\pi/2$, for glass-reinforced composite. Comment on whether the decrease of E_x as θ increases or decreases from $\theta = 0°$ is as extreme for glass-reinforced material is it is for graphite-reinforced material.

2. Plot the coefficients of mutual influence of the first and second kind for glass-reinforced composite. For each coefficient, what values of θ correspond to the regions of greatest influence? Are these regions the same as for a graphite-reinforced material?

5.6
FREE THERMAL AND FREE MOISTURE STRAINS

To study thermally induced deformations in an x-y-z global coordinate system, transformation from the 1-2-3 principal material coordinate system is necessary. Because they are legitimate strains, the transformations of equation (5.15) are valid for thermally induced, or moisture-induced, strains. Considering free thermal strains, the transformations become

$$\varepsilon_x^T(T, T_{ref}) = \cos^2\theta \varepsilon_1^T(T, T_{ref}) + \sin^2\theta \varepsilon_2^T(T, T_{ref})$$

$$\varepsilon_y^T(T, T_{ref}) = \sin^2\theta \varepsilon_1^T(T, T_{ref}) + \cos^2\theta \varepsilon_2^T(T, T_{ref})$$

$$\varepsilon_z^T(T, T_{ref}) = \varepsilon_3^T(T, T_{ref})$$

$$\gamma_{yz}^T(T, T_{ref}) = 0 \tag{5.140}$$

$$\gamma_{xz}^T(T, T_{ref}) = 0$$

$$\frac{1}{2}\gamma_{xy}^T(T, T_{ref}) = (\varepsilon_1^T(T, T_{ref}) - \varepsilon_2^T(T, T_{ref}))\sin\theta\cos\theta$$

Note, equation (5.140) uses the reverse of the transformation of equation (5.15). Equation (5.140) is rather simple because in the principal material coordinate system there are no free thermal shear strains. If the strains are assumed to be linearly dependent on the difference between a particular temperature, T, and the reference temperature, T_{ref}, then, if we use equations (2.98) and (2.99), equation (5.140) becomes

$$\varepsilon_x^T(T, T_{ref}) = (\cos^2\theta\alpha_1 + \sin^2\theta\alpha_2)\Delta T$$

$$\varepsilon_y^T(T, T_{ref}) = (\sin^2\theta\alpha_1 + \cos^2\theta\alpha_2)\Delta T$$

$$\varepsilon_z^T(T, T_{ref}) = \alpha_3\Delta T \tag{5.141}$$

$$\gamma_{xy}^T(T, T_{ref}) = 2(\alpha_1 - \alpha_2)\sin\theta\cos\theta\Delta T$$

If the *coefficients of thermal deformation* in the x-y-z system are defined to be such that, due to a temperature change ΔT,

$$\varepsilon_x^T(T, T_{ref}) = \alpha_x\Delta T$$

$$\varepsilon_y^T(T, T_{ref}) = \alpha_y\Delta T$$

$$\gamma_{xy}^T(T, T_{ref}) = \alpha_{xy}\Delta T \tag{5.142}$$

$$\varepsilon_z^T(T, T_{ref}) = \alpha_z\Delta T$$

then we can define the *coefficients of thermal deformation*, or CTD, in the x-y-z system to be

$$\boxed{\begin{aligned}\alpha_x &= \alpha_1\cos^2\theta + \alpha_2\sin^2\theta \\ \alpha_y &= \alpha_1\sin^2\theta + \alpha_2\cos^2\theta \\ \alpha_{xy} &= 2(\alpha_1 - \alpha_2)\cos\theta\sin\theta \\ \alpha_z &= \alpha_3\end{aligned}} \tag{5.143}$$

Needless to say, though there are no free thermal shear strains in the 1-2-3 system, this is not the case for the x-y-z system. Heating or cooling a small element of material with its fibers not aligned with the x or y axis results in a change in the right angle of the corners in the x-y plane. Using the values of α_1, α_2, and α_3 from Table 2.1, Figure 5.13 shows the variation of α_x, α_y, and α_{xy}, and α_z with θ for a graphite-reinforced material. Of course, α_z doesn't vary with θ and is equal to α_3. The coefficients of thermal deformation α_x and α_y essentially interchange roles as θ changes from $-90°$ to $+90°$. They are both even functions of θ. The coefficient of thermal deformation α_{xy} is an odd function of θ and attains a value as large as α_x and α_y (and α_z) at $\theta = \pm45°$.

A simple example will serve to underscore the importance of fiber orientation effects on free thermal strains. Consider, as in Figure 5.14(a), an unconstrained 50 mm × 50 mm × 50 mm off-axis element of graphite-reinforced material with its fibers oriented at 45° relative to the x-axis. Assume the temperature of the material is increased by 50°C and the deformed shape of the material and the lengths of the

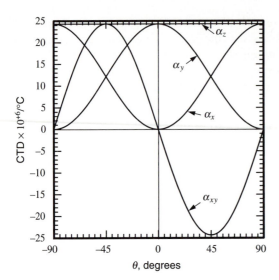

FIGURE 5.13
Variation of coefficients of thermal deformation with fiber angle θ for graphite-reinforced composite

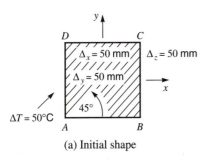

FIGURE 5.14
Thermal deformations of an unconstrained off-axis element

(a) Initial shape

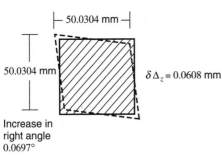

(b) Deformed shape

original 50 mm sides are of interest. From Table 2.1

$$\alpha_1 = -0.018 \times 10^{-6}/°C$$

$$\alpha_2 = 24.3 \times 10^{-6}/°C \tag{5.144}$$

and because the material shrinks in the fiber direction when heated, diagonal AC

contracts. On the other hand, diagonal BD expands. Intuitively, then, the corners A, B, C, and D cannot remain orthogonal when the material is heated. As Figure 5.14(b) shows, corners A and C must open, while corners D and B must close. Quantitative information regarding the shape changes can be obtained by using equation (5.143) as

$$\alpha_x = (-0.018 \cos^2 45° + 24.3 \sin^2 45°) \times 10^{-6}$$

$$\alpha_y = (-0.018 \sin^2 45° + 24.3 \cos^2 45°) \times 10^{-6} \qquad (5.145)$$

$$\alpha_{xy} = 2(-0.018 - 24.3) \sin 45° \cos 45° \times 10^{-6}$$

or
$$\alpha_x = \alpha_y = 12.14 \times 10^{-6}/°C$$
$$\alpha_{xy} = -24.3 \times 10^{-6}/°C \qquad (5.146)$$

(These quantities could also be read directly from Figure 5.13.) For $\Delta T = 50°C$, from equation (5.142),

$$\varepsilon_x^T = 607 \times 10^{-6} = \varepsilon_y^T$$
$$\gamma_{xy}^T = -1216 \times 10^{-6} \qquad (5.147)$$

From the definition of free thermal strain,

$$\delta\Delta_x^T = \varepsilon_x^T \Delta_x = (607 \times 10^{-6})(50) = 0.0304 \text{ mm}$$
$$\delta\Delta_y^T = \varepsilon_y^T \Delta_y = (607 \times 10^{-6})(50) = 0.0304 \text{ mm} \qquad (5.148)$$

and the change in the corner right angle in the x-y plane is

$$\gamma_{xy}^T = -1216 \times 10^{-6} \text{ rad } (-0.0697°) \qquad (5.149)$$

The free thermal strain in the z direction is, by equations (5.142) and (5.143),

$$\varepsilon_z = \varepsilon_3 = \alpha_3 \Delta T = (24.3 \times 10^{-6}) \times 50$$
$$\varepsilon_z = 1215 \times 10^{-6} \qquad (5.150)$$

resulting in

$$\delta\Delta_z = \varepsilon_z \Delta_z = (1215 \times 10^{-6})(50) = 0.0608 \text{ mm} \qquad (5.151)$$

This completes the problem.

By analogy to equation (5.143), the *coefficients of moisture deformation*, or CMD, in the x-y-z system are defined as

$$\boxed{\begin{aligned} \beta_x &= \beta_1 \cos^2 \theta + \beta_2 \sin^2 \theta \\ \beta_y &= \beta_1 \sin^2 \theta + \beta_2 \cos^2 \theta \\ \beta_{xy} &= 2\,(\beta_1 - \beta_2) \cos \theta \sin \theta \\ \beta_z &= \beta_3 \end{aligned}} \qquad (5.152)$$

By analogy to equation (5.142), then, the free moisture strains in the x-y-z system are given by

$$\varepsilon_x^M = \beta_x \Delta M$$

$$\varepsilon_y^M = \beta_y \Delta M$$

$$\gamma_{xy}^M = \beta_{xy} \Delta M \tag{5.153}$$

$$\varepsilon_z^M = \beta_3 \Delta M$$

Exercise for Section 5.6

Plot α_x, α_y, and α_{xy} versus θ, $-90° \le \theta \le 90°$ for glass-reinforced composite. Except for maximum and minimum values, the relations for α_x, α_y, and α_{xy} as a function of θ should be very similar to the graphite-reinforced case. The coefficients of thermal deformation α_x and α_y are even functions of θ, while α_{xy} is an odd function of θ.

5.7
PLANE-STRESS STRESS-STRAIN RELATIONS AND THE EFFECTS OF FREE THERMAL AND FREE MOISTURE STRAINS: GLOBAL COORDINATE SYSTEM

We are now in a position to derive the plane-stress stress-strain relations in an x-y-z system with the effects of free thermal and moisture strains included. The derivation is straightforward but the results are quite far-reaching. The implication of the relations, and the important influence of free thermal- and moisture-induced effects, will not really be evident until we discuss laminates and, in particular, the influence of temperature changes and moisture absorption on stresses in laminates. However, this chapter is the appropriate place to derive the plane-stress stress-strain relations with thermal and moisture deformation effects included. Equations (4.29) and (4.30) are the logical starting points. Considering equation (4.29) first, including the factor of 2 with the shear strain,

$$\left\{ \begin{array}{c} \varepsilon_1 - \alpha_1 \Delta T - \beta_1 \Delta M \\ \varepsilon_2 - \alpha_2 \Delta T - \beta_2 \Delta M \\ \dfrac{1}{2}\gamma_{12} \end{array} \right\} = \left[\begin{array}{ccc} S_{11} & S_{12} & 0 \\ S_{12} & S_{22} & 0 \\ 0 & 0 & \dfrac{1}{2}S_{66} \end{array} \right] \left\{ \begin{array}{c} \sigma_1 \\ \sigma_2 \\ \tau_{12} \end{array} \right\} \tag{5.154}$$

This can be expanded to the form

$$\left\{ \begin{array}{c} \varepsilon_1 \\ \varepsilon_2 \\ \dfrac{1}{2}\gamma_{12} \end{array} \right\} - \left\{ \begin{array}{c} \alpha_1 \Delta T \\ \alpha_2 \Delta T \\ \dfrac{0}{2} \end{array} \right\} - \left\{ \begin{array}{c} \beta_1 \Delta M \\ \beta_2 \Delta M \\ \dfrac{0}{2} \end{array} \right\} = \left[\begin{array}{ccc} S_{11} & S_{12} & 0 \\ S_{12} & S_{22} & 0 \\ 0 & 0 & \dfrac{1}{2}S_{66} \end{array} \right] \left\{ \begin{array}{c} \sigma_1 \\ \sigma_2 \\ \tau_{12} \end{array} \right\} \tag{5.155}$$

where the $0/2$ is a reminder we are dealing with tensor strains. The second term on the left hand side can be rewritten with the aid of equation (5.143) in the form

$$
\left\{
\begin{array}{c}
\alpha_x \Delta T \\
\alpha_y \Delta T \\
\frac{1}{2}\alpha_{xy}\Delta T
\end{array}
\right\}
=
\left[
\begin{array}{ccc}
\cos^2\theta & \sin^2\theta & -2\sin\theta\cos\theta \\
\sin^2\theta & \cos^2\theta & -2\sin\theta\cos\theta \\
\sin\theta\cos\theta & -\sin\theta\cos\theta & \cos^2\theta-\sin^2\theta
\end{array}
\right]
\left\{
\begin{array}{c}
\alpha_1\Delta T \\
\alpha_2\Delta T \\
\dfrac{0}{2}
\end{array}
\right\}
$$

(5.156)

Recognizing the matrix as $[T]^{-1}$, equation (5.12), inversion leads to

$$
\left\{
\begin{array}{c}
\alpha_1\Delta T \\
\alpha_2\Delta T \\
\dfrac{0}{2}
\end{array}
\right\}
= [T]
\left\{
\begin{array}{c}
\alpha_x\Delta T \\
\alpha_y\Delta T \\
\frac{1}{2}\alpha_{xy}\Delta T
\end{array}
\right\}
$$

(5.157)

Likewise

$$
\left\{
\begin{array}{c}
\beta_1\Delta M \\
\beta_2\Delta M \\
\dfrac{0}{2}
\end{array}
\right\}
= [T]
\left\{
\begin{array}{c}
\beta_x\Delta M \\
\beta_y\Delta M \\
\frac{1}{2}\beta_{xy}\Delta M
\end{array}
\right\}
$$

(5.158)

If we use equations (5.9), (5.20), (5.157), and (5.158), equation (5.155) becomes

$$
[T]
\left\{
\begin{array}{c}
\varepsilon_x \\
\varepsilon_y \\
\frac{1}{2}\gamma_{xy}
\end{array}
\right\}
- [T]
\left\{
\begin{array}{c}
\alpha_x\Delta T \\
\alpha_y\Delta T \\
\frac{1}{2}\alpha_{xy}\Delta T
\end{array}
\right\}
- [T]
\left\{
\begin{array}{c}
\beta_x\Delta M \\
\beta_y\Delta M \\
\frac{1}{2}\beta_{xy}\Delta M
\end{array}
\right\}
$$

(5.159)

$$
=
\left[
\begin{array}{ccc}
S_{11} & S_{12} & 0 \\
S_{12} & S_{22} & 0 \\
0 & 0 & \frac{1}{2}S_{66}
\end{array}
\right]
[T]
\left\{
\begin{array}{c}
\sigma_x \\
\sigma_y \\
\tau_{xy}
\end{array}
\right\}
$$

Multiplying both sides by $[T]^{-1}$, combining the terms on the left hand side, and multiplying the three matrices together on the right side, accounting for the factors of $1/2$, leads to

$$
\left\{
\begin{array}{c}
\varepsilon_x - \alpha_x\Delta T - \beta_x\Delta M \\
\varepsilon_y - \alpha_y\Delta T - \beta_y\Delta M \\
\gamma_{xy} - \alpha_{xy}\Delta T - \beta_{xy}\Delta M
\end{array}
\right\}
=
\left[
\begin{array}{ccc}
\bar{S}_{11} & \bar{S}_{12} & \bar{S}_{16} \\
\bar{S}_{12} & \bar{S}_{22} & \bar{S}_{26} \\
\bar{S}_{16} & \bar{S}_{26} & \bar{S}_{66}
\end{array}
\right]
\left\{
\begin{array}{c}
\sigma_x \\
\sigma_y \\
\tau_{xy}
\end{array}
\right\}
$$

(5.160)

where all terms in the equation have been previously defined. The α's, β's, and $\bar{S}_{ij}$'s are material properties and are known, while the strains ε_x, ε_y, and γ_{xy} are the total strains in the x-y-z system. They are related to the change in the geometry of an element of material by the usual expressions, namely,

$$\varepsilon_x = \frac{\delta \Delta_x}{\Delta_x} \qquad \varepsilon_y = \frac{\delta \Delta_y}{\Delta_y} \tag{5.161}$$

and γ_{xy} is the total shearing strain and is the change in the right corner angles of the element. Equation (5.160) is the off-axis counterpart to equation (4.29), where the mechanical strains are given by

$$\left\{ \begin{array}{c} \varepsilon_x^{mech} \\ \varepsilon_y^{mech} \\ \gamma_{xy}^{mech} \end{array} \right\} = \left\{ \begin{array}{c} \varepsilon_x - \alpha_x \Delta T - \beta_x \Delta M \\ \varepsilon_y - \alpha_y \Delta T - \beta_y \Delta M \\ \gamma_{xy} - \alpha_{xy} \Delta T - \beta_{xy} \Delta M \end{array} \right\} \tag{5.162}$$

Note that with the fibers oriented off axis there is a distinction between total shearing strain, γ_{xy}, and mechanical shearing strain. Recall, in the principal material coordinate system these two measures of strain are the same and the phrase "mechanical shearing strain" has no meaning.

By starting with the inverse of equation (4.29), namely, equation (4.30), and using the various definitions, the inverse of equation (5.160) can be derived as

$$\left\{ \begin{array}{c} \sigma_x \\ \sigma_y \\ \tau_{xy} \end{array} \right\} = \left[\begin{array}{ccc} \bar{Q}_{11} & \bar{Q}_{12} & \bar{Q}_{16} \\ \bar{Q}_{12} & \bar{Q}_{22} & \bar{Q}_{26} \\ \bar{Q}_{16} & \bar{Q}_{26} & \bar{Q}_{66} \end{array} \right] \left\{ \begin{array}{c} \varepsilon_x - \alpha_x \Delta T - \beta_x \Delta M \\ \varepsilon_y - \alpha_y \Delta T - \beta_y \Delta M \\ \gamma_{xy} - \alpha_{xy} \Delta T - \beta_{xy} \Delta M \end{array} \right\} \tag{5.163}$$

It would be a serious error not to remind ourselves of the presence of ε_z. Because $\varepsilon_z = \varepsilon_3$ throughout our work, ε_z can be obtained directly from equation (4.28), namely,

$$\varepsilon_z = \varepsilon_3 = \alpha_3 \Delta T + \beta_3 \Delta M + S_{13} \sigma_1 + S_{23} \sigma_2 \tag{5.164}$$

However, the stresses σ_1 and σ_2 can be written in terms of σ_x, σ_y, and τ_{xy} to obtain an expression for ε_z that represents the strain in the z direction expressed in terms of global system stresses. If we use the stress transformation equation (5.7), equation (5.164) becomes

$$\begin{aligned} \varepsilon_z = \alpha_3 \Delta T + \beta_3 \Delta M &+ (S_{13} \cos^2 \theta + S_{23} \sin^2 \theta) \sigma_x \\ &+ (S_{13} \sin^2 \theta + S_{23} \cos^2 \theta) \sigma_y + 2(S_{13} - S_{23}) \sin \theta \cos \theta \tau_{xy} \end{aligned} \tag{5.165}$$

The mechanical extensional strain in the z direction is given by

$$\begin{aligned} \varepsilon_z^{mech} = \varepsilon_z - \alpha_z \Delta T - \beta_z \Delta T &= S_{12} \sigma_1 + S_{23} \sigma_2 \\ &= (S_{13} \cos^2 \theta + S_{23} \sin^2 \theta) \sigma_x + (S_{13} \sin^2 \theta + S_{23} \cos^2 \theta) \sigma_y \\ &+ 2(S_{13} - S_{23}) \sin \theta \cos \theta \tau_{xy} \end{aligned} \tag{5.166}$$

Even in the presence of free thermal and free moisture strain effects, for the condition of plane stress there are no shear strains whatsoever in the y-z and x-z planes;

that is,

$$\gamma_{yz} = \gamma_{yz}^T = \gamma_{yz}^{mech} = \gamma_{xz} = \gamma_{xz}^T = \gamma_{xz}^{mech} = 0 \tag{5.167}$$

and similarly for free moisture strain effects.

The relations developed in the previous sections are important in the analysis of behavior of fiber-reinforced material in the presence of free thermal and free moisture strains. We shall see in the ensuing chapters the type of information that can be obtained from the various relations. However, before closing this chapter, we present two more simple examples to provide some insight into the mechanics of thermal effects and, more importantly, into the physical interpretation of the strains ε_x, ε_y, and γ_{xy} in the presence of free thermal or free moisture strains. As an example, assume that instead of being completely free to deform, the off-axis element in the example of Figure 5.14 is completely restrained in the x-y plane. There is no restraint in the z direction and thus the problem is one of plane stress. Because of the constraints,

$$\delta\Delta_x = 0 = \delta\Delta_y \tag{5.168}$$

and because the total strains are given by

$$\varepsilon_x = \frac{\delta\Delta_x}{\Delta_x} \quad \text{and} \quad \varepsilon_y = \frac{\delta\Delta_y}{\Delta_y} \tag{5.169}$$

then

$$\varepsilon_x = 0 \quad \text{and} \quad \varepsilon_y = 0 \tag{5.170}$$

Also, because the right corner angles in the x-y plane cannot change, the total shear strain is zero:

$$\gamma_{xy} = 0 \tag{5.171}$$

As moisture effects are not present, incorporating the restraint effects of equations (5.170) and (5.171), equation (5.163) becomes

$$\left\{\begin{array}{c} \sigma_x \\ \sigma_y \\ \tau_{xy} \end{array}\right\} = \left[\begin{array}{ccc} \bar{Q}_{11} & \bar{Q}_{12} & \bar{Q}_{16} \\ \bar{Q}_{12} & \bar{Q}_{22} & \bar{Q}_{26} \\ \bar{Q}_{16} & \bar{Q}_{26} & \bar{Q}_{66} \end{array}\right] \left\{\begin{array}{c} -\alpha_x\Delta T \\ -\alpha_y\Delta T \\ -\alpha_{xy}\Delta T \end{array}\right\} \tag{5.172}$$

These are the stresses required to keep the element from deforming. For $\theta = 45°$ and for the graphite-reinforced material,

$$\bar{Q}_{11} = 47.9 \text{ GPa} \qquad \bar{Q}_{12} = 39.1 \text{ GPa} \qquad \bar{Q}_{16} = 35.9 \text{ GPa}$$
$$\bar{Q}_{22} = 47.9 \text{ GPa} \qquad \bar{Q}_{26} = 35.9 \text{ GPa} \qquad \bar{Q}_{66} = 40.5 \text{ GPa} \tag{5.173}$$

and using the numerical values of α_x, α_y, and α_{xy} from equation (5.146), with $\Delta T = 50°C$, we find that equation (5.172) yields

$$\sigma_x = \sigma_y = -9.15 \text{ MPa}$$
$$\tau_{xy} = 5.62 \text{ MPa} \tag{5.174}$$

Because of the combined effects of $\alpha_3\Delta T$ and the stresses σ_x, σ_y, τ_{xy} generated, equation (5.165) (with $\beta_3\Delta M = 0$) gives us the value of the total through-thickness strain ε_z. With no constraints whatsoever on the element, the total through-thickness expansion in the example of Figure 5.14 was strictly $\alpha_3 \Delta T$. Equations (5.142) and (5.150) gave the value of ε_z for this situation. That value was $\varepsilon_z = 1215 \times 10^{-6}$. For the present constrained problem, using $\theta = 45°$, the values of the stresses from equation (5.174), and the values of S_{13} and S_{23} from equation (2.56), we compute ε_z from equation (5.165) to be

$$\varepsilon_z = 1780 \times 10^{-6} \tag{5.175}$$

This is significantly more than $\alpha_3 \Delta T$ and indicates the error in not remembering that the plane-stress condition retains important three-dimensional interactions between stresses and strains. Note that this is the same result obtained when specifying that the element was fully constrained in the 1-2 plane, as in equation (4.36). Specifying that the element is fully constrained in the 1-2 plane is equivalent to specifying that the element is fully constrained in the x-y plane. In fact, if we use the stresses σ_x, σ_y, and τ_{xy} from equation (5.174) to compute the stresses in the 1-2 system, namely, σ_1, σ_3, and τ_{12}, by employing equation (5.10), the result is the stresses of equation (4.36). For this problem the nonzero mechanical strains are given by

$$\begin{Bmatrix} \varepsilon_x^{mech} \\ \varepsilon_y^{mech} \\ \varepsilon_z^{mech} \\ \gamma_{xy}^{mech} \end{Bmatrix} = \begin{Bmatrix} -\alpha_x \Delta T \\ -\alpha_y \Delta T \\ \varepsilon_z - \alpha_z \Delta T \\ \alpha_{xy} \Delta T \end{Bmatrix} = \begin{Bmatrix} -607 \\ -607 \\ 565 \\ -1216 \end{Bmatrix} \times 10^{-6} \tag{5.176}$$

To solve the above problem we could have also started with equation (5.160), which would become

$$\begin{Bmatrix} -\alpha_x \Delta T \\ -\alpha_y \Delta T \\ -\alpha_{xy} \Delta T \end{Bmatrix} = \begin{bmatrix} \bar{S}_{11} & \bar{S}_{12} & \bar{S}_{16} \\ \bar{S}_{12} & \bar{S}_{22} & \bar{S}_{26} \\ \bar{S}_{16} & \bar{S}_{26} & \bar{S}_{66} \end{bmatrix} \begin{Bmatrix} \sigma_x \\ \sigma_y \\ \tau_{xy} \end{Bmatrix} \tag{5.177}$$

and solved the three equations for the three unknowns σ_x, σ_y, and τ_{xy}. This approach would have led to identical results. (Because $[\bar{Q}] = [\bar{S}]^{-1}$, this has to be the case.)

As a closing example to this chapter, let us examine another variant of the problem of Figure 5.14. As in Figure 5.15(a), let us assume that the 50 mm $\times$ 50 mm $\times$ 50 mm off-axis element of graphite-reinforced material is not completely constrained; rather, it is partially constrained by frictionless rollers from deformation in the y direction, but is otherwise free to deform. The problem is one of plane stress, and because of the constraints,

$$\varepsilon_y = 0 \tag{5.178}$$

Also, because of the rollers and the lack of contact on the edges perpendicular to the x axis,

$$\tau_{xy} = \sigma_x = 0 \tag{5.179}$$

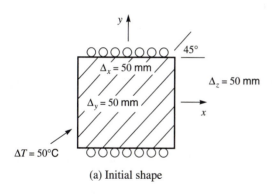

(a) Initial shape

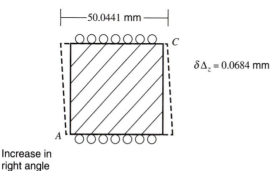

FIGURE 5.15
Thermal deformations of a
partially constrained off-axis
element of fiber-reinforced
material

Increase in
right angle
0.0527°

(b) Deformed shape

Using equation (5.160) with $\Delta M = 0$ yields

$$
\left\{ \begin{array}{c} \varepsilon_x - \alpha_x \Delta T \\ 0 - \alpha_y \Delta T \\ \gamma_{xy} - \alpha_{xy} \Delta T \end{array} \right\} = \left[\begin{array}{ccc} \bar{S}_{11} & \bar{S}_{12} & \bar{S}_{16} \\ \bar{S}_{12} & \bar{S}_{22} & \bar{S}_{26} \\ \bar{S}_{16} & \bar{S}_{26} & \bar{S}_{66} \end{array} \right] \left\{ \begin{array}{c} 0 \\ \sigma_y \\ 0 \end{array} \right\} \tag{5.180}
$$

which expands to

$$
\varepsilon_x - \alpha_x \Delta T = \bar{S}_{12}\sigma_y
$$
$$
-\alpha_y \Delta T = \bar{S}_{22}\sigma_y \tag{5.181}
$$
$$
\gamma_{xy} - \alpha_{xy} \Delta T = \bar{S}_{26}\sigma_y
$$

The stress required to constrain the deformation in the y direction is given by the second equation, namely,

$$
\sigma_y = \frac{-\alpha_y \Delta T}{\bar{S}_{22}} \tag{5.182}
$$

If we substitute this expression for σ_y into the other two equations, the two nonzero strains are given by

$$\varepsilon_x = \left(\alpha_x - \frac{\bar{S}_{12}}{\bar{S}_{22}}\alpha_y\right)\Delta T$$

$$\gamma_{xy} = \left(\alpha_{xy} - \frac{\bar{S}_{26}}{\bar{S}_{22}}\alpha_y\right)\Delta T \tag{5.183}$$

For $\theta = 45°$ and for graphite-reinforced composite,

$$\bar{S}_{11} = 78.3 \text{ (TPa)}^{-1} \qquad \bar{S}_{12} = -35.3 \text{ (TPa)}^{-1} \qquad \bar{S}_{16} = -38.1 \text{ (TPa)}^{-1}$$

$$\bar{S}_{22} = 78.3 \text{ (TPa)}^{-1} \qquad \bar{S}_{26} = -38.1 \text{ (TPa)}^{-1} \qquad \bar{S}_{66} = 92.3 \text{ (TPa)}^{-1} \tag{5.184}$$

Using the values for α_x, α_y, and α_{xy}, equation (5.146), we find

$$\sigma_y = -7.75 \text{ MPa}$$

$$\varepsilon_x = 881 \times 10^{-6} \tag{5.185}$$

$$\gamma_{xy} = -921 \times 10^{-6} \text{ rad } (-0.0527°)$$

Since

$$\frac{\delta\Delta_x}{\Delta_x} = \varepsilon_x \tag{5.186}$$

the change in length in the x direction is

$$\delta\Delta_x = \varepsilon_x\Delta_x = (881 \times 10^{-6}) \times 50 = 0.0441 \text{ mm} \tag{5.187}$$

and the right angles at corners A and C increase by 0.000 921 rad, or 0.0527°. Because σ_x and τ_{xy} are zero, the strain in the z direction for the situation in Figure 5.15, from equation (5.165), simplifies to

$$\varepsilon_z = \alpha_3\Delta T + (S_{13}\sin^2\theta + S_{23}\cos^2\theta)\sigma_y \tag{5.188}$$

or using numerical values,

$$\varepsilon_z = 1368 \times 10^{-6} \tag{5.189}$$

This through-thickness strain is smaller than the strain in the case with constraints in both the x and y directions, $\varepsilon_z = 1780 \times 10^{-6}$ in equation (5.175), and closer to the value given by $\alpha_3\Delta T$, 1215×10^{-6}. The lack of constraints allows the element of material to more closely approach the situation of free thermal expansion. If we use the value $\Delta_z = 50$ mm, the change in thickness of the element is given by

$$\delta\Delta_z = 0.0684 \text{ mm} \tag{5.190}$$

This problem could have been solved starting with equation (5.163) instead of equation (5.160), which would have resulted in equations of the form

$$0 = \bar{Q}_{11}(\varepsilon_x - \alpha_x\Delta T) + \bar{Q}_{12}(0 - \alpha_y\Delta T) + \bar{Q}_{16}(\gamma_{xy} - \alpha_{xy}\Delta T)$$

$$\sigma_y = \bar{Q}_{12}(\varepsilon_x - \alpha_x\Delta T) + \bar{Q}_{22}(0 - \alpha_y\Delta T) + \bar{Q}_{26}(\gamma_{xy} - \alpha_{xy}\Delta T) \tag{5.191}$$

$$0 = \bar{Q}_{16}(\varepsilon_x - \alpha_x\Delta T) + \bar{Q}_{26}(0 - \alpha_y\Delta T) + \bar{Q}_{66}(\gamma_{xy} - \alpha_{xy}\Delta T)$$

These equations could be solved for σ_y, ε_x, and γ_{xy} to yield the results of equation (5.185). Figure 5.15(b) illustrates the deformations due to a 50°C temperature change. For this problem, the nonzero mechanical strains are

$$
\left\{
\begin{array}{c}
\varepsilon_x^{mech} \\
\varepsilon_y^{mech} \\
\varepsilon_z^{mech} \\
\gamma_{xy}^{mech}
\end{array}
\right\}
=
\left\{
\begin{array}{c}
\varepsilon_x - \alpha_x \Delta T \\
-\alpha_y \Delta T \\
\varepsilon_z - \alpha_z \Delta T \\
\gamma_{xy} - \alpha_{xy} \Delta T
\end{array}
\right\}
=
\left\{
\begin{array}{c}
274 \\
-607 \\
153 \\
295
\end{array}
\right\}
\times 10^{-6}
\qquad (5.192)
$$

Before closing, we should expand upon and summarize the stresses and strains for the various thermal situations studied in examples. This summary will serve to illustrate the coupling of thermal strains and constraints. Recall that in each case we are considering a 50 mm × 50 mm × 50 mm element of fiber-reinforced graphite material with $\Delta T = 50°C$.

1. No constraints, $\theta = 0°$, fully three-dimensional, like the results of equation (2.101):

$$\text{Specified:} \quad \sigma_1 = \sigma_2 = \sigma_3 = \tau_{23} = \tau_{13} = \tau_{12} = 0$$

$$\text{Solve for:} \quad \varepsilon_1 = \alpha_1 \Delta T = -0.900 \times 10^{-6}$$

$$\varepsilon_2 = \alpha_2 \Delta T = 1215 \times 10^{-6} \qquad (5.193)$$

$$\varepsilon_3 = \alpha_3 \Delta T = 1215 \times 10^{-6}$$

$$\gamma_{23} = \gamma_{13} = \gamma_{12} = 0$$

2. No constraints, $\theta = 45°$, plane stress assumed, equations (5.149) and (5.150):

$$\text{Specified:} \quad \sigma_x = \sigma_y = \tau_{xy} = 0$$

$$(\sigma_z = \tau_{yz} = \tau_{xz} = 0, \text{ implied by plane-stress assumption})$$

$$\text{Solved for:} \quad \varepsilon_x = \alpha_x \Delta T = 607 \times 10^{-6}$$

$$\varepsilon_y = \alpha_y \Delta T = 607 \times 10^{-6} \qquad (5.194)$$

$$\varepsilon_z = \alpha_3 \Delta T = 1215 \times 10^{-6}$$

$$\gamma_{xy} = \alpha_{xy} \Delta T = -1216 \times 10^{-6} \text{ rad } (-0.0697°)$$

$$\gamma_{yz} = \gamma_{xz} = 0$$

The first and second cases are in reality the same problem. It is a question of studying the problem in either the 1-2-3 or the x-y-z system. Specifying no constraints in the 1-2-3 system or specifying no constraints in the x-y-z system are the same thing.

3. Constraints in the 1-2 plane, no constraints in 3 direction, $\theta = 0°$, plane stress assumed, equations (4.33) and (4.36):

Specified: $\varepsilon_1 = \varepsilon_2 = \gamma_{12} = 0$

$(\sigma_3 = \tau_{23} = \tau_{13} = 0,$ implied by plane-stress assumption)

Solved for: $\sigma_1 = -3.52$ MPa

$\sigma_2 = -14.77$ MPa   (5.195)

$\tau_{12} = 0$

$\varepsilon_3 = 1780 \times 10^{-6}$

$\gamma_{23} = \gamma_{13} = 0$

4. Constraints in the x-y plane, no constraints in z direction, $\theta = 45°$, plane stress assumed, equations (5.170), (5.171), (5.174), and (5.175):

Specified: $\varepsilon_x = \varepsilon_y = \gamma_{xy} = 0$

$(\sigma_z = \tau_{yz} = \tau_{xz} = 0,$ implied by plane-stress assumption)

Solved for: $\sigma_x = \sigma_y = -9.15$ MPa

$\tau_{xy} = 5.62$ MPa   (5.196)

$\varepsilon_z = 1780 \times 10^{-6}$

$\gamma_{yz} = \gamma_{xz} = 0$

The third and fourth cases are also the same problem, though it is not so obvious. Being completely constrained in the x-y plane and being completely constrained in the 1-2 plane are the same condition.

5. Constraint in y direction, no constraints in x or z directions $\theta = 45°$, plane stress assumed, equations (5.178), (5.179), (5.185), and (5.189):

Specified: $\sigma_x = \tau_{xy} = \varepsilon_y = 0$

$(\sigma_z = \tau_{yz} = \tau_{xz} = 0,$ implied by plane-stress assumption)

Solved for: $\sigma_y = -7.75$ MPa

$\varepsilon_x = 881 \times 10^{-6}$ (5.197)

$\varepsilon_z = 1368 \times 10^{-6}$

$\gamma_{xy} = -921 \times 10^{-6}$ rad $(0.0527°)$

$\gamma_{yz} = \gamma_{xz} = 0$

The third case should be contrasted with the fully constrained case with $\theta = 0°$, the numerical example presented in Chaper 2, equation (2.118). For that case,

Specified: $\varepsilon_1 = \varepsilon_2 = \varepsilon_3 = \gamma_{23} = \gamma_{13} = \gamma_{12} = 0$

Solved for: $\sigma_1 = -13.55$ MPa

$\sigma_2 = -27.6$ MPa   (5.198)

$\sigma_3 = -27.6$ MPa

$\tau_{23} = \tau_{13} = \tau_{12} = 0$

Comparison of the stress levels in equations (5.195) and (5.198) illustrates the influence of constraints, or lack there of, in the 3 direction on stress levels.

Exercises for Section 5.7

1. Transform the stresses of equation (5.174), the case of a 50 mm × 50 mm × 50 mm element with $\theta = 45°$, with no constraint in the z direction but completely constrained in the x and y directions, to the 1-2 system to show that the problem is equivalent to the case of an element with $\theta = 0°$ and with no constraint in the 3 direction but completely constrained in the 1 and 2 directions, equation (4.36).

2. Consider a 50 mm × 50 mm × 50 mm element of material with its fibers oriented at 45° and constrained against deformation in the x direction. The element is heated 50°C. What is the stress σ_x required to enforce this constraint and what are the strains ε_y, γ_{xy}, and ε_z? Are there any similarities to this problem and the problem of Figure 5.15? See equations (5.185) and (5.189).

5.8
SUMMARY

This concludes this chapter on the response of an off-axis element of fiber-reinforced material in a state of plane stress. The concepts are quite important, particularly the idea of coupling of the various stress and strain components. The addition of free thermal strain effects (and free moisture strain effects) causes some unusual and perhaps unexpected responses in a fiber-reinforced composite material. Though the influence on inplane stresses and deformations is important, the coupling of inplane thermal effects, through Poisson's ratios, and through-the-thickness thermal effects is even more important. This coupling can easily be overlooked, leading to errors in predicted response. Understanding these concepts is essential for understanding the response of a laminate, the subject of the next chapter.

CHAPTER 6

Classical Lamination Theory: The Kirchhoff Hypothesis

In the preceding chapters we developed the tools needed to understand the elastic response of a small volume of fiber-reinforced material under the assumption that the fibers and the matrix material were smeared into one equivalent homogeneous material. The simplifications of the plane-stress assumption were developed. The use of an x-y-z global coordinate system, as opposed to a 1-2-3 principal material coordinate system, to describe the stress-strain behavior for a plane-stress state was introduced. We examined some of the important differences and similarities between conventional isotropic materials and fiber-reinforced composite materials in a series of examples. These examples illustrated the implications of shear-extension coupling, and we considered the influence of thermal expansion. These discussions were all related to understanding the stress-strain relations of a fiber-reinforced material, relations that are, by definition, valid at a point in what we are considering to be the equivalent homogeneous material.

As was stated in Chapter 1, fiber-reinforced materials are most frequently used by employing multiple layers of material to form a laminate. Each layer is thin (see Table 2.1 for layer thickness) and may have a different fiber orientation, and in some cases, not all layers are of the same material. Some layers may use graphite fibers for reinforcement, while others may use glass fibers. Some laminates may consist of three or four layers, and some of several hundred layers. Two laminates may involve the same number of layers and the same set of fiber angles, but the two laminates can be different and most certainly can exhibit entirely different behavior because of the arrangement of the layers. For example, one four-layer laminate may have the fibers in the two outer layers oriented at $0°$ and the fibers in the two inner layers oriented at $90°$, while another four-layer laminate may have the fibers in the outer layers at $90°$ and those in the inner layers at $0°$. As Figure 6.1 shows, when subjected to the same level of bending moment, M, the first laminate will deform much less than the second. (The nomenclature $[0/90]_S$ and $[90/0]_S$ of Figure 6.1

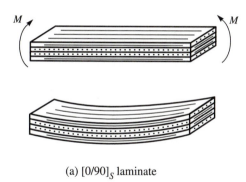

FIGURE 6.1

Differences in bending deformations between $[0/90]_S$ and $[90/0]_S$ laminates

(a) $[0/90]_S$ laminate

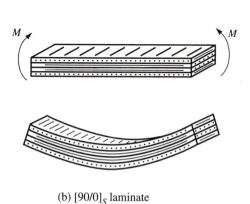

(b) $[90/0]_S$ laminate

will be explained shortly.) In the first case the fibers that resist bending are further apart, resulting in a larger bending stiffness. On the other hand, as in Figure 6.2, both laminates will stretch the same amount in the x direction when an inplane load is applied in that direction. Both laminates have two layers with their fibers parallel to the load and two layers with their fibers perpendicular to the load. The real issue, then, is to understand how laminates respond to loads, how the fiber angles of the individual layers influence laminate response, how the stacking arrangement of the layers influences the response, how changing material properties in a group of layers changes response, and so forth. Furthermore, the stresses within each layer, in addition to depending on the magnitude and character of the loading, must also depend on the arrangement of the layers, the fiber orientation in each layer, the material properties of each layer, and the like. How are the stresses influenced by these parameters? The number of variables that can be changed in a laminated fiber-reinforced composite structure, as well as the number of responses that can be studied, is immense. It is of prime concern to understand how changing these variables influences laminate response, and ultimately structural response. As an end product, it is important to be able to design laminates so that structures have a specific response, so that deformations are within certain limits and stress levels are below a given level.

This chapter will introduce simplifications in the analysis of fiber-reinforced composite materials that will allow us to obtain answers for a large class of problems.

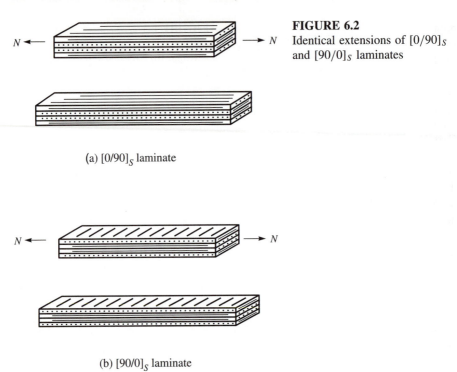

FIGURE 6.2
Identical extensions of $[0/90]_S$ and $[90/0]_S$ laminates

(a) $[0/90]_S$ laminate

(b) $[90/0]_S$ laminate

We can thus evaluate the influence of fiber directions, stacking arrangements, material properties, and so forth, on laminate and structural response. However, before we introduce the simplified theory, commonly called *classical lamination theory*, we will briefly discuss the nomenclature associated with laminates, including the manner in which fiber angles are specified and the stacking arrangement is identified. The appendix provides a brief overview of the all-important step of manufacturing composite laminates, and it includes a step-by-step description of fabricating a flat laminate by hand. This is particularly useful for small-scale university or laboratory settings, where fully automated high-volume production of composite laminates is not feasible.

6.1
LAMINATE NOMENCLATURE

Discussion of laminates requires, first, that we have a method of describing a laminate, particularly the fiber orientation of each layer. In addition, for purposes of analysis, we must establish a coordinate system for specifying locations through the thickness of the laminate as well as along the length and across the width. For flat plates, an x-y-z global Cartesian coordinate system, which we used in Chapter 5 for stress-strain behavior, is useful in describing a laminate. For cylindrical laminates, we use a global polar coordinate system. We can also use a spherical, elliptical, parabolic, or other global coordinate system germane to the geometry of the structure and laminate. The key feature is that whatever system we use, the origin of the through-thickness coordinate, designated z, is most conveniently located at the laminate geometric midplane. An important feature of what we are about to present is as follows: If you use any global orthogonal coordinate system, be it Cartesian, cylindrical, or spherical to analyze a laminate, then the figures we are about to present are applicable, even though they will be discussed in the context of a rectangular Cartesian system.

Figure 6.3 illustrates a global Cartesian coordinate system and a general laminate consisting of N layers. The upper portion of the figure is a cross-sectional view in the x-z plane ($y = 0$ plane), and the lower portion is a planform view. The laminate thickness is denoted by H and the thickness of an individual layer by h. Not all layers necessarily have the same thickness, so the thickness of the kth layer is denoted as h_k. The geometric midplane may be within a particular layer or at an interface between layers. What is important is that a geometric midplane can be defined. Herein the $+z$ axis will be downward and the laminate extends in the z direction from $-H/2$ to $+H/2$. We refer to the layer at the negativemost z location as layer 1, the next layer in as layer 2, the layer at an arbitrary location as layer k, and the layer at the positivemost z position as layer N. In the planform view, Figure 6.3(b), layer N is closest to the reader. The locations of the layer interfaces are denoted by a subscripted z; the first layer is bounded by locations z_0 and z_1, the second layer by z_1 and z_2, the kth layer by z_{k-1} and z_k, and the Nth layer by z_{N-1} and z_N.

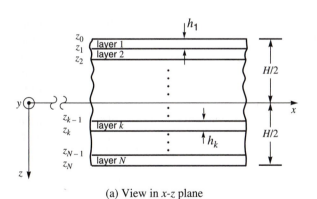

(a) View in x-z plane

FIGURE 6.3
Laminate nomenclature

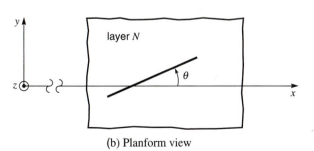

(b) Planform view

To identify the fiber angles of the various layers, the fiber angle relative to the $+x$ axis of each layer is specified. The specification starts with layer 1, the layer at the negativemost z location. For example, we denote the laminate in the upper portion of Figure 6.1 as a [0/90/90/0] laminate. We denote the laminate in the lower portion as a [90/0/0/90] laminate, where in each case we assume the x axis is oriented in the lengthwise direction of each laminate. The leftmost entry in the laminate notation refers to the orientation of layer 1. In cases where the stacking sequence to the one side of the $z = 0$ plane, the laminate geometric midplane, is a mirror image of the stacking sequence on the other side of the $z = 0$ plane, the stacking notation can be abbreviated by referring to only one-half of the laminate and subscripting the stacking notation with an S, which means symmetric. The upper and lower laminates of Figure 6.1 can thus be denoted by $[0/90]_S$ and $[90/0]_S$, respectively. With this notation, the leftmost entry in stacking specification is either layer 1 or layer N; that is, the stacking specification starts with the outer layer on each side of the laminate. To categorize a laminate as symmetric, it is imperative that the material properties, fiber orientation, and thickness of the layer at a specific location to one side of the geometric midplane be identical to the material properties, fiber orientation, and thickness of the layer at the mirror image location on the other side. Otherwise the laminate is not truly symmetric. The influence of symmetry, or its lack, will be discussed later. Suffice it to say that many aspects of the response of a laminate depend strongly on whether or not it is symmetric. In fact, symmetric laminates have been emphasized to such an extent that if one encounters the notation [0/90/90/0] in a discussion of composites, one usually assumes that an eight-layer laminate is being discussed, with stacking sequence [0/90/90/0/0/90/90/0]. To emphasize that indeed the complete laminate is being specified, the subscript T for total is sometimes used. Thus, the upper and lower laminates of Figure 6.1 could be denoted as $[0/90/90/0]_T$ and $[90/0/0/90]_T$, respectively.

When the stacking sequence involves adjacent layers of opposite orientation, as is often the case, shorthand notation is used. For example, if a six-layer laminate has the stacking sequence $[+45/-45/0/0/-45/+45]_T$, it would be abbreviated as $[\pm45/0]_S$. Here the $\pm$ is used to contract the notation and indicates there is a layer with its fibers oriented at $+45°$ with respect to the $+x$ axis and adjacent to it another layer with its fibers oriented at $-45°$ with relative to the $+x$ axis. Next to the $-45°$ layer is a layer with its fibers aligned with the x axis, as in Figure 6.4. When a stacking sequence of a subset consisting of several layers is repeated within a laminate, further

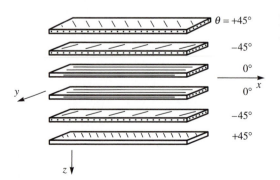

$\theta = +45°$

$-45°$

$0°$

$0°$

$-45°$

$+45°$

FIGURE 6.4
A $[\pm45/0]_S$ laminate

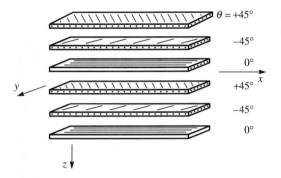

FIGURE 6.5
A $[(\pm 45/0)_2]_T$ laminate

shorthand notation is often used. If a 12-layer laminate has a stacking arrangement of $[+45/-45/0/+45/-45/0/0/-45/+45/0/-45/+45]_T$, it can be contracted to read $[(\pm 45/0)_2]_S$. Accordingly, a laminate denoted by $[(\pm 45/0)_2]_T$ would represent a six-layer laminate with an unsymmetric stacking arrangement of $[+45/-45/0/+45/-45/0]$, as in Figure 6.5. We will introduce other conventions and methods of describing laminates as specific problems are discussed. We shall now proceed with the development of classical lamination theory.

6.2
LAMINATE STRAINS AND DISPLACEMENTS: THE KIRCHHOFF HYPOTHESIS

One of the most important assumptions in the analysis of structures and of materials within structures was introduced in the mid 1800s by Kirchhoff. The assumption has greatly simplified analysis since that time and has permitted the analyst and designer to accurately predict the response of beams, plates, and shells. The hypothesis has been applied to metallic structures, wooden structures, concrete structures, and a wide variety of other materials. The assumption has very little to do with the material itself and much more to do with the geometry of the structure, the type of loadings it is designed to resist, and the conditions on the boundary of the structure. The assumption has been applied with a great deal of success to structures fabricated from composite materials. This has been verified experimentally and by comparison with analyses of the same problem conducted without employing the simplifying Kirchhoff hypothesis.

To begin consideration of the Kirchhoff hypothesis, consider an initially flat laminated plate acted upon by a variety of loads. The loads can consist of: applied moments, M; distributed applied loads, q; inplane loads, N; and point loads, P. The plate consists of multiple layers of fiber-reinforced materials, and the fibers in each layer are parallel to the plane of the plate. We assume that all layers are perfectly bonded together. There is no slippage between layers. Figure 6.6 illustrates the loaded plate, showing the x-y-z coordinate system. The Kirchhoff hypothesis focuses on the deformation of lines which before deformation are straight and normal to the laminate's geometric midsurface. In the figure one such line is drawn and it is denoted

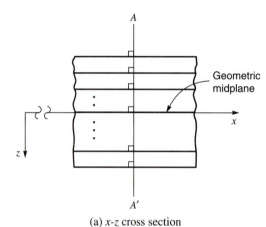

FIGURE 6.6
Laminated plate with undeformed normal AA' acted
upon by loads

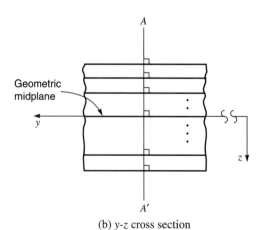

(a) x-z cross section

FIGURE 6.7
Cross sections showing normal
before deformations

(b) y-z cross section

as line AA'. Figure 6.7(a) shows the detail of an x-z cross section of the laminate,
and we see that line AA' passes through the laminate, and specifically through
each layer. Because before deformation the plate is flat and the layer interfaces are

parallel to each other and to the geometric midsurface of the plate, line AA' is normal to each interface; Figure 6.7(b) shows the details of this line viewed in a y-z cross section. The Kirchhoff hypothesis is simple: It assumes that despite the deformations caused by the applied loads, line AA' remains straight and normal to the deformed geometric midplane and does not change length. Specifically, the normal line does not deform; it simply translates and rotates as a consequence of the deformation, a very simple but yet very far-reaching assumption. Figure 6.8 shows the normal of Figure 6.6 having simply rotated and translated due to the deformations caused by the applied loads. That the line remains straight and normal to the geometric midplane after deformation is an important part of the assumption. Figure 6.9 illustrates the consequences of normal remaining straight by again showing the details of a laminate x-z cross section. Because the interfaces remain parallel with each other and with the geometric midplane, line AA' remains normal to each interface. The line is continuously straight through the thickness of the laminate, as in Figure 6.9(b). The line is not a series of straight line segments, as in Figure 6.9(c); rather, it is a single straight line. That the line does not change length is another important part of the assumption. For the length of the line to remain unchanged, the top and bottom surfaces of the laminate must remain the same distance apart in the thickness direction of the laminate. The distance between points t and t' in Figure 6.9(b), then, is the same as the distance between t and t' in Figure 6.9(a). According to the hypothesis, there is no through-thickness strain ε_z along line AA'. This is counter to what we know the stress-strain relations predict, namely stresses in the x direction causing strains in the z direction, as in equation (5.19). Generally there will be a through-thickness strain ε_z but the Kirchhoff hypothesis is inconsistent with this fact. Fortunately, the assumption that the normal remains fixed in length does not enter directly into the use of the hypothesis.

Because it is assumed that the line remains perfectly straight, normal to each interface, and does not change length, it is possible to express the displacement of material points on the line in terms of the displacement and rotation of the point on

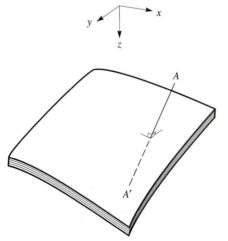

FIGURE 6.8
Normal remaining normal and simply translating and rotating

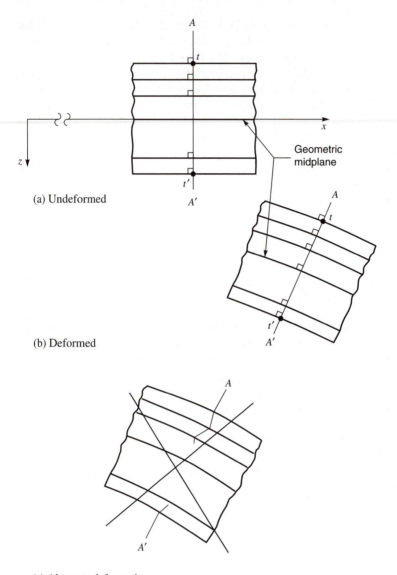

(a) Undeformed

(b) Deformed

(c) Alternate deformation

FIGURE 6.9
Consequences of Kirchhoff hypothesis

the line located at the laminate geometric midplane. In light of this, and because of the definition of stress resultants that will naturally arise at a later point in laminate analysis, it is convenient to think of the laminate geometric midplane as a *reference surface*. The mechanics of a laminate will be expressed in terms of what is happening at the reference surface. If the response of the reference surface is known, the strains, displacements, and stresses at each point along the normal line through the thickness of the laminate can be determined. This is an important advantage. Rather than treating a laminate as a three-dimensional domain and having to analyze it as such, the analysis of laminates degenerates to studying what is happening to the reference surface, a two-dimensional domain. With cylinders, for example, analysis degenerates to having to know what is happening to the surface of the cylinder located at the mean cylinder radius. As will be seen, understanding what is happening at the reference surface can be complicated enough. However, treating a plate or cylinder as a three-dimensional domain can become intractable. Hereafter, the laminate geometric midsurface will be referred to as the reference surface of the laminate.

6.3
IMPLICATIONS OF THE KIRCHHOFF HYPOTHESIS

We should stress that no mention has been made of material properties. The issue of the line remaining straight is strictly a kinematic and geometric issue. This is an important point; it implies that if we accept the validity of the Kirchhoff hypothesis, then we assume it is valid for the wide range of material properties that are available with fiber-reinforced composite materials. To study the implications of the Kirchhoff hypothesis, and to take advantage of it, let us examine the deformation of an x-z cross section of the plate being discussed. Figure 6.10 details the deformation of a cross section, and in particular the displacements of point P, a point located at an arbitrary distance z below point P^o, a point on the reference surface, points P and P^o both being on line AA'. Because line AA' remains straight, the deformation of the cross section as viewed in the x-z plane consists of three major components. There are two components of translation and one of rotation. As the laminate deforms, line AA' translates horizontally in the $+x$ direction; it translates down in the $+z$ direction; and in the process of translating downward, it rotates about the y axis. The superscript o will be reserved to denote the kinematics of point P^o on the reference surface. In particular, the horizontal translation of point P^o in the x direction will be denoted as u^o. The vertical translation will be denoted as w^o. The rotation of the reference surface about the y axis at point P^o is $\partial w^o / \partial x$. An important part of the Kirchhoff hypothesis is the assumption that line AA' remains perpendicular to the reference surface. Because of this, the rotation of line AA' is the same as the rotation of the reference surface, and thus the rotation of line AA', as viewed in the x-z plane, is $\partial w^o / \partial x$.

Since line AA' remains straight, the component of translation in the $+x$ direction of point P due to P^o translating horizontally an amount u^o is u^o. Downward translation and rotation of P^o cause additional movement at point P. For the present, we shall restrict our discussion to the case where points on the reference surface experience only small rotations in the x-z and y-z planes, the latter rotation not being apparent in Figure 6.10. In the context of Figure 6.10 this means

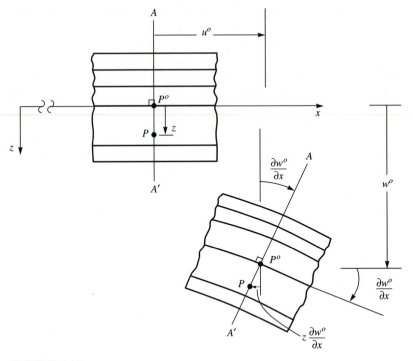

FIGURE 6.10
Kinematics of deformation as viewed in the x-z plane

$$\frac{\partial w^o}{\partial x} < 1 \tag{6.1}$$

By less than unity is meant that sines and tangents of angles of rotation are replaced by the rotations themselves, and cosines of the angles of rotation are replaced by 1. With this approximation, then, the rotation of point P^o causes point P to translate horizontally in the minus x direction by an amount

$$z\frac{\partial w^o}{\partial x} \tag{6.2}$$

This negative horizontal translation is denoted in Figure 6.10, and the total translation of point P in the x direction, denoted as $u(x, y, z)$, is thus the sum of two effects, namely,

$$u(x, y, z) = u^o(x, y) - z\frac{\partial w^o(x, y)}{\partial x} \tag{6.3}$$

It is important to note the notation used and the implications of the Kirchhoff hypothesis. Point P is located at (x, y, z), an arbitrary position within the laminate. The displacement of that point in the x direction is a function of all three coordinates, and thus the notation $u(x, y, z)$. The displacements and rotations of point P^o on the reference surface, however, depend only on x and y, and hence the notation $u^o(x, y)$ and $\partial w^o(x, y)/\partial x$. Clearly, due to the kinematics of the Kirchhoff hypothesis, the displacement of point (x, y, z) depends *linearly* on z, the distance the point is away from the reference surface.

Completing the picture of displacements of the x-z cross section, we see that as a result of the small rotation assumption, the vertical translation of point P is the same as the vertical translation of point P^o; that is, the vertical translation of point P is independent of z. As we shall see shortly, and as discussed just a few paragraphs ago, this leads to contradictory results for through-thickness extensional strains. With this independence of z,

$$w(x, y, z) = w^o(x, y) \tag{6.4}$$

where again the notation $w(x, y, z)$ indicates that the vertical displacement of point P at location (x, y, z) is, in general, a function of x, y, and z, but the hypothesis renders the vertical displacement independent of z and exactly equal to the reference surface displacement. Another interpretation of the independence of z is that all points on line AA' move vertically the same amount.

A similar picture emerges if the deformation is viewed in the y-z plane. As shown in Figure 6.11, the translation of point P in the $+y$ direction is

$$v(x, y, z) = v^o(x, y) - z\frac{\partial w^o(x, y)}{\partial y} \tag{6.5}$$

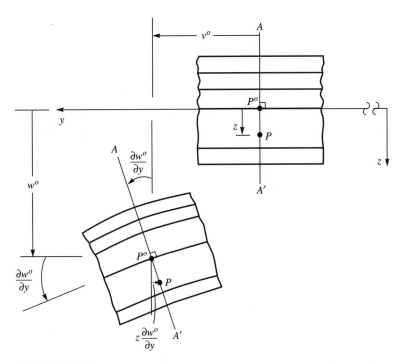

FIGURE 6.11
Kinematics of deformation as viewed in the y-z plane

In the above, v^o is the translation of point P^o on the reference surface in the $+y$ direction and $\partial w^o/\partial y$ is the rotation of that point about the x axis. In summary, then, the displacement of an arbitrary point P with coordinates (x, y, z) is given by

$$u(x, y, z) = u^o(x, y) - z\frac{\partial w^o(x, y)}{\partial x}$$

$$v(x, y, z) = v^o(x, y) - z\frac{\partial w^o(x, y)}{\partial y} \tag{6.6}$$

$$w(x, y, z) = w^o(x, y)$$

It is important to realize that generally the displacements and rotations of points on the reference surface vary from location to location within the plate. For the moment we shall not be concerned with how they vary. To determine this, laminated plate theories, as presented in Chapter 13, must be developed. To reemphasize: The important points to note from the Kirchhoff hypothesis are that the inplane displacements $u(x, y, z)$ and $v(x, y, z)$ everywhere within the laminate vary linearly with z, and the out-of-plane displacement $w(x, y, z)$ is independent of z. On the surface that may seem somewhat restrictive but it is an accurate approximation for a large class of problems.

6.4
LAMINATE STRAINS

With the assumptions regarding the displacement field established by way of the Kirchhoff hypothesis, the next step is to investigate the strains that result from the displacements. This can be done by using the strain-displacement relations from the theory of elasticity. Using these relations and equation (6.6), we can compute the strains at any point within the laminate, and by using these laminate strains in the stress-strain relations, we can compute the stresses at any point within the laminate. Thus, determining the expressions for the strains is important.

From the strain-displacement relations and equation (6.6), the extensional strain in the x direction, ε_x, is given by

$$\varepsilon_x(x, y, z) \equiv \frac{\partial u(x, y, z)}{\partial x} = \frac{\partial u^o(x, y)}{\partial x} - z\frac{\partial^2 w^o(x, y)}{\partial x^2} \tag{6.7}$$

where the triple horizontal bars are to be interpreted as "is defined as." We can see that the strain ε_x is composed of two parts. The first term, $\partial u^o(x, y)/\partial x$, is the extensional strain of the reference surface in the x direction. Because we are restricting our discussion to small rotations of the reference surface, the second term, $\partial^2 w^o(x, y)/\partial x^2$, is the curvature of the reference surface in the x direction. In general, the curvature, which is the inverse of the radius of curvature, involves more than just the second derivative of w. However, for the case of small rotations, the curvature and the second derivative are identical. Accordingly, the strain ε_x is written as

$$\varepsilon_x(x, y, z) = \varepsilon_x^o(x, y) + z\kappa_x^o(x, y) \tag{6.8}$$

where we use the notation

$$\varepsilon_x^o(x, y) = \frac{\partial u^o(x, y)}{\partial x} \qquad \kappa_x^o(x, y) = -\frac{\partial^2 w^o(x, y)}{\partial x^2} \tag{6.9}$$

Notice the minus sign associated with the definition of curvature. The quantity ε_x^o is referred to as the the extensional strain of the reference surface in the x direction, and κ_x^o is referred to as curvature of the reference surface in the x direction.

The other five strain components are given by

$$\varepsilon_y(x, y, z) \equiv \frac{\partial v(x, y, z)}{\partial y} = \varepsilon_y^o(x, y) + z\kappa_y^o(x, y)$$

$$\varepsilon_z(x, y, z) \equiv \frac{\partial w(x, y, z)}{\partial z} = \frac{\partial w^o(x, y)}{\partial z} = 0$$

$$\gamma_{yz}(x, y, z) \equiv \frac{\partial w(x, y, z)}{\partial y} + \frac{\partial v(x, y, z)}{\partial z} = \frac{\partial w^o(x, y)}{\partial y} - \frac{\partial w^o(x, y)}{\partial y} = 0 \tag{6.10}$$

$$\gamma_{xz}(x, y, z) \equiv \frac{\partial w(x, y, z)}{\partial x} + \frac{\partial u(x, y, z)}{\partial z} = \frac{\partial w^o(x, y)}{\partial x} - \frac{\partial w^o(x, y)}{\partial x} = 0$$

$$\gamma_{xy}(x, y, z) \equiv \frac{\partial v(x, y, z)}{\partial x} + \frac{\partial u(x, y, z)}{\partial y} = \gamma_{xy}^o + z\kappa_{xy}^o$$

where we can define

$$\varepsilon_y^o(x, y) = \frac{\partial v^o(x, y)}{\partial y} \quad \text{and} \quad \kappa_y^o(x, y) = -\frac{\partial^2 w^o(x, y)}{\partial y^2}$$

$$\gamma_{xy}^o(x, y) = \frac{\partial v^o(x, y)}{\partial x} + \frac{\partial u^o(x, y)}{\partial y} \quad \text{and} \quad \kappa_{xy}^o = -2\frac{\partial^2 w^o(x, y)}{\partial x \partial y} \tag{6.11}$$

The quantities ε_y^o, κ_y^o, γ_{xy}^o, and κ_{xy}^o are referred to as the reference surface extensional strain in the y direction, the reference surface curvature in the y direction, the reference surface inplane shear strain, and the reference surface twisting curvature, respectively. The notation used emphasizes again the fact that the reference surface strains and curvatures are functions only of x and y, while the strains are, in general, functions of x, y, and z. The term *inplane shear strain* in connection with γ_{xy}^o is used to indicate that it is related to changes in right angles between perpendicular line segments lying in the reference surface. Three of the six strain components are exactly zero. The two shear strains through the thickness are zero because the Kirchhoff hypothesis assumes that lines perpendicular to the reference surface before deformation remain perpendicular after the deformation; right angles in the thickness direction do not change when the laminate deforms (see Figures 6.10 and 6.11). By definition, then, the through-thickness shear strains must be exactly zero. If they were not computed to be zero in equation (6.10), there would be an inconsistency. The fact that the through-thickness extensional strain ε_z is predicted to be zero is the inconsistency we referred to previously that is due to the assumption that the length of the normal AA' does not change when the laminate deforms. Equations (2.45) and (2.47) indicate that extensional strains in all three directions are an integral part of the stress-strain relations. For a state of plane stress, equations (4.3) and (4.10) indicate that the extensional strain ε_z is not zero for this situation either. Hence, the third equation of equation (6.6) leads to an inconsistency. This is inherent in the Kirchhoff hypothesis and cannot be resolved within the context of the theory. The issue can be resolved by using the third term of equation (6.6) when the vertical displacement of a point in the cross section is needed, and using the stress-strain relations, such as equation (4.3) or equation (4.10), when the through-thickness extensional strain component ε_z is needed. The focus of Chapter 12 will be studying ε_z in laminates.

In summary, the displacements and the nonzero strains that result from the Kirchhoff hypothesis are as follows:

$$
\begin{aligned}
u(x, y, z) &= u^o(x, y) - z\frac{\partial w^o(x, y)}{\partial x} \\
v(x, y, z) &= v^o(x, y) - z\frac{\partial w^o(x, y)}{\partial y} \\
w(x, y, z) &= w^o(x, y)
\end{aligned}
\tag{6.12}
$$

$$
\begin{aligned}
\varepsilon_x(x, y, z) &= \varepsilon_x^o(x, y) + z\kappa_x^o(x, y) \\
\varepsilon_y(x, y, z) &= \varepsilon_y^o(x, y) + z\kappa_y^o(x, y) \\
\gamma_{xy}(x, y, z) &= \gamma_{xy}^o(x, y) + z\kappa_{xy}^o(x, y)
\end{aligned}
\tag{6.13}
$$

with

$$
\begin{aligned}
\varepsilon_x^o(x, y) &= \frac{\partial u^o(x, y)}{\partial x} \quad \text{and} \quad \kappa_x^o(x, y) = -\frac{\partial^2 w^o(x, y)}{\partial x^2} \\
\varepsilon_y^o(x, y) &= \frac{\partial v^o(x, y)}{\partial y} \quad \text{and} \quad \kappa_y^o(x, y) = -\frac{\partial^2 w^o(x, y)}{\partial y} \\
\gamma_{xy}^o(x, y) &= \frac{\partial v^o(x, y)}{\partial x} + \frac{\partial u^o(x, y)}{\partial y} \quad \text{and} \quad \kappa_{xy}^o = -2\frac{\partial^2 w^o(x, y)}{\partial x \partial y}
\end{aligned}
\tag{6.14}
$$

These are very important equations and are one of the important assumptions of classical lamination theory. The equations imply that the displacements in the x and y directions vary linearly through the thickness of the laminate. Also, they imply the strains vary linearly through the thickness of the laminate. This appears to be a rather simple solution to what could be considered a more complicated situation.

Equations (6.12)–(6.14) are important for another reason. They imply that if the reference surface strains and curvatures are known at every point within a laminate, then the strains at every point within the three-dimensional volume are known. For example, consider the situation in Figure 6.12(a), which illustrates a small segment of a four-layer laminate deformed such that at a point P^o on the reference surface the extensional strain in the x direction is 1000×10^{-6}. The radius of curvature of the reference surface, R^o, is 0.2 m, the latter corresponding to a curvature of 5 m^{-1}. The x-direction extensional strain as a function of the z coordinate through

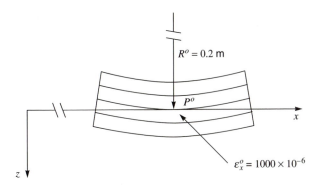

(a) ε_x^o and κ_x^o specified at a point on the reference surface of a laminate

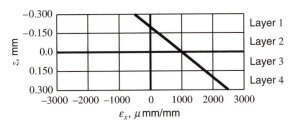

(b) Linear variation of ε_x with z through the laminate thickness

FIGURE 6.12
Example of predicted strain distribution through the laminate as a result of the Kirchhoff hypothesis

the thickness of the laminate above and below point P^o on the reference surface is given by the first relation of equation (6.13) as

$$\varepsilon_x = 0.001 + (z)(5) \tag{6.15}$$

Note the sign of κ_x^o relative to the deformation illustrated in Figure 6.12(a). If the laminate is 0.600 mm thick, the extensional strain in the x direction at the top of the laminate, where $z = -0.300$ mm, is

$$\varepsilon_x = 0.001 + (-300 \times 10^{-6})(5) = -500 \times 10^{-6} \text{ m/m}$$
$$= -500 \times 10^{-6} \text{ mm/mm} = -500 \ \mu \text{ mm/mm} \tag{6.16}$$

The nondimensional character of the strain becomes apparent. At the bottom of the laminate, where $z = +0.300$ mm,

$$\varepsilon_x = 0.001 + (300 \times 10^{-6})(5) = 2500 \times 10^{-6} \text{ m/m}$$
$$= 2500 \times 10^{-6} \text{ mm/mm} = 2500 \ \mu \text{ mm/mm} \tag{6.17}$$

Figure 6.12(b) shows the linear variation of strain with z through the laminate thickness.

Knowing the response of the reference surface is very valuable and useful. Unfortunately, determining the response of the reference surface everywhere within a laminate is not always easy. However, it is far easier to determine the response of the two-dimensional reference surface of the laminate and then use equations (6.12)–(6.14) than to determine the response at every point within the three-dimensional volume of the laminate without the benefits of these equations.

6.5
LAMINATE STRESSES

The second important assumption of classical lamination theory is that *each point within the volume of a laminate is in a state of plane stress.* Laminates used in the cases discussed in Figures 4.2–4.5 would clearly violate this assumption. However, in many situations the stresses in the plane of the laminate clearly dominate the stress state, and the out-of-plane components of stress can be assumed to zero with little loss of accuracy. If this is the case, then equation (5.83) is valid for every point in the laminate. However, recall that nothing has yet been said about material properties. If we are to discuss stresses, fiber orientation and material properties enter the picture through the $\bar{Q}_{ij}$. In the last chapter we learned how to compute the $\bar{Q}_{ij}$. We are thus in a position to compute the stresses if we know the strains and curvatures of the reference surface. Accordingly, using the strains that result from the Kirchhoff hypothesis, equation (6.13), we find that the stress-strain relations for a laminate become

$$
\left\{
\begin{array}{c}
\sigma_x \\
\sigma_y \\
\tau_{xy}
\end{array}
\right\}
=
\left[
\begin{array}{ccc}
\bar{Q}_{11} & \bar{Q}_{12} & \bar{Q}_{16} \\
\bar{Q}_{12} & \bar{Q}_{22} & \bar{Q}_{26} \\
\bar{Q}_{16} & \bar{Q}_{26} & \bar{Q}_{66}
\end{array}
\right]
\left\{
\begin{array}{c}
\varepsilon_x^o + z\kappa_x^o \\
\varepsilon_y^o + z\kappa_x^o \\
\gamma_{xy}^o + z\kappa_{xy}^o
\end{array}
\right\}
\tag{6.18}
$$

As can be seen, because the strains are functions of z, the stresses are functions of z. For a laminate, however, the stresses depend on z for another very important reason: The material properties in one layer, represented in equation (6.18) by the reduced stiffnesses $\bar{Q}_{ij}$, are generally different than the material properties of another layer, due both to varying fiber orientation from layer to layer, and to the fact that different materials can be used in different layers. Consequently, the transformed reduced stiffnesses are functions of location through the thickness and thus functions of z. The stresses therefore vary with z not only because the strains vary linearly with z *but also because the reduced stiffnesses vary with z.* In fact, the reduced stiffnesses vary in piecewise constant fashion, having one set of values through the thickness of one layer, another set of values through another layer, a third set of values through yet another layer, and so forth. The net result is that, in general, the variation of the stresses through the thickness of the laminate is discontinuous and very much *unlike* the variation through the thickness of an isotropic material.

6.6
STRESS DISTRIBUTIONS THROUGH THE THICKNESS

The most effective way to illustrate the implications on the stresses of the layer-by-layer change in material properties, coupled with the results of the Kirchhoff hypothesis, is with a series of simple examples. This will also allow us to introduce other important concepts and generally illustrate the implications of classical lamination theory (CLT). We start with a four-layer $[0/90]_S$ laminate; the $(0/90)$ construction is called a cross-ply construction. This laminate is an overly simplistic case, but we consider it so that the source of the response of more complicated laminates can be isolated by comparison with this case.

6.6.1 CLT Example 1: $[0/90]_S$ Laminate Subjected to Known ε_x^o

Consider a four-layer $[0/90]_S$ graphite-epoxy laminate, like the one shown in Figure 6.1(a), subjected to loads such that at a particular point (x, y) on the reference surface

$$\varepsilon_x^o(x, y) = 1000 \times 10^{-6} \qquad \kappa_x^o(x, y) = 0$$
$$\varepsilon_y^o(x, y) = 0 \qquad \kappa_y^o(x, y) = 0 \qquad (6.19)$$
$$\gamma_{xy}^o(x, y) = 0 \qquad \kappa_{xy}^o(x, y) = 0$$

Using equation (6.13), we find that the strain distribution through the thickness of the laminate is given by

$$\varepsilon_x(x, y, z) = \varepsilon_x^o(x, y) + z\kappa_x^o(x, y) = 1000 \times 10^{-6}$$
$$\varepsilon_y(x, y, z) = \varepsilon_y^o(x, y) + z\kappa_y^o(x, y) = 0 \qquad (6.20)$$
$$\gamma_{xy}(x, y, z) = \gamma_{xy}^o(x, y) + z\kappa_{xy}^o(x, y) = 0$$

Recall in conjunction with Table 2.1 that the thickness of a layer, h, will be assumed to be 0.150 mm. From equation (6.20), through the entire thickness of the laminate, above and below the point (x, y) on the reference surface, the only nonzero strain is the extensional strain in the x direction, as in Figure 6.13. This represents a laminate stretched, or extended, in the x direction but with no associated Poisson contraction in the y direction, and with no inplane shear strain and no curvature effects. For this $[0/90]_S$ laminate, we find, referring to the nomenclature of Figure 6.3, the layer interface locations are

$$z_o = -0.300\text{mm} \qquad z_1 = -0.150\text{mm} \qquad z_2 = 0$$
$$z_3 = 0.150\text{mm} \qquad z_4 = 0.300\text{mm} \qquad (6.21)$$

The stresses at each point through the thickness of the laminate are determined by using the above reference surface strains and curvatures in the stress-strain relation, equation (6.18); specifically:

$$\left\{ \begin{array}{c} \sigma_x \\ \sigma_y \\ \tau_{xy} \end{array} \right\} = \left[\begin{array}{ccc} \bar{Q}_{11} & \bar{Q}_{12} & \bar{Q}_{16} \\ \bar{Q}_{12} & \bar{Q}_{22} & \bar{Q}_{26} \\ \bar{Q}_{16} & \bar{Q}_{26} & \bar{Q}_{66} \end{array} \right] \left\{ \begin{array}{c} 1000 \times 10^{-6} \\ 0 \\ 0 \end{array} \right\} \qquad (6.22)$$

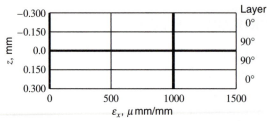

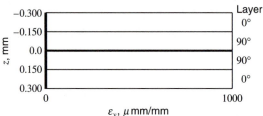

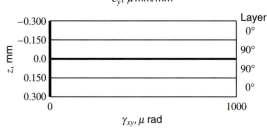

FIGURE 6.13
Strain distribution through the thickness of $[0/90]_S$ laminate subjected to $\varepsilon_x^o = 1000 \times 10^{-6}$

For the two 0° layers

$$\left\{ \begin{array}{c} \sigma_x \\ \sigma_y \\ \tau_{xy} \end{array} \right\} = \left[\begin{array}{ccc} \bar{Q}_{11}(0°) & \bar{Q}_{12}(0°) & \bar{Q}_{16}(0°) \\ \bar{Q}_{12}(0°) & \bar{Q}_{22}(0°) & \bar{Q}_{26}(0°) \\ \bar{Q}_{16}(0°) & \bar{Q}_{26}(0°) & \bar{Q}_{66}(0°) \end{array} \right] \left\{ \begin{array}{c} 1000 \times 10^{-6} \\ 0 \\ 0 \end{array} \right\} \quad (6.23)$$

while for the two 90° layers

$$\left\{ \begin{array}{c} \sigma_x \\ \sigma_y \\ \tau_{xy} \end{array} \right\} = \left[\begin{array}{ccc} \bar{Q}_{11}(90°) & \bar{Q}_{12}(90°) & \bar{Q}_{16}(90°) \\ \bar{Q}_{12}(90°) & \bar{Q}_{22}(90°) & \bar{Q}_{26}(90°) \\ \bar{Q}_{16}(90°) & \bar{Q}_{26}(90°) & \bar{Q}_{66}(90°) \end{array} \right] \left\{ \begin{array}{c} 1000 \times 10^{-6} \\ 0 \\ 0 \end{array} \right\} \quad (6.24)$$

From equation (5.86), the stresses in the 0° layers are

$$\left\{ \begin{array}{c} \sigma_x \\ \sigma_y \\ \tau_{xy} \end{array} \right\} = \left[\begin{array}{ccc} Q_{11} & Q_{12} & 0 \\ Q_{12} & Q_{22} & 0 \\ 0 & 0 & Q_{66} \end{array} \right] \left\{ \begin{array}{c} 1000 \times 10^{-6} \\ 0 \\ 0 \end{array} \right\} \quad (6.25)$$

For $\theta = 90°$, which means $m = 0$ and $n = 1$ in equation (5.84),

$$\begin{array}{lll} \bar{Q}_{11}(90°) = Q_{22} & \bar{Q}_{12}(90°) = Q_{12} & \bar{Q}_{16}(90°) = Q_{16} = 0 \\ \bar{Q}_{22}(90°) = Q_{11} & \bar{Q}_{26}(90°) = Q_{26} = 0 & \bar{Q}_{66}(90°) = Q_{66} \end{array} \quad (6.26)$$

From equation (6.24), then, for the two 90° layers,

$$\left\{ \begin{array}{c} \sigma_x \\ \sigma_y \\ \tau_{xy} \end{array} \right\} = \left[\begin{array}{ccc} Q_{22} & Q_{12} & 0 \\ Q_{12} & Q_{11} & 0 \\ 0 & 0 & Q_{66} \end{array} \right] \left\{ \begin{array}{c} 1000 \times 10^{-6} \\ 0 \\ 0 \end{array} \right\} \qquad (6.27)$$

Using numerical values for the reduced stiffnesses, equation (5.94), we find that for the 0° layers

$$\sigma_x = Q_{11} \times 1000 \times 10^{-6} = (155.7 \times 10^9)(1000 \times 10^{-6}) = 155.7 \text{ MPa}$$
$$\sigma_y = Q_{12} \times 1000 \times 10^{-6} = (3.02 \times 10^9)(1000 \times 10^{-6}) = 3.02 \text{ MPa} \qquad (6.28)$$
$$\tau_{xy} = 0$$

and for the 90° layers

$$\sigma_x = Q_{22} \times 1000 \times 10^{-6} = (12.16 \times 10^9)(1000 \times 10^{-6}) = 12.16 \text{ MPa}$$
$$\sigma_y = Q_{12} \times 1000 \times 10^{-6} = (3.02 \times 10^9)(1000 \times 10^{-6}) = 3.02 \text{ MPa} \qquad (6.29)$$
$$\tau_{xy} = 0$$

Figure 6.14 illustrates the distribution of the stresses through the thickness of the laminate. The character of the distribution of stress component σ_x illustrates the previously mentioned important difference between laminated fiber-reinforced composite materials and isotropic materials, namely, the discontinuous nature of the

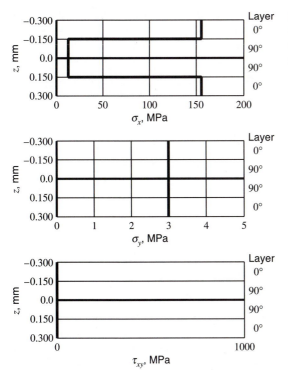

FIGURE 6.14
Stress distribution through the thickness of $[0/90]_S$ laminate subjected to $\varepsilon_x^o = 1000 \times 10^{-6}$

stress distribution through the thickness of the laminate. For the problem here, the stress component σ_x is constant within each layer, but varies from layer to layer. Formally, σ_x is said to be *piecewise constant* through the thickness of the laminate. On the other hand, the stress component σ_y is constant through the thickness of the laminate. The constancy of σ_y is a very special condition that can occur in cross-ply laminates or other special lamination arrangements. It can occur because $\bar{Q}_{12}(0°) = \bar{Q}_{12}(90°) = Q_{12}$. Note that a tensile value for σ_y is needed to overcome the natural tendency of the laminate to contract in the y direction. Due to the Poisson effect, stretching in the x direction causes contraction in the y direction. By the statement of the problem in equation (6.19), contraction in the y direction is here stipulated to be zero ($\varepsilon_y \equiv 0$), so a tensile stress is required to enforce this. Also special is the fact that the shear stress is zero at all z locations.

It is important to recall the example described in Figure 5.8(c), namely, an element of graphite-reinforced material with $\theta = 0°$ and stretched in the fiber direction. In that example the strain state of the $0°$ element was exactly the same as the strain state for the $0°$ layers of the $[0/90]_S$ laminate in the current example, namely, equation (6.19). The stress state for the $0°$ element in Figure 5.8(c) was given by equation (5.95), whereas the stress state for the $0°$ layers in the $[0/90]_S$ laminate is given by equation (6.28). *The stress states are the same.* Thus, if the state of strain of an element of material is specified, its stresses are uniquely determined, independently of whether the element is isolated, as in Figure 5.8(c), or whether it is part of a laminate. This is an important point.

If we consider a four-layer aluminum laminate made by perfecting bonding together four layers of aluminum, the distribution of strain through the thickness would be as in Figure 6.13. As we have emphasized, the reference surface strains and curvatures dictate the distribution of the strains through the thickness. Specified values of reference surface strains and curvatures produce the same distributions of ε_x, ε_y, and γ_{xy} through the thickness of aluminum as through the thickness of a composite. The stresses for each layer in the aluminum laminate are given by equation (6.18); the reduced stiffnesses for aluminum are used instead of the reduced stiffnesses of the composite. From our previous examples with aluminum, specifically equation (5.87), and the numerical values for aluminum from Table 2.1,

$$\sigma_x = (79.6 \times 10^9)(1000 \times 10^{-6}) = 79.6 \text{ MPa}$$
$$\sigma_y = (23.9 \times 10^9)(1000 \times 10^{-6}) = 23.9 \text{ MPa} \qquad (6.30)$$
$$\tau_{xy} = 0$$

Due to the assumption of perfection bonding between layers, the four layers of aluminum act as one; Figure 6.15 shows the distribution of the stresses through the thickness of the aluminum. The character of this figure should be compared with that of Figure 6.14, particularly the continuity of the stress component σ_x, and the magnitude of σ_y relative to σ_x.

Though the determination of the stresses through the thickness of the composite laminate is straightforward, given that we know the reference surface strains and curvatures, the really important issue is the determination of the stresses and strains in the principal material system of the individual layers. When we consider failure

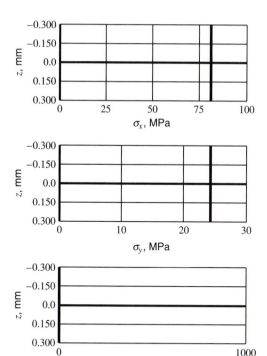

FIGURE 6.15
Stress distribution through the thickness of a four-layer aluminum laminate subjected to $\varepsilon_x^o = 1000 \times 10^{-6}$

of fiber-reinforced materials, what's important is the stresses and strains in the fiber direction, perpendicular to the fibers, and in shear. As far as the material is concerned, these are the basic responses. We introduced an x-y-z global coordinate system simply for convenience, specifically so we would not have to deal with a number of coordinate systems. However, we should address fundamental issues in the principal material system. To do this, we employ the transformation relations for stress and strain of Chapter 5, specifically equations (5.10) and (5.21). As they will be used frequently in this chapter, we here reproduce those relations:

$$\left\{ \begin{array}{c} \sigma_1 \\ \sigma_2 \\ \tau_{12} \end{array} \right\} = [T] \left\{ \begin{array}{c} \sigma_x \\ \sigma_y \\ \tau_{xy} \end{array} \right\} = \left[\begin{array}{ccc} m^2 & n^2 & 2mn \\ n^2 & m^2 & -2mn \\ -mn & mn & m^2 - n^2 \end{array} \right] \left\{ \begin{array}{c} \sigma_x \\ \sigma_y \\ \tau_{xy} \end{array} \right\} \quad (6.31)$$

and

$$\left\{ \begin{array}{c} \varepsilon_1 \\ \varepsilon_2 \\ \frac{1}{2}\gamma_{12} \end{array} \right\} = [T] \left\{ \begin{array}{c} \varepsilon_x \\ \varepsilon_y \\ \frac{1}{2}\gamma_{xy} \end{array} \right\} = \left[\begin{array}{ccc} m^2 & n^2 & 2mn \\ n^2 & m^2 & -2mn \\ -mn & mn & m^2 - n^2 \end{array} \right] \left\{ \begin{array}{c} \varepsilon_x \\ \varepsilon_y \\ \frac{1}{2}\gamma_{xy} \end{array} \right\} \quad (6.32)$$

For the particular problem here, no transformation is needed for the top and bottom layers. As their fibers are oriented at $\theta = 0°$, $\varepsilon_1 = \varepsilon_x$, $\varepsilon_2 = \varepsilon_y$, $\gamma_{xy} = \gamma_{12}$. Nevertheless, to remain formal, for the $0°$ layers, since $m = 1$ and $n = 0$, equation

(6.32) results in

$$\left\{ \begin{array}{c} \varepsilon_1 \\ \varepsilon_2 \\ \frac{1}{2}\gamma_{12} \end{array} \right\} = [T(0°)] \left\{ \begin{array}{c} \varepsilon_x \\ \varepsilon_y \\ \frac{1}{2}\gamma_{xy} \end{array} \right\} = \left[\begin{array}{ccc} 1 & 0 & 0 \\ 0 & 1 & 0 \\ 0 & 0 & 1 \end{array} \right] \left\{ \begin{array}{c} \varepsilon_x \\ \varepsilon_y \\ \frac{1}{2}\gamma_{xy} \end{array} \right\} \tag{6.33}$$

Substituting for the values of strain in the x-y-z system for the 0° layers, we find

$$\left\{ \begin{array}{c} \varepsilon_1 \\ \varepsilon_2 \\ \frac{1}{2}\gamma_{12} \end{array} \right\} = \left[\begin{array}{ccc} 1 & 0 & 0 \\ 0 & 1 & 0 \\ 0 & 0 & 1 \end{array} \right] \left\{ \begin{array}{c} 1000 \times 10^{-6} \\ 0 \\ 0 \end{array} \right\} \tag{6.34}$$

As expected, then, in the 0° layers,

$$\varepsilon_1 = 1000 \times 10^{-6}$$
$$\varepsilon_2 = 0 \tag{6.35}$$
$$\gamma_{12} = 0$$

For the 90° layers, $m = 0$ and $n = 1$ and equation (6.32) becomes

$$\left\{ \begin{array}{c} \varepsilon_1 \\ \varepsilon_2 \\ \frac{1}{2}\gamma_{12} \end{array} \right\} = [T(90°)] \left\{ \begin{array}{c} \varepsilon_x \\ \varepsilon_y \\ \frac{1}{2}\gamma_{xy} \end{array} \right\} = \left[\begin{array}{ccc} 0 & 1 & 0 \\ 1 & 0 & 0 \\ 0 & 0 & -1 \end{array} \right] \left\{ \begin{array}{c} 1000 \times 10^{-6} \\ 0 \\ 0 \end{array} \right\} \tag{6.36}$$

which results in

$$\varepsilon_1 = 0$$
$$\varepsilon_2 = 1000 \times 10^{-6} \tag{6.37}$$
$$\gamma_{12} = 0$$

Figure 6.16 shows the distribution of these principal material system strains ε_1, ε_2, and γ_{12}. The distribution of the principal material systems strains, unlike the laminate system strains, is discontinuous with z. The distributions are discontinuous because the value of a particular strain component, ε_1 for example, is the strain in a direction that changes from layer to layer. In the first and fourth layers, ε_1 is in the x direction, while in the second and third layers ε_1 is in the y direction. This is unlike the strains in the laminate x-y-z system where ε_x, for example, is the strain in the x direction, independent of the layer.

The stresses in the principal material system are computed using equation (6.31). As with the strains, the stresses σ_1, σ_2, and τ_{12} for the 0° layers are σ_x, σ_y, and τ_{xy}, respectively; so for the 0° layers,

$$\sigma_1 = 155.7 \text{ MPa}$$
$$\sigma_2 = 3.02 \text{ MPa} \tag{6.38}$$
$$\tau_{12} = 0$$

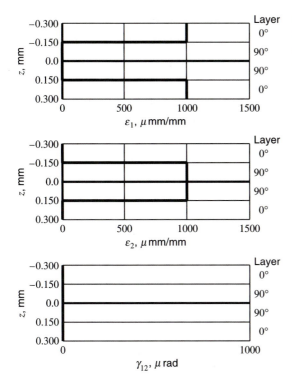

FIGURE 6.16
Principal material system strain
distribution through the thickness
of $[0/90]_S$ laminate subjected to
$\varepsilon_x^o = 1000 \times 10^{-6}$

while for the 90° layers,

$$\left\{ \begin{array}{c} \sigma_1 \\ \sigma_2 \\ \tau_{12} \end{array} \right\} = [T(90°)] \left\{ \begin{array}{c} \sigma_x \\ \sigma_y \\ \tau_{xy} \end{array} \right\} = \left[\begin{array}{ccc} 0 & 1 & 0 \\ 1 & 0 & 0 \\ 0 & 0 & -1 \end{array} \right] \left\{ \begin{array}{c} 12.16 \\ 3.02 \\ 0 \end{array} \right\} \quad (6.39)$$

or
$$\sigma_1 = 3.02 \text{ MPa}$$
$$\sigma_2 = 12.16 \text{ MPa} \quad (6.40)$$
$$\tau_{12} = 0$$

The thickness distribution of the principal material system stresses, in Figure 6.17,
like the distribution of the laminate system stresses, is discontinuous with z. The
discontinuity results not only because of the way the abrupt changes in the material
properties from layer to layer influence σ_x σ_y, and τ_{xy} but also, like the principal
material system strains, because the direction in which these stresses act changes
abruptly from layer to layer.

We should mention before moving to the second example that the stresses in
the principal material system could be computed by using the stress-strain relations
in the principal material system and the principal material strains, that is, equation
(4.14). This is opposed to transforming σ_x, σ_y, and τ_{xy}. The results would obviously
be the same.

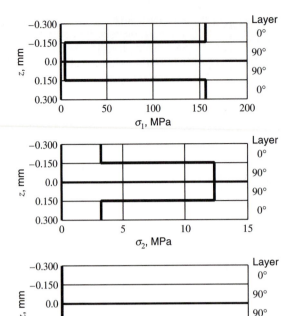

FIGURE 6.17
Principal material system stress distribution through the thickness of $[0/90]_S$ laminate subjected to $\varepsilon_x^o = 1000 \times 10^{-6}$

6.6.2 CLT Example 2: $[0/90]_S$ Laminate Subjected to Known κ_x^o

As a second example of determining the stress and strain response of a laminate, consider the same four-layer $[0/90]_S$ laminate. However, assume it is subjected to loads such that at a particular point (x, y) on the reference surface

$$\varepsilon_x^o(x, y) = 0 \qquad \kappa_x^o(x, y) = 3.33 \text{ m}^{-1}$$
$$\varepsilon_y^o(x, y) = 0 \qquad \kappa_y^o(x, y) = 0 \qquad (6.41)$$
$$\gamma_{xy}^o(x, y) = 0 \qquad \kappa_{xy}^o(x, y) = 0$$

Unlike CLT Example 1, the strain distribution through the thickness of the laminate is not independent of z; rather, it is linear in z and is given by

$$\varepsilon_x(x, y, z) = \varepsilon_x^o(x, y) + z\kappa_x^o(x, y) = 3.33z$$
$$\varepsilon_y(x, y, z) = \varepsilon_y^o(x, y) + z\kappa_y^o(x, y) = 0 \qquad (6.42)$$
$$\gamma_{xy}(x, y, z) = \gamma_{xy}^o(x, y) + z\kappa_{xy}^o(x, y) = 0$$

As with CLT Example 1, through the entire thickness of the laminate, above and below the point on the reference surface, the only nonzero strain is the extensional strain in the x direction. This nonzero strain is linear in z, as in Figure 6.18, and is zero on the reference surface, where $z = 0$; -1000×10^{-6} on the top surface, where $z = -0.300$ mm; and $+1000 \times 10^{-6}$ on the bottom surface, where $z = +0.300$ mm. The physical interpretation of this curvature-only deformation will be

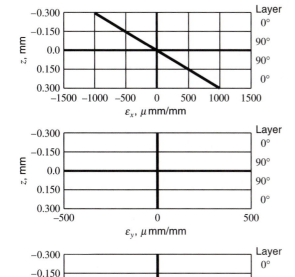

Layer
0°

90°

90°

0°

Layer
0°

90°

90°

0°

Layer
0°

90°

90°

0°

FIGURE 6.18
Strain distribution through the thickness of $[0/90]_s$ laminate subjected to $\kappa_x^o = 3.33 \text{m}^{-1}$

discussed shortly, but as we go through this present example keep in mind the minus sign in connection with the definitions of the curvatures, equation (6.14), and the fact that the $+z$ direction is downward. By the convention being used here, if the reference surface is deformed such that there is a negative second derivative of w^o with respect to x (i.e., positive κ_x^o), then there is the tendency toward a compressive strain in the upper surface of the laminate.

The stresses at each point through the thickness of the laminate are determined by using the strains from equation (6.42) in the stress-strain relations, equation (6.18), namely,

$$\left\{ \begin{array}{c} \sigma_x \\ \sigma_y \\ \tau_{xy} \end{array} \right\} = \left[\begin{array}{ccc} \bar{Q}_{11} & \bar{Q}_{12} & \bar{Q}_{16} \\ \bar{Q}_{12} & \bar{Q}_{22} & \bar{Q}_{26} \\ \bar{Q}_{16} & \bar{Q}_{26} & \bar{Q}_{66} \end{array} \right] \left\{ \begin{array}{c} 3.33z \\ 0 \\ 0 \end{array} \right\} \quad (6.43)$$

For the two 0° layers

$$\left\{ \begin{array}{c} \sigma_x \\ \sigma_y \\ \tau_{xy} \end{array} \right\} = \left[\begin{array}{ccc} \bar{Q}_{11}(0°) & \bar{Q}_{12}(0°) & \bar{Q}_{16}(0°) \\ \bar{Q}_{12}(0°) & \bar{Q}_{22}(0°) & \bar{Q}_{26}(0°) \\ \bar{Q}_{16}(0°) & \bar{Q}_{26}(0°) & \bar{Q}_{66}(0°) \end{array} \right] \left\{ \begin{array}{c} 3.33z \\ 0 \\ 0 \end{array} \right\} \quad (6.44)$$

while for the two 90° layers

$$\left\{ \begin{array}{c} \sigma_x \\ \sigma_y \\ \tau_{xy} \end{array} \right\} = \left[\begin{array}{ccc} \bar{Q}_{11}(90°) & \bar{Q}_{12}(90°) & \bar{Q}_{16}(90°) \\ \bar{Q}_{12}(90°) & \bar{Q}_{22}(90°) & \bar{Q}_{26}(90°) \\ \bar{Q}_{16}(90°) & \bar{Q}_{26}(90°) & \bar{Q}_{66}(90°) \end{array} \right] \left\{ \begin{array}{c} 3.33z \\ 0 \\ 0 \end{array} \right\} \quad (6.45)$$

For the 0° layers

$$\sigma_x = Q_{11} \times (3.33z) = (155.7 \times 10^9)(3.33z) = 519\,000z \text{ MPa}$$

$$\sigma_y = Q_{12} \times (3.33z) = (3.02 \times 10^9)(3.33z) = 10\,060z \text{ MPa} \quad (6.46)$$

$$\tau_{xy} = 0$$

These relations are valid for z in the range -0.300 mm $\leq z \leq -0.150$ mm and $+0.150$ mm $\leq z \leq +0.300$ mm. For the 90° layers,

$$\sigma_x = Q_{22} \times (3.33z) = (12.16 \times 10^9)(3.33z) = 40\,500z \text{ MPa}$$

$$\sigma_y = Q_{12} \times (3.33z) = (3.02 \times 10^9)(3.33z) = 10\,060z \text{ MPa} \quad (6.47)$$

$$\tau_{xy} = 0$$

These relations are valid for z in the range -0.150 mm $\leq z \leq +0.150$ mm. Figure 6.19 illustrates the stress distribution; the maximum tensile stress of $\sigma_x = 155.7$ MPa occurs in the 0° layer at the bottom of the laminate where $z = +0.300$ mm, and the maximum compressive stress of $\sigma_x = -155.7$ MPa occurs

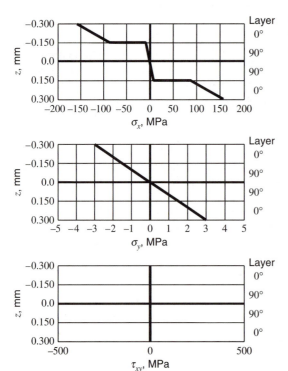

FIGURE 6.19

Stress distribution through the thickness of $[0/90]_S$ laminate subjected to $\kappa_x^o = 3.33\text{m}^{-1}$

at the top of the laminate, where $z = -0.300$ mm. In contrast to CLT Example 1, the distribution of the stresses through the thickness of each layer varies with z, but like CLT Example 1, the stress component σ_x is discontinuous from one layer to the next. At the interface between the first and second layers, σ_x jumps from 78.0 MPa compression in layer 1 to 6.09 MPa compression in layer 2. There is a similar jump between layers 3 and 4. The stress component σ_y is continuous; σ_y reaches a value of 3.02 MPa compression at the top of the laminate and varies linearly to 3.02 MPa tension at the bottom. This continuity in σ_y is strictly an artifact of the cross-ply construction, as it was in CLT Example 1, namely, $\bar{Q}_{12}(0°) = \bar{Q}_{12}(90°) = Q_{12}$. As will be seen, for laminates with fiber orientations other than $0°$ or $90°$, this will not necessarily be the case. In fact, a review of Chapter 5 should lead to the conclusion that if the strain states of these two examples were applied to a laminate with layers with fiber angles other than $0°$ or $90°$, in each case there would be a shear stress due to the existence of $\bar{Q}_{16}$ and, in general, none of the stresses would be continuous. The exception will be when adjacent layers have identical fiber orientations, or when adjacent layers have opposite fiber orientations (e.g., $\pm 30°$). In the latter case some components of stress are continuous despite the abrupt changes in fiber orientation.

Whatever the layer arrangement, the deformations specified by equation (6.41) require that there be a value for the stress component σ_y. This is because equation (6.41) specifies that at the particular point on the reference surface, where equation (6.41) is valid, there is no curvature in the y direction, κ_y^o. If the deformations specified in equation (6.41) were assumed to be valid at every point on the entire reference surface of a laminated plate, not just at a particular point, then the plate would appear as shown in Figure 6.20(a), namely, deformed into a cylindrical surface. The cylindrical shape would be such that there would be curvature in the x direction but not the y direction. With this deformation the upper surface of the plate would experience compressive strains in the x direction and the lower surface would experience tensile strains. If the curvature were not specified to be zero in the y direction, the compressive strains on the top surface in the x direction, through the natural tendency of the Poisson effect, would produce tensile strains in the y direction, and the tensile strains on the lower surface in the x direction would produce contraction strains in the y direction. As a result, the plate would be *saddle shaped*, as in Figure 6.20(b). This curvature in the y direction due to Poisson effects is referred to as *anticlastic curvature*. This is the natural tendency of the plate if it is given a curvature in the x direction and nothing is specified about the curvature in the y direction. However, here we have specified the curvature in the y direction to be zero; the anticlastic curvature is suppressed. To overcome the tendency to develop anticlastic curvature in the y direction, a stress component, σ_y, is required. The distribution of σ_y in Figure 6.19 is the distribution required to have this particular laminate remain flat in the y direction, yet cylindrical in the x direction.

A word of warning: It appears from Figure 6.19 that there is what could be called a neutral axis, a location of zero stress, at $z = 0$. This is the case here, but it will not always be the case. There may not always be a location of zero stress within the thickness of the laminate. Conversely, the location $z = 0$ will not always be a point of zero stress.

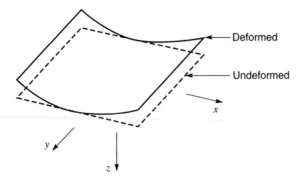

(a) Cylindrical shape defined by κ_x^o, with κ_y^o specified to be zero

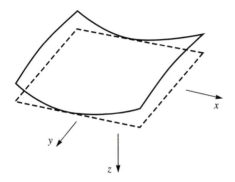

(b) Anticlastic curvature, κ_y^o not specified to be zero

FIGURE 6.20
Deformations of laminated plates

Again, we can determine the strain distribution in the principal material system by using transformation. The distribution of material system strains in this second example will be somewhat more complicated than in the first example. In the $0°$ layers, using the transformation matrix $[T]$ for the $0°$ layers yields

$$\left\{ \begin{array}{c} \varepsilon_1 \\ \varepsilon_2 \\ \frac{1}{2}\gamma_{12} \end{array} \right\} = \left[\begin{array}{ccc} 1 & 0 & 0 \\ 0 & 1 & 0 \\ 0 & 0 & 1 \end{array} \right] \left\{ \begin{array}{c} 3.33z \\ 0 \\ 0 \end{array} \right\} \tag{6.48}$$

As expected, then, in the $0°$ layers,

$$\varepsilon_1 = 3.33z$$
$$\varepsilon_2 = 0 \tag{6.49}$$
$$\gamma_{12} = 0$$

For the 90° layers

$$
\left\{
\begin{array}{c}
\varepsilon_1 \\
\varepsilon_2 \\
\dfrac{1}{2}\gamma_{12}
\end{array}
\right\}
=
\left[
\begin{array}{ccc}
0 & 1 & 0 \\
1 & 0 & 0 \\
0 & 0 & -1
\end{array}
\right]
\left\{
\begin{array}{c}
3.33z \\
0 \\
0
\end{array}
\right\}
\tag{6.50}
$$

or

$$
\varepsilon_1 = 0
$$
$$
\varepsilon_2 = 3.33z \tag{6.51}
$$
$$
\gamma_{12} = 0
$$

These principal material system strains, Figure 6.21, are piecewise linear; the discontinuities result because, as mentioned before, they represent strains in different directions at the different layer locations. We write the principal material system stresses in the 0° layers directly from equation (6.46) as

$$
\sigma_1 = 519\,000z \text{ MPa}
$$
$$
\sigma_2 = 10\,060z \text{ MPa} \tag{6.52}
$$
$$
\tau_{12} = 0
$$

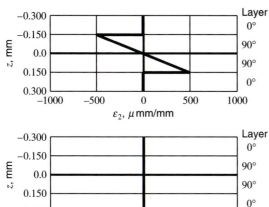

FIGURE 6.21
Principal material system strain distribution through the thickness of $[0/90]_S$ laminate subjected to $\kappa_x^o = 3.33\text{m}^{-1}$

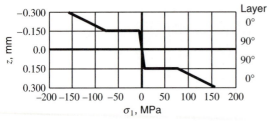

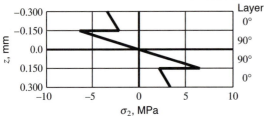

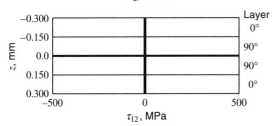

FIGURE 6.22
Principal material system stress distribution through the thickness of $[0/90]_S$ laminate subjected to $\kappa_x^o = 3.33 \text{m}^{-1}$

and for the 90° layers from equation (6.47) as

$$\left\{ \begin{array}{c} \sigma_1 \\ \sigma_2 \\ \tau_{12} \end{array} \right\} = \left[\begin{array}{ccc} 0 & 1 & 0 \\ 1 & 0 & 0 \\ 0 & 0 & -1 \end{array} \right] \left\{ \begin{array}{c} 40\,500z \\ 10\,060z \\ 0 \end{array} \right\} \tag{6.53}$$

which leads to

$$\sigma_1 = 10\,060z \text{ MPa}$$
$$\sigma_2 = 40\,500z \tag{6.54}$$
$$\tau_{12} = 0$$

Figure 6.22 shows the piecewise linear variations of the principal material system stresses, and the distributions are very dissimilar to any one would see with metals. However, the distributions shown in Figure 6.22 and in the previous figures are commonplace with composites, so one should get used to seeing them.

The utility of knowing the distribution of stresses through the thickness of the laminate in the principal material system is evident from Figure 6.22. The location and value of the maximum tensile stress perpendicular to the fibers, which is the stress that might be expected to cause failure, can immediately be determined. This stress is 6.09 MPa and it occurs at the bottom of the third layer. This exemplifies another characteristic of composite materials. Just because a laminate is subjected to bending doesn't mean that the stresses that might cause failure are at the extremities

For the 90° layers

$$
\left\{ \begin{array}{c} \varepsilon_1 \\ \varepsilon_2 \\ \dfrac{1}{2}\gamma_{12} \end{array} \right\} = \left[\begin{array}{ccc} 0 & 1 & 0 \\ 1 & 0 & 0 \\ 0 & 0 & -1 \end{array} \right] \left\{ \begin{array}{c} 3.33z \\ 0 \\ 0 \end{array} \right\}
\tag{6.50}
$$

or

$$
\varepsilon_1 = 0
$$
$$
\varepsilon_2 = 3.33z
\tag{6.51}
$$
$$
\gamma_{12} = 0
$$

These principal material system strains, Figure 6.21, are piecewise linear; the discontinuities result because, as mentioned before, they represent strains in different directions at the different layer locations. We write the principal material system stresses in the 0° layers directly from equation (6.46) as

$$
\sigma_1 = 519\ 000z \text{ MPa}
$$
$$
\sigma_2 = 10\ 060z \text{ MPa}
\tag{6.52}
$$
$$
\tau_{12} = 0
$$

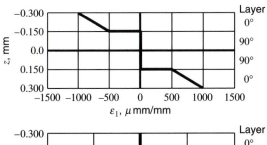

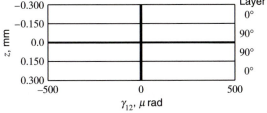

FIGURE 6.21
Principal material system strain distribution through the thickness of $[0/90]_S$ laminate subjected to $\kappa_x^o = 3.33\text{m}^{-1}$

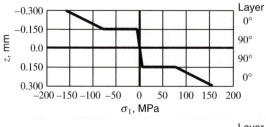

FIGURE 6.22

Principal material system stress distribution through the thickness of $[0/90]_S$ laminate subjected to $\kappa_x^o = 3.33\text{m}^{-1}$

and for the 90° layers from equation (6.47) as

$$
\left\{ \begin{array}{c} \sigma_1 \\ \sigma_2 \\ \tau_{12} \end{array} \right\} = \left[\begin{array}{ccc} 0 & 1 & 0 \\ 1 & 0 & 0 \\ 0 & 0 & -1 \end{array} \right] \left\{ \begin{array}{c} 40\,500z \\ 10\,060z \\ 0 \end{array} \right\}
\tag{6.53}
$$

which leads to

$$\sigma_1 = 10\,060z \text{ MPa}$$
$$\sigma_2 = 40\,500z \tag{6.54}$$
$$\tau_{12} = 0$$

Figure 6.22 shows the piecewise linear variations of the principal material system stresses, and the distributions are very dissimilar to any one would see with metals. However, the distributions shown in Figure 6.22 and in the previous figures are commonplace with composites, so one should get used to seeing them.

The utility of knowing the distribution of stresses through the thickness of the laminate in the principal material system is evident from Figure 6.22. The location and value of the maximum tensile stress perpendicular to the fibers, which is the stress that might be expected to cause failure, can immediately be determined. This stress is 6.09 MPa and it occurs at the bottom of the third layer. This exemplifies another characteristic of composite materials. Just because a laminate is subjected to bending doesn't mean that the stresses that might cause failure are at the extremities

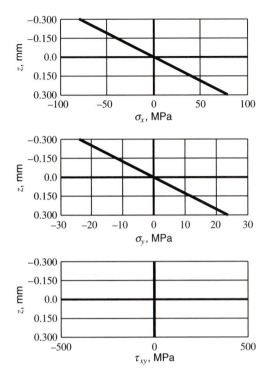

FIGURE 6.23
Stress distribution through the thickness of a four-layer aluminum laminate subjected to $\kappa_x^o = 3.33\text{m}^{-1}$

of the laminate. For an isotropic beam or plate subjected to bending, this is so; however, it is wrong to assume a priori that this is also the case for composites.

Let us again consider, as we did in the first example, a laminate constructed of four aluminum layers and subjected to the reference surface deformations given by equation (6.41). Equation (6.42) remains valid and the strain distribution for this case is that shown by Figure 6.18. The four layers of aluminum act as one, and thus equation (6.18) with the reduced stiffnesses of aluminum is valid for the entire thickness, -0.300 mm $\leq z \leq +0.300$ mm; as shown in Figure 6.23, that equation leads to

$$\sigma_x = (79.6 \times 10^9)(3.33z) = 265\,000z \text{ MPa}$$

$$\sigma_y = (23.9 \times 10^6)(3.33z) = 79\,600z \tag{6.55}$$

$$\tau_{xy} = 0$$

The simple linear variation of σ_x for the aluminum is a contrast to the piecewise-linear nature of the distribution of σ_x for the laminate, Figure 6.19, which is caused in the latter case by the discontinuous nature of the material properties.

These two simple examples with a rather elementary laminate illustrate the results predicted by the Kirchhoff hypothesis coupled with the plane-stress assumption. As is apparent from these examples, if we know the reference surface strains and curvatures, then calculating the distribution of the strains and stresses, both in the laminate x-y-z system and in the principal material 1-2-3 system, follows in a rather straightforward manner. Other examples will be presented shortly, but at this point it is appropriate to introduce the idea of force and moment resultants.

6.7
FORCE AND MOMENT RESULTANTS

As the discussion related to Figure 6.20 implies, to keep the reference surface of a laminated plate from exhibiting anticlastic curvature at the point (x, y), stresses σ_y are required, as in Figure 6.19. Equally important, though not discussed specifically, to produce the specified curvature in the x direction, σ_x stresses are required; these were also illustrated in Figure 6.19. From the perspective of an entire laminated plate, as in Figure 6.24, bending moments are required along the edge of the plate to produce the deformations of Figure 6.20(a). Similarly, to keep a laminated plate deformed as specified by the first example, equation (6.19), Figure 6.14 indicated that inplane stresses in the x and y directions are required. These stresses are manifested as inplane loads along the edges of the plate, as in Figure 6.25. The loads and moments required to produce the specified midplane deformations in any particular problem are actually *integrals* through the laminate thickness of the stresses. We refer to these integrals through the thickness as *stress resultants*. Specifically, the inplane force resultant in the x direction, N_x, is defined as

$$N_x \equiv \int_{-\frac{H}{2}}^{\frac{H}{2}} \sigma_x dz \qquad (6.56)$$

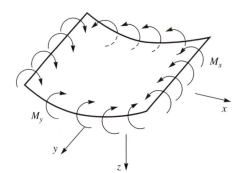

FIGURE 6.24
Moments required to deform laminated plate into a cylindrical shape

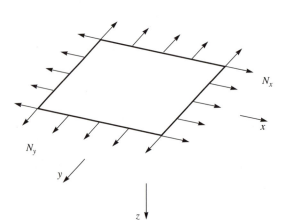

FIGURE 6.25
Forces required to stretch laminated plate in x direction

where, recall, H is the thickness of the laminate. We define the bending moment resultant due to σ_x as

$$M_x \equiv \int_{-\frac{H}{2}}^{\frac{H}{2}} \sigma_x z \, dz \qquad (6.57)$$

It is important to note the units associated with the force and moment resultants: The force resultant has the units of force per unit length, and the moment resultant has the units of moment per unit length. The unit of length is length in the y direction.

The force and moment resultant integrals are easy to evaluate since, in general, the stresses vary at most only in a stepwise linear fashion with z. For CLT Example 1, since σ_x is constant within a given layer, the integral for N_x is even simpler to evaluate. Specifically, we find, referring to Figure 6.14 for CLT Example 1 and formally going through the steps,

$$\begin{aligned} N_x &= \int_{-\frac{H}{2}}^{\frac{H}{2}} \sigma_x \, dz \\ &= \int_{z_0}^{z_1} \sigma_x \, dz + \int_{z_1}^{z_2} \sigma_x \, dz + \int_{z_2}^{z_3} \sigma_x \, dz + \int_{z_3}^{z_4} \sigma_x \, dz \end{aligned} \qquad (6.58)$$

Substituting from equations (6.28) and (6.29) the numerical values for the stresses that are valid over the appropriate range of z, and from equation (6.21) substituting for the values of the interface locations z_k, we find

$$\begin{aligned} N_x &= \int_{-300\times10^{-6}}^{-150\times10^{-6}} 155.7 \times 10^6 dz + \int_{-150\times10^{-6}}^{0} 12.16 \times 10^6 dz \\ &\quad + \int_{0}^{150\times10^{-6}} 12.16 \times 10^6 dz + \int_{150\times10^{-6}}^{300\times10^{-6}dz} 155.7 \times 10^6 dz \end{aligned} \qquad (6.59)$$

Because the stresses in each integrand are constant, they can be taken outside the integrals, resulting in

$$\begin{aligned} N_x &= 155.7 \times 10^6 \int_{-300\times10^{-6}}^{-150\times10^{-6}} dz + 12.16 \times 10^6 \int_{-150\times10^{-6}}^{0} dz \\ &\quad + 12.16 \times 10^6 \int_{0}^{150\times10^{-6}} dz + 155.7 \times 10^6 \int_{150\times10^{-6}}^{300\times10^{-6}} dz \end{aligned} \qquad (6.60)$$

The integrals on z are just the thickness of each layer, here 0.150 mm, and so

$$\begin{aligned} N_x &= [(155.7 + 12.16 + 12.16 + 155.7) \times 10^6](150 \times 10^{-6}) \\ &= 50\,400 \text{ N/m} \end{aligned} \qquad (6.61)$$

We determined N_x on a step-by-step basis for this simple example to emphasize the meaning of the integral through the thickness, namely, integration must be carried out through each layer. For this simple example, shortcuts in the integrations could obviously be made, but for more complicated situations, for which the stresses are not constant within each layer, the step-by-step approach is necessary.

Force resultants based on σ_y and τ_{xy} can be defined in a similar manner; specifically:

$$N_y \equiv \int_{-\frac{H}{2}}^{\frac{H}{2}} \sigma_y\,dz \qquad N_{xy} \equiv \int_{-\frac{H}{2}}^{\frac{H}{2}} \tau_{xy}\,dz \qquad (6.62)$$

where the latter is referred to as the shear force resultant. For CLT Example 1, again from equations (6.28) and (6.29),

$$N_y = (4)(3.02 \times 10^6)(150 \times 10^{-6}) = 1809 \text{ N/m} \qquad N_{xy} = 0 \qquad (6.63)$$

the latter stress resultant being zero because of the particular case. The integral for N_y is simple because the value of σ_y is the same in each layer and the four integrals can be lumped into one.

Because the definitions of the stress resultants are integrals with respect to z, and because z is measured from the reference surface, the stress resultants should be considered forces per unit inplane length acting at the reference surface. Hence for CLT Example 1, the proper interpretation is that if at a point (x, y) on the reference surface of a $[0/90]_S$ graphite-epoxy laminate the force resultants are

$$N_x = 50\,400 \text{ N/m} \qquad N_y = 1809 \text{ N/m} \qquad N_{xy} = 0 \qquad (6.64)$$

then that point on the reference surface will deform in the fashion given by equation (6.19), namely,

$$\begin{aligned}
\varepsilon_x^o(x, y) &= 1000 \times 10^{-6} & \kappa_x^o(x, y) &= 0 \\
\varepsilon_y^o(x, y) &= 0 & \kappa_y^o(x, y) &= 0 \\
\gamma_{xy}^o(x, y) &= 0 & \kappa_{xy}^o(x, y) &= 0
\end{aligned} \qquad (6.65)$$

The stress and strain distributions of Figures 6.13, 6.14, 6.16, and 6.17 result from the application of these stress resultants.

To further clarify the physical meaning of equations (6.64) and (6.65), consider, for example, a $[0/90]_S$ graphite-reinforced laminate with a length L_x in the x direction of 0.250 m and a width L_y in the y direction of 0.125 m. To have every point on the entire 0.0312 m^2 reference surface subjected to the deformations of equation (6.65) requires a load

$$N_x \times L_y = 50\,400 \text{ N/m} \times 0.125 \text{ m} = 6300 \text{ N} \qquad (6.66a)$$

in the x direction, and a load

$$N_y \times L_x = 1809 \text{ N/m} \times 0.250 \text{ m} = 452 \text{ N} \qquad (6.66b)$$

in the y direction. These loads, as in Figure 6.26, should be uniformly distributed along the appropriate edges. The displacements of the reference surface for this case can be determined by considering equations (6.14) and (6.65), namely,

$$\begin{aligned}
\varepsilon_x^o(x, y) &= \frac{\partial u^o(x, y)}{\partial x} = 1000 \times 10^{-6} \\
\varepsilon_y^o(x, y) &= \frac{\partial v^o(x, y)}{\partial y} = 0
\end{aligned} \qquad (6.67)$$

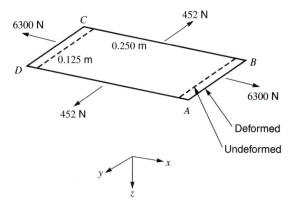

FIGURE 6.26
Forces required to produce state
of deformation $\varepsilon_x^o = 1000 \times 10^{-6}$
in $[0/90]_S$ laminate

Integrating these two equations results in

$$u^o(x, y) = 0.001x + g(y)$$
$$v^o(x, y) = h(x) \tag{6.68}$$

where $g(y)$ and $h(x)$ are functions of integration resulting from integrating partial derivatives. As a result, because the shear strain $\gamma_{xy}^o(x, y)$ is zero,

$$\gamma_{xy}^o(x, y) = \frac{\partial v^o(x, y)}{\partial x} + \frac{\partial u^o(x, y)}{\partial y} = \frac{dh(x)}{dx} + \frac{dg(y)}{dy} = 0 \tag{6.69}$$

The quantity $dg(y)/dy$ is a function of y only, and the quantity $dh(x)/dx$ is a function of x only, and equation (6.69) specifies that they add to yield a constant, namely zero. The only way that a function of x can add to a function of y to produce a constant is if the two functions are constants. Here they must be the same constant, differing by a sign. Specifically,

$$\frac{dg(y)}{dy} = C_1 \qquad \frac{dh(x)}{dx} = -C_1 \tag{6.70}$$

The result of integrating these two equations is

$$g(y) = C_1 y + C_2$$
$$h(x) = -C_1 x + C_3 \tag{6.71}$$

The constants C_2 and C_3 represent rigid body translations of the laminate in the x and y directions, respectively, and C_1 represents rigid body rotation about the z axis. Setting u^o and v^o to zero at the origin of coordinate system, $x = 0$ and $y = 0$, thereby suppressing rigid body translations, and arbitrarily suppressing rigid body rotation about the z axis by setting C_1 to zero, results in expressions for the displacements of the laminate reference surface, namely,

$$u^o(x, y) = 0.001x$$
$$v^o(x, y) = 0 \tag{6.72}$$
$$w^o(x, y) = 0$$

where the third equation results from the fact that curvatures do not develop due to inplane loads for this laminate. Though it is not needed for the determination of equation (6.72), we assume here that the origin of the x-y-z global coordinate system is at the geometric center of the laminate. With the displacements of equation (6.72), we can see that u^o varies linearly with x. The deformed length of the laminate is 250.25 mm, the width remains the same, and the corner right angles at A, B, C, and D remain right.

Returning to the resultants, the moment resultants are somewhat more complicated. For CLT Example 2, the bending moment resultant in the x direction is given by

$$M_x \equiv \int_{-\frac{H}{2}}^{\frac{H}{2}} \sigma_x z\, dz$$

$$= \int_{z_0}^{z_1} \sigma_x z\, dz + \int_{z_1}^{z_2} \sigma_x z\, dz + \int_{z_2}^{z_3} \sigma_x z\, dz + \int_{z_3}^{z_4} \sigma_x z\, dz \tag{6.73}$$

If we use the functional form of the stresses that are valid in the various ranges of the interface locations z, equations (6.46) and (6.47), and if we substitute for the values of z_k, then

$$M_x = \left\{ \int_{-300\times10^{-6}}^{-150\times10^{-6}} 519\,000 z^2\, dz + \int_{-150\times10^{-6}}^{0} 40\,500 z^2\, dz \right.$$

$$\left. + \int_{0}^{150\times10^{-6}} 40\,500 z^2\, dz + \int_{150\times10^{-6}}^{300\times10^{-6}} 519\,000 z^2\, dz \right\} \times 10^6 \tag{6.74}$$

Going one step further, we find

$$M_x = \left\{ 519\,000 \times 10^6 \int_{-300\times10^{-6}}^{-150\times10^{-6}} z^2\, dz + 40\,500 \times 10^6 \int_{-150\times10^{-6}}^{0} z^2\, dz \right.$$

$$\left. + 40\,500 \times 10^6 \int_{0}^{150\times10^{-6}} z^2\, dz + 519\,000 \times 10^6 \int_{150\times10^{-6}}^{300\times10^{-6}} z^2\, dz \right\} \tag{6.75}$$

with the integrals on z leading to differences in the cubes of the z_k; that is:

$$M_x = \frac{1}{3} \left\{ 519 \left((-150)^3 - (-300)^3\right) + 40.5 \left(0^3 - (-150)^3\right) \right.$$

$$\left. + 40.5 \left((150)^3 - 0^3\right) + 519 \left((300)^3 - (150)^3\right) \right\} \times \left(10^9\right) \left(10^{-6}\right)^3 \tag{6.76}$$

The final numerical result is

$$M_x = 8.27 \text{ N·m/m} \tag{6.77}$$

It is important to note the sign of the moment resultant: The sign is consistent with the sense of the moments shown in Figure 6.24; the moments there are shown in a positive sense. We will expand our discussion of the sign of the stress resultants shortly. Moment resultants associated with the other two stresses can be defined as

$$M_y \equiv \int_{-\frac{H}{2}}^{\frac{H}{2}} \sigma_y z\, dz \qquad M_{xy} \equiv \int_{-\frac{H}{2}}^{\frac{H}{2}} \tau_{xy} z\, dz \qquad (6.78)$$

The latter is referred to as the twisting moment resultant. For this problem

$$M_y = \int_{z_0}^{z_1} \sigma_y z\, dz + \int_{z_1}^{z_2} \sigma_y z\, dz + \int_{z_2}^{z_3} \sigma_y z\, dz + \int_{z_3}^{z_4} \sigma_y z\, dz$$

$$= (10\ 060 \times 10^6) \left\{ \left[\int_{-300 \times 10^{-6}}^{300 \times 10^{-6}} z^2\, dz \right] \right\} \qquad (6.79)$$

The latter step results because, by equations (6.46) and (6.47), the functional dependence of σ_y on z is the same for each layer and the four spatial integrals can be lumped into one, resulting in

$$M_y = 0.1809\ \text{N·m/m} \qquad (6.80)$$

By equations (6.46) and (6.47),

$$M_{xy} = 0 \qquad (6.81)$$

Again, take care to note the sign of the bending moment resultant M_y, as it is consistent with the sense shown in Figure 6.24. Finally, the definitions of the moment resultants imply that the moments are taken about the point $z = 0$; that is, the reference surface.

The proper interpretation of CLT Example 2 is that if at a point (x, y) on the reference surface of a $[0/90]_S$ laminate, the moment resultants are

$$M_x = 8.27\ \text{N·m/m} \qquad M_y = 0.1809\ \text{N·m/m} \qquad M_{xy} = 0 \qquad (6.82)$$

then at that point the deformation of the reference surface is given by equation 6.41, namely,

$$\begin{aligned}
\varepsilon_x^o(x, y) &= 0 & \kappa_x^o(x, y) &= 3.33\ \text{m}^{-1} \\
\varepsilon_y^o(x, y) &= 0 & \kappa_y^o(x, y) &= 0 \\
\gamma_{xy}^o(x, y) &= 0 & \kappa_{xy}^o(x, y) &= 0
\end{aligned} \qquad (6.83)$$

The stress and strain distributions of Figures 6.18, 6.19, 6.21, and 6.22 result from the application of these stress resultants. Consider again the 0.250 m long by 0.125 m wide $[0/90]_S$ laminate; if it is to have the deformations of equation (6.83) at every point on its entire 0.0312 m² reference surface, then along the 0.125 m widthwise side there must be a total bending moment of

$$M_x \times L_y = 8.27 \times 0.1255 = 1.033\ \text{N·m} \qquad (6.84a)$$

and along the 0.250 m lengthwise edge there must be a total bending moment of

$$M_y \times L_x = 0.1809 \times 0.250 = 0.0452\ \text{N·m} \qquad (6.84b)$$

Figure 6.27 illustrates these moments and it is assumed they are uniformly distributed along the edges. Here the double-headed arrows are used to indicate moments, as opposed to the curved arrows in Figure 6.24. We are introducing the double-headed arrow notation to avoid confusion in later figures. The sense of the moments along the 0.250 m lengthwise edges is such as to counter anticlastic curvature effects.

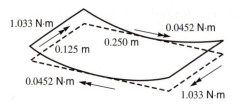

1.033 N·m   0.0452 N·m
0.250 m
0.125 m
0.0452 N·m
1.033 N·m

FIGURE 6.27
Moments required to produce state of deformation $\kappa_x^o = 3.33$ m^{-1} in $[0/90]_S$ laminate

We can determine the out-of-plane displacement of the reference surface by considering the definitions of the curvatures and equation (6.83), namely,

$$\kappa_x^o(x, y) = -\frac{\partial^2 w^o(x, y)}{\partial x^2} = 3.33$$

$$\kappa_y^o(x, y) = -\frac{\partial^2 w^o(x, y)}{\partial y^2} = 0 \tag{6.85}$$

$$\kappa_{xy}^o = -2\frac{\partial^2 w^o(x, y)}{\partial x \partial y} = 0$$

Integrating the first two equations results in two different expressions for $w^o(x, y)$:

$$w^o(x, y) = -\frac{1}{2}3.33x^2 + q(y)x + r(y)$$

$$w^o(x, y) = s(x)y + t(x) \tag{6.86}$$

where $q(y), r(y), s(x)$, and $t(x)$ are arbitrary functions of integration. Computing the reference surface twisting curvature, which is equal to zero, from both expressions leads to

$$\kappa_{xy}^o = -2\frac{\partial^2 w^o(x, y)}{\partial x \partial y} = -2\frac{dq(y)}{dy} = 0$$

$$\kappa_{xy}^o = -2\frac{\partial^2 w^o(x, y)}{\partial x \partial y} = -2\frac{ds(x)}{dx} = 0 \tag{6.87}$$

From these two equations we can conclude that $q(y)$ and $s(x)$ are constants; that is:

$$q(y) = K_1$$

$$s(x) = K_2 \tag{6.88}$$

Thus, the two expressions for $w^o(x, y)$ become

$$w^o(x, y) = -\frac{1}{2}3.33x^2 + K_1x + r(y)$$

$$w^o(x, y) = K_2y + t(x) \tag{6.89}$$

As there can be only one expression for $w^o(x, y)$, we conclude that

$$r(y) = K_2 y + K_3$$

$$t(x) = -\frac{1}{2}3.33x^2 + K_1 x + K_3 \tag{6.90}$$

with K_3 being a constant that is common to both functions. The final expression for $w^o(x, y)$ is

$$w^o(x, y) = -\frac{1}{2}3.33x^2 + K_1 x + K_2 y + K_3 \tag{6.91}$$

The constants K_1 and K_2 represent rigid body rotations of the laminate about the y and x axes, respectively, and K_3 represents rigid body translation in the z direction. If w^o, $\partial w^o/\partial x$, and $\partial w^o/\partial y$ are arbitrarily set equal to zero at the origin of the coordinate system, $x = 0$ and $y = 0$, then to supress rigid body motion results in requiring K_1, K_2, and K_3 to be zero. The deformed shape of the laminate is thus given by

$$u^o(x, y) = 0$$

$$v^o(x, y) = 0$$

$$w^o(x, y) = -\frac{1}{2}(3.33)x^2 \tag{6.92}$$

where the first two equations result from the fact that ε_x^o, ε_y^o, and γ_{xy}^o are zero for this CLT example, and we assume that the origin of the coordinate system is at the center of the laminate.

The following couplings are very important:

1. The coupling of the deformations of the reference surface with the distribution of strains through the thickness.
2. The coupling of the distribution of strains through the thickness with the distribution of stresses through the thickness.
3. The coupling of the distribution of stresses through the thickness with the stress resultants that act at the reference surface.

The couplings begin to tie together the analysis of laminates in a cause-and-effect relation. Generally the loads on a laminate are specified and we want to know the resulting stresses. The effect is the reference surface deformations and stresses, while the cause is the stress resultants, or loads. If we know the reference surface deformations, we can determine the stresses. At this point we cannot determine the reference surface deformations from the stress resultants; we can only compute the stress resultants, given that we know the reference surface deformations. Later we will be able to compute the reference surface deformations from the stress resultants, and thus determine the thickness distribution of the strains and stresses from the given loads.

Though we have not shown it directly, in the first example the moment resultants are identically zero, and in the second example the force resultants are identically zero. An examination of the distribution of the stresses through the thickness of the laminate in the first example indicates that both σ_x and σ_y are even functions of z. Multiplying the stresses for that example by z makes the integrands of the moment resultants odd functions of z, resulting in zeros for the integrals. In the second example, the distribution of σ_x and σ_y are linear functions of z, making the integrands of the force resultants odd functions of z, resulting in zeros for those integrals too.

Exercise for Section 6.7

Compute the force and moment resultants for the aluminum laminates of Figures 6.15 and 6.23, the counterparts to the $[0/90]_S$ graphite-reinforced laminates of CLT Examples 1 and 2. Compare these resultants with the resultants for the graphite-reinforced laminates. Is the graphite-reinforced laminate stiffer than the aluminum laminate?

6.8
FURTHER EXAMPLES

6.8.1 CLT Example 3: $[\pm30/0]_S$ Laminate Subjected to Known ε_x^o

As a third example of the response of fiber-reinforced composite laminates as predicted by classical lamination theory, consider a six-layer $[\pm30/0]_S$ laminate with a point (x, y) on the reference surface subjected to the extension-only conditions of equation (6.19), namely,

$$\varepsilon_x^o(x, y) = 1000 \times 10^{-6} \qquad \kappa_x^o(x, y) = 0$$
$$\varepsilon_y^o(x, y) = 0 \qquad\qquad \kappa_y^o(x, y) = 0 \qquad\qquad (6.93)$$
$$\gamma_{xy}^o(x, y) = 0 \qquad\qquad \kappa_{xy}^o(x, y) = 0$$

The distribution of the strains ε_x, ε_y, and γ_{xy} is similar to Figure 6.13, except the laminate is 0.900 mm thick instead of 0.600 mm thick. For this six-layer laminate, as in equation (6.20),

$$\varepsilon_x(x, y, z) = \varepsilon_x^o(x, y) + z\kappa_x^o(x, y) = 1000 \times 10^{-6}$$
$$\varepsilon_y(x, y, z) = \varepsilon_y^o(x, y) + z\kappa_y^o(x, y) = 0 \qquad\qquad (6.94)$$
$$\gamma_{xy}(x, y, z) = \gamma_{xy}^o(x, y) + z\kappa_{xy}^o(x, y) = 0$$

and the interface locations are

$$z_0 = -0.450 \text{ mm} \qquad z_1 = -0.300 \text{ mm} \qquad z_2 = -0.150 \text{ mm}$$
$$z_3 = 0 \qquad z_4 = 0.150 \text{ mm} \qquad z_5 = 0.300 \text{ mm} \qquad z_6 = 0.450 \text{ mm}$$

(6.95)

with the strain distribution being shown in Figure 6.28. By the stress-strain relations the stresses in the $+30°$ layers are given by

$$\left\{ \begin{array}{c} \sigma_x \\ \sigma_y \\ \tau_{xy} \end{array} \right\} = \left[\begin{array}{ccc} \bar{Q}_{11}(30°) & \bar{Q}_{12}(30°) & \bar{Q}_{16}(30°) \\ \bar{Q}_{12}(30°) & \bar{Q}_{22}(30°) & \bar{Q}_{26}(30°) \\ \bar{Q}_{16}(30°) & \bar{Q}_{26}(30°) & \bar{Q}_{66}(30°) \end{array} \right] \left\{ \begin{array}{c} 1000 \times 10^{-6} \\ 0 \\ 0 \end{array} \right\} \qquad (6.96)$$

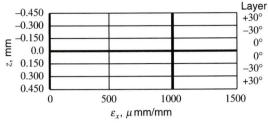

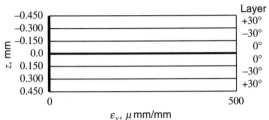

Layer

FIGURE 6.28
Strain distribution through the thickness of $[\pm30/0]_S$ laminate subjected to $\varepsilon_x^o = 1000 \times 10^{-6}$

or

$$\sigma_x = \bar{Q}_{11}(30°) \times 1000 \times 10^{-6}$$
$$\sigma_y = \bar{Q}_{12}(30°) \times 1000 \times 10^{-6} \qquad (6.97)$$
$$\tau_{xy} = \bar{Q}_{16}(30°) \times 1000 \times 10^{-6}$$

If we use the values of the reduced stiffness matrix for $\theta = +30°$, equation (5.97), the stresses are

$$\sigma_x = (92.8 \times 10^9)(1000 \times 10^{-6}) = 92.8 \text{ MPa}$$
$$\sigma_y = (30.1 \times 10^9)(1000 \times 10^{-6}) = 30.1 \text{ MPa} \qquad (6.98)$$
$$\tau_{xy} = (46.7 \times 10^9)(1000 \times 10^{-6}) = 46.7 \text{ MPa}$$

while for the $-30°$ layers,

$$\left\{ \begin{array}{c} \sigma_x \\ \sigma_y \\ \tau_{xy} \end{array} \right\} = \left[\begin{array}{ccc} \bar{Q}_{11}(-30°) & \bar{Q}_{12}(-30°) & \bar{Q}_{16}(-30°) \\ \bar{Q}_{12}(-30°) & \bar{Q}_{22}(-30°) & \bar{Q}_{26}(-30°) \\ \bar{Q}_{16}(-30°) & \bar{Q}_{26}(-30°) & \bar{Q}_{66}(-30°) \end{array} \right] \left\{ \begin{array}{c} 1000 \times 10^{-6} \\ 0 \\ 0 \end{array} \right\}$$

$$(6.99)$$

Or, if we use the values of the reduced stiffness matrix for $\theta = -30°$, equation (5.100),

$$\sigma_x = (92.8 \times 10^9)(1000 \times 10^{-6}) = 92.8 \text{ MPa}$$

$$\sigma_y = (30.1 \times 10^9)(1000 \times 10^{-6}) = 30.1 \text{ MPa} \tag{6.100}$$

$$\tau_{xy} = (-46.7 \times 10^9)(1000 \times 10^{-6}) = -46.7 \text{ MPa}$$

For the 0° layers the stresses are the same as in the first example with the $[0/90]_S$ laminate, namely,

$$\left\{ \begin{array}{c} \sigma_x \\ \sigma_y \\ \tau_{xy} \end{array} \right\} = \left[\begin{array}{ccc} \bar{Q}_{11}(0°) & \bar{Q}_{12}(0°) & \bar{Q}_{16}(0°) \\ \bar{Q}_{12}(0°) & \bar{Q}_{22}(0°) & \bar{Q}_{26}(0°) \\ \bar{Q}_{16}(0°) & \bar{Q}_{26}(0°) & \bar{Q}_{66}(0°) \end{array} \right] \left\{ \begin{array}{c} 1000 \times 10^{-6} \\ 0 \\ 0 \end{array} \right\} \tag{6.101}$$

$$\left\{ \begin{array}{c} \sigma_x \\ \sigma_y \\ \tau_{xy} \end{array} \right\} = \left[\begin{array}{ccc} Q_{11} & Q_{12} & 0 \\ Q_{12} & Q_{22} & 0 \\ 0 & 0 & Q_{66} \end{array} \right] \left\{ \begin{array}{c} 1000 \times 10^{-6} \\ 0 \\ 0 \end{array} \right\} \tag{6.102}$$

or

$$\sigma_x = 155.7 \text{ MPa}$$

$$\sigma_y = 3.02 \text{ MPa} \tag{6.103}$$

$$\tau_{xy} = 0$$

Figure 6.29 shows the stress distribution for this third example. Unlike the previous examples, τ_{xy} is not zero throughout the thickness. The stress components σ_x and

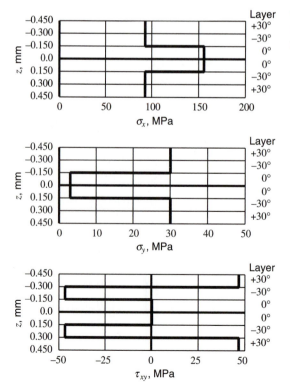

FIGURE 6.29
Stress distribution through the thickness of $[\pm 30/0]_S$ laminate subjected to $\varepsilon_x^o = 1000 \times 10^{-6}$

σ_y are continuous with z between the $\pm30°$ layers despite the abrupt change in fiber angle, whereas the component τ_{xy} is not continuous. These characteristics are a reflection of the nature of the variation of the reduced stiffnesses, the $\bar{Q}_{ij}$, with θ. This problem again emphasizes the fact that if the state of strain is given, the resulting stresses are independent of whether the element of material is isolated, as in equation (5.98), or whether it is part of a laminate, as in equation (6.98). The equivalence of these two results for the $+30°$ layers for these two different problems is attributed to the uniqueness of the strains in defining the stresses.

The force resultants are computed directly as follows:

$$N_x = (92.8 + 92.8 + 155.7 + 155.7 + 92.8 + 92.8) \times 10^6 (150 \times 10^{-6})$$
$$= 0.1024 \text{ MN/m}$$
$$N_y = (30.1 + 30.1 + 3.02 + 3.02 + 30.1 + 30.1) \times 10^6 (150 \times 10^{-6})$$
$$= 0.018\ 94 \text{ MN/m}$$
$$N_{xy} = (46.7 - 46.7 + 0 + 0 - 46.7 + 46.7) \times 10^6 (150 \times 10^{-6})$$
$$= 0$$

$$(6.104a)$$

The zero value of the shear force resultant N_{xy} is important to note. Even though there are shear stresses in the $\pm30°$ layers, the effect of these shear stresses on N_{xy} cancel each other when integrated through the thickness. There are no moment resultants for this problem; that is:

$$M_x = M_y = M_{xy} = 0 \qquad (6.104b)$$

The proper interpretation of this third example is: If the force resultants given by equations (6.104a) and (6.104b) are applied at a point (x, y) on the reference surface of a $[\pm30/0]_S$ laminate, then at that point the reference surface will deform as given by equation (6.93) and the stresses given by equations (6.98), (6.100), and (6.103) result. A 0.250 m long by 0.125 m wide $[\pm30/0]_S$ laminate with the entire reference surface deformed as given in equation (6.93) would require the forces illustrated in Figure 6.30, assuming they are uniformly distributed along the edges. These forces are determined from the force resultants of equation (6.104) and the dimensions of

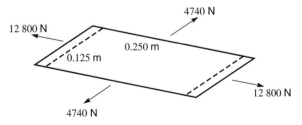

FIGURE 6.30
Forces required to produce state of deformation $\varepsilon_x^o = 1000 \times 10^{-6}$ in $[\pm30/0]_S$ laminate

the laminate; this situation is similar to that of Figure 6.26 for the $[0/90]_S$ laminate. Equation (6.72) is valid for this $[\pm30/0]_S$ case, and again the deformed length of the plate is 250.25 mm long; the width and right corner angles remain unchanged. Note that considerably more force is required to deform the $[\pm30/0]_S$ laminate than to deform the $[0/90]_S$ laminate in exactly the same manner. There are more layers in the $[\pm30/0]_S$ laminate than in the $[0/90]_S$ laminate, but more importantly, the $\pm30°$ layers have a greater stiffness in the x direction than the $90°$ layers. The requirement of having a larger N_y for the $[\pm30/0]_S$ case is a reflection of the greater tendency of the $[\pm30/0]_S$ laminate to contract more in the y direction. This is also due to the influence of the $\pm30°$ layers.

The strains in the principal material system can be computed from the transformation equations equation (6.32). For the $+30°$ layers,

$$
\left\{ \begin{array}{c} \varepsilon_1 \\ \varepsilon_2 \\ \dfrac{1}{2}\gamma_{12} \end{array} \right\} = [T(+30°)] \left\{ \begin{array}{c} \varepsilon_x \\ \varepsilon_y \\ \dfrac{1}{2}\gamma_{xy} \end{array} \right\}
$$

$$
= \begin{bmatrix} 0.750 & 0.250 & 0.867 \\ 0.250 & 0.750 & -0.867 \\ -0.433 & 0.433 & 0.500 \end{bmatrix} \left\{ \begin{array}{c} 1000 \times 10^{-6} \\ 0 \\ 0 \end{array} \right\}
$$

(6.105)

or
$$
\begin{aligned}
\varepsilon_1 &= 750 \times 10^{-6} \\
\varepsilon_2 &= 250 \times 10^{-6} \\
\gamma_{12} &= -867 \times 10^{-6}
\end{aligned}
$$
(6.106)

For the $-30°$ layers,

$$
\left\{ \begin{array}{c} \varepsilon_1 \\ \varepsilon_2 \\ \dfrac{1}{2}\gamma_{12} \end{array} \right\} = \big[T(-30°)\big] \left\{ \begin{array}{c} \varepsilon_x \\ \varepsilon_y \\ \dfrac{1}{2}\gamma_{xy} \end{array} \right\}
$$

$$
= \begin{bmatrix} 0.750 & 0.250 & -0.867 \\ 0.250 & 0.750 & 0.867 \\ 0.433 & -0.433 & 0.500 \end{bmatrix} \left\{ \begin{array}{c} 1000 \times 10^{-6} \\ 0 \\ 0 \end{array} \right\}
$$

(6.107)

or
$$
\begin{aligned}
\varepsilon_1 &= 750 \times 10^{-6} \\
\varepsilon_2 &= 250 \times 10^{-6} \\
\gamma_{12} &= 867 \times 10^{-6}
\end{aligned}
$$
(6.108)

For the $0°$ layers

$$\varepsilon_1 = \varepsilon_x = 1000 \times 10^{-6}$$
$$\varepsilon_2 = \varepsilon_y = 0 \tag{6.109}$$
$$\gamma_{12} = \gamma_{xy} = 0$$

The stresses in the principal material system are computed from the transformation equations equation (6.31). For the $+30°$ layers

$$
\begin{Bmatrix} \sigma_1 \\ \sigma_2 \\ \tau_{12} \end{Bmatrix} = [T(+30°)] \begin{Bmatrix} \sigma_x \\ \sigma_y \\ \tau_{xy} \end{Bmatrix}
$$

$$\tag{6.110}$$

$$
= \begin{bmatrix} 0.750 & 0.250 & 0.867 \\ 0.250 & 0.750 & -0.867 \\ -0.433 & 0.433 & 0.500 \end{bmatrix} \begin{Bmatrix} 92.8 \text{ MPa} \\ 30.1 \text{ MPa} \\ 46.7 \text{ MPa} \end{Bmatrix}
$$

or

$$\sigma_1 = 117.6 \text{ MPa}$$
$$\sigma_2 = 5.30 \text{ MPa} \tag{6.111}$$
$$\tau_{12} = -3.81 \text{ MPa}$$

For the $-30°$ layers,

$$
\begin{Bmatrix} \sigma_1 \\ \sigma_2 \\ \tau_{12} \end{Bmatrix} = [T(-30°)] \begin{Bmatrix} \sigma_x \\ \sigma_y \\ \tau_{xy} \end{Bmatrix}
$$

$$\tag{6.112}$$

$$
= \begin{bmatrix} 0.750 & 0.250 & -0.867 \\ 0.250 & 0.750 & 0.867 \\ 0.433 & -0.433 & 0.500 \end{bmatrix} \begin{Bmatrix} 92.8 \text{ MPa} \\ 30.1 \text{ MPa} \\ -46.7 \text{ MPa} \end{Bmatrix}
$$

or

$$\sigma_1 = 117.6 \text{ MPa}$$
$$\sigma_2 = 5.30 \text{ MPa} \tag{6.113}$$
$$\tau_{12} = 3.81 \text{ MPa}$$

For the $0°$ layers

$$\sigma_1 = \sigma_x = 155.7 \text{ MPa}$$
$$\sigma_2 = \sigma_y = 3.02 \text{ MPa} \tag{6.114}$$
$$\tau_{12} = \tau_{xy} = 0$$

Figures 6.31 and 6.32 illustrate distributions of the principal material system strains and stresses; the discontinuities from layer to layer have now become familiar.

6.8.2 CLT Example 4: $[\pm 30/0]_S$ Laminate Subjected to Known κ_x^o

As a fourth example, consider the bending of the six-layer $[\pm 30/0]_S$ laminate. This is the counterpart to the bending of the $[0/90]_S$ laminate, and upon completion of

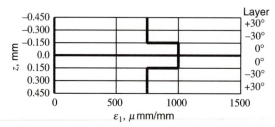

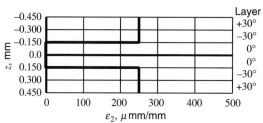

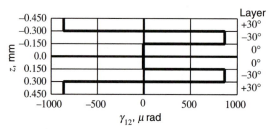

FIGURE 6.31

Principal material system strain distribution through the thickness of $[\pm30/0]_S$ laminate subjected to $\varepsilon_x^o = 1000 \times 10^{-6}$

this example, we will have examined the inplane stretching and the out-of-plane bending of a $[0/90]_S$ and a $[\pm30/0]_S$ laminate. This quartet of examples provides insight into some of the important characteristics of stresses in composite materials.

Consider a six-layer $[\pm30/0]_S$ laminate with a point (x, y) on the reference surface subjected to the curvature-only condition

$$\varepsilon_x^o(x, y) = 0 \qquad \kappa_x^o(x, y) = 2.22 \text{ m}^{-1}$$
$$\varepsilon_y^o(x, y) = 0 \qquad \kappa_y^o(x, y) = 0 \qquad (6.115)$$
$$\gamma_{xy}^o(x, y) = 0 \qquad \kappa_{xy}^o(x, y) = 0$$

Equation (6.95) gives the z_k for the laminate, and Figure 6.33 shows the distribution of the strains ε_x, ε_y, and γ_{xy} through the thickness of the laminate. Functionally, the strain distributions are given as

$$\varepsilon_x(x, y, z) = \varepsilon_x^o(x, y) + z\kappa_x^o(x, y) = 2.22z$$
$$\varepsilon_y(x, y, z) = \varepsilon_y^o(x, y) + z\kappa_y^o(x, y) = 0 \qquad (6.116)$$
$$\gamma_{xy}(x, y, z) = \gamma_{xy}^o(x, y) + z\kappa_{xy}^o(x, y) = 0$$

Because this six-layer $[\pm30/0]_S$ laminate is thicker than the four-layer $[0/90]_S$ laminate, we take the curvature of the $[\pm30/0]_S$ laminate to be smaller than the curvature of the $[0/90]_S$ laminate of CLT Example 3; thus, the maximum strains at the outer extremities of the two laminates are the same. This step is strictly based

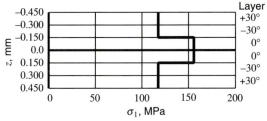

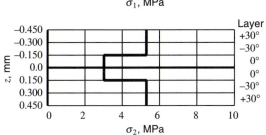

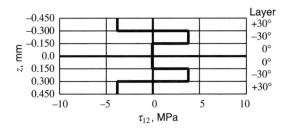

FIGURE 6.32
Principal material system stress
distribution through the thickness
of $[\pm 30/0]_S$ laminate subjected to
$\varepsilon_x^o = 1000 \times 10^{-6}$

on the desire to be consistent in these examples. The deformation represented by
equation (6.115) is as it appears in Figure 6.20(a).

We compute the stresses in the various layers by the now-familiar approach of
applying the stress-strain relations on a layer-by-layer basis. Specifically, for the
$+30°$ layers,

$$\left\{ \begin{array}{c} \sigma_x \\ \sigma_y \\ \tau_{xy} \end{array} \right\} = \left[\begin{array}{ccc} \bar{Q}_{11}(30°) & \bar{Q}_{12}(30°) & \bar{Q}_{16}(30°) \\ \bar{Q}_{12}(30°) & \bar{Q}_{22}(30°) & \bar{Q}_{26}(30°) \\ \bar{Q}_{16}(30°) & \bar{Q}_{26}(30°) & \bar{Q}_{66}(30°) \end{array} \right] \left\{ \begin{array}{c} 2.22z \\ 0 \\ 0 \end{array} \right\} \quad (6.117a)$$

For the $-30°$ layers,

$$\left\{ \begin{array}{c} \sigma_x \\ \sigma_y \\ \tau_{xy} \end{array} \right\} = \left[\begin{array}{ccc} \bar{Q}_{11}(-30°) & \bar{Q}_{12}(-30°) & \bar{Q}_{16}(-30°) \\ \bar{Q}_{12}(-30°) & \bar{Q}_{22}(-30°) & \bar{Q}_{26}(-30°) \\ \bar{Q}_{16}(-30°) & \bar{Q}_{26}(-30°) & \bar{Q}_{66}(-30°) \end{array} \right] \left\{ \begin{array}{c} 2.22z \\ 0 \\ 0 \end{array} \right\} \quad (6.117b)$$

For the $0°$ layers,

$$\left\{ \begin{array}{c} \sigma_x \\ \sigma_y \\ \tau_{xy} \end{array} \right\} = \left[\begin{array}{ccc} \bar{Q}_{11}(0°) & \bar{Q}_{12}(0°) & \bar{Q}_{16}(0°) \\ \bar{Q}_{12}(0°) & \bar{Q}_{22}(0°) & \bar{Q}_{26}(0°) \\ \bar{Q}_{16}(0°) & \bar{Q}_{26}(0°) & \bar{Q}_{66}(0°) \end{array} \right] \left\{ \begin{array}{c} 2.22z \\ 0 \\ 0 \end{array} \right\} \quad (6.117c)$$

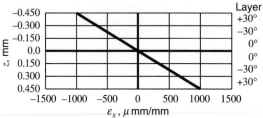

Layer
+30°
−30°
0°
0°
−30°
+30°

FIGURE 6.33

Strain distribution through the thickness of $[\pm 30/0]_S$ laminate subjected to $\kappa_x^o = 2.22$ m^{-1}

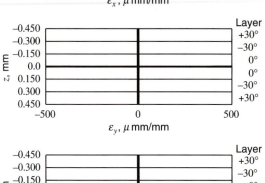

Layer
+30°
−30°
0°
0°
−30°
+30°

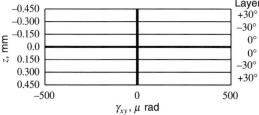

Layer
+30°
−30°
0°
0°
−30°
+30°

Expanding these expressions for the stresses and using numerical values for the $\bar{Q}_{ij}$ results in, for the $+30°$ layers,

$$\sigma_x = (92.8 \times 10^9)(2.22z) = 206\,000z \text{ MPa}$$
$$\sigma_y = (30.1 \times 10^9)(2.22z) = 66\,800z \text{ MPa} \qquad (6.118a)$$
$$\tau_{xy} = (46.7 \times 10^9)(2.22z) = 103\,800z \text{ MPa}$$

for the $-30°$ layers,

$$\sigma_x = (92.8 \times 10^9)(2.22z) = 206\,000z \text{ MPa}$$
$$\sigma_y = (30.1 \times 10^9)(2.22z) = 66\,800z \qquad (6.118b)$$
$$\tau_{xy} = (-46.7 \times 10^9)(2.22z) = -103\,800z$$

and for the $0°$ layers,

$$\sigma_x = (155.7 \times 10^9)(2.22z) = 346\,000z \text{ MPa}$$
$$\sigma_y = (3.02 \times 10^9)(2.22z) = 6700z \qquad (6.118c)$$
$$\tau_{xy} = (0)(2.22z) = 0$$

Figure 6.34 illustrates the variation of σ_x, σ_y, and τ_{xy} with z. Though the functional dependence on z of τ_{xy} is the same in the $-30°$ layers as in the $+30°$ layers, except for the sign, the magnitude of the shear stress is smaller in the $-30°$ layers because

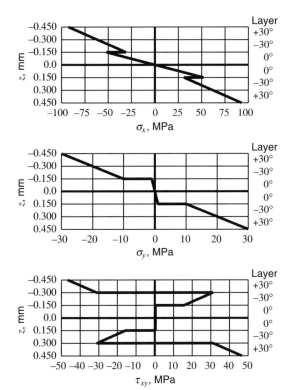

FIGURE 6.34

Stress distribution through the thickness of a $[\pm 30/0]_S$ laminate subjected to $\kappa_x^o = 2.22$ m^{-1}

these layers are closer to the reference surface (smaller values of z) and experience less strain for the given value of curvature κ_x^o.

Comparing CLT Examples 3 and 4 yields an important observation. Specifically, σ_x is considerably higher in the 0° layers of the $[\pm 30/0]_S$ laminate of CLT Example 3 than in the 0° layers of the $[\pm 30/0]_S$ laminate of CLT Example 4; refer to Figures 6.29 and 6.34. In fact, in CLT Example 4 the maximum value of σ_x in the 0° layers is one-third as large. Yet, this is the direction of bending. In CLT Example 4 the 0° layers are contributing very little to resisting bending in the x direction, whereas in CLT Example 3 they contribute considerably to resisting extension in the x direction. This lack of effectiveness of the 0° layers in the bending problem results because they are located near the reference surface of the laminate, that is, near what would be the "neutral axis" if this were an isotropic material problem. The stiffness and strength of the fibers will not be utilized if they are located near the reference surface and if the bending response is the primary response. On the other hand, for resisting inplane deformations, the 0° layers can be anywhere through the thickness of the laminate and have the same effect. For this reason, it is best to design composite structures so laminates are utilized primarily to resist tension and compression, as opposed to resisting bending.

The force resultants for CLT Example 4 are zero because the stresses σ_x, σ_y, and τ_{xy} are odd functions of z. The moment resultants are computed as

$$M_x \equiv \int_{-\frac{H}{2}}^{\frac{H}{2}} \sigma_x z \, dz$$

$$= \int_{z_0}^{z_1} \sigma_x z \, dz + \int_{z_1}^{z_2} \sigma_x z \, dz + \int_{z_2}^{z_3} \sigma_x z \, dz + \int_{z_3}^{z_4} \sigma_x z \, dz \qquad (6.119)$$

$$+ \int_{z_4}^{z_5} \sigma_x z \, dz + \int_{z_5}^{z_6} \sigma_x z \, dz$$

Using the functional forms of σ_x that are valid in the various ranges of z, and substituting for the values of z_k, results in

$$M_x = \left\{ \int_{-450\times10^{-6}}^{-300\times10^{-6}} 206z^2 dz + \int_{-300\times10^{-6}}^{-150\times10^{-6}} 206z^2 dz + \int_{-150\times10^{-6}}^{0} 346z^2 dz \right.$$

$$+ \int_{0}^{150\times10^{-6}} 346z^2 dz + \int_{150\times10^{-6}}^{300\times10^{-6}} 206z^2 dz + \left. \int_{300\times10^{-6}}^{450\times10^{-6}} 206z^2 dz \right\} \times 10^9$$

$$(6.120)$$

or
$$M_x = 12.84 \text{ N·m/m} \qquad (6.121)$$

Carrying out similar computations for M_y and M_{xy} leads to

$$M_y = 3.92 \text{ N·m/m} \qquad (6.122)$$

$$M_{xy} = 2.80 \text{ N·m/m} \qquad (6.123)$$

Note well that the twisting moment M_{xy} is not zero. This is because the $-30°$ layers are closer to the reference surface than the $+30°$ layers. As a result, the shear stresses τ_{xy} in the $-30°$ layers are smaller in magnitude than in the $+30°$ layers. Also, because they are closer, the contribution from the $-30°$ layers to the integral for M_{xy} is less, thereby not cancelling the contribution from the $+30°$ layers to the integral. The overall effect is that there is a net value of M_{xy} from the integral.

The proper interpretation of the moment resultants for this example, then, is that if for a point (x, y) on the reference surface of a $[\pm30/0]_S$ laminate the moment resultants are

$$M_x = 12.84 \text{ N·m/m} \qquad M_y = 3.92 \text{ N·m/m} \qquad M_{xy} = 2.80 \text{ N·m/m} \qquad (6.124)$$

then the deformation at that point on the reference surface is given by

$$\begin{aligned}
\varepsilon_x^o(x, y) &= 0 & \kappa_x^o(x, y) &= 2.22 \text{ m}^{-1} \\
\varepsilon_y^o(x, y) &= 0 & \kappa_y^o(x, y) &= 0 \\
\gamma_{xy}^o(x, y) &= 0 & \kappa_{xy}^o(x, y) &= 0
\end{aligned} \qquad (6.125)$$

and the distributions of the stresses and strains shown in Figures 6.33 and 6.34 result. To have *only* curvature in the x direction, κ_x^o, a twisting moment resultant M_{xy}, in addition to bending moment resultants M_x and M_y, must be applied at this point on the reference surface. If a $[\pm 30/0]_S$ laminate that is 0.250 m long by 0.125 m wide is to have the deformations of equation (6.125) at each point on its reference surface, then there must be a total bending moment of

$$M_x \times L_y = 12.84 \times 0.125 = 1.605 \text{ N·m} \tag{6.126a}$$

plus a total twisting moment of

$$M_{xy} \times L_y = 2.80 \times 0.125 = 0.350 \text{ N·m} \tag{6.126b}$$

along the 0.125 m edges, and a total bending moment of

$$M_y \times L_x = 3.92 \times 0.250 = 0.981 \text{ N·m} \tag{6.126c}$$

plus a total twisting moment of

$$M_{xy} \times L_x = 2.80 \times 0.250 = 0.701 \text{ N·m} \tag{6.126d}$$

along the 0.250 m edges. Figure 6.35 depicts these moments, and the sense is correctly indicated. Referring to the development of equation (6.92), we see that the deformed shape of the laminate is given by

$$u^o(x, y) = 0$$
$$v^o(x, y) = 0 \tag{6.127}$$
$$w^o(x, y) = -\frac{1}{2}(2.22)x^2$$

If the twisting moments were not present along the edges, then there would be a twisting curvature $\kappa_{xy}^o(x, y)$. The notion of needing a twisting moment to produce a deformation that consists *only* of a curvature in the x direction, and conversely, as we shall see, the notion of having a twisting curvature generated even if there is no twisting moment, are unique to composite materials. In general, when one considers pairs of off-axis angles (e.g., $\pm\theta$ pairs), it is physically impossible to have these pairs be at the same distance from the reference surface and this $M_{xy} - \kappa_x^o$ effect will always be present. Of course, if a laminate consists only of 0° and 90°

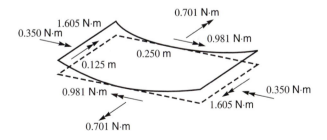

FIGURE 6.35
Moments required to produce state of deformation $\kappa_x^o = 2.22 \text{ m}^{-1}$ in $[\pm 30/0]_S$ laminate

layers, then there are no shear stresses and the effect is not present, as in CLT Example 2. With only $0°$ or $90°$ layers, $\bar{Q}_{16}$ and $\bar{Q}_{26}$ are zero for every layer and the extensional strains due to bending do not cause shear stresses. Finally, the presence of M_y for the $[\pm 30/0]_S$ laminate is necessary to counter anticlastic curvature effects.

We can compute the strains in the principal material system from the transformation relations. For the $+30°$ layers,

$$\left\{\begin{array}{c} \varepsilon_1 \\ \varepsilon_2 \\ \frac{1}{2}\gamma_{12} \end{array}\right\} = [T(+30°)] \left\{\begin{array}{c} \varepsilon_x \\ \varepsilon_y \\ \frac{1}{2}\gamma_{xy} \end{array}\right\}$$

(6.128)

$$= \left[\begin{array}{ccc} 0.750 & 0.250 & 0.867 \\ 0.250 & 0.750 & -0.867 \\ -0.433 & 0.433 & 0.500 \end{array}\right] \left\{\begin{array}{c} 2.22z \\ 0 \\ 0 \end{array}\right\}$$

or

$$\varepsilon_1 = 1.667z$$
$$\varepsilon_2 = 0.556z$$
$$\gamma_{12} = -1.924z$$

(6.129)

For the $-30°$ layers,

$$\left\{\begin{array}{c} \varepsilon_1 \\ \varepsilon_2 \\ \frac{1}{2}\gamma_{12} \end{array}\right\} = [T(-30°)] \left\{\begin{array}{c} \varepsilon_x \\ \varepsilon_y \\ \frac{1}{2}\gamma_{xy} \end{array}\right\}$$

(6.130)

$$= \left[\begin{array}{ccc} 0.750 & 0.250 & -0.867 \\ 0.250 & 0.750 & 0.867 \\ 0.433 & -0.433 & 0.500 \end{array}\right] \left\{\begin{array}{c} 2.22z \\ 0 \\ 0 \end{array}\right\}$$

or

$$\varepsilon_1 = 1.667z$$
$$\varepsilon_2 = 0.556z$$
$$\gamma_{12} = 1.924z$$

(6.131)

For the $0°$ layers, by direct observation,

$$\varepsilon_1 = \varepsilon_x = 2.22z$$
$$\varepsilon_2 = \varepsilon_y = 0$$
$$\gamma_{12} = \gamma_{xy} = 0$$

(6.132)

Figure 6.36 shows these strain distributions graphically. The stresses in the principal material system are given by transformation. For the $+30°$ layers,

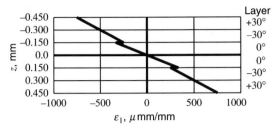

FIGURE 6.36
Principal material system strain distribution through the thickness of $[\pm 30/0]_S$ laminate subjected to $\kappa_x^o = 2.22$ m^{-1}

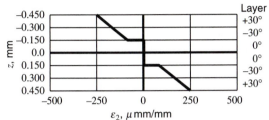

$$
\left\{
\begin{array}{c}
\sigma_1 \\
\sigma_2 \\
\tau_{12}
\end{array}
\right\}
=
\left[T(+30°) \right]
\left\{
\begin{array}{c}
\sigma_x \\
\sigma_y \\
\tau_{xy}
\end{array}
\right\}
$$

$$
=
\begin{bmatrix}
0.750 & 0.250 & 0.867 \\
0.250 & 0.750 & -0.867 \\
-0.433 & 0.433 & 0.500
\end{bmatrix}
\left\{
\begin{array}{c}
206\ 000z \\
66\ 800z \\
103\ 800z
\end{array}
\right\}
$$

(6.133)

or

$$
\sigma_1 = 261\ 000z \text{ MPa}
$$

$$
\sigma_2 = 11\ 780z
$$

(6.134)

$$
\tau_{12} = -8470z
$$

For the $-30°$ layers,

$$
\left\{
\begin{array}{c}
\sigma_1 \\
\sigma_2 \\
\tau_{12}
\end{array}
\right\}
=
\left[T(-30°) \right]
\left\{
\begin{array}{c}
\sigma_x \\
\sigma_y \\
\tau_{xy}
\end{array}
\right\}
$$

(6.135)

$$
=
\begin{bmatrix}
0.750 & 0.250 & -0.867 \\
0.250 & 0.750 & 0.867 \\
0.433 & -0.433 & 0.500
\end{bmatrix}
\left\{
\begin{array}{c}
206\ 000z \\
66\ 800z \\
-103\ 800z
\end{array}
\right\}
$$

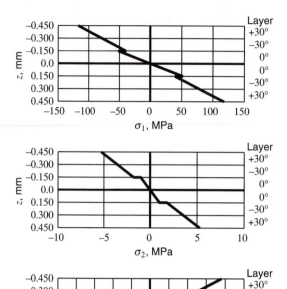

FIGURE 6.37
Principal material system stress distribution through the thickness of $[\pm 30/0]_S$ laminate subjected to $\kappa_x^o = 2.22$ m^{-1}

or

$$\sigma_1 = 261\ 000z \text{ MPa}$$
$$\sigma_2 = 11\ 780z \tag{6.136}$$
$$\tau_{12} = 8470z$$

For the 0° layers,

$$\sigma_1 = \sigma_x = 346\ 000z \text{ MPa}$$
$$\sigma_2 = \sigma_y = 6700z \tag{6.137}$$
$$\tau_{12} = \tau_{xy} = 0$$

Figure 6.37 illustrates these stress distributions.

From these examples, we can conclude that within a layer the stresses and strains vary linearly, or are a constant, through the thickness. For the most general deformation of the most general laminate, this will *always* be the case. It is a direct result of the Kirchhoff hypothesis, that is, the linear variation of the displacements through the thickness as described by equation (6.12). Graphically, because of this linear character, it is really only necessary to compute the stresses, or strains, at the top and bottom of each layer and then connect these values with a straight line. This will give an accurate description of the stresses and strains through the thickness.

6.8.3 CLT Example 5: $[\pm30/0]_T$ Laminate Subjected to Known ε_x^o

As a fifth and final example, consider a laminate that consists of one-half of the laminate in CLT Examples 3 and 4. The laminate for this fifth example will be a $[+30/-30/0]_T$ laminate, an unsymmetric laminate. We shall examine the strains, stresses, and force and moment resultants for this laminate assuming previously used extension-only deformations of

$$\varepsilon_x^o(x, y) = 1000 \times 10^{-6} \qquad \kappa_x^o(x, y) = 0$$
$$\varepsilon_y^o(x, y) = 0 \qquad \kappa_y^o(x, y) = 0 \qquad (6.138)$$
$$\gamma_{xy}^o(x, y) = 0 \qquad \kappa_{xy}^o(x, y) = 0$$

Except for the fact that it only involves three layers, the strain distribution for this example is like the strain distribution of CLT Examples 1 and 3. For this three-layer laminate

$$z_0 = -0.225 \text{ mm} \quad z_1 = -0.075 \text{ mm} \quad z_2 = +0.075 \text{ mm} \quad z_3 = +0.225 \text{ mm}$$

$$(6.139)$$

and the strain distribution is shown in Figure 6.38.

As you may have anticipated by now, the stresses σ_x, σ_y, and τ_{xy} in the $+30°$, $-30°$, and $0°$ layers are identical to the case of the $[\pm30/0]_S$ laminate of Example 3.

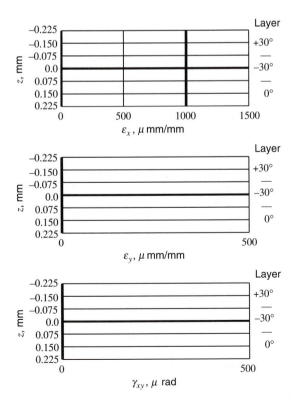

FIGURE 6.38
Strain distribution through the thickness of $[\pm30/0]_T$ laminate subjected to $\varepsilon_x^o = 1000 \times 10^{-6}$

The strain distribution of equation (6.138), when applied to +30° layers, −30° layers, and 0° layers, produces the same stresses independently of the laminate the layers are within. Thus the calculations from equations (6.98), (6.100), and (6.103) are valid for this laminate for the +30°, −30°, and 0° layers, respectively. For the +30° layer

$$\sigma_x = 92.8 \text{ MPa}$$
$$\sigma_y = 30.1 \text{ MPa}$$
$$\tau_{xy} = 46.7 \text{ MPa}$$

(6.140a)

For the −30° layer

$$\sigma_x = 92.8 \text{ MPa}$$
$$\sigma_y = 30.1 \text{ MPa}$$
$$\tau_{xy} = -46.7 \text{ MPa}$$

(6.140b)

For the 0° layer

$$\sigma_x = 155.7 \text{ MPa}$$
$$\sigma_y = 3.02 \text{ MPa}$$
$$\tau_{xy} = 0$$

(6.140c)

Figure 6.39 illustrates the stress distribution described by the above equations, and it is clear that for this unsymmetric laminate the distribution of stresses is not symmetric

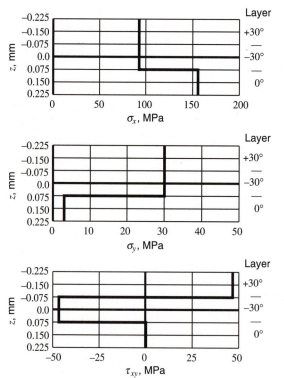

FIGURE 6.39
Stress distribution through the thickness of $[\pm 30/0]_T$ laminate subjected to $\varepsilon_x^o = 1000 \times 10^{-6}$

with respect to $z = 0$. This lack of symmetry occurs despite the symmetric and rather simple strain distribution through the thickness. The effects of this lack of symmetry on the stresses will be evident in the moment resultants.

The force resultants for this problem are computed as

$$
\begin{aligned}
N_x &= (92.8 + 92.8 + 155.7) \times 10^6 (150 \times 10^{-6}) \\
&= 51\,200 \text{ N/m} \\
N_y &= (30.1 + 30.1 + 3.02) \times 10^6 (150 \times 10^{-6}) \\
&= 9470 \text{ N/m} \\
N_{xy} &= (46.7 - 46.7 + 0) \times 10^6 (150 \times 10^{-6}) \\
&= 0
\end{aligned}
\tag{6.141}
$$

The bending moment resultant M_x is

$$
M_x = \int_{-\frac{H}{2}}^{\frac{H}{2}} \sigma_x z\,dz = \int_{z_0}^{z_1} \sigma_x z\,dz + \int_{z_1}^{z_2} \sigma_x z\,dz + \int_{z_2}^{z_3} \sigma_x z\,dz
\tag{6.142}
$$

$$
M_x = \left\{ \int_{-225 \times 10^{-6}}^{-75 \times 10^{-6}} 92.8z\,dz + \int_{-75 \times 10^{-6}}^{75 \times 10^{-6}} 92.8z\,dz + \int_{75 \times 10^{-6}}^{225 \times 10^{-6}} 155.7z\,dz \right\} \times 10^6
\tag{6.143}
$$

$$
\begin{aligned}
M_x &= \frac{1}{2} \left\{ 92.8 \left((-75)^2 - (-225)^2 \right) + 92.8 \left((75)^2 - (-75)^2 \right) \right. \\
&\quad \left. + 155.7 \left((225)^2 - (75)^2 \right) \right\} \times 10^6 \times (10^{-6})^2
\end{aligned}
\tag{6.144}
$$

$$
M_x = 1.416 \text{ N·m/m}
\tag{6.145a}
$$

The bending moment resultant M_y is

$$
M_y = -0.609 \text{ N·m/m}
\tag{6.145b}
$$

while the twisting moment resultant M_{xy} is

$$
M_{xy} = -1.051 \text{ N·m/m}
\tag{6.145c}
$$

In summary, therefore,

$$
\begin{aligned}
N_x &= 51\,200 \text{ N/m} \\
N_y &= 9470 \text{ N/m} \\
N_{xy} &= 0 \\
M_x &= 1.416 \text{ N·m/m} \\
M_y &= -0.609 \text{ N·m/m} \\
M_{xy} &= -1.051 \text{ N·m/m}
\end{aligned}
\tag{6.146}
$$

The proper interpretation of this fifth example is as follows: If the force and moment resultants given by equation (6.146) are applied at a point (x, y) on the reference surface of a $[+30/-30/0]_T$ laminate, then at that point the reference surface will deform as in equation (6.138), and the strains and stresses of Figures 6.38 and 6.39 will

result. Note that to produce only an extensional strain ε_x with this unsymmetric laminate, *both force and moment resultants are necessary.* With a symmetric laminate, to produce only an extensional strain, only force resultants are required. This is an important difference between symmetric and unsymmetric laminates. The fact that both force resultants and moment resultants are required to produce pure extension of an unsymmetric laminate is called the *bending-stretching coupling effect.* Symmetric laminates do not exhibit bending-stretching coupling effects. Unsymmetric laminates always do, and more will be said of this later.

This completes our five examples designed to demonstrate the implications of the Kirchhoff hypothesis, and to demonstrate how the stresses in a laminate are computed in the context of this hypothesis and the plane-stress assumption. We have learned that by knowing the strains and curvatures at the reference surface, by the Kirchhoff hypothesis the strains at any point through the thickness can be determined. Through the stress-strain relations for each layer, we can determine the stress distribution through the thickness. By transformation, we can compute the strains and stresses in the principal material direction of each layer. We have also learned that within each layer the stresses are constant, or vary linearly, and that, in general, the stress components exhibit discontinuities from one layer to the next. Force and moment resultants can be quite naturally defined as integrals of the stresses through the thickness of the laminate. These resultants can be thought of as acting at the reference surface and, in fact, are responsible for the strains at the reference surface. Finally, we saw in CLT Example 5 that unsymmetric laminates require moment resultants, as well as force resultants, to deform in simple extension in the x direction. Although it was not demonstrated, to deform with just pure curvature and with no extension as the laminates of CLT Examples 2 and 4 did, an unsymmetric laminate requires force resultants as well as moment resultants. A symmetric laminate requires only moment resultants.

You are now ready to compute stresses in laminates, and you will have the opportunity to do so in the exercises for Section 6.8. In addition, you will be asked to modify one of your existing computer programs. When you finish the exercises, you will have a thorough understanding of the important steps in the stress analysis of composite laminates.

6.8.4 A Note on the Kirchhoff Hypothesis

One of the key assumptions of the Kirchhoff hypothesis is that all points through the thickness of a laminate with coordinates (x, y) displace the same amount in the z direction. The displacement is given by the displacement of the reference surface in the z direction:

$$w(x, y, z) = w^o(x, y) \tag{6.147}$$

In the context of Figure 6.9, this means points t and t' are the same distance apart after deformation as they are before deformation. That distance is H, the thickness of the laminate. As mentioned, this contradicts the fact that because of Poisson effects, an ε_x causes an ε_z and an ε_y causes an ε_z. Hence, $w(x, y, z)$ cannot be the same at each

z location through the thickness. Kirchhoff's hypothesis was originally introduced in the 1800s for the purpose of studying bending response; that is, $u^o(x, y)$ and $v^o(x, y)$ were both equal to zero. For nonlayered single-material beams, such as aluminum or steel, or for symmetric laminates subjected to bending, points t and t' *do* indeed remain a fixed distance apart. For the situation shown in Figure 6.9, due to Poisson effects, point t moves *toward* the reference surface and point t' moves *away* from the reference surface the same amount t moves towards the reference surface. Therefore, on a point-by-point basis through the thickness, equation (6.147) is not valid. However, on an *average* basis, the equation is valid. If the context of equation (6.147) is kept in mind, and either equation (4.3) or equation (4.10) are used in conjunction with the predictions of classical lamination theory for ε_1, ε_2, and σ_1, σ_2, then accurate and legitimate information regarding $\varepsilon_z (= \varepsilon_3)$ can be obtained.

Exercises for Section 6.8

1. A six-layer $[\pm 15/0]_S$ graphite-reinforced composite laminate is deformed so that at a point (x, y) on the reference surface

$$\varepsilon_x^o = 1000 \times 10^{-6} \qquad \kappa_x^o = 0$$
$$\varepsilon_y^o = 0 \qquad \kappa_y^o = 0$$
$$\gamma_{xy}^o = 0 \qquad \kappa_{xy}^o = 0$$

This is the same deformation given the $[\pm 30/0]_S$ laminate of CLT Example 3. Here, however, the off-axis fiber angles are more closely aligned with the direction of the applied deformation. The value of N_x to produce these reference surface deformations should be larger than for the $[\pm 30/0]_S$ case, reflecting the rapid change of $\bar{Q}_{11}$ with θ, as in Figure 5.7. (*a*) Compute and plot, as a function of z, the three components of stress and three components of strain in the x-y and 1-2 systems. (*b*) Compute the values of N_x, N_y, N_{xy}, M_x, M_y, and M_{xy} required at that point to produce these reference surface deformations. (*c*) Compare the values of N_x and N_y with those of CLT Example 3.

2. Suppose a $[\pm 30/0]_T$ unsymmetric laminate that is 0.250 m long by 0.125 m wide, as in CLT Example 5, has the deformations of equation (6.138) at every point on its reference surface. (*a*) Sketch the force and moment resultants required along the edges to produce this deformation. (*b*) Determine the numerical values of these resultants.

Exercises 3, 4, and 5 should be studied carefully. The laminate is the same in each problem, but the applied deformation state is quite different. In each case, however, the maximum strain produced has a magnitude of 1000×10^{-6}. Note, you may want to complete the Computer Exercise before proceeding with Exercises 3 through 6.

3. A six-layer $[\pm 45/0]_S$ graphite-reinforced composite laminate is deformed so that at a point (x, y) the reference surface

$$\varepsilon_x^o = 1000 \times 10^{-6} \qquad \kappa_x^o = 0$$
$$\varepsilon_y^o = 1000 \times 10^{-6} \qquad \kappa_y^o = 0$$
$$\gamma_{xy}^o = 0 \qquad \kappa_{xy}^o = 0$$

This deformation is referred to as biaxial stretching (i.e., stretching in two directions). (*a*) Compute and plot, as a function of z, the three components of strain and the three components of stress in the x-y and 1-2 systems. (b) Compute the values of N_x, N_y, N_{xy}, M_x, M_y, M_{xy} required at that point to produce these reference surface deformations. This laminate is slightly orthotropic in its inplane properties. Therefore, it takes more force in the x direction than in the y direction to achieve the given deformation state. Also note that every layer is in the same state of stress in the principal material system.

4. A six-layer $[\pm 45/0]_S$ graphite-reinforced composite laminate is deformed so that at a point (x, y) the reference surface

$$\varepsilon_x^o = 0 \qquad \kappa_x^o = 2.22 \text{ m}^{-1}$$
$$\varepsilon_y^o = 0 \qquad \kappa_y^o = -2.22 \text{ m}^{-1}$$
$$\gamma_{xy}^o = 0 \qquad \kappa_{xy}^o = 0$$

(a) Compute and plot, as a function of z, the three components of strain and the three components of stress in the x-y and 1-2 systems. (b) Compute the values of N_x, N_y, N_{xy}, M_x, M_y, M_{xy} required at that point to produce these reference surface deformations. (c) Note that a nonzero value of M_{xy} is required to cause this symmetric deformation state. Note also that the value of M_x required is slightly greater than the value of M_y required. Why is this so? (d) Suppose a $[\pm 45/0]_S$ laminate that is 0.250 m long by 0.125 m wide has the curvatures given above at every point on its reference surface. Sketch the deformed shape of the laminate, including the x-y-z coordinate system and the correct sense of the out-of-plane displacements w^o. (e) What moments are required along the edges of the rectangular laminate to produce the curvatures?

5. A six-layer $[\pm 45/0]_S$ graphite composite laminate is deformed so that at a point (x, y) on the reference surface

$$\varepsilon_x^o = 0 \qquad \kappa_x^o = 0$$
$$\varepsilon_y^o = 0 \qquad \kappa_y^o = 0$$
$$\gamma_{xy}^o = 0 \qquad \kappa_{xy}^o = 2.22 \text{ m}^{-1}$$

For this situation answer questions (a), (b), (d), and (e) from Exercise 4. Note that the values of M_x and M_y required to produce just a twisting curvature are about one-half the value of M_{xy}, a large percentage.

6. In CLT Example 5 we found that to stretch an unsymmetric laminate, equation (6.138), requires both force resultants and moment resultants. This is unlike the stetching of a symmetric laminate (e.g., CLT Examples 1 and 3), where stretching a symmetric laminate does not require moment resultants. Like stretching, bending an unsymmetric laminate requires both force resultants and moment resultants. To show this, consider the $[\pm 30/0]_T$ laminate of CLT Example 5 deformed such that at a point (x, y) on the reference surface

$$\varepsilon_x^o(x, y) = 0 \qquad \kappa_x^o(x, y) = 4.44 \text{ m}^{-1}$$
$$\varepsilon_y^o(x, y) = 0 \qquad \kappa_y^o(x, y) = 0$$
$$\gamma_{xy}^o(x, y) = 0 \qquad \kappa_{xy}^o(x, y) = 0$$

This curvature produces $\varepsilon_x = 1000 \times 10^{-6}$ at $z = \pm H/2$. (a) Compute and plot the distribution through the thickness of stresses σ_x, σ_y, and τ_{xy}. (b) Calculate the force and moment resultants at that point to show that bending an unsymmetric laminate requires both force and moment resultants.

7. Suppose in Exercise 6 the deformations given were assumed to hold for every point on the reference surface of a $[\pm 30/0]_T$ laminate of dimensions 0.250 m in the x direction

and 0.125 m in the y direction. (a) What would be the forces and moments acting on each edge? Sketch the 0.250×0.125 m laminate to show the sense of these forces and moments. (b) Is the deformed shape described by the following?

$$u^o(x, y) = 0$$
$$v^o(x, y) = 0$$
$$w^o(x, y) = (2.22)x^2$$

If the above description of the deformed shape is not valid, how would the deformed shape be described?

8. In the next chapter we shall study the extensional and bending stiffnesses of laminates. For nonlayered single materials, the extensional stiffness of a beam is given by the product of extensional modulus, E, and the cross-sectional area, A, namely EA. The bending stiffness of a beam is given by the product of extensional modulus and the cross-sectional second moment of area, I, namely EI. These are concepts learned in elementary strength-of-materials courses. For simple laminates such as a $[0/90]_S$ laminate, the extensional and bending stiffnesses of a composite beam can be approximated by computing the effective EA and effective EI. Consider a beam 10 mm wide and of arbitrary length with a lamination sequence of $[0/90]_S$. By considering only the contribution of the $0°$ layers to the cross-sectional stiffness, compute the effective EA and effective EI of a $[0/90]_S$ beam. Hint: For bending, consider the cross section of the composite to be like the cross section of an I-beam. With an I-beam, the contribution of the web to the second moment of area is negligible compared to the contribution of the flanges. For the $[0/90]_S$ composite, consider the $90°$ layers to have negligible contribution to the effective EI. These approximations are depicted in the accompanying figure:

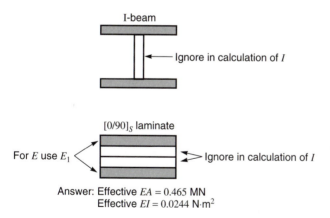

Answer: Effective $EA = 0.465$ MN
Effective $EI = 0.0244$ N·m²

Computer Exercise

You are now ready to make an important addition to your existing computer program involving the $\bar{Q}$. Modify and expand the program to compute the stresses through the thickness of a laminate, given that we know the engineering properties, fiber orientations, layer thickness, layer stacking sequence, and $\varepsilon_x^o, \ldots, \kappa_{xy}^o$ for the laminate. For simplicity, assume that all layers are made of the same material and all are the same thickness. The program should be user-friendly, prompting the user for information, and printing out, either on the screen or on a printer, all the information regarding the laminate. The program should flow roughly as follows:

1. Read in the number of layers.

2. Read in the engineering properties of a layer, and the thickness.

3. Compute the Q_{ij}.

4. Print the number of layers, the engineering properties, and the Q_{ij}.

5. Compute and print the z_k.

6. Read in and print the fiber orientations in each layer.

7. Compute and print the $\bar{Q}_{ij}$ for each layer.

8. Read in and print the values of $\varepsilon_x^o, \ldots, \kappa_{xy}^o$.

9. Compute and print the values of ε_x, ε_y, and γ_{xy} at the top and bottom of each layer. (These are the extreme values in each layer.)

10. Compute and print the values of ε_1, ε_2, and γ_{12} at the top and bottom of each layer.

11. Compute and print the values of σ_x, σ_y, and τ_{xy} at the top and bottom of each layer.

12. Compute and print the values of σ_1, σ_2, and τ_{12} at the top and bottom of each layer.

13. As a check at a later stage when we add further to this program, you might want to compute and print the values of $N_x, \ldots, M_{xy}$ as calculated from the values of σ_x, σ_y, and τ_{xy}, and the z_k.

    Use the stress and strain calculations of CLT Examples 1 through 5 to check your program for errors.

SUGGESTED READINGS

For a discussion of the Kirchhoff hypothesis, particularly in the context of classical isotropic materials, see:

1. Timoshenko, S., and S. Winowsky-Krieger. *Theory of Plates and Shells.* 2nd ed. New York: McGraw-Hill, 1959.

2. Fung, Y. C. *Foundations of Solid Mechanics.* Englewood Cliffs, NJ: Prentice Hall, 1969.

For a discussion of the Kirchhoff hypothesis in the context of variational and energy methods and isotropic materials, see:

3. Shames, I. H., and C. L. Dym. *Energy and Finite Element Methods in Structural Mechanics.* New York: Hemisphere Publishing, 1985.

One of the earliest books on applying the concepts of isotropic plate theory to laminates is:

4. Ashton, J. E., and J. M. Whitney. *Theory of Laminated Plates.* Lancaster, PA: Technomic Publishing Co., 1970.

Other books on the topic are:

5. Ambartsumyan, S. A. *Theory of Anisotropic Plates.* Lancaster, PA: Technomic Publishing Co., 1969.

6. Whitney, J. M. *Structural Analysis of Laminated Anisotropic Plates*. Lancaster, PA: Technomic Publishing Co., 1987.

7. Jones, R. M. *Mechanics of Composite Materials*. New York: McGraw-Hill, 1975.

8. Calcote, L. R. *The Analysis of Laminated Composite Structures*. New York: Van Nostrand Reinhold, 1969.

For a discussion of the relaxation of the Kirchhoff hypothesis, see:

9. Reddy, J. N. "A Simple Higher-Order Theory for Laminated Composite Plates." *Transactions of the ASME, Journal of Applied Mechanics* 51 (1984), pp. 745–52.

CHAPTER 7

Classical Lamination Theory: Laminate Stiffness Matrix

To this point in the development of classical lamination theory it is clear that if we are given the strains and curvatures at a point (x, y) on the reference surface— ε_x^o, ε_y^o, γ_{xy}^o, κ_x^o, κ_y^o, and κ_{xy}^o—then we can compute the strain distributions through the thickness of the laminate. By using the stress-strain relations, we can compute the stress distributions. By using the transformation relations, we can determine the stresses and strains in the principal material system, and by using the definitions of the stress resultants, we can compute the force and moment resultants acting at that point. If we specify that the given strains and curvatures are the same at every point on the reference surface of a laminate of a given length and width, then we can determine the total forces and moments acting on the edges of the laminate. These steps are all a result of the plane stress assumption and the Kirchhoff hypothesis. Figure 7.1 illustrates the connection between these steps. What remains in the development of classical lamination theory is to be able to specify the force and moment resultants acting at a point (x, y) on the reference surface, and then to be able to compute the strains and curvatures of that point on the reference surface that these resultants cause. We want to fill in the missing link in the upper left portion of Figure 7.1, namely, compute the reference surface strains and curvatures, knowing the stress resultants. With these computed reference surface strains and curvatures, we can then, as in the previous examples, compute the strain and stress distributions through the thickness of the laminate. Relating the stress resultants to the reference surface strains and curvatures is an important step. In the application of composite materials to structures, we are often given the forces and moments that act on a laminate, and we want to know the stresses and strains that are caused by these loads. Having a relation between the

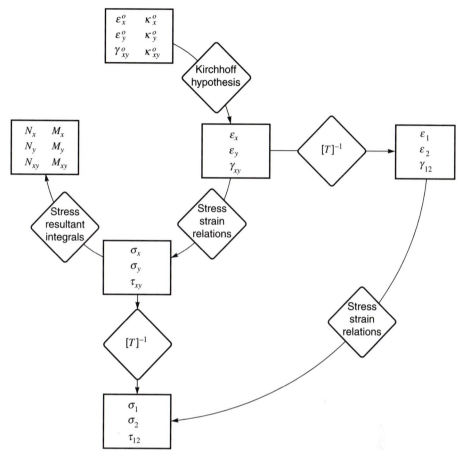

FIGURE 7.1
Interrelations among concepts developed so far

force and moment resultants and the strains and curvatures at a point (x, y) on the reference surface will allow us to do this. In what follows we shall develop this key relationship.

7.1
FORMAL DEFINITION OF FORCE AND MOMENT RESULTANTS

Though we have introduced and computed them in the examples, we have not formally defined in one location the three force resultants and three moment resultants that are a result of integrating the stresses through the thickness of the laminate. The stress resultants in the x direction, N_x, in the y direction, N_y, and in shear, N_{xy}, are defined as

$$N_x \equiv \int_{-\frac{H}{2}}^{\frac{H}{2}} \sigma_x dz$$

$$N_y \equiv \int_{-\frac{H}{2}}^{\frac{H}{2}} \sigma_y dz \qquad (7.1)$$

$$N_{xy} \equiv \int_{-\frac{H}{2}}^{\frac{H}{2}} \tau_{xy} dz$$

The resultants N_x and N_y will be referred to as the *normal* force resultants, and N_{xy} will be referred to as the *shear* force resultant.

Figure 7.2 illustrates a small element of laminate surrounding a point (x, y) on the geometric midplane, and the figure indicates the directions of these three stress resultants. The units of the stress resultants are force per unit length; the unit of length is a unit of length in the x direction or in the y direction. If the small element of laminate has dimensions dx by dy, then the *force* in the x direction due to N_x, the *force* in the x direction due to N_{xy}, the *force* in the y direction due to N_y, and the *force* in the y direction due to N_{xy} are, respectively,

$$N_x dy \qquad N_{xy} dx \qquad N_y dx \qquad N_{xy} dy \qquad (7.2)$$

The moment resultants M_x, M_y, and M_{xy} are defined as

$$M_x \equiv \int_{-\frac{H}{2}}^{\frac{H}{2}} \sigma_x z dz$$

$$M_y \equiv \int_{-\frac{H}{2}}^{\frac{H}{2}} \sigma_y z dz \qquad (7.3)$$

$$M_{xy} \equiv \int_{-\frac{H}{2}}^{\frac{H}{2}} \tau_{xy} z dz$$

The resultants M_x and M_y will be referred to as *bending* moment resultants and M_{xy} will be referred to as the *twisting* moment resultant. In Figure 7.3 the sense of these three moment resultants is illustrated, and the senses shown are important.

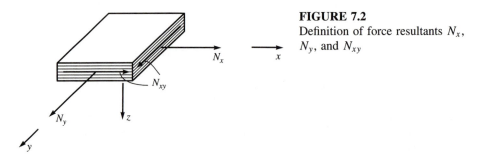

FIGURE 7.2
Definition of force resultants N_x, N_y, and N_{xy}

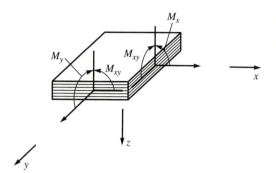

FIGURE 7.3
Definition of moment resultants M_x, M_y, and M_{xy}

Shortly we shall be relating a positive M_x, for example, with a positive curvature in the x direction, κ_x^o. If you refer to CLT Examples 2 and 4 in the previous chapter, you will see that the moment M_x required to produce the positive curvature in those examples was indeed positive. Knowing the correct sense of positive quantities is critical. The sense illustrated in Figure 7.3 is consistent with the cross-product definition of moments learned in elementary mechanics courses (i.e., $\bar{M} = \bar{r} \times \bar{F}$). The units of the moment resultants are moment per unit length, again the unit of length being a unit of length in either the x direction or y direction.

If a small element of laminate has dimensions dx by dy, then the *moments* about the positive y and the positive x axis due to M_x, M_y, and M_{xy} are, respectively,

$$M_x dy \qquad M_{xy} dx \qquad -M_y dx \qquad -M_{xy} dy \qquad (7.4)$$

We will use these formal definitions of $N_x, \ldots, M_{xy}$ in computing the laminate stiffness matrix and they will be used in a later chapter when we study composite plates. Specifically, we will use the force and moment resultants when studying the equilibrium of a small element of laminated plate. As the definitions of the moment resultants involve integrals with respect to the coordinate z, and because z is measured relative to the reference surface, the moment resultants have to be thought of as producing moments on the reference surface. The force resultants can be considered to act anywhere through the thickness of the laminate. But when force resultants are used in conjunction with moment resultants, it is most consistent to think of the force resultants as producing forces on the reference surface. This will also be consistent when we begin discussing the relationship between the resultants and reference surface response, namely, $\varepsilon_x^o, \ldots, \kappa_{xy}^o$. From this point forward, then, the force and moment resultants will be considered to act on the reference surface.

7.2
LAMINATE STIFFNESS: THE *ABD* MATRIX

Equations (7.1) and (7.3) give the definitions of the force and moment resultants. These definitions involve the stresses σ_x, σ_y, and τ_{xy}. From the stress-strain relations, the stresses can be written in terms of the strains by the now-familiar relation

$$
\left\{
\begin{array}{c}
\sigma_x \\
\sigma_y \\
\tau_{xy}
\end{array}
\right\}
=
\left[
\begin{array}{ccc}
\bar{Q}_{11} & \bar{Q}_{12} & \bar{Q}_{16} \\
\bar{Q}_{12} & \bar{Q}_{22} & \bar{Q}_{26} \\
\bar{Q}_{16} & \bar{Q}_{26} & \bar{Q}_{66}
\end{array}
\right]
\left\{
\begin{array}{c}
\varepsilon_x \\
\varepsilon_y \\
\gamma_{xy}
\end{array}
\right\}
\tag{7.5}
$$

and the strains can, in turn, be written in terms of the strains and curvatures of the reference surface to produce another familiar relation, namely,

$$
\left\{
\begin{array}{c}
\sigma_x \\
\sigma_y \\
\tau_{xy}
\end{array}
\right\}
=
\left[
\begin{array}{ccc}
\bar{Q}_{11} & \bar{Q}_{12} & \bar{Q}_{16} \\
\bar{Q}_{12} & \bar{Q}_{22} & \bar{Q}_{26} \\
\bar{Q}_{16} & \bar{Q}_{26} & \bar{Q}_{66}
\end{array}
\right]
\left\{
\begin{array}{c}
\varepsilon_x^o + z\kappa_x^o \\
\varepsilon_y^o + z\kappa_x^o \\
\gamma_{xy}^o + z\kappa_{xy}^o
\end{array}
\right\}
\tag{7.6}
$$

We have emphasized repeatedly that in a laminate the stresses are functions of z because the transformed reduced stiffnesses $\bar{Q}_{ij}$ change from layer to layer, and also because of the linear dependence on z. The reference surface strains and curvatures, by definition, do not depend on z. Let us examine the expression for the normal force resultant N_x by substituting for σ_x from equation (7.6) into the integrand of the first equation of equation (7.1), specifically:

$$
N_x = \int_{-\frac{H}{2}}^{\frac{H}{2}} \{\bar{Q}_{11}(\varepsilon_x^o + z\kappa_x^o) + \bar{Q}_{12}(\varepsilon_y^o + z\kappa_y^o) + \bar{Q}_{16}(\gamma_{xy}^o + z\kappa_{xy}^o)\}dz \tag{7.7}
$$

and expanding the integrand leads to

$$
N_x = \int_{-\frac{H}{2}}^{\frac{H}{2}} \{\bar{Q}_{11}\varepsilon_x^o + \bar{Q}_{11}z\kappa_x^o + \bar{Q}_{12}\varepsilon_y^o + \bar{Q}_{12}z\kappa_y^o + \bar{Q}_{16}\gamma_{xy}^o + \bar{Q}_{16}z\kappa_{xy}^o\}dz \tag{7.8}
$$

The integration can be distributed over the six terms to give

$$
N_x = \int_{-\frac{H}{2}}^{\frac{H}{2}} \bar{Q}_{11}\varepsilon_x^o dz + \int_{-\frac{H}{2}}^{\frac{H}{2}} \bar{Q}_{11}z\kappa_x^o dz + \int_{-\frac{H}{2}}^{\frac{H}{2}} \bar{Q}_{12}\varepsilon_y^o dz
$$
$$
+ \int_{-\frac{H}{2}}^{\frac{H}{2}} \bar{Q}_{12}z\kappa_y^o dz + \int_{-\frac{H}{2}}^{\frac{H}{2}} \bar{Q}_{16}\gamma_{xy}^o dz + \int_{-\frac{H}{2}}^{\frac{H}{2}} \bar{Q}_{16}z\kappa_{xy}^o dz
\tag{7.9}
$$

Because the reference surface deformations are not functions of z, they can be taken outside of the integrals, resulting in

$$
N_x = \left\{ \int_{-\frac{H}{2}}^{\frac{H}{2}} \bar{Q}_{11}dz \right\}\varepsilon_x^o + \left\{ \int_{-\frac{H}{2}}^{\frac{H}{2}} \bar{Q}_{11}zdz \right\}\kappa_x^o + \left\{ \int_{-\frac{H}{2}}^{\frac{H}{2}} \bar{Q}_{12}dz \right\}\varepsilon_y^o
$$
$$
+ \left\{ \int_{-\frac{H}{2}}^{\frac{H}{2}} \bar{Q}_{12}zdz \right\}\kappa_y^o + \left\{ \int_{-\frac{H}{2}}^{\frac{H}{2}} \bar{Q}_{16}dz \right\}\gamma_{xy}^o + \left\{ \int_{-\frac{H}{2}}^{\frac{H}{2}} \bar{Q}_{16}zdz \right\}\kappa_{xy}^o
\tag{7.10}
$$

The integrals through the thickness now involve just the material properties, specifically, the reduced stiffnesses. We need to look at these integrals in detail.

The first integral in equation (7.10) is

$$
\int_{-\frac{H}{2}}^{\frac{H}{2}} \bar{Q}_{11}dz \tag{7.11}
$$

Because the reduced stiffnesses are materials properties, they are constant within a layer. However, from layer to layer the values of the reduced stiffnesses change. Because of this piecewise-constant property of the reduced stiffnesses, the integral through the thickness can be expanded to give

$$\int_{-\frac{H}{2}}^{\frac{H}{2}} \bar{Q}_{11} dz = \int_{z_0}^{z_1} \bar{Q}_{11_1} dz + \int_{z_1}^{z_2} \bar{Q}_{11_2} dz + \int_{z_2}^{z_3} \bar{Q}_{11_3} dz$$
$$+ \cdots + \int_{z_{k-1}}^{z_k} \bar{Q}_{11_k} dz + \cdots + \int_{z_{N-1}}^{z_N} \bar{Q}_{11_N} dz \tag{7.12}$$

In the above expansion the locations of the layer interfaces have been used and the subscript on $\bar{Q}_{11}$ indicates layer number. The systematic scheme introduced earlier to identify the layer numbers and the z location of the layer interfaces will be used to advantage in what follows.

Because within each layer $\bar{Q}_{11}$ is a constant; in the above series of integrals it can be removed from each integral and the series of integrals written as

$$\int_{-\frac{H}{2}}^{\frac{H}{2}} \bar{Q}_{11} dz = \bar{Q}_{11_1} \int_{z_0}^{z_1} dz + \bar{Q}_{11_2} \int_{z_1}^{z_2} dz + \bar{Q}_{11_3} \int_{z_2}^{z_3} dz$$
$$+ \cdots + \bar{Q}_{11_k} \int_{z_{k-1}}^{z_k} dz + \cdots + \bar{Q}_{11_N} \int_{z_{N-1}}^{z_N} dz \tag{7.13}$$

But the integrations with respect to z are quite simple, equation (7.13) becoming

$$\int_{-\frac{H}{2}}^{\frac{H}{2}} \bar{Q}_{11} dz = \bar{Q}_{11_1}(z_1 - z_0) + \bar{Q}_{11_2}(z_2 - z_1) + \bar{Q}_{11_3}(z_3 - z_2)$$
$$+ \cdots + \bar{Q}_{11_k}(z_k - z_{k-1}) + \cdots + \bar{Q}_{11_N}(z_N - z_{N-1}) \tag{7.14}$$

or

$$\int_{-\frac{H}{2}}^{\frac{H}{2}} \bar{Q}_{11} dz = \sum_{k=1}^{N} \bar{Q}_{11_k}(z_k - z_{k-1}) \tag{7.15}$$

The difference in the z values is recognized as simply the thickness of the kth layer, h_k, so equation (7.15) becomes

$$\int_{-\frac{H}{2}}^{\frac{H}{2}} \bar{Q}_{11} dz = \sum_{k=1}^{N} \bar{Q}_{11_k} h_k \tag{7.16}$$

The integral of $\bar{Q}_{11}$ through the thickness is denoted by A_{11}; that is:

$$\int_{-\frac{H}{2}}^{\frac{H}{2}} \bar{Q}_{11} dz = A_{11} = \sum_{k=1}^{N} \bar{Q}_{11_k} h_k \tag{7.17}$$

In a similar manner, we find, referring to the third and fifth terms in equation (7.10), that

$$\int_{-\frac{H}{2}}^{\frac{H}{2}} \bar{Q}_{12} dz = A_{12} = \sum_{k=1}^{N} \bar{Q}_{12_k} h_k \tag{7.18}$$

and
$$\int_{-\frac{H}{2}}^{\frac{H}{2}} \bar{Q}_{16} dz = A_{16} = \sum_{k=1}^{N} \bar{Q}_{16_k} h_k \tag{7.19}$$

Thus the first, third, and fifth terms on the right side of equation (7.10) become

$$A_{11}\varepsilon_x^o + A_{12}\varepsilon_y^o + A_{16}\gamma_{xy}^o \tag{7.20}$$

The second, fourth, and sixth terms on the right side of equation (7.10) involve integrals of $\bar{Q}_{11} \times z$, $\bar{Q}_{12} \times z$, and $\bar{Q}_{16} \times z$, rather than just $\bar{Q}_{11}$, $\bar{Q}_{12}$, and $\bar{Q}_{16}$. Specifically, the second term is

$$\int_{-\frac{H}{2}}^{\frac{H}{2}} \bar{Q}_{11} z \, dz \tag{7.21}$$

Again using the fact that $\bar{Q}_{11}$ is constant within each layer but changes from layer to layer, we find that

$$\int_{-\frac{H}{2}}^{\frac{H}{2}} \bar{Q}_{11} z \, dz = \bar{Q}_{11_1} \int_{z_0}^{z_1} z \, dz + \bar{Q}_{11_2} \int_{z_1}^{z_2} z \, dz + \bar{Q}_{11_3} \int_{z_2}^{z_3} z \, dz$$
$$+ \cdots + \bar{Q}_{11_k} \int_{z_{k-1}}^{z_k} z \, dz + \cdots + \bar{Q}_{11_N} \int_{z_{N-1}}^{z_N} z \, dz \tag{7.22}$$

Here the integrals on z are also quite simple. Equation (7.22) becomes

$$\int_{-\frac{H}{2}}^{\frac{H}{2}} \bar{Q}_{11} z \, dz = \frac{1}{2}\{\bar{Q}_{11_1}(z_1^2 - z_0^2) + \bar{Q}_{11_2}(z_2^2 - z_1^2) + \bar{Q}_{11_3}(z_3^2 - z_2^2)$$
$$+ \cdots + \bar{Q}_{11_k}(z_k^2 - z_{k-1}^2) + \cdots + \bar{Q}_{11_N}(z_N^2 - z_{N-1}^2)\} \tag{7.23}$$

or
$$\int_{-\frac{H}{2}}^{\frac{H}{2}} \bar{Q}_{11} z \, dz = \frac{1}{2} \sum_{k=1}^{N} \bar{Q}_{11_k}(z_k^2 - z_{k-1}^2) \tag{7.24}$$

This integral is denoted as B_{11}; that is:

$$\int_{-\frac{H}{2}}^{\frac{H}{2}} \bar{Q}_{11} z \, dz = B_{11} = \frac{1}{2} \sum_{k=1}^{N} \bar{Q}_{11_k}(z_k^2 - z_{k-1}^2) \tag{7.25}$$

In like manner, referring to the fourth and sixth terms in equation (7.10), we find

$$\int_{-\frac{H}{2}}^{\frac{H}{2}} \bar{Q}_{12} z \, dz = B_{12} = \frac{1}{2} \sum_{k=1}^{N} \bar{Q}_{12_k}(z_k^2 - z_{k-1}^2) \tag{7.26}$$

and
$$\int_{-\frac{H}{2}}^{\frac{H}{2}} \bar{Q}_{16} z \, dz = B_{16} = \frac{1}{2} \sum_{k=1}^{N} \bar{Q}_{16_k}(z_k^2 - z_{k-1}^2) \tag{7.27}$$

The expression for N_x in equation (7.10) becomes, after slight rearrangement,

$$N_x = A_{11}\varepsilon_x^o + A_{12}\varepsilon_y^o + A_{16}\gamma_{xy}^o + B_{11}\kappa_x^o + B_{12}\kappa_y^o + B_{16}\kappa_{xy}^o \tag{7.28}$$

Because they are summations of material and geometric properties, A_{11}, A_{12}, A_{16}, B_{11}, B_{12}, and B_{16} are easy to compute and are an inherent property of the laminate. Because the A_{ij} and B_{ij} are integrals, they represent *smeared* or *integrated* properties of the laminate. Equation (7.28) represents a relation between the normal force resultant N_x and the six reference surface deformations. This is part of the missing link in Figure 7.1. We will develop five other equations that will lead to a total of six relations between the six stress resultants and the six reference surface deformations. The missing link will then be complete. However, before proceeding to complete the link, let us compute A_{11}, A_{12}, A_{16}, B_{11}, B_{12} and B_{16} for the laminates discussed in CLT Examples 1 through 5 of the previous chapter.

For the four-layer $[0/90]_S$ laminate of CLT Examples 1 and 2, as each layer is the same thickness h, by equation (7.17),

$$A_{11} = \sum_{k=1}^{N} \bar{Q}_{11_k} h_k$$

$$= \{\bar{Q}_{11_1} + \bar{Q}_{11_2} + \bar{Q}_{11_3} + \bar{Q}_{11_4}\} \times h \tag{7.29}$$

or $\qquad A_{11} = \{\bar{Q}_{11}(0°) + \bar{Q}_{11}(90°) + \bar{Q}_{11}(90°) + \bar{Q}_{11}(0°)\} \times h \tag{7.30}$

Substituting numerical values from CLT Examples 1 and 2 for the appropriate $\bar{Q}_{ij}$ into equation (7.30) yields

$$A_{11} = \{155.7 + 12.16 + 12.16 + 155.7\} \times 10^9 \times 150 \times 10^{-6}$$

$$A_{11} = 50.4 \times 10^6 \text{ N/m} \tag{7.31}$$

Likewise,

$$A_{12} = \{\bar{Q}_{12}(0°) + \bar{Q}_{12}(90°) + \bar{Q}_{12}(90°) + \bar{Q}_{12}(0°)\} \times h,$$

$$A_{12} = \{3.02 + 3.02 + 3.02 + 3.02\} \times 10^9 \times 150 \times 10^{-6}$$

$$A_{12} = 1.809 \times 10^6 \text{ N/m}$$

$$A_{16} = \{\bar{Q}_{16}(0°) + \bar{Q}_{16}(90°) + \bar{Q}_{16}(90°) + \bar{Q}_{16}(0°)\} \times h, \tag{7.32}$$

$$A_{16} = \{0 + 0 + 0 + 0\} \times 150 \times 10^{-6}$$

$$A_{16} = 0$$

For all cross-ply laminates A_{16} will be zero because for each layer $\bar{Q}_{16} = 0$.

The B_{ij} can be computed from equations (7.25)–(7.27) as

$$B_{11} = \frac{1}{2}\{\bar{Q}_{11_1}(z_1^2 - z_0^2) + \bar{Q}_{11_2}(z_2^2 - z_1^2) + \bar{Q}_{11_3}(z_3^2 - z_2^2) + \bar{Q}_{11_4}(z_4^2 - z_3^2)\} \tag{7.33}$$

or $\qquad B_{11} = \frac{1}{2}\{\bar{Q}_{11}(0°)(z_1^2 - z_0^2) + \bar{Q}_{11}(90°)(z_2^2 - z_1^2)$

$$+ \bar{Q}_{11}(90°)(z_3^2 - z_2^2) + \bar{Q}_{11}(0°)(z_4^2 - z_3^2)\} \tag{7.34}$$

We find, using the numerical values of the interface locations for the $[0/90]_S$ laminate, equation (6.21),

$$B_{11} = \frac{1}{2}\{155.7[(-150)^2 - (-300)^2] + 12.16[0^2 - (-150)^2]$$

$$+ 12.16[(150)^2 - 0^2] + 155.7[(300)^2 - (150)^2]\} \times 10^9 \times (10^{-6})^2 \tag{7.35}$$

where carrying out the algebra leads to

$$B_{11} = 0 \tag{7.36}$$

The quantity B_{11} is exactly zero because the contributions to B_{11} from the layers at negative z locations exactly cancel the contributions to B_{11} from the layers at the positive z locations. This is a characteristic of a symmetric laminate. The values of all the B_{ij} will be zero for all symmetric laminates because of this cancellation effect. For symmetric laminates there is no need to go through the algebra to compute the values of the B_{ij}. They are *all* exactly zero; thus,

$$B_{12} = 0 \quad \text{and} \quad B_{16} = 0 \tag{7.37}$$

B_{16} is zero not only because it is a symmetric laminate, but also because the $\bar{Q}_{16}$ are zero for every layer.

For the six-layer $[\pm 30/0]_S$ laminate of CLT Examples 3 and 4,

$$A_{1j} = \{\bar{Q}_{1j_1} + \bar{Q}_{1j_2} + \bar{Q}_{1j_3} + \bar{Q}_{1j_4} + \bar{Q}_{1j_5} + \bar{Q}_{1j_6}\} \times h, \ j = 1, 2, 6 \tag{7.38}$$

Using layer orientation, we find that

$$A_{1j} = \{\bar{Q}_{1j}(30°) + \bar{Q}_{1j}(-30°) + \bar{Q}_{1j}(0°) + \bar{Q}_{1j}(0°)$$
$$+ \bar{Q}_{1j}(-30°) + \bar{Q}_{1j}(30°)\} \times h, \ j = 1, 2, 6 \tag{7.39}$$

For $j = 1$, substituting numerical values for the appropriate $\bar{Q}_{ij}$, we find

$$A_{11} = \{92.8 + 92.8 + 155.7 + 155.7 + 92.8 + 92.8\} \times 10^9 \times 150 \times 10^{-6} \tag{7.40}$$

or

$$A_{11} = 102.4 \times 10^6 \text{ N/m} \tag{7.41}$$

while for $j = 2$,

$$A_{12} = \{30.1 + 30.1 + 3.02 + 3.02 + 30.1 + 30.1\} \times 10^9 \times 150 \times 10^{-6}$$
$$A_{12} = 18.94 \times 10^6 \text{ N/m} \tag{7.42}$$

and for $j = 6$,

$$A_{16} = \{46.7 - 46.7 + 0 + 0 - 46.7 + 46.7\} \times 10^9 \times 150 \times 10^{-6}$$
$$A_{16} = 0 \tag{7.43}$$

The term A_{16} is exactly zero because the contributions from the $+30°$ layers exactly cancel the contributions from the $-30°$ layers. This is characteristic of the 16-component of the A_{ij} for what we will come to call a *balanced laminate*. Balanced laminates are characterized by $\pm\theta$ pairs of layers. More will be said of this later.

Because the laminate of CLT Examples 3 and 4 is symmetric, the components B_{11}, B_{12}, and B_{16} are exactly zero.

Turning to the $[\pm 30/0]_T$ laminate of CLT Example 5, because relative to the six-layer laminate of CLT Examples 3 and 4 there are exactly one-half the number of layers in any particular direction, the A_{ij} for the $[\pm 30/0]_T$ laminate are one-half the values of the A_{ij} for the $[\pm 30/0]_S$ laminate. However, as the $[\pm 30/0]_T$ is not a symmetric laminate, the B_{ij} will not be zero. Specifically, referring to equations

(7.41), (7.42), and (7.43), we find that

$$A_{11} = 51.2 \times 10^6 \text{ N/m}$$
$$A_{12} = 9.47 \times 10^6 \text{ N/m} \qquad (7.44)$$
$$A_{16} = 0 \text{ N/m}$$

For the B_{ij},

$$B_{1j} = \frac{1}{2}\{\bar{Q}_{1j_1}(z_1^2 - z_0^2) + \bar{Q}_{1j_2}(z_2^2 - z_1^2) + \bar{Q}_{1j_3}(z_3^2 - z_2^2)\}, j = 1, 2, 6$$

$$= \frac{1}{2}\{\bar{Q}_{1j}(30°)(z_1^2 - z_0^2) + Q_{1j}(-30°)(z_2^2 - z_1^2) \qquad (7.45)$$

$$+ Q_{1j}(0°)(z_3^2 - z_2^2)\}, j = 1, 2, 6$$

For $j = 1$ and using the interface locations, equation (6.138), we find

$$B_{11} = \frac{1}{2}\{92.8[(-75)^2 - (-225)^2] + 92.8[(75)^2 - (-75)^2]$$

$$+ 155.7[(225)^2 - (75)^2] \times 10^9 \times (10^{-6})^2\} \qquad (7.46)$$

resulting in
$$B_{11} = 1416 \text{ N} \qquad (7.47)$$

For $j = 2$ and 6,

$$B_{12} = -609 \quad \text{and} \quad B_{16} = -1051 \text{ N} \qquad (7.48)$$

To continue with the development of the laminate stiffness matrix, we can substitute the expressions for σ_y and τ_{xy} into the definitions in equation (7.1) for N_y and N_{xy}, respectively, to yield

$$N_y = \int_{-\frac{H}{2}}^{\frac{H}{2}} \{\bar{Q}_{12}\varepsilon_x^o + \bar{Q}_{12}z\kappa_x^o + \bar{Q}_{22}\varepsilon_y^o + \bar{Q}_{22}z\kappa_y^o + \bar{Q}_{26}\gamma_{xy}^o + \bar{Q}_{26}z\kappa_{xy}^o\}dz \quad (7.49)$$

and

$$N_{xy} = \int_{-\frac{H}{2}}^{\frac{H}{2}} \{\bar{Q}_{16}\varepsilon_x^o + \bar{Q}_{16}z\kappa_x^o + \bar{Q}_{26}\varepsilon_y^o + \bar{Q}_{26}z\kappa_y^o + \bar{Q}_{66}\gamma_{xy}^o + \bar{Q}_{66}z\kappa_{xy}^o\}dz \quad (7.50)$$

As in going from equation (7.8) to (7.9), equation (7.9) to (7.10), and finally to equation (7.28), and continuing to define quantities A_{ij} and B_{ij} and rearranging, we find that

$$N_y = A_{12}\varepsilon_x^o + A_{22}\varepsilon_y^o + A_{26}\gamma_{xy}^o + B_{12}\kappa_x^o + B_{22}\kappa_y^o + B_{26}\kappa_{xy}^o \qquad (7.51)$$

and
$$N_{xy} = A_{16}\varepsilon_x^o + A_{26}\varepsilon_y^o + A_{66}\gamma_{xy}^o + B_{16}\kappa_x^o + B_{26}\kappa_y^o + B_{66}\kappa_{xy}^o \qquad (7.52)$$

If we combine these definitions with the definitions associated with N_x, we can write the results in matrix notation as

$$\begin{Bmatrix} N_x \\ N_y \\ N_{xy} \end{Bmatrix} = \begin{bmatrix} A_{11} & A_{12} & A_{16} \\ A_{12} & A_{22} & A_{26} \\ A_{16} & A_{26} & A_{66} \end{bmatrix} \begin{Bmatrix} \varepsilon_x^o \\ \varepsilon_y^o \\ \gamma_{xy}^o \end{Bmatrix} + \begin{bmatrix} B_{11} & B_{12} & B_{16} \\ B_{12} & B_{22} & B_{26} \\ B_{16} & B_{26} & B_{66} \end{bmatrix} \begin{Bmatrix} \kappa_x^o \\ \kappa_y^o \\ \kappa_{xy}^o \end{Bmatrix} \qquad (7.53)$$

The above matrix equation constitutes three of the six relations between the six stress resultants and the six reference surface deformations. Three equations have yet to be derived, but with the above notation, it is clear that the A_{ij} and B_{ij} are components of three-by-three matrices. We can compute the numerical values of the additional A_{ij} and B_{ij} for the laminates of CLT Examples 1–5 in a manner similar to the previous calculations of A_{11}, A_{12}, A_{16}, B_{11}, B_{12}, and B_{16}.

To derive the remaining three equations, we use the moment expressions, equation (7.3), by substituting for σ_x from equation (7.6) into the first equation of equation (7.3) to yield

$$M_x = \int_{-\frac{H}{2}}^{\frac{H}{2}} \{\bar{Q}_{11}\varepsilon_x^o + \bar{Q}_{11}z\kappa_x^o + \bar{Q}_{12}\varepsilon_y^o + \bar{Q}_{12}z\kappa_y^o + \bar{Q}_{16}\gamma_{xy}^o + \bar{Q}_{16}z\kappa_{xy}^o\}z\,dz \quad (7.54)$$

where, as before, the integration can be distributed over the six terms to give

$$M_x = \int_{-\frac{H}{2}}^{\frac{H}{2}} \bar{Q}_{11}\varepsilon_x^o z\,dz + \int_{-\frac{H}{2}}^{\frac{H}{2}} \bar{Q}_{11}z\kappa_x^o z\,dz + \int_{-\frac{H}{2}}^{\frac{H}{2}} \bar{Q}_{12}\varepsilon_y^o z\,dz$$
$$+ \int_{-\frac{H}{2}}^{\frac{H}{2}} \bar{Q}_{12}z\kappa_y^o z\,dz + \int_{-\frac{H}{2}}^{\frac{H}{2}} \bar{Q}_{16}\gamma_{xy}^o z\,dz + \int_{-\frac{H}{2}}^{\frac{H}{2}} \bar{Q}_{16}z\kappa_{xy}^o z\,dz \quad (7.55)$$

Removing the reference surface deformations from the integrals, we find that

$$M_x = \left\{\int_{-\frac{H}{2}}^{\frac{H}{2}} \bar{Q}_{11}z\,dz\right\}\varepsilon_x^o + \left\{\int_{-\frac{H}{2}}^{\frac{H}{2}} \bar{Q}_{11}z^2\,dz\right\}\kappa_x^o + \left\{\int_{-\frac{H}{2}}^{\frac{H}{2}} \bar{Q}_{12}z\,dz\right\}\varepsilon_y^o$$
$$+ \left\{\int_{-\frac{H}{2}}^{\frac{H}{2}} \bar{Q}_{12}z^2\,dz\right\}\kappa_y^o + \left\{\int_{-\frac{H}{2}}^{\frac{H}{2}} \bar{Q}_{16}z\,dz\right\}\gamma_{xy}^o + \left\{\int_{-\frac{H}{2}}^{\frac{H}{2}} \bar{Q}_{16}z^2\,dz\right\}\kappa_{xy}^o \quad (7.56)$$

The first integral on the right hand side of equation (7.56) is

$$\int_{-\frac{H}{2}}^{\frac{H}{2}} \bar{Q}_{11}z\,dz \quad (7.57)$$

and this is immediately recognizable, from equation (7.25), as B_{11}. In fact, the first, third, and fifth terms on the right side of equation (7.56) become

$$B_{11}\varepsilon_x^o + B_{12}\varepsilon_y^o + B_{16}\gamma_{xy}^o \quad (7.58)$$

The B_{ij} appear in both the force resultant and the moment resultant expressions; this is important to remember.

The second, fourth, and sixth terms on the right hand side of equation (7.56) are new; the second term is

$$\int_{-\frac{H}{2}}^{\frac{H}{2}} \bar{Q}_{11}z^2\,dz \quad (7.59)$$

which can be expanded to

$$\int_{-\frac{H}{2}}^{\frac{H}{2}} \bar{Q}_{11} z^2 dz = \bar{Q}_{11_1} \int_{z_0}^{z_1} z^2 dz + \bar{Q}_{11_2} \int_{z_1}^{z_2} z^2 dz + \bar{Q}_{11_3} \int_{z_2}^{z_3} z^2 dz$$
$$+ \cdots + \bar{Q}_{11_k} \int_{z_{k-1}}^{z_k} z^2 dz + \cdots + \bar{Q}_{11_N} \int_{z_{N-1}}^{z_N} z^2 dz \tag{7.60}$$

Because of the constancy of the reduced stiffnesses within layers,

$$\int_{-\frac{H}{2}}^{\frac{H}{2}} \bar{Q}_{11} z^2 dz = \frac{1}{3} \{ \bar{Q}_{11_1}(z_1^3 - z_0^3) + \bar{Q}_{11_2}(z_2^3 - z_1^3) + \bar{Q}_{11_3}(z_3^3 - z_2^3)$$
$$+ \cdots + \bar{Q}_{11_k}(z_k^3 - z_{k-1}^3) + \cdots + \bar{Q}_{11_N}(z_N^3 - z_{N-1}^3) \} \tag{7.61}$$

or
$$\int_{-\frac{H}{2}}^{\frac{H}{2}} \bar{Q}_{11} z^2 dz = \frac{1}{3} \sum_{k=1}^{N} \bar{Q}_{11_k}(z_k^3 - z_{k-1}^3) \tag{7.62}$$

This integral is denoted as D_{11}; that is:

$$\int_{-\frac{H}{2}}^{\frac{H}{2}} \bar{Q}_{11} z^2 dz = D_{11} = \frac{1}{3} \sum_{k=1}^{N} \bar{Q}_{11_k}(z_k^3 - z_{k-1}^3) \tag{7.63}$$

Referring to the fourth and sixth terms in equation (7.56), we find that

$$\int_{-\frac{H}{2}}^{\frac{H}{2}} \bar{Q}_{12} z^2 dz = D_{12} = \frac{1}{3} \sum_{k=1}^{N} \bar{Q}_{12_k}(z_k^3 - z_{k-1}^3) \tag{7.64}$$

and
$$\int_{-\frac{H}{2}}^{\frac{H}{2}} \bar{Q}_{16} z^2 dz = D_{16} = \frac{1}{3} \sum_{k=1}^{N} \bar{Q}_{16_k}(z_k^3 - z_{k-1}^3) \tag{7.65}$$

With the definitions of the D_{ij}, and the earlier B_{ij}, the expression for the bending moment resultant M_x in equation (7.56) becomes

$$M_x = B_{11}\varepsilon_x^o + B_{12}\varepsilon_y^o + B_{16}\gamma_{xy}^o + D_{11}\kappa_x^o + D_{12}\kappa_y^o + D_{16}\kappa_{xy}^o \tag{7.66}$$

The D_{ij} are similar to the B_{ij} in that they involve powers of the z locations of the layer interfaces. Although the A_{ij} also involved z locations of the layer interfaces, it was really only the thickness of each layer that was involved. The actual location of the layers through the thickness is not important. However, with the B_{ij} and D_{ij}, the locations through the thickness of the laminate, as well as the thickness of the layers, are important.

   To continue with the numerical examples, the values of D_{11}, D_{12}, and D_{16} for the example laminates will be computed for illustration. For the four-layer $[0/90]_S$ laminate of CLT Examples 1 and 2,

$$D_{11} = \frac{1}{3} \{ \bar{Q}_{11_1}(z_1^3 - z_0^3) + \bar{Q}_{11_2}(z_2^3 - z_1^3) + \bar{Q}_{11_3}(z_3^3 - z_2^3) + \bar{Q}_{11_4}(z_4^3 - z_3^3) \} \tag{7.67}$$

or
$$D_{11} = \frac{1}{3} \{ \bar{Q}_{11}(0°)(z_1^3 - z_0^3) + \bar{Q}_{11}(90°)(z_2^3 - z_1^3)$$
$$+ \bar{Q}_{11}(90°)(z_3^3 - z_2^3) + \bar{Q}_{11}(0°)(z_4^3 - z_3^3) \} \tag{7.68}$$

The ensuing algebra leads to

$$D_{11} = 2.48 \text{ N·m} \tag{7.69}$$

For D_{12}

$$D_{12} = \frac{1}{3}\{\bar{Q}_{12_1}(z_1^3 - z_0^3) + \bar{Q}_{12_2}(z_2^3 - z_1^3) \tag{7.70}$$
$$+ \bar{Q}_{12_3}(z_3^3 - z_2^3) + \bar{Q}_{12_4}(z_4^3 - z_3^3)\}$$

or

$$D_{12} = \frac{1}{3}\{\bar{Q}_{12}(0°)(z_1^3 - z_0^3) + \bar{Q}_{12}(90°)(z_2^3 - z_1^3) \tag{7.71}$$
$$+ \bar{Q}_{12}(90°)(z_3^3 - z_2^3) + \bar{Q}_{12}(0°)(z_4^3 - z_3^3)\}$$

Numerically,

$$D_{12} = 0.0543 \text{ N·m} \tag{7.72}$$

Because Q_{16} is zero for each layer,

$$D_{16} = 0 \tag{7.73}$$

For the $[\pm 30/0]_S$ laminate

$$D_{1j} = \frac{1}{3}\{\bar{Q}_{1j_1}(z_1^3 - z_0^3) + \bar{Q}_{1j_2}(z_2^3 - z_1^3) + \bar{Q}_{1j_3}(z_3^3 - z_2^3) + \bar{Q}_{1j_4}(z_4^3 - z_3^3) \tag{7.74}$$
$$+ \bar{Q}_{1j_5}(z_5^3 - z_4^3) + \bar{Q}_{1j_6}(z_6^3 - z_5^3)\}, \ j = 1, 2, 6$$

or

$$D_{1j} = \frac{1}{3}\{\bar{Q}_{1j}(30°)(z_1^3 - z_0^3) + \bar{Q}_{1j}(-30°)(z_2^3 - z_1^3)$$
$$+ \bar{Q}_{1j}(0°)(z_3^3 - z_2^3) + \bar{Q}_{1j}(0°)(z_4^3 - z_3^3) \tag{7.75}$$
$$+ \bar{Q}_{1j}(-30°)(z_5^3 - z_4^3) + \bar{Q}_{1j}(30°)(z_6^3 - z_5^3)\}, \ j = 1, 2, 6$$

We find, referring to the interface locations for this laminate, equation (6.95), that the algebra leads to

$$D_{11} = 5.78 \text{ N·m}$$
$$D_{12} = 1.78 \text{ N·m} \tag{7.76}$$
$$D_{16} = 1.26 \text{ N·m}$$

Note that the value of D_{16} is not zero. Unlike the situation with A_{16}, where the contribution of the $-30°$ layers to the A_{16} cancels the contribution of the $+30°$ layers, the contributions of the layers with opposite orientation do not cancel in the calculation of D_{16}. This is because the locations of the layers, in addition to the layer thicknesses, are involved in the computation of the D_{ij}. We will see later in this chapter that the existence of a nonzero D_{16} term has important consequences. Also, we will see the dependence of the A_{26} on the $\pm\theta$ layer arrangement is identical to the dependence of A_{16}, and the dependence of D_{26} is identical to the dependence of D_{16}.

Finally, for the three-layer $[\pm 30/0]_T$ laminate,

$$D_{1j} = \frac{1}{3}\{\bar{Q}_{1j_1}(z_1^3 - z_0^3) + \bar{Q}_{1j_2}(z_2^3 - z_1^3) + \bar{Q}_{1j_3}(z_3^3 - z_2^3)\}, \ j = 1, 2, 6 \tag{7.77}$$

or

$$D_{1j} = \frac{1}{3}\{\bar{Q}_{1j}(30°)(z_1^3 - z_0^3) + \bar{Q}_{1j}(-30°)(z_2^3 - z_1^3) \tag{7.78}$$
$$+ \bar{Q}_{1j}(0°)(z_3^3 - z_2^3)\}, \ j = 1, 2, 6$$

Using numerical values leads to

$$D_{11} = 0.935 \text{ N·m}$$
$$D_{12} = 0.1294 \text{ N·m} \qquad (7.79)$$
$$D_{16} = 0.1576 \text{ N·m}$$

To finish the development of the laminate stiffness matrix, we can substitute the expressions for σ_y and τ_{xy} into the definitions in equation (7.3) for M_y and M_{xy}. The result is

$$M_y = \int_{-\frac{H}{2}}^{\frac{H}{2}} \{\bar{Q}_{12}\varepsilon_x^o + \bar{Q}_{12}z\kappa_x^o + \bar{Q}_{22}\varepsilon_y^o + \bar{Q}_{22}z\kappa_y^o$$
$$+ \bar{Q}_{26}\gamma_{xy}^o + \bar{Q}_{26}z\kappa_{xy}^o\}zdz \qquad (7.80)$$

and

$$M_{xy} = \int_{-\frac{H}{2}}^{\frac{H}{2}} \{\bar{Q}_{16}\varepsilon_x^o + \bar{Q}_{16}z\kappa_x^o + \bar{Q}_{26}\varepsilon_y^o$$
$$+ \bar{Q}_{26}z\kappa_y^o + \bar{Q}_{66}\gamma_{xy}^o + \bar{Q}_{66}z\kappa_{xy}^o\}zdz \qquad (7.81)$$

As in going from equation (7.54) to (7.55), equation (7.55) to (7.56), and finally to equation (7.66), and continuing to define additional D_{ij}, we find that

$$M_y = B_{12}\varepsilon_x^o + B_{22}\varepsilon_y^o + B_{26}\gamma_{xy}^o + D_{12}\kappa_x^o + D_{22}\kappa_y^o + D_{26}\kappa_{xy}^o \qquad (7.82)$$

and

$$M_{xy} = B_{16}\varepsilon_x^o + B_{26}\varepsilon_y^o + B_{66}\gamma_{xy}^o + D_{16}\kappa_x^o + D_{26}\kappa_y^o + D_{66}\kappa_{xy}^o \qquad (7.83)$$

We can write the results for the three moments M_x, M_y, and M_{xy} in matrix notation as

$$\begin{Bmatrix} M_x \\ M_y \\ M_{xy} \end{Bmatrix} = \begin{bmatrix} B_{11} & B_{12} & B_{16} \\ B_{12} & B_{22} & B_{26} \\ B_{16} & B_{26} & B_{66} \end{bmatrix} \begin{Bmatrix} \varepsilon_x^o \\ \varepsilon_y^o \\ \gamma_{xy}^o \end{Bmatrix} + \begin{bmatrix} D_{11} & D_{12} & D_{16} \\ D_{12} & D_{22} & D_{26} \\ D_{16} & D_{26} & D_{66} \end{bmatrix} \begin{Bmatrix} \kappa_x^o \\ \kappa_y^o \\ \kappa_{xy}^o \end{Bmatrix} \qquad (7.84)$$

The above analysis constitutes the remaining three relations of the six relations between the six stress resultants and the six reference surface deformations.

We can combine the six relations just derived to form one matrix relation between the six stress resultants and the six reference surface deformations. The combined matrix relation is

$$\begin{Bmatrix} N_x \\ N_y \\ N_{xy} \\ M_x \\ M_y \\ M_{xy} \end{Bmatrix} = \begin{bmatrix} A_{11} & A_{12} & A_{16} & B_{11} & B_{12} & B_{16} \\ A_{12} & A_{22} & A_{26} & B_{12} & B_{22} & B_{26} \\ A_{16} & A_{26} & A_{66} & B_{16} & B_{26} & B_{66} \\ B_{11} & B_{12} & B_{16} & D_{11} & D_{12} & D_{16} \\ B_{12} & B_{22} & B_{26} & D_{12} & D_{22} & D_{26} \\ B_{16} & B_{26} & B_{66} & D_{16} & D_{26} & D_{66} \end{bmatrix} \begin{Bmatrix} \varepsilon_x^o \\ \varepsilon_y^o \\ \gamma_{xy}^o \\ \kappa_x^o \\ \kappa_y^o \\ \kappa_{xy}^o \end{Bmatrix} \qquad (7.85)$$

Of course, in the above

$$A_{ij} = \sum_{k=1}^{N} \bar{Q}_{ij_k}(z_k - z_{k-1})$$

$$B_{ij} = \frac{1}{2} \sum_{k=1}^{N} \bar{Q}_{ij_k}(z_k^2 - z_{k-1}^2) \qquad (7.86)$$

$$D_{ij} = \frac{1}{3} \sum_{k=1}^{N} \bar{Q}_{ij_k}(z_k^3 - z_{k-1}^3)$$

The six-by-six matrix in equation (7.85) consisting of the components A_{ij}, B_{ij}, and D_{ij}, $i = 1, 2, 6$; $j = 1, 2, 6$, is called the *laminate stiffness matrix*. For obvious reasons it is also known as the *ABD* matrix. The *ABD* matrix defines a relationship between the stress resultants (i.e., loads) applied to a laminate, and the reference surface strains and curvatures (i.e., deformations). This form is a direct result of the Kirchhoff hypothesis, the plane-stress assumption, and the definition of the stress resultants. The laminate stiffness matrix involves *everything* that is used to define the laminate—layer material properties, fiber orientation, thickness, and location.

The relation of equation (7.85) can be inverted to give

$$\begin{Bmatrix} \varepsilon_x^o \\ \varepsilon_y^o \\ \gamma_{xy}^o \\ \kappa_x^o \\ \kappa_y^o \\ \kappa_{xy}^o \end{Bmatrix} = \begin{bmatrix} a_{11} & a_{12} & a_{16} & b_{11} & b_{12} & b_{16} \\ a_{12} & a_{22} & a_{26} & b_{21} & b_{22} & b_{26} \\ a_{16} & a_{26} & a_{66} & b_{61} & b_{62} & b_{66} \\ b_{11} & b_{21} & b_{61} & d_{11} & d_{12} & d_{16} \\ b_{12} & b_{22} & b_{62} & d_{12} & d_{22} & d_{26} \\ b_{16} & b_{26} & b_{66} & d_{16} & d_{26} & d_{66} \end{bmatrix} \begin{Bmatrix} N_x \\ N_y \\ N_{xy} \\ M_x \\ M_y \\ M_{xy} \end{Bmatrix} \qquad (7.87a)$$

where

$$\begin{bmatrix} a_{11} & a_{12} & a_{16} & b_{11} & b_{12} & b_{16} \\ a_{12} & a_{22} & a_{26} & b_{21} & b_{22} & b_{26} \\ a_{16} & a_{26} & a_{66} & b_{61} & b_{62} & b_{66} \\ b_{11} & b_{21} & b_{61} & d_{11} & d_{12} & d_{16} \\ b_{12} & b_{22} & b_{62} & d_{12} & d_{22} & d_{26} \\ b_{16} & b_{26} & b_{66} & d_{16} & d_{26} & d_{66} \end{bmatrix} = \begin{bmatrix} A_{11} & A_{12} & A_{16} & B_{11} & B_{12} & B_{16} \\ A_{12} & A_{22} & A_{26} & B_{12} & B_{22} & B_{26} \\ A_{16} & A_{26} & A_{66} & B_{16} & B_{26} & B_{66} \\ B_{11} & B_{12} & B_{16} & D_{11} & D_{12} & D_{16} \\ B_{12} & B_{22} & B_{26} & D_{12} & D_{22} & D_{26} \\ B_{16} & B_{26} & B_{66} & D_{16} & D_{26} & D_{66} \end{bmatrix}^{-1}$$

$$(7.87b)$$

Equation (7.85) and its inverse equation (7.87) are very important in the analysis of composite structures, particularly the inverse form. Knowing the loads, we can determine the reference surface strains and curvatures uniquely from equation (7.87). If we know the reference surface strains and curvatures, as we have seen in the previous examples, then we can determine the strains and stresses in each layer. Figure 7.4 illustrates the importance of the *ABD* matrix, namely the fact that it closes the loop on the analysis at composite laminates (see Figure 7.1).

It is appropriate at this point to present the numerical values of the remaining components of the *ABD* matrix for each of the three laminates we have been discussing. So far only limited numerical values have been illustrated. It will be left as an exercise to verify the numerical values presented for all the components of the *ABD* matrix for the three laminates. It is important to note the units on the various components of the *ABD* matrix. The units are consistent with the units of the reference surface strains and curvatures and the units of the stress resultants.

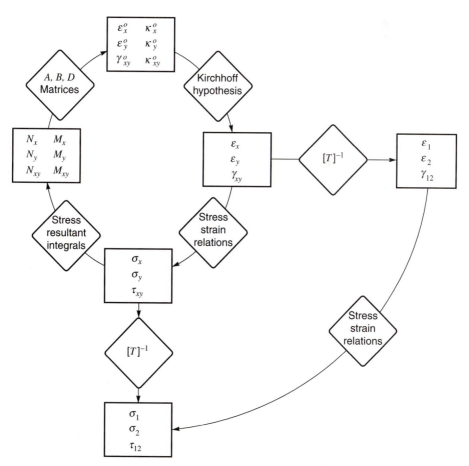

FIGURE 7.4
Closing the loop on concepts developed

For the $[0/90]_S$ laminate:

$$[A] = \begin{bmatrix} 50.4 & 1.809 & 0 \\ 1.809 & 50.4 & 0 \\ 0 & 0 & 2.64 \end{bmatrix} \times 10^6 \text{ N/m} \qquad (7.88a)$$

$$[B] = \begin{bmatrix} 0 & 0 & 0 \\ 0 & 0 & 0 \\ 0 & 0 & 0 \end{bmatrix} \text{ N} \qquad (7.88b)$$

$$[D] = \begin{bmatrix} 2.48 & 0.0543 & 0 \\ 0.0543 & 0.542 & 0 \\ 0 & 0 & 0.0792 \end{bmatrix} \text{ N·m} \qquad (7.88c)$$

For the $[\pm 30/0]_S$ laminate:

$$[A] = \begin{bmatrix} 102.4 & 18.94 & 0 \\ 18.94 & 16.25 & 0 \\ 0 & 0 & 20.2 \end{bmatrix} \times 10^6 \text{ N/m} \qquad (7.89a)$$

$$[B] = \begin{bmatrix} 0 & 0 & 0 \\ 0 & 0 & 0 \\ 0 & 0 & 0 \end{bmatrix} \text{ N} \qquad (7.89b)$$

$$[D] = \begin{bmatrix} 5.78 & 1.766 & 1.261 \\ 1.766 & 1.256 & 0.418 \\ 1.261 & 0.418 & 1.850 \end{bmatrix} \text{ N·m} \qquad (7.89c)$$

For the $[\pm 30/0]_T$ laminate:

$$[A] = \begin{bmatrix} 51.2 & 9.47 & 0 \\ 9.47 & 8.12 & 0 \\ 0 & 0 & 10.10 \end{bmatrix} \times 10^6 \text{ N·m} \qquad (7.90a)$$

$$[B] = \begin{bmatrix} 1416 & -609 & -1051 \\ -609 & -199.0 & -348 \\ -1051 & -348 & -609 \end{bmatrix} \text{ N} \qquad (7.90b)$$

$$[D] = \begin{bmatrix} 0.935 & 0.1294 & 0.1576 \\ 0.1294 & 0.1272 & 0.0522 \\ 0.1576 & 0.0522 & 0.1400 \end{bmatrix} \text{ N·m} \qquad (7.90c)$$

For later use it is convenient to record the numerical values of the inverse quantities, defined by equation (7.87b), for the three laminates. For the $[0/90]_S$ laminate:

$$\begin{bmatrix} a_{11} & a_{12} & a_{16} \\ a_{12} & a_{22} & a_{26} \\ a_{16} & a_{26} & a_{66} \end{bmatrix} = \begin{bmatrix} 19.87 & -0.714 & 0 \\ -0.714 & 19.87 & 0 \\ 0 & 0 & 379 \end{bmatrix} \times 10^{-9} \text{ m/N} \qquad (7.91a)$$

$$\begin{bmatrix} b_{11} & b_{12} & b_{16} \\ b_{21} & b_{22} & b_{26} \\ b_{61} & b_{62} & b_{66} \end{bmatrix} = \begin{bmatrix} 0 & 0 & 0 \\ 0 & 0 & 0 \\ 0 & 0 & 0 \end{bmatrix} \frac{1}{N} \qquad (7.91b)$$

$$\begin{bmatrix} d_{11} & d_{12} & d_{16} \\ d_{12} & d_{22} & d_{26} \\ d_{16} & d_{26} & d_{66} \end{bmatrix} = \begin{bmatrix} 0.404 & -0.0405 & 0 \\ -0.0405 & 1.849 & 0 \\ 0 & 0 & 12.63 \end{bmatrix} 1/\text{N·m} \qquad (7.91c)$$

For the $[\pm 30/0]_S$ laminate:

$$\begin{bmatrix} a_{11} & a_{12} & a_{16} \\ a_{12} & a_{22} & a_{26} \\ a_{16} & a_{26} & a_{66} \end{bmatrix} = \begin{bmatrix} 12.45 & -14.52 & 0 \\ -14.52 & 78.5 & 0 \\ 0 & 0 & 49.5 \end{bmatrix} \times 10^{-9} \text{ m/N} \qquad (7.92a)$$

$$\begin{bmatrix} b_{11} & b_{12} & b_{16} \\ b_{21} & b_{22} & b_{26} \\ b_{61} & b_{62} & b_{66} \end{bmatrix} = \begin{bmatrix} 0 & 0 & 0 \\ 0 & 0 & 0 \\ 0 & 0 & 0 \end{bmatrix} \frac{1}{\text{N}} \qquad (7.92b)$$

$$\begin{bmatrix} d_{11} & d_{12} & d_{16} \\ d_{12} & d_{22} & d_{26} \\ d_{16} & d_{26} & d_{66} \end{bmatrix} = \begin{bmatrix} 0.330 & -0.420 & -0.1300 \\ -0.420 & 1.397 & -0.0287 \\ -0.1300 & -0.0287 & 0.636 \end{bmatrix} 1/\text{N·m} \qquad (7.92c)$$

For the $[\pm 30/0]_T$ laminate:

$$\begin{bmatrix} a_{11} & a_{12} & a_{16} \\ a_{12} & a_{22} & a_{26} \\ a_{16} & a_{26} & a_{66} \end{bmatrix} = \begin{bmatrix} 41.5 & -39.8 & 12.19 \\ -39.8 & 185.5 & 22.3 \\ 12.19 & 22.3 & 150.4 \end{bmatrix} \times 10^{-9} \text{ m/N} \qquad (7.93a)$$

$$\begin{bmatrix} b_{11} & b_{12} & b_{16} \\ b_{21} & b_{22} & b_{26} \\ b_{61} & b_{62} & b_{66} \end{bmatrix} = \begin{bmatrix} -164.6 & 179.4 & 384 \\ 207 & -71.8 & 52.5 \\ 18.80 & 195.4 & 707 \end{bmatrix} \times 10^{-6} \text{ m/N} \qquad (7.93b)$$

$$\begin{bmatrix} d_{11} & d_{12} & d_{16} \\ d_{12} & d_{22} & d_{26} \\ d_{16} & d_{26} & d_{66} \end{bmatrix} = \begin{bmatrix} 2.10 & -1.553 & -2.42 \\ -1.553 & 10.84 & -0.276 \\ -2.42 & -0.276 & 16.07 \end{bmatrix} 1/\text{N·m} \qquad (7.93c)$$

These results can also be written in the full form of equation (7.85). For the $[0/90]_S$ laminate,

$$\begin{Bmatrix} N_x \\ N_y \\ N_{xy} \\ M_x \\ M_y \\ M_{xy} \end{Bmatrix} = \begin{bmatrix} 50.4 \times 10^6 & 1.809 \times 10^6 & 0 & 0 & 0 & 0 \\ 1.809 \times 10^6 & 50.4 \times 10^6 & 0 & 0 & 0 & 0 \\ 0 & 0 & 2.64 \times 10^6 & 0 & 0 & 0 \\ 0 & 0 & 0 & 2.48 & 0.0543 & 0 \\ 0 & 0 & 0 & 0.0543 & 0.542 & 0 \\ 0 & 0 & 0 & 0 & 0 & 0.0792 \end{bmatrix} \begin{Bmatrix} \varepsilon_x^o \\ \varepsilon_y^o \\ \gamma_{xy}^o \\ \kappa_x^o \\ \kappa_y^o \\ \kappa_{xy}^o \end{Bmatrix} \qquad (7.94)$$

For the $[\pm 30/0]_S$ laminate:

$$\begin{Bmatrix} N_x \\ N_y \\ N_{xy} \\ M_x \\ M_y \\ M_{xy} \end{Bmatrix} = \begin{bmatrix} 102.4 \times 10^6 & 18.94 \times 10^6 & 0 & 0 & 0 & 0 \\ 18.94 \times 10^6 & 16.25 \times 10^6 & 0 & 0 & 0 & 0 \\ 0 & 0 & 20.2 \times 10^6 & 0 & 0 & 0 \\ 0 & 0 & 0 & 5.78 & 1.766 & 1.261 \\ 0 & 0 & 0 & 1.766 & 1.256 & 0.418 \\ 0 & 0 & 0 & 1.261 & 0.418 & 1.850 \end{bmatrix} \begin{Bmatrix} \varepsilon_x^o \\ \varepsilon_y^o \\ \gamma_{xy}^o \\ \kappa_x^o \\ \kappa_y^o \\ \kappa_{xy}^o \end{Bmatrix} \qquad (7.95)$$

For the $[\pm30/0]_T$ laminate:

$$
\begin{Bmatrix} N_x \\ N_y \\ N_{xy} \\ M_x \\ M_y \\ M_{xy} \end{Bmatrix} =
\begin{bmatrix}
51.2 \times 10^6 & 9.47 \times 10^6 & 0 & 1416 & -609 & -1051 \\
9.47 \times 10^6 & 8.12 \times 10^6 & 0 & -609 & -199.0 & -348 \\
0 & 0 & 10.10 \times 10^6 & -1051 & -348 & -609 \\
1416 & -609 & -1051 & 0.935 & 0.1294 & 0.1576 \\
-609 & -199.0 & -348 & 0.1294 & 0.1272 & 0.0522 \\
-1051 & -348 & -609 & 0.1576 & 0.0522 & 0.1400
\end{bmatrix}
\begin{Bmatrix} \varepsilon_x^o \\ \varepsilon_y^o \\ \gamma_{xy}^o \\ \kappa_x^o \\ \kappa_y^o \\ \kappa_{xy}^o \end{Bmatrix}
\qquad (7.96)
$$

The full form of the inverse relations, equation (7.87), for the $[0/90]_S$ laminate is

$$
\begin{Bmatrix} \varepsilon_x^o \\ \varepsilon_y^o \\ \gamma_{xy}^o \\ \kappa_x^o \\ \kappa_y^o \\ \kappa_{xy}^o \end{Bmatrix} =
\begin{bmatrix}
19.87 \times 10^{-9} & -0.714 \times 10^{-9} & 0 & 0 & 0 & 0 \\
-0.714 \times 10^{-9} & 19.87 \times 10^{-9} & 0 & 0 & 0 & 0 \\
0 & 0 & 379 \times 10^{-9} & 0 & 0 & 0 \\
0 & 0 & 0 & 0.404 & -0.0405 & 0 \\
0 & 0 & 0 & -0.0405 & 1.849 & 0 \\
0 & 0 & 0 & 0 & 0 & 12.63
\end{bmatrix}
\begin{Bmatrix} N_x \\ N_y \\ N_{xy} \\ M_x \\ M_y \\ M_{xy} \end{Bmatrix}
\qquad (7.97)
$$

For the $[\pm30/0]_S$ laminate:

$$
\begin{Bmatrix} \varepsilon_x^o \\ \varepsilon_y^o \\ \gamma_{xy}^o \\ \kappa_x^o \\ \kappa_y^o \\ \kappa_{xy}^o \end{Bmatrix} =
\begin{bmatrix}
12.45 \times 10^{-9} & -14.52 \times 10^{-9} & 0 & 0 & 0 & 0 \\
-14.52 \times 10^{-9} & 78.5 \times 10^{-9} & 0 & 0 & 0 & 0 \\
0 & 0 & 49.5 \times 10^{-9} & 0 & 0 & 0 \\
0 & 0 & 0 & 0.330 & -0.420 & -0.1300 \\
0 & 0 & 0 & -0.420 & 1.397 & -0.0287 \\
0 & 0 & 0 & -0.1300 & -0.0287 & 0.636
\end{bmatrix}
\begin{Bmatrix} N_x \\ N_y \\ N_{xy} \\ M_x \\ M_y \\ M_{xy} \end{Bmatrix}
$$

$$(7.98)$$

For the $[\pm30/0]_T$ laminate:

$$
\begin{Bmatrix} \varepsilon_x^o \\ \varepsilon_y^o \\ \gamma_{xy}^o \\ \kappa_x^o \\ \kappa_y^o \\ \kappa_{xy}^o \end{Bmatrix} =
\begin{bmatrix}
41.5 \times 10^{-9} & -39.8 \times 10^{-9} & 12.19 \times 10^{-9} & -164.6 \times 10^{-6} & 179.4 \times 10^{-6} & 384 \times 10^{-6} \\
-39.8 \times 10^{-9} & 185.5 \times 10^{-9} & 22.3 \times 10^{-9} & 207 \times 10^{-6} & -71.8 \times 10^{-6} & 52.5 \times 10^{-6} \\
12.19 \times 10^{-9} & 22.3 \times 10^{-9} & 150.4 \times 10^{-9} & 18.80 \times 10^{-6} & 195.4 \times 10^{-6} & 707 \times 10^{-6} \\
-164.6 \times 10^{-6} & 207 \times 10^{-6} & 18.80 \times 10^{-6} & 2.10 & -1.553 & -2.42 \\
1.794 \times 10^{-6} & -71.8 \times 10^{-6} & 195.4 \times 10^{-6} & -1.553 & 10.84 & -0.276 \\
384 \times 10^{-6} & 52.5 \times 10^{-6} & 707 \times 10^{-6} & -2.42 & -0.276 & 16.07
\end{bmatrix}
\begin{Bmatrix} N_x \\ N_y \\ N_{xy} \\ M_x \\ M_y \\ M_{xy} \end{Bmatrix}
$$

$$(7.99)$$

Writing the numerical values in full form is quite revealing; the coupling, or lack thereof, between the various stress resultants and reference deformations becomes quite obvious. The coupling is important in determining how much interaction there will be between the specific loads and specific reference surface strains and curvatures. Essentially, the more zeros there are in the ABD matrix, or its inverse, the less coupling there is between the stress resultants and reference surface deformations. For example, for the $[0/90]_S$ and $[\pm 30/0]_S$ laminates, because all components of the B matrix are zero, the reference surface curvatures are decoupled from the force resultants, and the reference surface strains are decoupled from the moment resultants. The six-by-six relation of equation (7.85) reduces to two three-by-three relations, namely,

$$\left\{ \begin{array}{c} N_x \\ N_y \\ N_{xy} \end{array} \right\} = \left[\begin{array}{ccc} A_{11} & A_{12} & A_{16} \\ A_{12} & A_{22} & A_{26} \\ A_{16} & A_{26} & A_{66} \end{array} \right] \left\{ \begin{array}{c} \varepsilon_x^o \\ \varepsilon_y^o \\ \gamma_{xy}^o \end{array} \right\} \qquad (7.100a)$$

and

$$\left\{ \begin{array}{c} M_x \\ M_y \\ M_{xy} \end{array} \right\} = \left[\begin{array}{ccc} D_{11} & D_{12} & D_{16} \\ D_{12} & D_{22} & D_{26} \\ D_{16} & D_{26} & D_{66} \end{array} \right] \left\{ \begin{array}{c} \kappa_x^o \\ \kappa_y^o \\ \kappa_{xy}^o \end{array} \right\} \qquad (7.100b)$$

Such a reduction is not possible when there are elements of the B matrix present (i.e., when the laminate is unsymmetric). We will have more to say regarding the couplings within the ABD matrix in the next section, where the degree of coupling, and its effects on laminate response, are categorized in a systematic way. Moreover, we can do the categorization *a priori*, before any analysis is started, and we can use the advantages of knowing the level of coupling to simplify the analysis procedure.

Exercises for Section 7.2

1. Verify that the numerical values of the components of the A, B, and D matrices in equations (7.88), (7.89), and (7.90) are correct.

2. Use your computer program from the exercise below to compute the A, B, and D matrices of a $[+45/0/-30]_T$ graphite-reinforced composite laminate.

Computer Exercise

Below is a modification of the steps outlined in the Computer Exercise for Section 6.8. New steps have been added to include the calculation of A, B, and D matrices. These additional steps are underlined. The steps to compute the inverse of the ABD matrix are also included. Finally, steps are added to provide the user with the option to enter either the equivalent loads, or the reference surface strains and curvatures, and then proceed to compute the stresses as before. Of course, if the user enters the stress resultants, then the inverse of the ABD matrix must be used to compute the reference surface strains and curvatures.

1. Read in the number of layers.

2. Read in the engineering properties of a layer, and the thickness.

3. Compute the Q_{ij}.

4. Print the number of layers, the engineering properties, and the Q_{ij}.

5. Compute and print the z_k.

6. Read in and print the fiber orientations in each layer.

7. Compute and print the $\bar{Q}_{ij}$ for each layer.

7a. Compute and print the components of the A, B, and D matrices.

7b. Compute and print the inverse of ABD.

8a. EITHER read in and print the values of $\varepsilon_x^o, \ldots, \kappa_{xy}^o$.

8b. OR Read in and print the values of $N_x, \ldots, M_{xy}$, and compute and print values of $\varepsilon_x^o, \ldots, \kappa_{xy}^o$.

8. Compute and print the values of ε_x, ε_y, and γ_{xy} at the top and bottom of each layer. (These are the extreme values in each layer.)

9. Compute and print the values of ε_1, ε_2, and γ_{12} at the top and bottom of each layer.

10. Compute and print the values of σ_x, σ_y, and τ_{xy} at the top and bottom of each layer.

11. Compute and print the values of σ_1, σ_2, and τ_{12} at the top and bottom of each layer.

12. As a check at a later stage when we add further information to this program, you might want to compute and print the values of $N_x, \ldots, M_{xy}$ as calculated from the values of σ_x, σ_y, and τ_{xy}, and the z_k.

7.3
CLASSIFICATION OF LAMINATES AND THEIR EFFECT ON THE *ABD* MATRIX

We have seen in the numerical results presented in the previous section that the form of the *ABD* matrix depends strongly on whether the laminate is what we referred to as symmetric, whether it consists of just 0° and 90° layers, or whether for every layer at orientation $+\theta$ there is a layer at $-\theta$. In this section we would like to formally classify laminates as to their stacking arrangement, and indicate the effects of the various classifications on the *ABD* matrix. In a subsequent section, physical interpretations will be given to particular components of the *ABD* matrix. As a sidenote, in connection with equations (2.45) and (2.47), it was stated that the presence of the zeros in the compliance and stiffness matrices in the principal material coordinate system was associated with a fiber-reinforced material being orthotropic. Similarly, with the lack of any zero terms in equation (7.85), the elastic behavior of a general laminate is sometimes termed nonorthotropic, or anisotropic.

7.3.1 Symmetric Laminates

A laminate is said to be *symmetric* if for every layer to one side of the laminate reference surface with a specific thickness, specific material properties, and specific fiber orientation, there is another layer the identical distance on the *opposite* side of the reference surface with the identical thickness, material properties, and fiber orientation. If a laminate is not symmetric, then it is referred to as an *unsymmetric* laminate. Note that this pairing on opposite sides of the reference surface must occur for *every* layer. Although we have restricted our examples to laminates with layers of the same material and same thickness, this does not have to be the case for a laminate to be symmetric. A four-layer laminate with the two outer layers made of glass-reinforced material oriented at $+45°$ and the two inner layers made of graphite-reinforced material oriented at $-30°$ is a symmetric laminate.

For a symmetric laminate all the components of the B matrix are identically zero. Consequently, the full six-by-six set of equations in equation (7.85) decouples into two three-by-three sets of equations, namely,

$$\begin{Bmatrix} N_x \\ N_y \\ N_{xy} \end{Bmatrix} = \begin{bmatrix} A_{11} & A_{12} & A_{16} \\ A_{12} & A_{22} & A_{26} \\ A_{16} & A_{26} & A_{66} \end{bmatrix} \begin{Bmatrix} \varepsilon_x^o \\ \varepsilon_y^o \\ \gamma_{xy}^o \end{Bmatrix} \qquad (7.101a)$$

and

$$\begin{Bmatrix} M_x \\ M_y \\ M_{xy} \end{Bmatrix} = \begin{bmatrix} D_{11} & D_{12} & D_{16} \\ D_{12} & D_{22} & D_{26} \\ D_{16} & D_{26} & D_{66} \end{bmatrix} \begin{Bmatrix} \kappa_x^o \\ \kappa_y^o \\ \kappa_{xy}^o \end{Bmatrix} \qquad (7.101b)$$

With symmetric laminates, the inverse relation, equation (7.87), also degenerates into two three-by-three relations, specifically,

$$\begin{Bmatrix} \varepsilon_x^o \\ \varepsilon_y^o \\ \gamma_{xy}^o \end{Bmatrix} = \begin{bmatrix} a_{11} & a_{12} & a_{16} \\ a_{12} & a_{22} & a_{26} \\ a_{16} & a_{26} & a_{66} \end{bmatrix} \begin{Bmatrix} N_x \\ N_y \\ N_{xy} \end{Bmatrix} \qquad (7.102a)$$

$$\begin{Bmatrix} \kappa_x^o \\ \kappa_y^o \\ \kappa_{xy}^o \end{Bmatrix} = \begin{bmatrix} d_{11} & d_{12} & d_{16} \\ d_{12} & d_{22} & d_{26} \\ d_{16} & d_{26} & d_{66} \end{bmatrix} \begin{Bmatrix} M_x \\ M_y \\ M_{xy} \end{Bmatrix} \qquad (7.102b)$$

In the above the a matrix is the inverse of the A matrix and has components

$$a_{11} = \frac{A_{22}A_{66} - A_{26}^2}{\det[A]}$$

$$a_{12} = \frac{A_{26}A_{16} - A_{12}A_{66}}{\det[A]}$$

$$a_{16} = \frac{A_{12}A_{26} - A_{22}A_{16}}{\det[A]}$$

$$a_{22} = \frac{A_{11}A_{66} - A_{16}^2}{\det[A]} \qquad (7.103a)$$

$$a_{26} = \frac{A_{12}A_{16} - A_{11}A_{26}}{\det[A]}$$

$$a_{66} = \frac{A_{11}A_{22} - A_{12}^2}{\det[A]}$$

and the d matrix is the inverse of the D matrix and has components

$$d_{11} = \frac{D_{22}D_{66} - D_{26}^2}{\det[D]}$$

$$d_{12} = \frac{D_{26}D_{16} - D_{12}D_{66}}{\det[D]}$$

$$d_{16} = \frac{D_{12}D_{26} - D_{22}D_{16}}{\det[D]}$$

$$d_{22} = \frac{D_{11}D_{66} - D_{16}^2}{\det[D]} \qquad (7.103b)$$

$$d_{26} = \frac{D_{12}D_{16} - D_{11}D_{26}}{\det[D]}$$

$$d_{66} = \frac{D_{11}D_{22} - D_{12}^2}{\det[D]}$$

where the definitions

$$\det[A] = A_{11}(A_{22}A_{66} - A_{26}^2) - A_{12}(A_{12}A_{66} - A_{26}A_{16})$$
$$+ A_{16}(A_{12}A_{26} - A_{22}A_{16}) \qquad (7.103c)$$

and $\quad \det[D] = D_{11}(D_{22}D_{66} - D_{26}^2) - D_{12}(D_{12}D_{66} - D_{26}D_{16})$
$$+ D_{16}(D_{12}D_{26} - D_{22}D_{16}) \qquad (7.103d)$$

have been used. Because neither the A nor D matrices have any zero terms, a general symmetric laminate is anisotropic in both inplane and bending behavior.

7.3.2 Balanced Laminates

A laminate is said to be *balanced* if for every layer with a specified thickness, specific material properties, and specific fiber orientation, there is another layer with the identical thickness, material properties, but opposite fiber orientation somewhere in the laminate. The layer with opposite fiber orientation does not have to be on the opposite side of the reference surface, nor immediately adjacent to the other layer, nor anywhere in particular. The other layer can be anywhere within the thickness. A laminate does not have to be symmetric to be balanced. The symmetric $[\pm 30/0]_S$ laminate and the unsymmetric $[\pm 30/0]_T$ laminate are both balanced laminates. If a laminate is balanced, then the stiffness matrix components A_{16} and A_{26} are always zero because the $\bar{Q}_{16}$ and $\bar{Q}_{26}$ from the layer pairs with opposite orientation are of opposite sign, and the net contribution to A_{16} and A_{26} from these layer pairs is then zero. To classify as balanced, all off-axis layers must occur in pairs. As a special case, all laminates consisting entirely of layers with their fibers oriented at either 0 or 90°, to be discussed shortly, are balanced, as $\bar{Q}_{16}$ and $\bar{Q}_{26}$ are zero for every layer, resulting in A_{16} and A_{26} being zero. The ABD matrix of a balanced but otherwise general laminate is not that much simpler than the ABD matrix of a general unsymmetric, unbalanced laminate. The full six-by-six form, equation (7.85), applies but with the A_{16} and A_{26} components set to zero. To obtain the inverse relation, equation (7.87a), the full six-by-six with the zero entries for A_{16} and A_{26} must be inverted.

7.3.3 Symmetric Balanced Laminates

A laminate is said to be a *symmetric balanced* laminate if it meets both the criterion for being symmetric and the criterion for being balanced. If this is the case, then equation (7.85) takes the decoupled form

$$\left\{ \begin{array}{c} N_x \\ N_y \end{array} \right\} = \left[\begin{array}{cc} A_{11} & A_{12} \\ A_{12} & A_{22} \end{array} \right] \left\{ \begin{array}{c} \varepsilon_x^o \\ \varepsilon_y^o \end{array} \right\} \tag{7.104a}$$

with
$$N_{xy} = A_{66}\gamma_{xy}^o$$

and
$$\left\{ \begin{array}{c} M_x \\ M_y \\ M_{xy} \end{array} \right\} = \left[\begin{array}{ccc} D_{11} & D_{12} & D_{16} \\ D_{12} & D_{22} & D_{26} \\ D_{16} & D_{26} & D_{66} \end{array} \right] \left\{ \begin{array}{c} \kappa_x^o \\ \kappa_y^o \\ \kappa_{xy}^o \end{array} \right\} \tag{7.104b}$$

The inverse form involving the force resultants is

$$\left\{ \begin{array}{c} \varepsilon_x^o \\ \varepsilon_y^o \end{array} \right\} = \left[\begin{array}{cc} a_{11} & a_{12} \\ a_{12} & a_{22} \end{array} \right] \left\{ \begin{array}{c} N_x \\ N_y \end{array} \right\} \tag{7.105a}$$

with
$$\gamma_{xy}^o = a_{66}N_{xy}$$

The inverse form involving the moments is no different than for symmetric but unbalanced laminates, namely,

$$\left\{ \begin{array}{c} \kappa_x^o \\ \kappa_y^o \\ \kappa_{xy}^o \end{array} \right\} = \left[\begin{array}{ccc} d_{11} & d_{12} & d_{16} \\ d_{12} & d_{22} & d_{26} \\ d_{16} & d_{26} & d_{66} \end{array} \right] \left\{ \begin{array}{c} M_x \\ M_y \\ M_{xy} \end{array} \right\} \tag{7.105b}$$

For the symmetric balanced case, the third of equation (7.85) decouples from all other equations. The components of the a matrix in equation (7.105a) are

$$a_{11} = \frac{A_{22}}{A_{11}A_{22} - A_{12}^2}$$

$$a_{12} = \frac{-A_{12}}{A_{11}A_{22} - A_{12}^2}$$

$$a_{22} = \frac{A_{11}}{A_{11}A_{22} - A_{12}^2} \tag{7.106a}$$

$$a_{66} = \frac{1}{A_{66}}$$

and the components of the d matrix are the same as the symmetric but balanced case, equation (7.103b), namely,

$$d_{11} = \frac{D_{22}D_{66} - D_{26}^2}{\det[D]}$$

$$d_{12} = \frac{D_{26}D_{16} - D_{12}D_{66}}{\det[D]}$$

$$d_{16} = \frac{D_{12}D_{26} - D_{22}D_{16}}{\det[D]}$$

$$d_{22} = \frac{D_{11}D_{66} - D_{16}^2}{\det[D]}$$

$$d_{26} = \frac{D_{12}D_{16} - D_{11}D_{26}}{\det[D]}$$

$$d_{66} = \frac{D_{11}D_{22} - D_{12}^2}{\det[D]}$$

(7.106b)

with $\det[D]$ being given by equation (7.103d). Because A_{16} and A_{26} are zero, a symmetric balanced laminate is orthotropic with respect to inplane behavior.

7.3.4 Cross-Ply Laminates

A laminate is said to be a *cross-ply* laminate if every layer has its fibers oriented at either $0°$ or $90°$. If this is the case, because $\bar{Q}_{16}$ and $\bar{Q}_{26}$ are zero for every layer, then A_{16}, A_{26}, B_{16}, B_{26}, D_{16}, and D_{26} are zero, and there is some decoupling of the six equations. In particular, the six equations decouple to a set of four equations and a set of two equations, namely,

$$\begin{Bmatrix} N_x \\ N_y \\ M_x \\ M_y \end{Bmatrix} = \begin{bmatrix} A_{11} & A_{12} & B_{11} & B_{12} \\ A_{12} & A_{22} & B_{12} & B_{22} \\ B_{11} & B_{12} & D_{11} & D_{12} \\ B_{12} & B_{22} & D_{12} & D_{22} \end{bmatrix} \begin{Bmatrix} \varepsilon_x^o \\ \varepsilon_y^o \\ \kappa_x^o \\ \kappa_y^o \end{Bmatrix}$$

(7.107a)

$$\begin{Bmatrix} N_{xy} \\ M_{xy} \end{Bmatrix} = \begin{bmatrix} A_{66} & B_{66} \\ B_{66} & D_{66} \end{bmatrix} \begin{Bmatrix} \gamma_{xy}^o \\ \kappa_{xy}^o \end{Bmatrix}$$

(7.107b)

7.3.5 Symmetric Cross-Ply Laminates

A laminate is said to be a *symmetric cross-ply* laminate if it meets the criterion for being symmetric and the criterion for being cross-ply. This results in the simplest form of the ABD matrix. All the B_{ij} are zero, and A_{16}, A_{26}, D_{16}, and D_{26} are zero.

The six equations of equation (7.85) decouple significantly. Specifically,

$$\left\{ \begin{array}{c} N_x \\ N_y \end{array} \right\} = \left[\begin{array}{cc} A_{11} & A_{12} \\ A_{12} & A_{22} \end{array} \right] \left\{ \begin{array}{c} \varepsilon_x^o \\ \varepsilon_y^o \end{array} \right\} \tag{7.108a}$$

with
$$N_{xy} = A_{66}\gamma_{xy}^o$$

and
$$\left\{ \begin{array}{c} M_x \\ M_y \end{array} \right\} = \left[\begin{array}{cc} D_{11} & D_{12} \\ D_{12} & D_{22} \end{array} \right] \left\{ \begin{array}{c} \kappa_x^o \\ \kappa_y^o \end{array} \right\} \tag{7.108b}$$

with
$$M_{xy} = D_{66}\kappa_{xy}^o$$

The inverted form is

$$\left\{ \begin{array}{c} \varepsilon_x^o \\ \varepsilon_y^o \end{array} \right\} = \left[\begin{array}{cc} a_{11} & a_{12} \\ a_{12} & a_{22} \end{array} \right] \left\{ \begin{array}{c} N_x \\ N_y \end{array} \right\} \tag{7.109a}$$

with
$$\gamma_{xy}^o = a_{66}N_{xy}$$

and
$$\left\{ \begin{array}{c} \kappa_x^o \\ \kappa_y^o \end{array} \right\} = \left[\begin{array}{cc} d_{11} & d_{12} \\ d_{12} & d_{22} \end{array} \right] \left\{ \begin{array}{c} M_x \\ M_y \end{array} \right\} \tag{7.109b}$$

with
$$\kappa_{xy}^o = d_{66}M_{xy}$$

The components of the a matrix are the same as the symmetric balanced case,

$$a_{11} = \frac{A_{22}}{A_{11}A_{22} - A_{12}^2}$$

$$a_{12} = \frac{-A_{12}}{A_{11}A_{22} - A_{12}^2}$$

$$a_{22} = \frac{A_{11}}{A_{11}A_{22} - A_{12}^2} \tag{7.110a}$$

$$a_{66} = \frac{1}{A_{66}}$$

and the components of the d matrix are

$$d_{11} = \frac{D_{22}}{D_{11}D_{22} - D_{12}^2}$$

$$d_{12} = \frac{-D_{12}}{D_{11}D_{22} - D_{12}^2}$$

$$d_{22} = \frac{D_{11}}{D_{11}D_{22} - D_{12}^2} \tag{7.110b}$$

$$d_{66} = \frac{1}{D_{66}}$$

Compared to the most general case, the symmetric cross-ply laminate is trivial. Because A_{16}, A_{26}, D_{16}, and D_{26} are zero, a symmetric cross-ply laminate is orthotropic with respect to both inplane and bending behavior.

7.3.6 Single Isotropic Layer

As a comparison, it is of interest to record the ABD matrix for the case of a single isotropic layer, the subject of studies in classical plate theory. What we have presented in the development of classical lamination theory is a generalization of classical plate theory. The isotropic case is a special case of this generalization, and for a single isotropic layer of thickness H,

$$A_{11} = A_{22} = \frac{EH}{1 - v^2} = A \qquad A_{12} = v\frac{EH}{1 - v^2} = vA$$

$$A_{66} = \frac{EH}{2(1 + v)} = \frac{1 - v}{2}A \qquad A_{16} = A_{26} = 0$$

$$D_{11} = D_{22} = \frac{EH^3}{12(1 - v^2)} = D \qquad D_{12} = v\frac{EH^3}{12(1 - v^2)} = vD$$

$$D_{66} = \frac{EH^3}{24(1 + v)} = \frac{1 - v}{2}D \qquad D_{16} = D_{26} = 0$$

$$(7.111)$$

In the above equation, E is the extensional modulus, referred to as Young's modulus when dealing with isotropic materials, and v is Poisson's ratio of the single layer of material. Because of equation (7.111), the ABD matrix for a single isotropic layer decouples and greatly simplifies, and equation (7.85) becomes

$$\left\{ \begin{array}{c} N_x \\ N_y \end{array} \right\} = \frac{EH}{1 - v^2} \left[\begin{array}{cc} 1 & v \\ v & 1 \end{array} \right] \left\{ \begin{array}{c} \varepsilon_x^o \\ \varepsilon_y^o \end{array} \right\}$$

$$(7.112a)$$

with

$$N_{xy} = \frac{EH}{2(1 + v)}\gamma_{xy}^o$$

and

$$\left\{ \begin{array}{c} M_x \\ M_y \end{array} \right\} = \frac{EH^3}{12(1 - v^2)} \left[\begin{array}{cc} 1 & v \\ v & 1 \end{array} \right] \left\{ \begin{array}{c} \kappa_x^o \\ \kappa_y^o \end{array} \right\}$$

$$(7.112b)$$

with

$$M_{xy} = \frac{EH^3}{24(1 + v)}\kappa_{xy}^o$$

The inverse relation is

$$\left\{ \begin{array}{c} \varepsilon_x^o \\ \varepsilon_y^o \end{array} \right\} = \frac{1}{EH} \left[\begin{array}{cc} 1 & -v \\ -v & 1 \end{array} \right] \left\{ \begin{array}{c} N_x \\ N_y \end{array} \right\}$$

$$(7.113a)$$

with

$$\gamma_{xy}^o = \frac{2(1 + v)}{EH}N_{xy}$$

and

$$\left\{ \begin{array}{c} \kappa_x^o \\ \kappa_y^o \end{array} \right\} = \frac{12}{EH^3} \left[\begin{array}{cc} 1 & -v \\ -v & 1 \end{array} \right] \left\{ \begin{array}{c} M_x \\ M_y \end{array} \right\}$$

$$(7.113b)$$

with

$$\kappa_{xy} = \frac{24(1 + v)}{EH^3}M_{xy}$$

Finally, as we have implied, there are no simplifications in the ABD matrix if the laminate is unsymmetric, if it is unbalanced, and if it is not a cross-ply. One must work with the full six-by-six matrix of equation (7.85). In equation (7.87), one must invert the full six-by-six relation.

7.4
REFERENCE SURFACE STRAINS AND CURVATURES OF EXAMPLE LAMINATES

We shall return to the laminates of our example problems one more time to verify an earlier claim. In Chapter 6, Section 6.7, in conjunction with equations (6.64) and (6.65), we stated that the proper interpretation of CLT Example 1 is that if a point (x, y) on the reference surface of a $[0/90]_S$ graphite-reinforced laminate is subjected to stress resultants

$$N_x = 50\,400 \text{ N/m}$$
$$N_y = 1809 \text{ N/m} \qquad (7.114)$$
$$N_{xy} = 0$$

and no moment resultants, then at that point the reference surface would deform in an extension-only fashion given by

$$\varepsilon_x^o(x, y) = 1000 \times 10^{-6} \qquad \kappa_x^o(x, y) = 0$$
$$\varepsilon_y^o(x, y) = 0 \qquad\qquad \kappa_y^o(x, y) = 0 \qquad (7.115)$$
$$\gamma_{xy}^o(x, y) = 0 \qquad\qquad \kappa_{xy}^o(x, y) = 0$$

With the introduction of the ABD matrix and its inverse, this can be verified directly. Specifically, because the laminate of CLT Example 1 is a symmetric cross-ply laminate with no moment resultants involved, equation (7.109) applies. We find that, using the numerical values of the a_{ij} for this laminate, equation (7.91a), equation (7.109a) yields

$$\varepsilon_x^o = 1000 \times 10^{-6}$$
$$\varepsilon_y^o = 0 \qquad (7.116)$$
$$\gamma_{xy}^o = 0$$

Because the moments are zero and the laminate is symmetric, equation (7.109b) indicates the curvatures are identically zero. We have thus closed the loop on this problem, ending with the conditions we started with.

Similarly, for CLT Example 2, we stated in conjunction with equations (6.82) and (6.83) that the proper interpretation of CLT Example 2 was that if the moment resultants at a point (x, y) on the reference surface of the $[0/90]_S$ graphite-reinforced laminate are prescribed as

$$M_x = 8.27 \text{ N·m/m}$$
$$M_y = 0.1809 \text{ N·m/m} \qquad (7.117)$$
$$M_{xy} = 0$$

then at that point the reference surface would deform in a curvature-only fashion given by

$$\varepsilon_x^o(x, y) = 0 \qquad \kappa_x^o(x, y) = 3.33 \text{ m}^{-1}$$
$$\varepsilon_y^o(x, y) = 0 \qquad \kappa_y^o(x, y) = 0 \qquad\qquad (7.118)$$
$$\gamma_{xy}^o(x, y) = 0 \qquad \kappa_{xy}^o(x, y) = 0$$

Equation (7.109a) indicates that because there are no force resultants involved, the reference surface strains are identically zero. Using the values of the d_{ij} for this laminate, equation (7.91c), we find that equation (7.109b) provides the curvatures as

$$\kappa_x^o = 3.33 \text{ m}^{-1}$$
$$\kappa_y^o = 0 \qquad\qquad (7.119)$$
$$\kappa_{xy}^o = 0$$

Finally, consider CLT Example 4, where we stated in equations (6.124) and (6.125) that if we prescribe the moment resultants at a point (x, y) on the reference surface of a $[\pm30/0]_S$ of a graphite-reinforced laminate as

$$M_x = 12.84 \text{ N·m/m}$$
$$M_y = 3.92 \text{ N·m/m} \qquad\qquad (7.120)$$
$$M_{xy} = 2.80 \text{ N·m/m}$$

then at that point the reference surface would deform in a curvature-only fashion given by

$$\varepsilon_x^o(x, y) = 0 \qquad \kappa_x^o(x, y) = 2.22 \text{ m}^{-1}$$
$$\varepsilon_y^o(x, y) = 0 \qquad \kappa_y^o(x, y) = 0 \qquad\qquad (7.121)$$
$$\gamma_{xy}^o(x, y) = 0 \qquad \kappa_{xy}^o(x, y) = 0$$

As the $[\pm30/0]_S$ laminate is symmetric and balanced, equation (7.105) applies, specifically equation (7.105b). Using the values of the d_{ij}, equation (7.92c), in equation (7.105b) results in

$$\kappa_x^o = 2.22 \text{ m}^{-1} \qquad \kappa_y^o = 0 = \kappa_{xy}^o \qquad\qquad (7.122)$$

Exercises for Section 7.4

1. Use the numerical values in CLT Example 3 and the various components of the inverse of the ABD matrix for the $[\pm30/0]_S$ laminate to verify that the stress resultants in equations (6.104(a) and (b)) result in the reference surface deformations given by equation (6.93).

2. Use the numerical values in CLT Example 5 for the various components of the inverse of the ABD matrix for the $[\pm30/0]_T$ laminate to verify that the stress resultants in equations (6.146) result in the reference surface deformations given by equation (6.138).

7.5
COMMENTS

Whether we invert the full six-by-six ABD matrix or take advantage of the simplifications just discussed, it is clear we are now in a position to start at any point in the loop of Figure 7.4 and calculate our way to any other point. Despite this fact, it is the loads that are usually known, and these are converted to force and moment resultants by dividing by the appropriate lengths. We use the laminate material properties (materials, fiber orientations, etc.) to compute the ABD matrix. We can then compute the reference surface strains and curvatures and we can determine the stress and strain response, in the x-y-z global coordinate system, or the 1-2-3 principal material coordinate system. An important alternative to this process is determining the laminate material properties that will produce a specific response to a known set of loads. This process is called *laminate design*.

Optimal laminate design goes one step further and seeks the laminate material properties that minimize, for example, laminate stresses in the presence of a given set of loads. Design and optimization are important facets of composite analysis, and in many instances they are the primary activities. However, they both depend on having a thorough understanding of the mechanics of composite materials, and all of the steps that go into closing the loop of Figure 7.4. Certain subtleties are involved in some of the steps of Figure 7.4, and one subtlety, the coupling effects that occur in laminates, is discussed next. These couplings are reflected in the various components of the ABD matrix and its inverse. Because the ABD matrix is a stiffness matrix, these couplings are properly referred to as *elastic couplings*. We have discussed two of these important couplings. One was the coupling that occurs in an unsymmetric laminate due to the presence of the B matrix. The other occurred in CLT Example 4, where, due to the presence of the D_{16} stiffness term, a twisting moment M_{xy} was required if the $[\pm 30/0]_S$ laminate was to have only curvature in the x direction—an M_{xy}-κ_x^o coupling. Other couplings that influence response are also present, and these are just as important. These couplings do not occur with metallic materials and hence are not fully appreciated by many who do not work with composites.

7.6
ELASTIC COUPLINGS

7.6.1 Influence of D_{16} and D_{26}

We have just revisited CLT Example 4 of Chapter 6, and equation (7.120) indicates that three components of moment are required to produce a deformation in a $[\pm 30/0]_S$ laminate consisting of just a simple curvature in the x direction. The fact that a twisting moment resultant M_{xy} is required is a unique characteristic of laminated fiber-reinforced materials. To study the twisting moment effect more closely, consider the following: The $[\pm 30/0]_S$ laminate is symmetric and balanced and thus equation (7.104*b*) applies. In particular, with the deformations of equation (7.121) and the numerical values for the D_{ij} from equation (7.89c), equation (7.104b) leads to the moment resultants previously computed, namely,

$$M_x = D_{11}\kappa_x^o = 5.78 \times 2.22 = 12.84 \text{ N·m/m}$$

$$M_y = D_{12}\kappa_x^o = 1.766 \times 2.22 = 3.92 \text{ N·m/m} \qquad (7.123)$$

$$M_{xy} = D_{16}\kappa_{xy}^o = 1.261 \times 2.22 = 2.80 \text{ N·m/m}$$

From equaton (7.123) we see that bending stiffness component D_{16} is directly responsible for the requirement that there be a twisting moment component M_{xy}. For this case, the sense and magnitude of this twisting moment for a 0.250 m × 0.125 m laminate were shown in Figure 6.35. For the cross-ply laminate of CLT Example 2 the component D_{16} is zero, and thus to deform this laminate as given by equation (7.121), no M_{xy} is needed, as in Figure 6.27. Hence, D_{16}, and likewise D_{26}, are responsible for the coupling of moments and deformations not normally associated with each other. An examination of our aluminum laminate would indicate that a twisting moment would not be required in this case either. (See the results of Exercise 1 in the Exercises for Section 6.7.)

As another example of the effects of D_{16}, consider the situation wherein at every point on the reference surface of a rectangular $[\pm30/0]_S$ laminate

$$M_x > 0 \text{ and all other stress resultants are zero} \qquad (7.124)$$

where it is assumed that M_x is the same at every point. This represents a symmetric balanced laminate subjected to a simple bending moment, and for this situation the reference surface deformations would be, from equation (7.105),

$$\varepsilon_x^o(x, y) = 0 \qquad \kappa_x^o(x, y) = d_{11}M_x = 0.330$$

$$\varepsilon_y^o(x, y) = 0 \qquad \kappa_y^o(x, y) = d_{12}M_x = -0.420 \qquad (7.125)$$

$$\gamma_{xy}^o(x, y) = 0 \qquad \kappa_{xy}^o(x, y) = d_{16}M_x = -0.1300 \text{ m}^{-1}$$

where we have used the numerical values from equation (7.92c) for the d_{ij} and, as an example, we considered a unit value of $M_x = 1$ N·m/m. If we assumed that every point on the reference surface has the deformations of equation (7.125), the deformed laminate would be as shown in Figure 7.5. Because the curvatures are

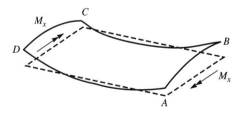

FIGURE 7.5
Bending and twisting deformations of $[\pm30/0]_S$ laminate due to bending moment M_x

constant over the entire reference surface, equation (6.14) can be integrated, as was done for equation (6.91), to provide the out-of-plane deflections. Here the situation is more complicated. Specifically, from the definitions of the curvatures,

$$\frac{\partial^2 w^o(x, y)}{\partial x^2} = -\kappa_x^o$$

$$\frac{\partial^2 w^o(x, y)}{\partial y^2} = -\kappa_y^o \qquad (7.126)$$

$$\frac{\partial^2 w^o(x, y)}{\partial x \partial y} = -\frac{1}{2}\kappa_{xy}^o$$

where it is understood that the right sides are not functions of x and y. Integrating the first two equations results in two different expressions for $w^o(x, y)$:

$$w^o(x, y) = -\frac{1}{2}\kappa_x^o x^2 + q(y)x + r(y)$$

$$w^o(x, y) = -\frac{1}{2}\kappa_y^o y^2 + s(x)y + t(x) \qquad (7.127)$$

where, as before, $q(y)$, $r(y)$, $s(x)$, and $t(x)$ are arbitrary functions of integration. The twisting curvature is given by two expressions

$$\frac{\partial^2 w^o(x, y)}{\partial x \partial y} = \frac{dq(y)}{dy} = -\frac{1}{2}\kappa_{xy}^o$$

$$\frac{\partial^2 w^o(x, y)}{\partial x \partial y} = \frac{ds(x)}{dx} = -\frac{1}{2}\kappa_{xy}^o \qquad (7.128)$$

Each of these equations can be integrated to yield

$$q(y) = -\frac{1}{2}\kappa_{xy}^o y + K_1$$

$$s(x) = -\frac{1}{2}\kappa_{xy}^o x + K_2 \qquad (7.129)$$

Using these, we find that the two expressions for $w^o(x, y)$ in equation (7.127) become

$$w^o(x, y) = -\frac{1}{2}\kappa_x^o x^2 - \frac{1}{2}\kappa_{xy}^o yx + K_1 x + r(y)$$

$$w^o(x, y) = -\frac{1}{2}\kappa_y^o y^2 - \frac{1}{2}\kappa_{xy}^o xy + K_2 y + t(x) \qquad (7.130)$$

leading to the conclusions that

$$r(y) = -\frac{1}{2}\kappa_y^o y^2 + K_2 y + K_3$$

$$t(x) = -\frac{1}{2}\kappa_x^o x^2 + K_1 x + K_3 \qquad (7.131)$$

As a result,

$$u^o(x, y) = 0$$

$$v^o(x, y) = 0 \tag{7.132}$$

$$w^o(x, y) = -\frac{1}{2}(\kappa_x^o x^2 + \kappa_y^o y^2 + \kappa_{xy}^o xy) + K_1 x + K_2 y + K_3$$

where the constants of integration associated with rigid body rotations and translations, K_1, K_2, and K_3, can be arbitrarily set to zero to give

$$w^o(x, y) = -\frac{1}{2}(d_{11}x^2 + d_{12}y^2 + d_{16}xy)M_x \tag{7.133a}$$

$$= -0.165x^2 + 0.210y^2 + 0.0650xy \tag{7.133b}$$

It must be reemphasized that equation (7.132) is valid only if κ_x^o, κ_y^o, and κ_{xy}^o are constant over the area of the laminate. The term $d_{11}M_x$ is responsible for the primary curvature in the x direction. The term $d_{12}M_x$ is responsible for the anticlastic curvature in the y direction. The third term, $d_{16}M_x$, is due to the twisting curvature κ_{xy}^o. For this situation with the $[\pm 30/0]_S$ laminate, the twisting curvature is manifest by the fact that in Figure 7.5 points A and C have a greater displacement in the $+z$ direction than do points B and D. If there was no twisting curvature, then the four points would have the same displacement in the z direction. As we see from the definition of d_{16}, equation (7.106b), the presence of D_{16} and D_{26} is ultimately responsible for the bending moment M_x causing a twisting curvature. Such an effect will not occur for cross-ply laminates, since D_{16} and D_{26} are zero for this class of laminates, resulting in $d_{16} = 0$.

As another example of the coupling due to the components of the D matrix, consider a $[\pm 30/0]_S$ laminate with the following deformations of the reference surface at a point:

$$\varepsilon_x^o(x, y) = 0 \qquad \kappa_x^o(x, y) = 0$$

$$\varepsilon_y^o(x, y) = 0 \qquad \kappa_y^o(x, y) = 0 \tag{7.134}$$

$$\gamma_{xy}^o(x, y) = 0 \qquad \kappa_{xy}^o(x, y) \neq 0 \text{ (say 2.22 m}^{-1})$$

This means that the point on the reference surface is experiencing only a twisting curvature; the displacement in the z direction is given by just the third term on the right hand side of equation (7.132). (See Exercise 5 in the Exercises for Section 6.8 for an example of this type of deformation.) From equation (7.104b),

$$M_x = D_{16}\kappa_{xy}^o = 1.261 \times 2.22 = 2.80 \text{ N·m/m}$$

$$M_y = D_{26}\kappa_{xy}^o = 0.418 \times 2.22 = 0.928 \text{ N·m/m} \tag{7.135}$$

$$M_{xy} = D_{66}\kappa_{xy}^o = 1.850 \times 2.22 = 4.11 \text{ N·m/m}$$

The presence of D_{16} and D_{26} dictates that to have only a twisting curvature, bending moments in both the x and y directions are required, in addition to the twisting moment. Figure 7.6 illustrates the deformed shape and required moment resultants if every point on the reference surface of a 0.250 m by 0.125 m laminate experiences this twisting curvature. If a laminate experiences only twisting curvature, then the edges of the laminate remain straight, simply rotating but not curving.

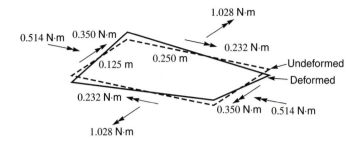

FIGURE 7.6
Moments required to deform $[\pm 30/0]_S$ laminate to have only twisting curvature $\kappa_{xy}^o = 2.22$ m^{-1}

What makes the influence of D_{16} and D_{26} even more interesting is the fact that the signs of these stiffnesses are sensitive to the signs of the fiber angles of the various layers. This sign sensitivity can be traced back to the sign sensitivity of $\bar{Q}_{16}$ and $\bar{Q}_{26}$ with fiber angle θ. The D matrix for the $[\pm 30/0]_S$ laminate is repeated here from earlier as

$$[D] = \begin{bmatrix} 5.78 & 1.766 & 1.261 \\ 1.766 & 1.256 & 0.418 \\ 1.261 & 0.418 & 1.850 \end{bmatrix} \text{N·m} \tag{7.136}$$

For a $(\mp 30/0)_S$ laminate, the D matrix is given by

$$[D] = \begin{bmatrix} 5.78 & 1.766 & -1.261 \\ 1.766 & 1.256 & -0.418 \\ -1.261 & -0.418 & 1.850 \end{bmatrix} \text{N·m} \tag{7.137}$$

Then to have a point on the reference surface of a $(\mp 30/0)_S$ laminate deformed in accordance with

$$\begin{aligned} \varepsilon_x^o(x, y) &= 0 & \kappa_x^o(x, y) &= 2.22 \text{ m}^{-1} \\ \varepsilon_y^o(x, y) &= 0 & \kappa_y^o(x, y) &= 0 \\ \gamma_{xy}^o(x, y) &= 0 & \kappa_{xy}^o(x, y) &= 0 \end{aligned} \tag{7.138}$$

a now familiar deformation state, requires

$$\begin{aligned} M_x &= D_{11}\kappa_x^o = 5.78 \times 2.22 = 12.84 \text{ N·m/m} \\ M_y &= D_{12}\kappa_x^o = 1.766 \times 2.22 = 3.92 \text{ N·m/m} \\ M_{xy} &= D_{16}\kappa_x^o = -1.261 \times 2.22 = -2.80 \text{ N·m/m} \end{aligned} \tag{7.139}$$

The value of M_{xy} is opposite in sign to the value for the $[\pm 30/0]_S$ case. Figure 7.7

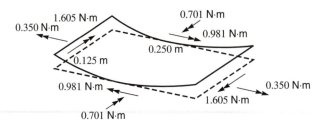

FIGURE 7.7
Moments required to produce state of deformation $\kappa_x^o = 2.22$ m^{-1} in $[\mp 30/0]_S$ laminate

shows the moments required on the edges of a 0.250 m × 0.125 m ($\mp 30/0)_S$ laminate if the deformations of equation (7.138) are valid for every point on the reference surface. This figure should be contrasted with Figure 6.35 for the $[\pm 30/0]_S$ case, and the senses of the twisting moments should be noted.

Consider next the situation where there is a unit M_x applied at every point on the reference surface of a $[\mp 30/0]_S$ laminate. For the $[\mp 30/0]_S$ laminate,

$$d_{11} = 0.330 \ 1/\text{N·m} \qquad d_{12} = -0.420 \ 1/\text{N·m} \qquad d_{16} = 0.1300 \ 1/\text{N·m}$$
$$d_{22} = 1.397 \ 1/\text{N·m} \qquad d_{26} = 0.0287 \ 1/\text{N·m} \qquad d_{66} = 0.636 \ 1/\text{N·m} \tag{7.140}$$

with the signs of D_{16} and D_{26} causing the signs of d_{16} and d_{26} to be opposite those for a $[\pm 30/0]_S$ laminate. The curvatures the $[\mp 30/0]_S$ laminate are, for a unit moment,

$$\kappa_x^o(x, y) = d_{11}M_x = 0.330 \text{ m}^{-1}$$
$$\kappa_y^o(x, y) = d_{12}M_x = -0.420 \text{ m}^{-1} \tag{7.141}$$
$$\kappa_{xy}^o(x, y) = d_{16}M_x = 0.1300 \text{ m}^{-1}$$

The sign of the twisting curvature is opposite the sign of the $[\pm 30/0]_S$ case. Figure 7.8 illustrates the deformations for this situation. The opposite sign of the twisting curvature is manifest in the fact that now points A and C have less displacement in the $+z$ direction than do points B and D. Comparison with Figure 7.5 contrasts these situations, and both figures illustrate the fact that due to D_{16} being nonzero, a simple bending moment results in a distorted reference surface shape.

We should mention that the influence of D_{16} and D_{26}, of course, depends on their magnitude relative to the primary bending stiffnesses, D_{11} and D_{22}. When D_{16} and D_{26} are small, the distortions of Figures 7.5 and 7.8 are not perceptible. When D_{16} and D_{26} are comparable to D_{12}, the anticlastic curvature term, then the distortions are noticeable and cannot be ignored. In the above examples, the magnitude of the twisting curvature is about one-third the magnitude of the anticlastic curvature effect, a noticable level.

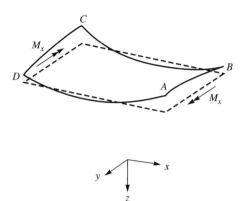

FIGURE 7.8
Bending and twisting deformations
of $[\mp 30/0]_S$ laminate due to bending
moment M_x

7.6.2 Influence of A_{16} and A_{26}

Two other important coupling terms in the laminate stiffness matrix are A_{16} and A_{26}. These two components produce effects in laminates much like the effects $\bar{Q}_{16}$ and $\bar{Q}_{26}$ produce in a single layer. Though the effects of $\bar{Q}_{16}$ and $\bar{Q}_{26}$ may be present in an individual layer in a laminate, if the laminate is balanced, we have seen that A_{16} and A_{26} are zero. In particular, consider the $[\pm 30/0]_S$ laminate. Although this laminate is balanced, there could be applications where the laminate is rotated slightly, or skewed, in its plane by an angle ϕ. Consider the case of $\phi = 10°$ such that the laminate is actually a $[+40/-20/+10]_S$ laminate. This laminate is not balanced and thus A_{16} and A_{26} are not zero. The A matrix for a $[\pm 30/0]_S$ graphite-reinforced laminate rotated 10° is

$$[A] = \begin{bmatrix} 99.8 & 18.94 & 7.37 \\ 18.94 & 18.85 & 7.37 \\ 7.37 & 7.37 & 20.2 \end{bmatrix} \times 10^6 \text{ N/m} \qquad (7.142)$$

If at a point on the reference surface the deformations are given by

$$\varepsilon_x^o(x, y) = 1000 \times 10^{-6} \qquad \kappa_x^o(x, y) = 0$$
$$\varepsilon_y^o(x, y) = 0 \qquad \kappa_y^o(x, y) = 0 \qquad (7.143)$$
$$\gamma_{xy}^o(x, y) = 0 \qquad \kappa_{xy}^o(x, y) = 0$$

then at that point the stress resultants required are, from equation (7.101a),

$$N_x = A_{11}\varepsilon_x^o = 99\,800 \text{ N/m}$$
$$N_y = A_{12}\varepsilon_x^o = 18\,940 \text{ N/m}$$
$$N_{xy} = A_{16}\varepsilon_x^o = 7370 \text{ N/m} \qquad (7.144)$$
$$M_x = M_y = M_{xy} = 0$$

The N_y is required to overcome the Poisson effects at this point on the reference surface; the laminate wants to contract in the y direction if loaded in the x direction. The N_{xy} is necessary to prevent shear strain, similar to the situation discussed in

Chapter 5 in connection with Figure 5.8(e) and (f). In that figure an element of fiber-reinforced material was subjected to just such an extension in the x direction. To effect this elongation, and have neither contraction in the y direction nor shear deformation, it was necessary to impose a tensile stress in the y direction *and* a shear stress on the element. The shear stress was necessary due to the influence of $\bar{Q}_{16}$. For the $[\pm 30/0]_S$ laminate rotated $10°$, the cumulative effect of the $\bar{Q}_{16}$ in the layers, which results in the nonzero value of A_{16}, leads to the requirement of having to have an N_{xy} present to restrain shear deformation. If every point on the reference surface of a 0.250 m by 0.125 m plate is to have the deformations given by equation (7.143), then the forces uniformly distributed along the edges of the plate as shown in Figure 7.9 are required. Even though the shear forces are the smallest of the forces, they are not negligible. For this situation, because no shearing or Poisson contraction is allowed, the plate simply gets longer, the width remains the same, and the right corner angles remain right. The deformed length of the plate is 250.25 m.

As we continue to study the deformations of the rotated $[\pm 30/0]_S$ laminate, consider a point on the reference surface to be loaded in tension with

$$N_x > 0 \qquad N_y = N_{xy} = M_x = M_y = M_{xy} = 0 \qquad (7.145)$$

At that point the reference surface strains are given by

$$\varepsilon_x^o = a_{11} N_x$$
$$\varepsilon_y^o = a_{12} N_x \qquad (7.146)$$
$$\gamma_{xy}^o = a_{16} N_x$$

where the a matrix is determined from equations (7.103a) and (7.142) as

$$\begin{bmatrix} a_{11} & a_{12} & a_{16} \\ a_{12} & a_{22} & a_{26} \\ a_{16} & a_{26} & a_{66} \end{bmatrix} = \begin{bmatrix} 12.38 & -12.46 & 0.0271 \\ -12.46 & 74.4 & -22.6 \\ 0.0271 & -22.6 & 57.8 \end{bmatrix} \times 10^{-9} \text{ N/m} \quad (7.147)$$

For this situation, after noting the signs of a_{11}, a_{12}, and a_{16}, we find that the point on the reference surface experiences a tensile strain in the x direction, a contraction

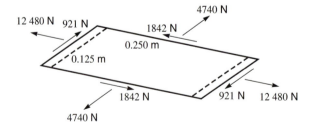

12 480 N   921 N   1842 N   4740 N   0.250 m   0.125 m   1842 N   921 N   12 480 N   4740 N

FIGURE 7.9
Forces required to produce $\varepsilon_x^o = 1000 \times 10^{-6}$ in a $[\pm 30/0]_S$ laminate rotated $10°$

strain in the y direction, and a positive shear strain. An examination of a_{16} in equation (7.103a) shows that it depends on A_{16} and A_{26}. The cumulative values of $\bar{Q}_{16}$ and $\bar{Q}_{26}$, which lead to nonzero values of A_{16} and A_{26}, are responsible for the tensile loading producing a shearing strain, behavior that is unique to fiber-reinforced materials. We also saw this behavior in Chapter 5 in connection with Figure 5.4(e) and (f) where an element of fiber-reinforced material was being subjected to a single tensile stress in the x direction. Such coupled behavior in laminates between tensile stresses and shear strains can be viewed as a favorable characteristic for tailoring the response of structures, tailoring that would not be possible with metallic materials. Using the numerical values of the a_{ij} and considering, for example, a value of $N_x = 10^5$ N/m, we find that the deformations at the point on the reference surface are given by

$$\varepsilon_x^o = 1238 \times 10^{-6}$$
$$\varepsilon_y^o = -1246 \times 10^{-6} \qquad (7.148)$$
$$\gamma_{xy}^o = 2.71 \times 10^{-6}$$

If the strains are the same at every point on the reference surface, then we can determine the displacement in a manner similar to deriving equation (6.72), although the present case is somewhat different. From equation (6.14)

$$\frac{\partial u^o(x, y)}{\partial x} = \varepsilon_x^o$$
$$\frac{\partial v^o(x, y)}{\partial y} = \varepsilon_y^o \qquad (7.149)$$
$$\frac{\partial u^o(x, y)}{\partial y} + \frac{\partial v^o(x, y)}{\partial x} = \gamma_{xy}^o$$

where it is understood the right sides are constants. Integrating the first two equations leads to

$$u^o(x, y) = \varepsilon_x^o x + g(y)$$
$$v^o(x, y) = \varepsilon_y^o y + h(x) \qquad (7.150)$$

which are then substituted into the third equation to give

$$\frac{dg(y)}{dy} + \frac{dh(x)}{dx} = \gamma_{xy}^o \qquad (7.151)$$

If a function of x and a function of y sum to a constant, then the two functions themselves must be constants, or

$$\frac{dg(y)}{dy} = C_1$$
$$\frac{dh(x)}{dx} = \gamma_{xy}^o - C_1 \qquad (7.152)$$

which integrate to

$$g(y) = C_1 y + C_2$$
$$h(x) = (\gamma_{xy}^o - C_1)x + C_3 \qquad (7.153)$$

The result is

$$u^o(x, y) = \varepsilon_x^o x + C_1 y + C_2$$
$$v^o(x, y) = \varepsilon_y^o y + (\gamma_{xy}^o - C_1)x + C_3 \qquad (7.154)$$
$$w^o(x, y) = 0$$

with C_1 being associated with rigid body rotation about the z axis, and C_2 and C_3 being associated with rigid body translations. Again, we stress that equation (7.154) is valid only if ε_x^o, ε_y^o, and γ_{xy}^o are constant over the area of the laminate. For the situation here

$$u^o(x, y) = a_{11} N_x x + C_1 y + C_2$$
$$v^o(x, y) = a_{12} N_x y + (a_{16} N_x - C_1)x + C_3 \qquad (7.155)$$

The deformations of a rectangular $[\pm 30/0]_S$ laminate rotated with $\phi = 10°$ and loaded such that equation (7.155) is valid everywhere on its reference surface are illustrated in Figure 7.10, where the figure, which is drawn somewhat differently than past similar figures in order to better illustrate the important features, is drawn with the conditions

$$C_2 = C_3 = 0$$
$$C_1 = \gamma_{xy}^o = a_{16} N_x \qquad (7.156)$$

This choice of the rigid body constants is completely arbitrary and is made strictly for the purpose of having Figure 7.10 be as simple as possible. Note the decrease in right angles at corners A and C of the laminate because γ_{xy}^o is positive.

As with the dual role of $\bar{Q}_{16}$ and $\bar{Q}_{26}$, A_{16} and A_{26} play a dual role. In addition to relating shear deformations to normal force resultants, A_{16} and A_{26} relate extensional deformations to shear force resultants. If a point on the reference surface of a $[\pm 30/0]_S$ laminate rotated $10°$ is loaded in shear such that

$$N_{xy} > 0 \qquad N_x = N_y = M_x = M_y = M_{xy} = 0 \qquad (7.157)$$

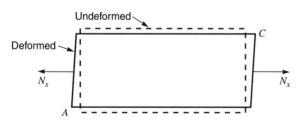

FIGURE 7.10
Deformations due to N_x of $[\pm 30/0]_S$ laminate rotated $10°$

then at that point the reference surface deformations are given by

$$\varepsilon_x^o = a_{16}N_{xy}$$
$$\varepsilon_y^o = a_{26}N_{xy} \qquad (7.158)$$
$$\gamma_{xy}^o = a_{66}N_{xy}$$

By examining the signs of a_{16} and a_{26}, we can see that through the cumulative influences of the $\bar{Q}_{16}$ and $\bar{Q}_{26}$, the positive shear force resultant causes, in addition to the expected shear strain, a slight extension strain in the x direction and a significant contraction strain in the y direction at the reference surface of the laminate. The ability to cause an extension or contraction strain with a shear resultant is a characteristic that can be used to advantage in structural applications. Considering the numerical values, and again assuming a value for $N_{xy} = 10^5$ N/m, we find that the reference surface strains at that point are given by

$$\varepsilon_x^o = 2.71 \times 10^{-6}$$
$$\varepsilon_y^o = -2260 \times 10^{-6} \qquad (7.159)$$
$$\gamma_{xy}^o = 5780 \times 10^{-6}$$

Figure 7.11 illustrates the deformations of a rectangular laminate experiencing these strains over the entire reference surface. Again the constants C_1-C_3 in equation (7.154) are chosen to make Figure 7.11 simple.

To study the sign sensitivity of A_{16} and A_{26}, consider the $[\pm 30/0]_S$ laminate rotated off axis by $\phi = -10°$. The laminate has a stacking arrangement of $[+20/-40/-10]_S$ and the A matrix is given by

$$[A] = \begin{bmatrix} 99.8 & 18.94 & -7.37 \\ 18.94 & 18.85 & -7.37 \\ -7.37 & -7.37 & 20.2 \end{bmatrix} \times 10^6 \text{ N/m} \qquad (7.160)$$

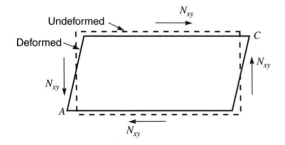

FIGURE 7.11
Deformation due to N_{xy} of $[\pm 30/0]_S$ laminate rotated $10°$

The difference between the $[\pm30/0]_S$ laminate's being rotated off axis by $+10°$ and its being rotated off axis by $-10°$ is in the A_{16} and A_{26} terms. By examining equation (7.142), we can see that these components of the A matrix are of opposite sign for the two laminate rotations. If at a point on the reference surface the deformations are given by

$$\varepsilon_x^o(x, y) = 1000 \times 10^{-6} \qquad \kappa_x^o(x, y) = 0$$
$$\varepsilon_y^o(x, y) = 0 \qquad \kappa_y^o(x, y) = 0 \qquad (7.161)$$
$$\gamma_{xy}^o(x, y) = 0 \qquad \kappa_{xy}^o(x, y) = 0$$

then the stress resultants required at that point are

$$N_x = A_{11}\varepsilon_x^o = 99\,800 \text{ N/m}$$
$$N_y = A_{12}\varepsilon_x^o = 18\,940 \text{ N/m}$$
$$\qquad (7.162)$$
$$N_{xy} = A_{16}\varepsilon_x^o = -7370 \text{ N/m}$$
$$M_x = M_y = M_{xy} = 0$$

If all points on the reference surface of a $0.250 \text{ m} \times 0.125 \text{ m}$ $[\pm30/0]_S$ laminate rotated $-10°$ are to have the above deformations, then, as in Figure 7.12, the forces required are identical to the case of the laminate rotated $+10°$, Figure 7.9, except the sign of the shear force resultant is reversed.

The a matrix of the $[\pm30/0]_S$ laminate rotated $-10°$ is

$$\begin{bmatrix} a_{11} & a_{12} & a_{16} \\ a_{12} & a_{22} & a_{26} \\ a_{16} & a_{26} & a_{66} \end{bmatrix} = \begin{bmatrix} 12.38 & -12.46 & -0.0271 \\ -12.46 & 74.4 & 22.6 \\ -0.0271 & 22.6 & 57.8 \end{bmatrix} \times 10^{-9} \text{ N/m} \qquad (7.163)$$

with the difference being in the a_{16} and a_{26} terms. Thus, a laminate loaded in tension in the x direction, again assuming a value of $N_x = 10^5 \text{ N/m}$, experiences the reference

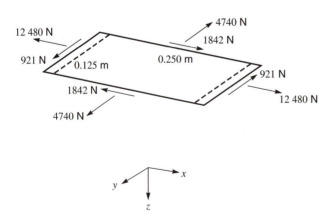

FIGURE 7.12
Forces required to produce $\varepsilon_x^o = 1000 \times 10^{-6}$ in a $[\pm30/0]_S$ laminate rotated $-10°$

surface deformations of

$$\varepsilon_x^o = 1238 \times 10^{-6}$$

$$\varepsilon_y^o = -1246 \times 10^{-6} \tag{7.164}$$

$$\gamma_{xy}^o = -2.71 \times 10^{-6}$$

A 0.250 m by 0.125 m laminate stretches in the x direction and contracts in the y direction, the same as if rotated $+10°$. However, the change in corner right angle is not the same; the change is opposite in sign. Figure 7.13 illustrates the deformed laminate; the right angle at corners A and C increases, in contrast to Figure 7.10 for the laminate rotated $+10°$.

Finally, consider the laminate rotated $-10°$ and loaded with a positive shear stress resultant $N_{xy} > 0$; the reference surface strains again are given by equation (7.158). By examining the signs of a_{16} and a_{26} in equation (7.163), we see clearly that the laminate now contracts in the x direction and extends in the y direction, as well as experiencing the expected shear strain. For a value of $N_{xy} = 10^5$ N/m,

$$\varepsilon_x^o = a_{16}N_{xy} = -2.71 \times 10^{-6}$$

$$\varepsilon_y^o = a_{26}N_{xy} = 2260 \times 10^{-6} \tag{7.165}$$

$$\gamma_{xy}^o = a_{66}N_{xy} = 5780 \times 10^{-6}$$

Figure 7.14 illustrates these deformations, with Figure 7.11 providing a contrast.

When studying the rotated laminate, we should keep in mind that the definition of the ABD matrix, in particular the A matrix, is valid at a point (x, y). A rectangular 0.250 m by 0.125 m laminate, with the ABD matrix and the strains and curvatures being the same at every point on the reference, is being used simply to dramatize the physical meaning of certain effects. The definition and numerical values of the components of the A matrix do not depend on the rectangular shape or the dimensions. In essence, the components of the ABD matrix are not influenced by any

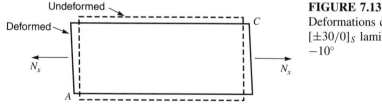

FIGURE 7.13
Deformations due to N_x of $[\pm 30/0]_S$ laminate rotated $-10°$

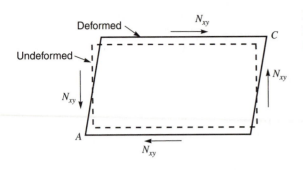

FIGURE 7.14
Deformations due to N_{xy} of $[\pm30/0]_S$ laminate rotated $-10°$

characteristic of the laminate except the specific stacking arrangement and specific material properties at point (x, y).

7.6.3 Influence of the B_{ij}

The elastic couplings due to the elements of the B matrix are very complex. For a general unsymmetric laminate loaded with all six stress resultants, the coupling effects are numerous. By examining equations (7.85) and (7.87), we see that each component of deformation requires the presence of all six stress resultants. Conversely, each stress resultant contributes to all six components of deformations. For a general unsymmetric laminate to deform in even a simple fashion with one component of deformation—for example, the now-familiar case of having only an extensional strain in the x direction—all three force resultants and all three moment resultants are required. For the case of only an x-direction extensional strain, equation (7.85) reduces to

$$N_x = A_{11}\varepsilon_x^o$$
$$N_y = A_{12}\varepsilon_x^o$$
$$N_{xy} = A_{16}\varepsilon_x^o$$
$$M_x = B_{11}\varepsilon_x^o \tag{7.166}$$
$$M_y = B_{12}\varepsilon_x^o$$
$$M_{xy} = B_{16}\varepsilon_x^o$$

The requirement for having moment resultants, both bending and twisting, to produce simple extension is what makes unsymmetric laminates unique. The $[\pm30/0]_T$ laminate of CLT Example 5 in Chapter 6 is an example of this situation. Using the values of the required components of the A and B matrices for this laminate from equation (7.90) and the value of $\varepsilon_x^o = 1000 \times 10^{-6}$, we find from the above equations that the stress resultants needed for the $[\pm 30/0]_T$ laminate to have only extension in the x direction are

$$
\begin{aligned}
N_x &= 51\,200 \text{ N/m} \\
N_y &= 9470 \text{ N/m} \\
N_{xy} &= 0 \\
M_x &= 1.416 \text{ N·m/m} \\
M_y &= -0.609 \text{ N·m/m} \\
M_{xy} &= -1.051 \text{ N·m/m}
\end{aligned}
\tag{7.167}
$$

These are the values computed in Chapter 6, equation (6.146), by integrating the stresses through the thickness of the three-layer laminate. The forces and moments required along the edges of a 0.250 m by 0.125 m $[\pm 30/0]_T$ laminate to produce $\varepsilon_x^o = 1000 \times 10^{-6}$ everywhere on its reference surface, all other deformations being zero, are illustrated in Figure 7.15. Because the laminate is balanced, no N_{xy} is required. Nonetheless, the necessity of having this many forces and moments to effect simple extension in the x direction certainly categorizes unsymmetric laminates as being unique.

Equally interesting is the fact that subjecting a general unsymmetric laminate to just a simple loading, such as having only N_x acting on the laminate, produces all three reference surface strains and all three reference surface curvatures. This can be seen by examining equation (7.87). For loading only in the x direction, the reference

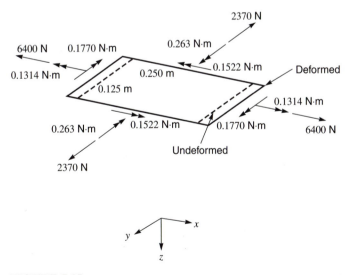

FIGURE 7.15
Forces and moments required to produce $\varepsilon_x^o = 1000 \times 10^{-6}$ in a $[\pm 30/0]_T$ laminate

surface deformations are given by

$$\varepsilon_x^o = a_{11} N_x$$
$$\varepsilon_y^o = a_{12} N_x$$
$$\gamma_{xy}^o = a_{16} N_x$$
$$\kappa_x^o = b_{11} N_x$$
$$\kappa_y^o = b_{12} N_x$$
$$\kappa_{xy}^o = b_{16} N_x$$

(7.168)

For the $[\pm 30/0]_T$ laminate, we find using the numerical values for the a_{ij} and b_{ij}, equation (7.93), and considering a unit load level, that

$$\varepsilon_x^o = 41.5 \times 10^{-9}$$
$$\varepsilon_y^o = -39.8 \times 10^{-9}$$
$$\gamma_{xy}^o = 12.19 \times 10^{-9}$$
$$\kappa_x^o = -164.6 \times 10^{-6} \text{ m}^{-1}$$
$$\kappa_y^o = 179.4 \times 10^{-6} \text{ m}^{-1}$$
$$\kappa_{xy}^o = 384 \times 10^{-6} \text{ m}^{-1}$$

(7.169)

It is difficult to envision the deformed shape of a rectangular laminate having equation (7.169) valid everywhere on its reference surface. However, the displacements can be written in terms of past results, namely, equations (7.132) and (7.154), specifically:

$$u^o(x, y) = \varepsilon_x^o x + C_1 y + C_2$$
$$v^o(x, y) = \varepsilon_y^o y + (\gamma_{xy}^o - C_1) x + C_3$$
$$w^o(x, y) = -\frac{1}{2}(\kappa_x^o x^2 + \kappa_y^o y^2 + \kappa_{xy}^o xy)$$
$$+ K_1 x + K_2 y + K_3$$

(7.170)

For this particular unsymmetric laminate and loading

$$u^o(x, y) = a_{11} N_x x + C_1 y + C_2$$
$$v_o(x, y) = a_{12} N_x y + (a_{16} N_x - C_1) x + C_3$$
$$w^o(x, y) = -\frac{1}{2}(b_{11} x^2 + b_{12} y^2 + b_{16} xy) N_x$$
$$+ K_1 x + K_2 y + K_3$$

(7.171)

The deformed laminate is longer in the x direction and narrower in the y direction, the right corner angles have changed, there is bending curvature in both the x and y directions, and there is twisting curvature. What is interesting is that the twisting curvature is larger than either of the two bending curvatures. This is due to the presence of the $\pm 30°$ layers.

As we can see by examining the details of the ABD matrix, *all* the components of the B matrix have dual roles. Every component B_{ij} can be thought of as coupling a moment resultant to an extensional strain, or coupling a force resultant to a curvature. This is a bending-stretching duality. In addition to the bending-stretching duality, with the components of the B matrix there is the level of duality that is also associated with the components of the A and D matrices. In particular, B_{16} couples the twisting curvature κ_{xy}^o with the normal force resultant N_x, and also the bending curvature κ_x^o with the shear force resultant N_{xy}. With this double level of duality, the components of the B matrix can produce interesting and complex effects. As a simple example of the bending-stretching duality, consider an unsymmetric cross-ply laminate $[0_2/90_2]_T$. First we will subject this laminate to a force resultant N_x and examine the resulting deformations, and then we will subject the laminate to a bending moment M_x and examine those deformations. In either case, equation (7.87) is applicable. For the $[0_2/90_2]_T$ laminate,

$$
\begin{bmatrix}
a_{11} & a_{12} & b_{11} & b_{12} \\
a_{12} & a_{22} & b_{21} & b_{22} \\
b_{11} & b_{21} & d_{11} & d_{12} \\
b_{12} & b_{22} & d_{12} & d_{22}
\end{bmatrix}
$$

$$
=
\begin{bmatrix}
44.1 \times 10^{-9} & -1.584 \times 10^{-9} & 188.5 \times 10^{-6} & 0 \\
-1.584 \times 10^{-9} & 44.1 \times 10^{-9} & 0 & -188.5 \times 10^{-6} \\
188.5 \times 10^{-6} & 0 & 1.470 & -0.0528 \\
0 & -188.5 \times 10^{-6} & -0.0528 & 1.470
\end{bmatrix}
$$

$$(7.172a)$$

$$
\begin{bmatrix}
a_{66} & b_{66} \\
b_{66} & d_{66}
\end{bmatrix}
=
\begin{bmatrix}
379 \times 10^{-9} & 0 \\
0 & 12.63
\end{bmatrix}
\tag{7.172b}
$$

$$a_{16} = a_{26} = b_{16} = b_{26} = b_{61} = b_{62} = d_{16} = d_{26} = 0 \tag{7.172c}$$

If a point on the reference surface of the $[0_2/90_2]_T$ laminate is subjected to a force resultant in the x direction, N_x, then at that point the reference surface deformations are given by

$$\varepsilon_x^o = a_{11} N_x$$

$$\varepsilon_y^o = a_{12} N_x$$

$$\gamma_{xy}^o = 0$$

$$\kappa_x^o = b_{11} N_x \tag{7.173}$$

$$\kappa_y^o = 0$$

$$\kappa_{xy}^o = 0$$

As we can see, the presence of b_{11} leads to a stretching-induced curvature in the x direction. There is no curvature in the y direction or any twisting curvature. Recall, this is not a situation where there is necessarily anticlastic curvature. If a stress resultant $N_x = 10^3$ N/m is considered, then the deformations are given by

$$\varepsilon_x^o = 44.1 \times 10^{-6}$$
$$\varepsilon_y^o = -1.584 \times 10^{-6}$$
$$\kappa_x^o = 0.1885 \text{ m}^{-1}$$
$$\gamma_{xy}^o = \kappa_y^o = \kappa_{xy}^o = 0$$

(7.174)

If every point on the reference surface has these deformations, then the out-of-plane shape of a laminate is given simply by

$$w(x, y)^o = -\frac{1}{2}\kappa_x^o x^2$$

(7.175)

A curvature of $\kappa_x^o = 0.1885$ m^{-1} means the reference surface of a 0.250 m by 0.125 m laminate would deform out of plane, as in Figure 7.16, such that the ends at $x = \pm 0.125$ m are deflected in the z direction an amount

$$w^o(\pm 0.125, y) = -\frac{1}{2}(0.1885 \text{ m}^{-1})(\pm 0.125)^2 = -1.473 \text{ mm} \quad (7.176)$$

This is a significant out-of-plane deflection for such a small load, specifically more than two times the laminate thickness. Bending-stretching effects are strong! This laminate represents an extreme level of asymmetry and it demonstrates the power of bending-stretching coupling.

Conversely, if a point on the reference surface of the $[0_2/90_2]_T$ laminate is subjected to just a unit bending moment in the x direction, then at that point reference surface deformations are given by

$$\varepsilon_x^o = b_{11}M_x = 188.5 \times 10^{-6}$$
$$\varepsilon_y^o = b_{21}M_x = 0$$
$$\gamma_{xy}^o = b_{61}M_x = 0$$
$$\kappa_x^o = d_{11}M_x = 1.470 \text{ m}^{-1}$$
$$\kappa_y^o = d_{12}M_x = -0.0528 \text{ m}^{-1}$$
$$\kappa_{xy}^o = d_{16}M_x = 0$$

(7.177)

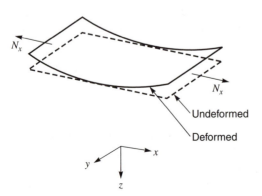

FIGURE 7.16
Deformations due to N_x of a $[0_2/90_2]_T$ laminate

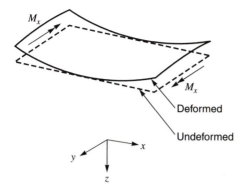

FIGURE 7.17
Deformations due to M_x of a $[0_2/90_2]_T$
laminate

where in this case the presence of b_{11} leads to bending-induced stretching in the x direction. There is no strain in the y direction but, as shown in Figure 7.17, because it is a bending problem there is an anticlastic curvature. There is no twisting curvature. If at every point on the reference of a rectangular laminate the curvatures are given by the above, then the out-of-plane deflection of the reference surface is given by

$$w^o(x, y) = -\frac{1}{2}(1.470x^2 - 0.0528y^2) \text{ m} \tag{7.178}$$

where rigid body translations and rotations in equation (7.170) have been suppressed, and the change in length of a 0.250 m by 0.125 m laminate is 0.0471 mm and there is no change of width.

This completes our discussion of the elastic couplings present in fiber-reinforced composite laminates. This discussion has centered on the deformations and stress resultants that occur because of these couplings. We have not addressed the stresses caused by these couplings. However, once we know the strains and curvatures at a point on the reference surface, we can compute the stresses. Because the couplings lead to unexpected and perhaps unusual strains and curvatures of the reference surface, the emphasis in this section has been on that aspect of the problem. However, the stresses should not be ignored. We will say more on this topic in later chapters.

Exercises for Section 7.6

1. A $[\pm45/0]_S$ graphite-reinforced laminate is to be used in a situation where the laminate may actually be rotated in plane by an amount ϕ. The lamination sequence is then $[45 + \phi/-45 + \phi/\phi]_S$. For $\phi = 15°$, for example, the lamination sequence is $[60/-30/15]_S$. The laminate is to be subjected to a simple bending moment M_x that is the same at every point on the reference surface. There is concern that the twisting curvature, κ_{xy}^o, will be excessive for certain ranges of ϕ. (a) Develop an expression for the ratio of d_{16} to d_{11}; d_{11} relates the primary x-direction bending curvature to the applied moment M_x, and d_{16} relates the twisting curvature to M_x. (b) Plot the ratio d_{16}/d_{11} versus ϕ, $-30° \le \phi \le +30°$. (c) Determine the range of ϕ such that the twisting curvature is less than 25 percent of the bending curvature. (d) Is it possible to have no twisting curvature?

2. Assume that the undeformed dimensions of the laminate in Figure 7.10 are 0.250 m in the x direction and 0.125 m in the y direction. (a) What are the changes in length and width

of the laminate? (b) What is the change in the corner right angle, in degrees? Assume $N_x = 10^5$ N/m and take note of equation (7.156).

3. Assume that the undeformed dimensions of the laminate in Figure 7.11 are 0.250 m in the x direction and 0.125 m in the y direction. (a) What are the changes in length and width of the laminate? (b) What is the change in the corner right angle, in degrees? Assume $N_{xy} = 10^5$ N/m.

4. Consider the unsymmetric $[0_2/90_2]_T$ laminate discussed at the end of this section. In particular, consider the case where a point on the reference surface of the laminate is subjected to just a force resultant in the x direction, namely, $N_x = 10^5$ N/m, equations (7.173) and (7.174). (a) Compute and plot the stresses $\sigma_x, \sigma_y, \tau_{xy}$ at this point as a function of z. (b) Note that although this is a problem where the laminate is subjected to only an inplane load, the stress distribution reflects the fact that bending-stretching coupling causes the stress distribution to appear as though the laminate is being subjected to a bending moment also. Oddly enough, there is a z location where $\sigma_x = 0$, that is, where there is no stress in the direction the laminate is being loaded. At what value of z is $\sigma_x = 0$?

5. Again consider the unsymmetric $[0_2/90_2]_T$ laminate discussed at the end of this section. In particular, consider the case where a point on the reference surface of the laminate is subjected to just a moment resultant in the x direction, namely, 10 N·m/m, equation (7.177). (a) Compute and plot the stresses $\sigma_x, \sigma_y, \tau_{xy}$ at this point as a function of z. (b) At the location $z = 0$ is there what might be referred to as a neutral plane, that is, a plane of zero stress that occurs in the bending of, say, an aluminum laminate? (c) At what value of z is $\sigma_x = 0$? (d) Is $\sigma_y = 0$ there also?

7.7
EFFECTIVE ENGINEERING PROPERTIES OF A LAMINATE

It is often convenient to have what can be referred to as *effective engineering properties* of a laminate. These consist of the effective extensional modulus in the x direction, the effective extensional modulus in the y direction, the effective Poisson's ratios, and the effective shear modulus in the x-y plane. These particular effective properties can be defined for general laminates but they make the most sense when considering the inplane loading of symmetric balanced laminates. Let us restrict our discussion to the inplane loading of symmetric balanced or symmetric cross-ply laminates. In this case the identical equations (7.105a) and (7.109a) are applicable. To introduce effective engineering properties, define the *average laminate stress* in the x direction, the *average laminate stress* in the y direction, and the *average laminate shear stress* to be, respectively,

$$\bar{\sigma}_x \equiv \frac{1}{H} \int_{-\frac{H}{2}}^{\frac{H}{2}} \sigma_x dz$$

$$\bar{\sigma}_y \equiv \frac{1}{H} \int_{-\frac{H}{2}}^{\frac{H}{2}} \sigma_y dz \qquad (7.179)$$

$$\bar{\tau}_{xy} \equiv \frac{1}{H} \int_{-\frac{H}{2}}^{\frac{H}{2}} \tau_{xy} dz$$

The average stresses, like all average quantities, do not really exist. They are simply a definition. In the definitions of the average stresses, the integrals are actually the definitions of the normal and shear force resultants, equation (7.1); that is:

$$\bar{\sigma}_x = \frac{1}{H} N_x$$

$$\bar{\sigma}_y = \frac{1}{H} N_y \qquad (7.180)$$

$$\bar{\tau}_{xy} = \frac{1}{H} N_{xy}$$

or

$$N_x = H\bar{\sigma}_x$$

$$N_y = H\bar{\sigma}_y \qquad (7.181)$$

$$N_{xy} = H\bar{\tau}_{xy}$$

with H being the thickness of the laminate. Knowing the force resultants and the laminate thickness, we can compute the average stresses easily. Substituting the above form into equation (7.105a), we find

$$
\left\{ \begin{array}{c} \varepsilon_x^o \\ \varepsilon_y^o \\ \gamma_{xy}^o \end{array} \right\} =
\left[\begin{array}{ccc} a_{11}H & a_{12}H & 0 \\ a_{12}H & a_{22}H & 0 \\ 0 & 0 & a_{66}H \end{array} \right]
\left\{ \begin{array}{c} \bar{\sigma}_x \\ \bar{\sigma}_y \\ \bar{\tau}_{xy} \end{array} \right\} \qquad (7.182)
$$

If we compare the form of this equation with the form of the compliance for a single layer in a state of plane stress, equation (4.4), it is possible to define the above three-by-three matrix as the *laminate compliance matrix*. Accordingly, by analogy with equation (4.5), the laminate's effective extensional modulus in the x direction, effective extensional modulus in the y direction, effective shear modulus, and two Poisson's ratios are given by

$$\bar{E}_x \equiv \frac{1}{a_{11}H}$$

$$\bar{E}_y \equiv \frac{1}{a_{22}H}$$

$$\bar{G}_{xy} \equiv \frac{1}{a_{66}H} \qquad (7.183)$$

$$\bar{\nu}_{xy} \equiv -\frac{a_{12}}{a_{11}}$$

$$\bar{\nu}_{yx} \equiv -\frac{a_{12}}{a_{22}}$$

In terms of the elements of the A matrix, from equation (7.106a),

$$\bar{E}_x \equiv \frac{A_{11}A_{22} - A_{12}^2}{A_{22}H}$$

$$\bar{E}_y \equiv \frac{A_{11}A_{22} - A_{12}^2}{A_{11}H}$$

$$\bar{G}_{xy} \equiv \frac{A_{66}}{H}$$ (7.184)

$$\bar{\nu}_{xy} \equiv \frac{A_{12}}{A_{22}}$$

$$\bar{\nu}_{yx} \equiv \frac{A_{12}}{A_{11}}$$

Of course, the reciprocity relation

$$\frac{\bar{\nu}_{xy}}{\bar{E}_x} = \frac{\bar{\nu}_{yx}}{\bar{E}_y}$$ (7.185)

is valid and so $\bar{\nu}_{yx}$ and $\bar{\nu}_{xy}$ are not independent.

As an example, for the four-layer $[0/90]_S$ laminate for which $H = 4 \times 150 \times 10^{-6} = 0.600$ mm, substituting into equation (7.184) from equation (7.88a), we find that the effective engineering properties are

$$\bar{E}_x = 83.8 \text{ GPa}$$

$$\bar{E}_y = 83.8 \text{ GPa}$$

$$\bar{G}_{xy} = 4.40 \text{ GPa}$$ (7.186)

$$\bar{\nu}_{xy} = 0.0359$$

$$\bar{\nu}_{yx} = 0.0359$$

For this particular laminate, the x and y directions have identical effective extensional moduli, and the effective shear modulus is the same as the shear modulus for a single layer, G_{12} (see Table 2.1). This is not surprising, considering that there are two layers with their fibers in the x direction and two with their fibers in the y direction, and that the two directions respond the same to an applied load. Furthermore, because both the $0°$ and the $90°$ layers have their fibers aligned with the global coordinate directions, the shear modulus for the laminate is identical to the shear modulus for a single layer.

Exercises for Section 7.7

1. Consider the $[\pm 30/0]_S$ graphite-reinforced laminate. (a) Compute the effective engineering properties $\bar{E}_x$, $\bar{E}_y$, $\bar{G}_{xy}$, $\bar{\nu}_{xy}$, and $\bar{\nu}_{yx}$. (b) Comment on the relative magnitudes of $\bar{E}_x$ and $\bar{E}_y$. (c) Why is the effective shear modulus greater than the value of G_{12} for a single layer? (d) What is the physical interpretation of $\bar{\nu}_{xy}$ such that its value is greater than 1?

You may want to complete the Computer Exercise before proceeding with Exercises 2, 3, and 4.

2. Consider a $[\pm\theta/0]_S$ graphite-reinforced laminate. For purposes of design, it is of interest to generate figures that show how $\bar{E}_x$, $\bar{E}_y$, $\bar{\nu}_{xy}$, $\bar{\nu}_{yx}$, and $\bar{G}_{xy}$ vary with θ. (a) Generate such design figures by computing these engineering properties as a function of θ, for $0° \leq \theta \leq 90°$. (b) Is there a range or value of θ for which $\bar{E}_x = \bar{E}_y$? (c) What about $\bar{\nu}_{xy}$ and $\bar{\nu}_{yx}$ for this range or value of θ? For what range is $\bar{\nu}_{xy} \geq 1$?

3. In Chapter 5 we defined coefficients of mutual influence for an element of fiber-reinforced material. Under conditions of load and deformation uniformity, these definitions can be thought of as applying to a single layer, as opposed to just a small element of material. Theoretically, like most of the definitions discussed, they are defined at a point. Two of those coefficients of mutual influence were defined as $\eta_{xy,x}$ and $\eta_{x,xy}$. Suppose those two effective coefficients of mutual influence for a laminate are given the notation $\bar{\eta}_{xy,x}$ and $\bar{\eta}_{x,xy}$. (a) Based on analogy with the coefficients of mutual influence for a single layer, state clearly the definition of these two coefficients of mutual influence for a laminate in terms of A_{ij}. Indicate with the definitions which stress resultants are zero and which are not zero, and what responses are being considered. (b) Compute and plot the two effective coefficients associated with a $[\pm30/0]_S$ laminate as this laminate is rotated by off-axis angle ϕ, $-10° < \phi < +10°$. (c) At what off-axis angles do the coefficients have maximum amplitude? (d) What is the physical interpretation of the dependency of the signs of $\bar{\eta}_{xy,x}$ and $\bar{\eta}_{x,xy}$ on ϕ?

4. Consider an eight-layer $[\pm45/0/90]_S$ graphite-reinforced laminate. Compute and plot the components of the A matrix of this laminate as it is rotated an angle ϕ in its plane. Consider the range $-\pi/2 \leq \phi \leq +\pi/2$. We refer to this laminate is as a quasi-isotropic laminate. We use the term *isotropic* because there is a single set of values for the elements of the A matrix, independent of ϕ. As a result, the engineering properties are independent of ϕ. The term *quasi* is used because the effective shear modulus, G_{xy}, is not related to the effective extensional modulus and effective Poisson's ratio by the relation

$$\bar{G}_{xy} = \frac{\bar{E}_x}{2(1 + \bar{\nu}_{xy})}$$

as they are for a truly isotropic material. The value of $\bar{G}_{xy}$ is *numerically* close to the right hand side of the above equation but the equality is simply not valid.

Computer Exercise

It will be useful to add a few lines to your computer program to compute and print the engineering properties as given by equation (7.184), specifically, from the Computer Exercise in the Exercises for Section 7.2: "7c. Compute and print the laminate engineering properties $\bar{E}_x$, $\bar{E}_y$, $\bar{G}_{xy}$, $\bar{\nu}_{xy}$, $\bar{\nu}_{yx}$." Then every time you use your program to solve a problem, the engineering properties will be printed. It would be best to print out a statement near the printed values of the engineering properties which indicates the definitions apply only to symmetric balanced or cross-ply laminates. It is possible to apply the definitions to general laminates; however, you must exercise care in using the numbers resulting from the definitions.

7.8
SUMMARY

Recall in Exercise 8 in the Exercises for Section 6.8 that the effective EA and effective EI of a $[0/90]_S$ beam 10 mm wide were computed to be

$$EA = 0.465 \times 10^6 \text{ N}$$
$$EI = 0.0244 \text{ N·m}^2 \tag{7.187}$$

The beam was 10 mm wide, and thus the effective EA and EI are actually per unit width, or per meter:

$$EA = 46.5 \times 10^6 \text{ N/m}$$
$$EI = 2.44 \text{ N·m}^2/\text{m} \tag{7.188}$$

For the $[0/90]_S$ laminate, the extensional and bending stiffnesses in the x direction are given by equation (7.88) as

$$A_{11} = 50.4 \times 10^6 \text{ N/m}$$
$$D_{11} = 2.48 \text{ N·m} \tag{7.189}$$

Hence, the rough strength-of-materials estimates of equation (7.188) for extensional and bending stiffnesses are close to the correct values. However, this good comparison must be viewed in two ways: (1) clearly fundamental notions such as effective EA and effective EI can be used with laminated composite materials; (2) only in certain simple cases can the fundamental notions be used with any accuracy. Crossply laminates are one example where they can be used with accuracy; equation (7.31) indicates the contribution of the 90° layers to A_{11} is negligible, and the details of equations (7.68) and (7.69) also show their contribution to D_{11} negligible. Thus, ignoring these layers and making strength-of-materials calculations will work. On the other hand, equation (7.40) for A_{11} of the $[\pm 30/0]_S$ shows no layer is negligible in that calculation. Also, what if the laminate was $[\pm 75/0]_S$? Would the contribution of the $\pm 75°$ layers be negligible? One has to work with laminated composites for some time to be able to make judgments. Once one is sure when approximations can be used, shortcuts are possible. However, with a readily available computer program that computes laminate properties and laminate response, resorting to approximations is not really necessary. When approximations can be and are indeed made, they provide insight into the mechanics of the problem—which should be viewed as the purpose of using approximations. However, they are not substitutions for the more rigorous and accurate approach.

7.9
SUGGESTED READINGS

The influences of A_{16} and A_{26} are discussed in:

1. Young, R. D., J. H. Starnes, Jr., and M. W. Hyer. "The Effects of Skewed Stiffeners and Anisotropic Skins on the Response of Compression-Loaded Composite Panels." *Proceedings of the Tenth DoD/NASA/FAA Conference on Fibrous Composites in Structural Design,*

1994, Report No. NAWCADWAR-94096-60, Naval Air Warfare Center, Warminster, PA, pp. II-109–II-123.

The implications of nonzero values of D_{16} and D_{26} for a problem of considerable importance are illustrated in:

2. Nemeth, M. P. "The Importance of Anisotropy on Buckling of Compression-Loaded Symmetric Composite Plates." *AIAA Journal* 24, no. 11 (1986), pp. 1831–35.

The following papers discuss the influence of terms in the *B* matrix:

3. Armanios, E. A., A. Makeev, and D. A. Hooke. "Finite-Displacement Analysis of Laminated Composite Strips with Extension-Twist Coupling." *Journal of Aerospace Engineering* (ASCE) 9, no. 3 (1996), pp. 80–91.
4. Dano, M.-L., and M. W. Hyer. "The Response of Unsymmetric Laminates to Simple Applied Forces." *Mechanics of Composite Materials and Structures* 3 (1996), pp. 65–80.

The following article discusses laminates that are specifically designed to show no coupling:

5. Bartholomew, P. "Ply Stacking Sequences for Laminated Plates Having In-Plane and Bending Orthotropy." *Fiber Science and Technology* 10 (1977), pp. 237–53.

CHAPTER 8

Classical Lamination Theory: Additional Examples

In the last two chapters we probed deeply into the assumptions and ramifications of classical lamination theory, examined implications of the theory, and reached a fundamental understanding of the response of fiber-reinforced composite laminates. In this chapter we use additional applications and discussions to provide deeper insight into the theory and to further our understanding of the response of fiber-reinforced composite laminates. The examples we will present are related to those in Chapter 6, namely, CLT Examples 1–5. However, the examples here are fundamentally different. In CLT Examples 1–5 the reference surface strains and curvatures were specified and the force resultants necessary to produce these deformations were computed; these resultants, of course, were due to the stresses. In the examples to follow, the force resultants will be specified, and the reference surface strains and curvatures, and resulting stresses, will be computed. Specifying the force resultants and computing the resulting deformations are not the same as specifying the deformations and computing the necessary force resultants. They are completely different problem statements.

Before proceeding with the force resultant specification examples, we must stress one point: In our discussions of classical lamination theory we have always specified that "if at *a point* on the reference surface" certain conditions hold, then such and such a response will occur *at that point*. Alternatively, we have specified that "if at *every point* on the reference surface of a laminate" certain conditions hold, then such and such a response will occur *for the entire laminate*. We have been emphatic about using this terminology because classical lamination theory focuses on what is happening at a point on the reference surface. The dependence of the laminate response on the z coordinate above and below this point on the reference surface has been removed from the problem. The dependence of response on the z direction is determined by the Kirchhoff hypothesis and by definitions. In this context, there were three key steps to the formulation of classical lamination theory that eliminated z dependence. First, by the Kirchhoff hypothesis the displacements and strains at any point through the thickness of the laminate were assumed to depend linearly on z; thus, it was only necessary to have an understanding of what is happening at the reference surface. Second, by the definition of the stress resultants, the stresses have been integrated through the thickness, eliminating z from that portion of the problem. Because the resultants are integrals with respect to the variable z, and since $z = 0$ at the reference surface, the resultants are the effective forces and moments that act at the reference surface. Finally, the laminate elastic properties have been defined by the A, B, and D matrices as integrals through the thickness of the laminate of the individual layer elastic properties. The z locations of the individual layers are embedded in the definitions of the matrices. Because the A, B, and D matrices relate the reference surface deformations to the stress resultants that act at the reference surface, they are easily interpreted as layer stiffness effects lumped at the reference surface.

Because of these assumptions and definitions, classical lamination theory clearly defines how laminate response depends on the variation of layer elastic properties and layer thicknesses in the z direction, and how displacements, strains, and stresses vary in the z direction. However, classical lamination theory does not define the manner in which laminate response varies in the x and y directions. We must rely on the laws of equilibrium to study how the response varies in these directions; we will do so in Chapter 13. For now we must be content to study the response at a particular point on the reference surface, or accept the statement that, for the purposes of the example, the response is the same at every point on the reference surface.

8.1
CLT EXAMPLE 6: $[0/90]_S$ LAMINATE SUBJECTED TO KNOWN N_x — COUNTERPART TO CLT EXAMPLE 1

To continue with our study of classical lamination theory, consider a problem that is the counterpart to CLT Example 1 of Chapter 6. In CLT Example 1 we studied the stresses and strains through the thickness of a $[0/90]_S$ laminate that had the following strains and curvatures at a point on the reference surface:

$$\varepsilon_x^o(x, y) = 1000 \times 10^{-6} \qquad \kappa_x^o(x, y) = 0$$
$$\varepsilon_y^o(x, y) = 0 \qquad \kappa_y^o(x, y) = 0 \qquad (8.1)$$
$$\gamma_{xy}^o(x, y) = 0 \qquad \kappa_{xy}^o(x, y) = 0$$

To effect this deformation, the following stress resultants were required at that point:

$$N_x = 50\,400 \text{ N/m} \qquad N_y = 1809 \text{ N/m}$$

and
$$N_{xy} = M_x = M_y = M_{xy} = 0$$

$$(8.2)$$

These results led to the interpretation that if every point on a 0.250 m long by 0.125 m wide $[0/90]_S$ laminate were to experience the deformations of equation (8.1), then the system of edge forces, uniformly distributed along the edges, shown in Figure 6.26, would be required. The new length of the laminate would be 250.25 mm, with the width remaining the same and the corner right angles remaining right. In CLT

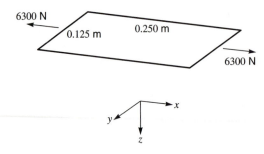

FIGURE 8.1
$[0/90]_S$ laminate subjected to specified force in x direction

Example 1 the deformations were specified and, through classical lamination theory, the force resultants were computed. However, in practice, the converse problem of specifying the force resultants and wanting to know the strains, curvatures, and stresses is as important, if not more so. Generally loads are known and it is of interest to determine deformations and stress levels. In that regard, consider, as in Figure 8.1, a 0.250 m by 0.125 m laminate subjected to a load in the x direction, at each end, of 6300 N. If the load is uniformly distributed along the 0.125 m edge, then by force equilibrium arguments it can be shown that for every point on the reference surface

$$N_x = 50\,400 \text{ N/m}$$

and
$$N_y = N_{xy} = M_x = M_y = M_{xy} = 0 \tag{8.3}$$

This loading is identical to the situation described in equations (8.1) and (8.2) except for the lack of N_y. In the present example we are specifying the *loading* in the x direction, while in CLT Example 1 we specified the *deformation* in the x direction. This is an important distinction; they are two different problems that only *appear* similar. Here we are purposely studying this counterpart to CLT Example 1, and later the counterparts to CLT Examples 3 and 4, to illustrate similarities and differences between problems that, at least on the surface, appear the same.

Using the values of the a_{ij} for this laminate from equation (7.91a), we find

$$\varepsilon_x^o = a_{11}N_x = 1001 \times 10^{-6}$$
$$\varepsilon_y^o = a_{12}N_x = -36.0 \times 10^{-6} \tag{8.4}$$
$$\gamma_{xy}^o = a_{16}N_{xy} = 0$$

Obviously the curvatures are zero. This state of deformation given by equation (8.4) is valid at every point on the 0.250 m by 0.125 m reference surface. Note that the elongation in the x direction is very close to the value in CLT Example 1, but in the present example there is a contraction strain in the y direction; the laminate is contracting 0.004 50 mm. Since no N_y is present, the laminate can contract in the y direction, through Poisson effects, and can elongate slightly more in the x direction. The solid lines in Figure 8.2 show the distributions of these strains through the thickness, where the strains from the original example, equation (8.1), are shown for comparison as dashed lines. For ε_x, the dashed line is indistinguishable from the solid line due to the similarity of ε_x for these two examples. Using the stress-strain

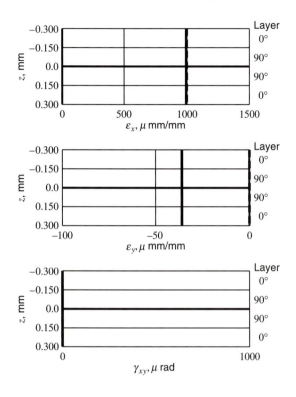

FIGURE 8.2
Strain distribution through
the thickness of $[0/90]_S$
laminate subjected to
$N_x = 50\ 400$ N/m

relations, for the 0° layers the stresses are

$$\sigma_x = 155.8 \text{ MPa}$$

$$\sigma_y = 2.58 \text{ MPa} \tag{8.5a}$$

$$\tau_{xy} = 0$$

while for the two 90° layers the stresses are

$$\sigma_x = 12.07 \text{ MPa}$$

$$\sigma_y = -2.58 \text{ MPa} \tag{8.5b}$$

$$\tau_{xy} = 0$$

These are similar to the stresses for Example 1, equations (6.28) and (6.29), except that here the values of σ_y for the two layers are equal in magnitude but opposite in sign. The opposite sign occurs because here the value of N_y is specified to be zero and the integral of σ_y through the thickness (i.e., the definition of N_y) must be exactly zero. Figure 8.3 illustrates these stress distributions through the thickness for CLT Example 1 and the current example. The distributions of σ_x for the two examples are, to within the scale of the figure, indistinguishable; however, the distribution of σ_y for the current problem reflects the equal and opposite character required for

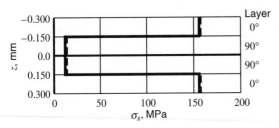

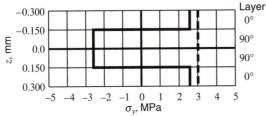

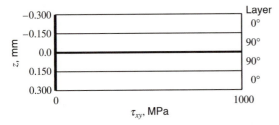

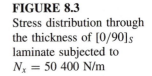

FIGURE 8.3
Stress distribution through the thickness of $[0/90]_s$ laminate subjected to $N_x = 50\,400$ N/m

———— Resultant specified   ▬ ▬ ▬ Displacement specified

N_y to be zero. Like CLT Example 1, there are no shear stresses in this problem. By inspection, the principal material system stresses are, for the 0° and 90° layers, respectively,

$$\sigma_1 = 155.8 \text{ MPa}$$
$$\sigma_2 = 2.58 \text{ MPa} \tag{8.6a}$$
$$\tau_2 = 0$$

and

$$\sigma_1 = -2.58 \text{ MPa}$$
$$\sigma_2 = 12.07 \text{ MPa} \tag{8.6b}$$
$$\tau_2 = 0$$

Again by inspection, the principal material system strains are, for the 0° and 90° layers, respectively,

$$\varepsilon_1 = 1001 \times 10^{-6}$$
$$\varepsilon_2 = -36.0 \times 10^{-6} \tag{8.7a}$$
$$\gamma_{12} = 0$$

and

$$\varepsilon_1 = -36.0 \times 10^{-6}$$

$$\varepsilon_2 = 1001 \times 10^{-6} \qquad (8.7b)$$

$$\gamma_{12} = 0$$

These distributions, illustrated in Figures 8.4 and 8.5, are very similar to the distributions of CLT Example 1. Despite the overall minimal differences, these two problems are distinctly different. The fact that the laminate is of cross-ply construction is partly responsible for the similar results. The strain and load levels chosen also contribute to the similarity of results. From the figures, it can be concluded that the response of a $[0/90]_S$ laminate subjected to the condition

$$N_x = 50\ 400 \text{ N/m} \qquad (8.8)$$

is very similar to the response of that laminate subjected to the condition

$$\varepsilon_x^o = 1000 \times 10^{-6} \qquad (8.9)$$

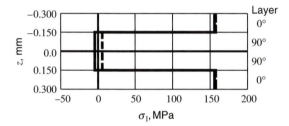

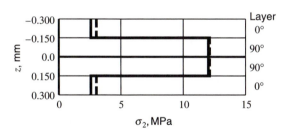

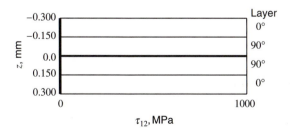

FIGURE 8.4
Principal material system stress distribution through the thickness of $[0/90]_S$ laminate subjected to $N_x = 50\ 400$ N/m

—— Resultant specified   — — — Displacement specified

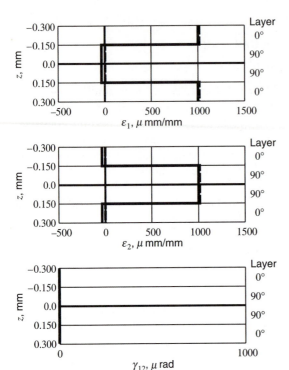

FIGURE 8.5
Principal material system strain distribution through the thickness of $[0/90]_S$ laminate subjected to $N_x = 50\,400$ N/m

——— Resultant specified ▬ ▬ ▬ Displacement specified

8.2
CLT EXAMPLE 7: $[\pm 30/0]_S$ LAMINATE SUBJECTED TO KNOWN N_x — COUNTERPART TO CLT EXAMPLE 3

CLT Example 3 considered the situation where a point on the reference surface of $[\pm30/0]_S$ laminate was subjected to the deformations given by equation (8.1). We concluded that to effect this deformation it was necessary to subject this point to the load conditions

$$N_x = 0.1024 \text{ MN/m} \qquad N_y = 0.01894 \text{ MN/m}$$

and $$N_{xy} = M_x = M_y = M_{xy} = 0$$

(8.10)

The alternative interpretation to that problem was that if every point on the reference surface of a 0.250 m by 0.125 m $[\pm30/0]_S$ laminate was to have the deformations of equation (8.1), then the system of uniformly distributed edge forces shown in Figure 6.30 would be necessary. As a related problem, consider a 0.250 m by 0.125 m laminate subjected to a loading in the x direction of 12 800 N. If the load is uniformly distributed along the ends, then by equilibrium arguments it can be shown that everywhere on the reference surface of the laminate the stress resultants are

$$N_x = 0.1024 \text{ MN/m}$$

and
$$N_y = N_{xy} = M_x = M_y = M_{xy} = 0$$
(8.11)

Using the values of the a_{ij} for this $[\pm 30/0]_S$ laminate from equation (7.92a), we find that the deformations at every point on the reference surface are given by

$$\varepsilon_x^o = a_{11}N_x = 1275 \times 10^{-6}$$

$$\varepsilon_y^o = a_{12}N_x = -1486 \times 10^{-6}$$
(8.12)

$$\gamma_{xy}^o = a_{16}N_x = 0$$

Figure 8.6 illustrates the distribution of these strains and the strains for CLT Example 3. Note that for this situation the laminate elongates about 30 percent more in the x direction than in CLT Example 3 even though the values of N_x are identical for the two problems. What is very important to note is that the contraction in the y direction is large. In fact, it is larger than the elongation in the x direction. This implies that the effective Poisson's ratio for the laminate $\bar{\nu}_{xy}$, from equation (7.184), is greater than unity! Indeed, for this laminate $\bar{\nu}_{xy} = 1.166$. The seemingly subtle difference between specifying the loading in the x direction that resulted from the specified

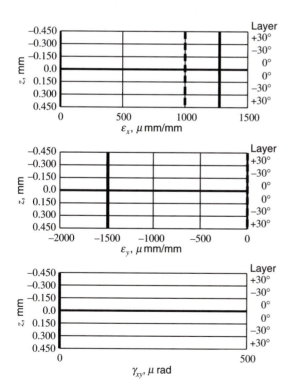

FIGURE 8.6
Strain distribution through the thickness of $[\pm 30/0]_S$ laminate subjected to $N_x = 0.1024$ MN/m

deformation in the x direction, and specifying the deformation in the x direction leads to distinctly different responses, particularly in the y direction. Using the stress-strain relations, we find that the stresses are, for the 30° layers:

$$\sigma_x = 73.6 \text{ MPa}$$

$$\sigma_y = 7.11 \tag{8.13a}$$

$$\tau_{xy} = 36.6$$

and for the −30° layers:

$$\sigma_x = 73.6 \text{ MPa}$$

$$\sigma_y = 7.11 \tag{8.13b}$$

$$\tau_{xy} = -36.6$$

while for the 0° layers,

$$\sigma_x = 194.1 \text{ MPa}$$

$$\sigma_y = -14.23 \tag{8.13c}$$

$$\tau_{xy} = 0$$

These are different from the stresses for CLT Example 3, equations (6.98), (6.100), and (6.103), particularly the σ_y stresses. In the +30° and −30° layers the σ_y stresses are considerably less for the current example, while for the 0° layers the sign of σ_y is different, as well as the magnitude. Figure 8.7 illustrates the through-thickness distributions of these stresses for the two examples. It is clear that the constraint $\varepsilon_y^o = 0$ for CLT Example 3 leads to larger stresses in the y direction. When unconstrained, the stresses are lower and, of course, $N_y = 0$. The principal material system stresses are as follows:

For the 30° layers:

$$\sigma_1 = 88.7 \text{ MPa}$$

$$\sigma_2 = -7.92 \tag{8.14a}$$

$$\tau_{12} = -10.52$$

For the −30° layers:

$$\sigma_1 = 88.7 \text{ MPa}$$

$$\sigma_2 = -7.92 \tag{8.14b}$$

$$\tau_{12} = 10.52$$

For the 0° layers:

$$\sigma_1 = 194.1 \text{ MPa}$$

$$\sigma_2 = -14.23 \tag{8.14c}$$

$$\tau_{12} = 0$$

The principal material system strains are,

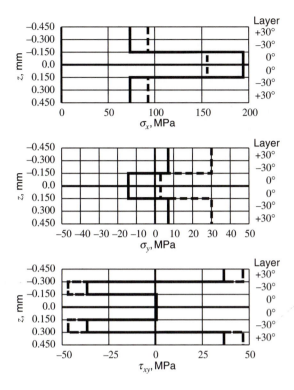

FIGURE 8.7
Stress distribution through the thickness of $[\pm30/0]_S$ laminate subjected to $N_x = 0.1024$ MN/m

—— Resultant specified — — — Displacement specified

For the $+30°$ layers:

$$\varepsilon_1 = 585 \times 10^{-6}$$
$$\varepsilon_2 = -796 \times 10^{-6} \qquad (8.15a)$$
$$\gamma_{12} = -2390 \times 10^{-6}$$

For the $-30°$ layers:

$$\varepsilon_1 = 585 \times 10^{-6}$$
$$\varepsilon_2 = -796 \times 10^{-6} \qquad (8.15b)$$
$$\gamma_{12} = 2390 \times 10^{-6}$$

For the $0°$ layers:

$$\varepsilon_1 = 1275 \times 10^{-6}$$
$$\varepsilon_2 = -1486 \times 10^{-6} \qquad (8.15c)$$
$$\gamma_{12} = 0$$

Figures 8.8 and 8.9 illustrate the through-thickness distributions for the principal material system stresses and strains for both examples.

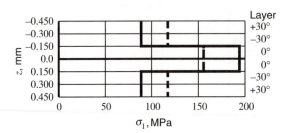

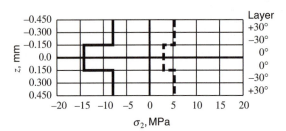

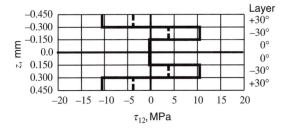

Resultant specified ▬ ▬ ▬ Displacement specified

FIGURE 8.8

Principal material system stress distribution through the thickness of $[\pm 30/0]_S$ laminate subjected to $N_x = 0.1024$ MN/m

We must note a difference in this example. When the loading is specified, the stresses in the 0° layers are higher than if the deformations are specified, that is, 191.4 MPa versus 155.7 MPa. With the load specified, the distribution of the stresses in the layers is governed solely by the directions of the fibers in the layers; the layers with fibers aligned with the load bear most of the load. If the laminate is artificially constrained (in this case in the y direction), then Poisson effects and other aspects of the constraints govern the distribution of the stresses in the individual layers. In summary, we can say that the response of a $[\pm 30/0]_S$ laminate subjected to the condition

$$N_x = 0.1024 \text{ MN/m} \tag{8.16}$$

is quite different from the response of that laminate to the condition

$$\varepsilon_x^o = 1000 \times 10^{-6} \tag{8.17}$$

The conclusions for this laminate are more the rule than the conclusions for the $[0/90]_S$ laminate.

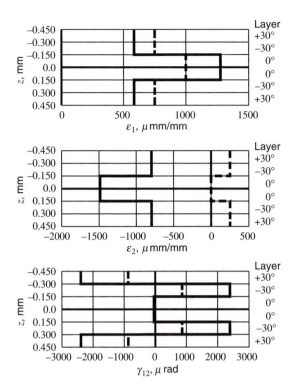

FIGURE 8.9

Principal material system strain distribution through the thickness of $[\pm30/0]_S$ laminate subjected to $N_x = 0.1024$ MN/m

——— Resultant specified — — — Displacement specified

8.3
CLT EXAMPLE 8: $[\pm30/0]_S$ LAMINATE SUBJECTED TO KNOWN M_x — COUNTERPART TO CLT EXAMPLE 4

In CLT Example 4 we specified that a point on the reference surface of a $[\pm30/0]_S$ laminate was subjected to the deformations

$$\varepsilon_x^o(x, y) = 0 \qquad \kappa_x^o(x, y) = 2.22 \text{ m}^{-1}$$
$$\varepsilon_y^o(x, y) = 0 \qquad \kappa_y^o(x, y) = 0 \qquad (8.18)$$
$$\gamma_{xy}^o(x, y) = 0 \qquad \kappa_{xy}^o(x, y) = 0$$

which led to the conclusion that the stress resultants

$$M_x = 12.84 \text{ N·m/m} \qquad M_y = 3.92 \text{ N·m/m} \qquad M_{xy} = 2.80 \text{ N·m/m}$$
$$(8.19)$$
and
$$N_x = N_y = N_{xy} = 0$$

were required at this point. In Chapter 7 we discussed the fact that D_{16} is responsible for the requirement of having a twisting moment present to produce just a simple bending curvature. Of course, M_y is necessary to resist anticlastic curvature effects.

Figure 6.35 provided an interpretation of the edge moments required for a rectangular 0.250 m by 0.125 m laminate to have these deformations everywhere on its reference surface. In Chapter 7 we introduced the counter to this problem in Figure 7.5, namely, subjecting the $[\pm 30/0]_S$ laminate to a specified bending moment in the x direction rather than a specified curvature. From that discussion it was clear that the application of a simple bending moment results in twisting curvature, as well as bending and anticlastic curvatures. The twisting curvature was again attributed to the D_{16} and D_{26} components of the D matrix. Let us revisit this counter problem and study the response of a 0.250 m by 0.125 m laminate with a bending moment of 1.605 N·m applied uniformly along the 0.125-m edge. By equilibrium arguments it can be shown that at every point on the reference surface of the laminate

$$M_x = 12.84 \text{ N·m/m}$$

and $$M_y = M_{xy} = N_x = N_y = N_{xy} = 0 \qquad (8.20)$$

This moment resultant is the value of the moment resultant required to produce $\kappa_x^o = 2.22 \text{ m}^{-1}$ in CLT Example 4. Missing is the value of bending moment required to resist the anticlastic curvature, M_y, and the twisting moment, M_{xy}; these are all specified in equation (8.19). As before, the d_{ij} can be used to compute the curvatures produced by the application of the simple loading of equation (8.20). These curvatures are, from equation (7.92c),

$$\kappa_x^o = d_{11} M_x = 4.24 \text{ m}^{-1}$$
$$\kappa_y^o = d_{12} M_x = -5.40 \text{ m}^{-1} \qquad (8.21)$$
$$\kappa_{xy}^o = d_{16} M_x = -1.669 \text{ m}^{-1}$$

Note that the curvature in the x direction is larger than the value of equation (8.18). More importantly, we observe that the anticlastic curvature is *larger* than the bending curvature, and the twisting curvature is substantial. The large anticlastic curvature is in keeping with the large Poisson contractions discussed in the previous section when this laminate was subjected to a load N_x. The bending curvature in the x direction is larger because the presence of M_y in CLT Example 4, acting through Poisson effects in the form of bending coefficient d_{12}, reduces curvature in the x direction. The presence of a twisting moment in CLT Example 4 also contributes to this effect through bending coefficient d_{26}. Of course, for the present case,

$$\varepsilon_x^o = \varepsilon_y^o = \gamma_{xy}^o = 0 \qquad (8.22)$$

Figure 7.5 shows the deformed shape of this laminate. Because the curvature is the same at every point of the reference surface, equation (7.132) gives the deformed shape of the reference surface,

$$w^o(x, y) = -\frac{1}{2}\left(4.24x^2 - 5.40y^2 - 1.669xy\right) \qquad (8.23)$$

As a result of equation (8.21), and using equation (6.13), we find that the strains through the thickness of the laminate are given by

$$\varepsilon_x = 4.24z$$

$$\varepsilon_y = -5.40z \tag{8.24}$$

$$\gamma_{xy} = -1.669z$$

These are illustrated in Figure 8.10, where they are compared with the strains from CLT Example 4. The stresses are determined by using the stress-strain relations; specifically:

For the $+30°$ layers:

$$\left\{ \begin{array}{c} \sigma_x \\ \sigma_y \\ \tau_{xy} \end{array} \right\} = \left[\begin{array}{ccc} \bar{Q}_{11}(30°) & \bar{Q}_{12}(30°) & \bar{Q}_{16}(30°) \\ \bar{Q}_{12}(30°) & \bar{Q}_{22}(30°) & \bar{Q}_{26}(30°) \\ \bar{Q}_{16}(30°) & \bar{Q}_{26}(30°) & \bar{Q}_{66}(30°) \end{array} \right] \left\{ \begin{array}{c} 4.24z \\ -5.40z \\ -1.669z \end{array} \right\} \tag{8.25a}$$

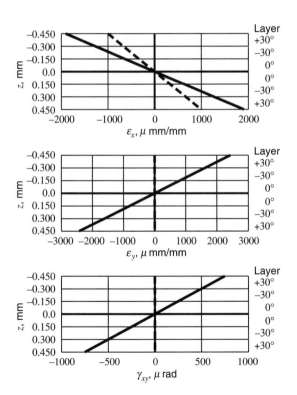

FIGURE 8.10
Strain distribution through the thickness of $[\pm 30/0]_S$ laminate subjected to $M_x = 12.84$ N·m/m

For the $-30°$ layers:

$$\left\{\begin{array}{c} \sigma_x \\ \sigma_y \\ \tau_{xy} \end{array}\right\} = \left[\begin{array}{ccc} \bar{Q}_{11}(-30°) & \bar{Q}_{12}(-30°) & \bar{Q}_{16}(-30°) \\ \bar{Q}_{12}(-30°) & \bar{Q}_{22}(-30°) & \bar{Q}_{26}(-30°) \\ \bar{Q}_{16}(-30°) & \bar{Q}_{26}(-30°) & \bar{Q}_{66}(-30°) \end{array}\right] \left\{\begin{array}{c} 4.24z \\ -5.40z \\ -1.669z \end{array}\right\}$$

$$(8.25b)$$

For the $0°$ layers:

$$\left\{\begin{array}{c} \sigma_x \\ \sigma_y \\ \tau_{xy} \end{array}\right\} = \left[\begin{array}{ccc} \bar{Q}_{11}(0°) & \bar{Q}_{12}(0°) & \bar{Q}_{16}(0°) \\ \bar{Q}_{12}(0°) & \bar{Q}_{22}(0°) & \bar{Q}_{26}(0°) \\ \bar{Q}_{16}(0°) & \bar{Q}_{26}(0°) & \bar{Q}_{66}(0°) \end{array}\right] \left\{\begin{array}{c} 4.24z \\ -5.40z \\ -1.669z \end{array}\right\} \qquad (8.25c)$$

Expanding these and using numerical values for the $\bar{Q}_{ij}$ leads to the following:
For the $+30°$ layers:

$$\sigma_x = 152\,800z \text{ MPa}$$
$$\sigma_y = -11\,870z \qquad (8.26a)$$
$$\tau_{xy} = 6180z$$

For the $-30°$ layers:

$$\sigma_x = 309\,000z \text{ MPa}$$
$$\sigma_y = 39\,800z \qquad (8.26b)$$
$$\tau_{xy} = -166\,800z$$

For the $0°$ layers:

$$\sigma_x = 643\,000z \text{ MPa}$$
$$\sigma_y = -52\,900z \qquad (8.26c)$$
$$\tau_{xy} = 7340z$$

These stress distributions should be compared with the stress distributions for CLT Example 4 given by equation (6.118); Figure 8.11 shows the comparison. While the distributions of σ_x are similar for the two cases, the distributions of σ_y are significantly different. What is most striking is that in CLT Example 4 the maximum values of σ_x, σ_y, and τ_{xy} occur in the outer $+30°$ layer, whereas with the current example the maximum values of these stresses occur in the $-30°$ layer, one layer in from the outer layer.

The stresses in the principal material system are determined by the usual transformations, accordingly:
For the $+30°$ layers:

$$\sigma_1 = 165\,200z \text{ MPa}$$
$$\sigma_2 = -24\,200z \qquad (8.27a)$$
$$\tau_{12} = -40\,400z$$

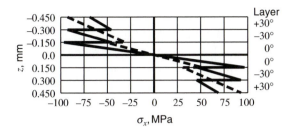

FIGURE 8.11
Stress distribution through
the thickness of $[\pm30/0]_S$
laminate subjected to
$M_x = 12.84$ N·m/m

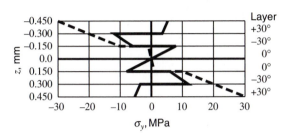

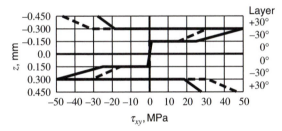

——— Resultant specified — — — Displacement specified

For the $-30°$ layers:

$$\sigma_1 = 386\,000z \text{ MPa}$$

$$\sigma_2 = -37\,500z \tag{8.27b}$$

$$\tau_{12} = 33\,000z$$

For the $0°$ layers:

$$\sigma_1 = 643\,000z \text{ MPa}$$

$$\sigma_2 = -52\,900z \tag{8.27c}$$

$$\tau_{12} = -7340z$$

As illustrated in Figure 8.12, the principal material system stresses in the outer
$+30°$ layers also tend to be lower than the stresses in the inner $-30°$ layers.

Finally, the strains in the principal material system are again given by the appropriate transformations.

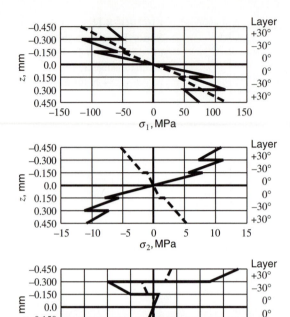

FIGURE 8.12
Principal material system stress distribution through the thickness for $[\pm30/0]_S$ laminate subjected to $M_x = 12.84$ N·m/m

—— Resultant specified — — Displacement specified

For the $+30°$ layers:

$$\varepsilon_1 = 1.105z$$

$$\varepsilon_2 = -2.27z \qquad (8.28a)$$

$$\gamma_{12} = -9.18z$$

For the $-30°$ layers:

$$\varepsilon_1 = 2.55z$$

$$\varepsilon_2 = -3.71z \qquad (8.28b)$$

$$\gamma_{12} = 7.51z$$

For the $0°$ layers:

$$\varepsilon_1 = 4.24z$$

$$\varepsilon_2 = -5.40z \qquad (8.28c)$$

$$\gamma_{12} = -1.669z$$

Figure 8.13 illustrates these principal material system strains, and the higher strain levels in the $-30°$ layers, as opposed to the outer $+30°$ layers, are evident.

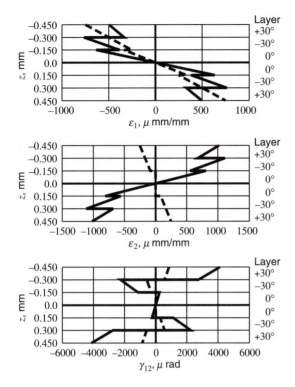

FIGURE 8.13
Principal material system
strain distribution through
the thickness for $[\pm30/0]_S$
laminate subjected to
$M_x = 12.84$ N·m/m

—— Resultant specified — — — Displacement specified

8.4
SUMMARY

These last three chapters have focused on classical lamination theory, a very broad topic with many ramifications. In these chapters we presented the basic assumptions, the implications of these assumptions, and numerical examples. As a result, it is possible to analyze the stress state in a laminate, given as a starting point either the reference surface deformations or the stress resultants. Next we need to introduce some of the basic concepts regarding failure of fiber-reinforced composite materials. After all, we compute stresses so we know something of the load-carrying capacity of a structure. The next two chapters are devoted to failure theories for composite laminates.

CHAPTER 9

Failure Theories for Fiber-Reinforced Materials: Maximum Stress Criterion

$\mathbf{F}$ailure of a structural component can be defined as the inability of the component to carry load. Though excessive deformation with the material still intact, as is the case for buckling, can certainly be considered failure in many situations, here failure will be considered to be the loss of integrity of the material itself. In the most basic sense, molecular bonds have been severed. If a fail-safe philosophy has been employed in the design of the structural component, then failure is not necessarily a catastrophic event. Rather, failure causes a load redistribution within the structure, a permanent deformation, or some other evidence that load levels have become excessive. The structure is still functional to a limited degree, but steps must be taken if continued use is to be considered.

Failure of fiber-reinforced materials is a complex and important topic, and studies of failure are an ongoing activity. For polymer-matrix composites, because the fiber direction is so strong relative to the other directions, it is clear that failure must be a function of the direction of the applied stress relative to the direction of the fibers. Causing failure of an element of material in the fiber direction requires significantly more stress than causing failure perpendicular to the fibers. Tensile failure in the fiber direction is controlled by fiber strength, while tensile failure perpendicular to the fibers is controlled by the strength of the bond between the fiber and matrix, and by the strength of the matrix itself. But what about the case of a tensile stress oriented at 30° relative to the fibers? We know that for this situation the stress component in the fiber direction, σ_1, the stress component perpendicular to the fibers, σ_2, and the shear stress, τ_{12}, can be determined using the stress transformation relations. Which stress component controls failure in this case? The stress component in the fiber direction? The stress component perpendicular to the fibers? The shear stress? Or is it a combination of all three? Because we are now in a position to calculate the stresses in the class of composite structures that satisfy the assumptions of classical lamination theory, it is appropriate to turn to the subject of failure and ask these questions.

There are many issues and controversies surrounding the subject of failure of composite materials. The matrix material of polymer matrix composites may be ductile and exhibit substantial yielding when subjected to high stress levels, and this yielding weakens support of the fiber, or degrades the mechanisms that transfer load into the fibers. On the other hand, the matrix material may be brittle and exhibit significant amounts of cracking around and between fibers as the stress level increases. This cracking will strongly influence the manner and efficiency with which load is transferred into the fibers, and strongly influence the performance of the material. In contrast, failure may be due to the fibers breaking or the fibers debonding and separating from the matrix. Subjected to a compressive load in the fiber direction, the fibers may buckle and deform excessively.

Clearly, we must consider many mechanisms when studying failure. In reality, failure is often a combination of several of these mechanisms, or modes. Failure can simply be the final event in a complex and difficult-to-understand process of damage initiation and accumulation within the material. A structure consists of multiple layers of fiber-reinforced materials, and even multiple materials, and there are multiple fiber directions and a range of load levels and load types. Consequently it is easy to understand why failure of fiber-reinforced composite structures is a difficult topic. Even with a single layer of material, the issues can be quite complicated. As a result, there have been many studies of failure. Each serious user of composite materials tends to develop their own philosophy about failure, based on the application, the material system, and their experience with testing and experimentation. Each large-scale commercial user of composite materials spends much time and capital gathering data to develop criteria and establish design stress levels.

While it is important to understand the mechanisms of failure, for many applications it is impossible to detail each step of the failure process. In the interest of utility, a failure criterion should be reducible to a level that can provide a means of judging whether or not a structure is safe from failure by knowing that a particular stress or combinations of stresses, or combination of strains, is less than some predetermined critical value. The failure criterion should be accurate without being overly conservative, it should be understandable by those using it, and it should be substantiated by experiment. A number of criteria have been proposed; some are rather straightforward and some are quite involved. The maximum stress criterion, maximum strain criterion, and failure criteria that account for interaction among the stress components are commonly used. This is because of the physical bases that underlie these criteria, particularly the maximum stress criterion and the maximum strain criterion. In addition, many of the criteria are simple variations of these, and the variations are based on experimental observation or on slightly different physical arguments put forth by the individuals identified with the criterion.

A legitimate question to ask at this point is, why are there a number of criteria? Isn't one sufficient? The answer is that no one criterion can accurately predict failure for all loading conditions and all composite materials. This is true for isotropic materials—some fail by yielding, others fail by brittle fracture. If we view failure criteria as indicators of failure rather than as predictors in an absolute sense, then having a number of criteria available, none of which covers all situations, becomes an acceptable situation.

In this book we will examine the maximum stress and the Tsai-Wu criteria. These two are chosen because they are among those commonly used for polymer-matrix composites. They represent a divergence of philosophies as to whether or not interaction between stress components is important in predicting failure. Also, by examining any particular criterion, we can present the issues that must be addressed when discussing failure of fiber-reinforced material. By incorporating a particular criterion into a stress analysis of a laminate, failure predictions are possible. The fact that several criteria are commonly used introduces the possibility of determining if using different criteria results in contradictory or similar predictions as to the stress levels permitted and the failure modes expected. Also, a detailed discussion of one or two basic criteria will allow you to form your own opinions regarding failure criteria.

As with the study of the stress-strain behavior of fiber-reinforced material, we shall approach the study of failure of a fiber-reinforced material by examining what happens to a small volume element of material when it is subjected to various components of stress. This is in keeping with the fact that stress is defined at a point, so logically we must assume that failure begins when certain conditions prevail at a point. We will continue to consider the fibers and matrix smeared into an equivalent homogeneous material when we are computing stresses, but to gain insight into the mechanisms that cause failure, it is useful to keep the separate constituents in mind. To that end, consider Figure 9.1; in the fiber direction, as a tensile load is applied, failure is due to fiber tensile fracture. One fiber breaks and the load is transferred through the matrix to the neighboring fibers. These fibers are overloaded, and with the small increase in load, they fail. As the load is increased, more fibers fail and more load is transferred to the unfailed fibers, which take a disproportionate share of the load. The surrounding matrix material certainly cannot sustain the load and so fibers begin to fail in succession; the failure propagates rapidly with increasing load. As with many fracture processes, the tensile strength of graphite fiber, for example, varies from fiber to fiber; and along the length of a fiber. The tensile strength of a fiber is a probabilistic quantity, and the mean value and its variance are important statistics. It is possible to study the failure of fiber-reinforced composites from this

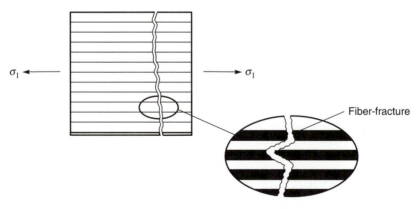

FIGURE 9.1
Failure in tension in the 1 direction

viewpoint, or to simply use a value of failure stress in the fiber direction that includes a high percentage of the fiber failure strengths. That will be the approach here. The tensile strength in the fiber direction will be denoted as σ_1^T.

With a compressive stress in the fiber direction, contemporary polymer-matrix composites fail by fiber kinking, or microbuckling, as in Figure 9.2. Kinking occurs among localized bands, or groups, of fibers, and the fibers in the band fracture at both ends of the kink; the fracture inclination angle is denoted as β, which varies from 10 to 30° in most composites. The width of the kink band, W, varies from 10 to 15 fiber diameters. The primary mechanism responsible for this behavior is yielding, or softening, of the matrix as the stresses within it increase to suppress fiber buckling. As might be suspected, any initial fiber waviness or misalignment, denoted

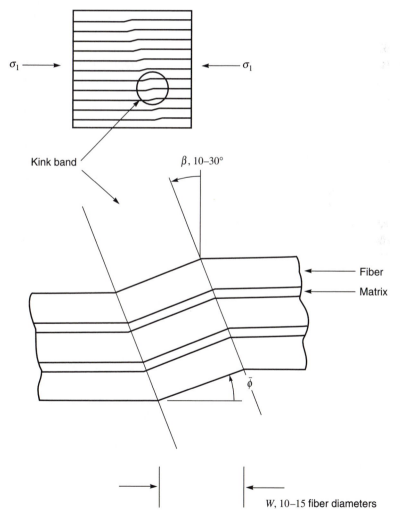

FIGURE 9.2
Failure in compression in the 1 direction

as $\bar{\phi}$ in Figure 9.2, greatly enhances kinking and reduces fiber-direction compressive strength, σ_1^C. In well-made composites, $\bar{\phi}$ is typically between 1 and $4°$ (0.017–0.070 rads). The study of kinking is an ongoing topic of research, where such issues as the magnitude of the fiber modulus relative to the magnitude of the matrix modulus, bending effects in the fiber, the variance of the misalignment angle, and the influence of other stresses (say, a compressive σ_2), are being studied. It is generally accepted, however, that fiber misalignment and yielding of the matrix influence composite compressive strength in the fiber direction, σ_1^C, by way of the relation

$$\sigma_1^C = \frac{G_{12}}{1 + \dfrac{\bar{\phi}}{\gamma_{12}^Y}} \tag{9.1}$$

where γ_{12}^Y is the shear strain at which the composite shear stress–shear strain relation loses validity due to softening, or yielding, effects in the matrix. Values of $\bar{\phi}/\gamma_{12}^Y$ range from 2 to 6, depending on the material.

Often polymeric fibers in a polymeric matrix fail in compression due to fiber crushing rather than fiber kinking. The compressive stresses in the fiber cause the fiber to fail before the matrix softens enough to allow kinking.

As illustrated in Figure 9.3, perpendicular to the fiber, say, in the 2 direction, failure could be due to a variety of mechanisms, depending on the exact matrix material and the exact fiber. Generally a tensile failure perpendicular to the fiber is due to a combination of three possible micromechanical failures: tensile failure of the matrix material; tensile failure of the fiber across its diameter, and failure of the interface between the fiber and matrix. The latter failure is more serious and indicates that the fiber and matrix are not well bonded. However, due to the chemistry of bonding, it is not always possible to have complete control of this bond.

Failure in compression perpendicular to the fibers, as in Figure 9.4, is generally due to material crushing, the fibers and matrix crushing and interacting. The com-

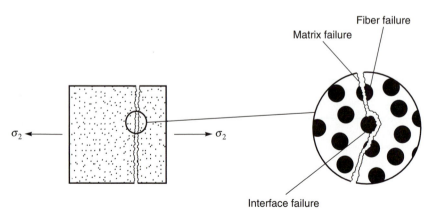

FIGURE 9.3
Failure in tension in the 2 direction

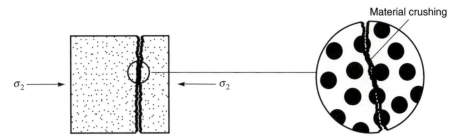

FIGURE 9.4
Failure in compression in the 2 direction

pressive failure stress perpendicular to the fibers is higher than the tensile failure stress in that direction. Herein the tensile failure stress perpendicular to the fibers will be denoted as σ_2^T, while the compressive failure stress will be denoted as σ_2^C. Failure in the 3 direction is similar to failure in the 2 direction and the failure stresses will be denoted as σ_3^T and σ_3^C.

The shear strength in the 2-3 plane, denoted as τ_{23}^F, is limited by the same mechanisms that govern tensile strengths perpendicular to the fibers, namely, matrix tensile failure, failure across the diameter of the fiber, and interfacial strength. Because a shear stress produces a tensile stress on a plane oriented at 45°, these tensile micromechanisms again limit the performance of the material. Figure 9.5 illustrates these mechanisms as viewed in this shear mode. Because of these mechanisms that control shear strength, the shear strength in the 2-3 plane is independent of the sign of the shear stress.

The shear strength in the 1-2 plane is limited by the shear strength of the matrix, the shear strength of the fiber, and the interfacial shear strength between the fiber and matrix. Figure 9.6 depicts failure in shear in the 1-2 plane due to a shear separation of the fiber from the matrix along the length of the interface. Failure in the 1-3 plane follows similar reasoning, and these shear strengths are denoted as τ_{12}^F and τ_{13}^F. As

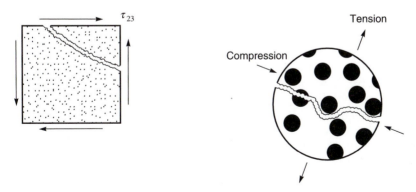

FIGURE 9.5
Failure in shear in the 2-3 plane

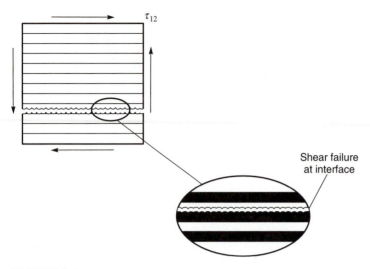

FIGURE 9.6
Failure in shear in the 1-2 plane

expected, the failure strength in this plane is independent of the sign of the shear stress.

In summary, then, for a fiber-reinforced composite material there are nine fundamental failure stresses to be concerned with, six normal stresses, three tensile and three compression, and three shear stresses. Each failure stress represents distinct microfailure mechanisms, and each results in a failure that is somewhat unique to that loading situation. Measuring these failure stresses is difficult. Many issues are involved with the testing of fiber-reinforced composites to determine failure stresses; most of them focus on the fixturing and specimen shape and dimensions. There is also the issue of interaction among stress components. We have indicated that compression failure in the fiber direction is due to kinking and microbuckling, and it stands to reason that a compressive stress perpendicular to the fibers—that is, a σ_2 or σ_3—might help support the fibers and prevent, or at least delay, microbuckling and increase the compressive load capacity in the fiber direction. However, testing to study interaction between stress components is difficult. To determine if a compressive stress perpendicular to the fibers increases the compressive strength in the fiber direction, a fixture to vary the level of compressive σ_2 and then the level of compressive σ_1 would have to be constructed and a range of load levels used. The construction of such a biaxial compressive test fixture is difficult. And what of triaxial compression? If a compressive σ_2 might increase the compressive capacity in the fiber direction, what about a combination of compressive σ_3 and a compressive σ_2? And what about the possible interaction of tension perpendicular to the fibers and shear? From Figures 9.3 and 9.5 we saw that the same mechanisms that are responsible for limiting the value of τ_{23} limit the value of a tensile σ_2, so it is conceivable that these two stress components interact to influence failure. The possibilities are enormous. Unfortunately, it is quite easy to generate failure criteria that are too complex to be verified experimentally. An important requirement of any

TABLE 9.1
Failure stresses (MPa) for graphite and glass composites

	Graphite-reinforced	Glass-reinforced
σ_1^C	−1250	−600
σ_1^T	1500	1000
σ_2^C	−200	−120
σ_2^T	50	30
τ_{12}^F	100	70

failure criterion is to be able to conduct failure tests on simple specimens subjected to fundamental stress states, and then be able to predict the load levels required to produce failure in more complicated structures with more complex stress states. The two criteria considered here rely on the fundamental failure strengths discussed above. The maximum stress criterion is a noninteractive failure theory, while the Tsai-Wu criterion is an interactive theory, and the influence of stress component interaction will be observed.

Because the majority of what we have discussed so far has been devoted to situations where the plane-stress assumption has been used, we shall limit our discussion of failure to those cases also. Hence we shall be interested in the following five failure stresses:

$$\sigma_1^C \;:\; \text{compression failure stress in the 1 direction}$$

$$\sigma_1^T \;:\; \text{tensile failure stress in the 1 direction}$$

$$\sigma_2^C \;:\; \text{compressive failure stresses in the 2 direction} \qquad (9.2)$$

$$\sigma_2^T \;:\; \text{tensile failure stress in the 2 direction}$$

$$\tau_{12}^F \;:\; \text{shear failure stress in the 12 plane}$$

Table 9.1 gives typical values of these stresses for a graphite-fiber composite and a glass-fiber composite. It stands to reason that the failure levels associated with σ_3 and τ_{13} can be equated to the levels for σ_2 and τ_{12}, respectively, and that the failure stress τ_{23}^F is similar to the other shear failure stresses.

We will devote the remainder of this chapter to the maximum stress criterion, while the next chapter addresses the Tsai-Wu criterion.

9.1
MAXIMUM STRESS FAILURE CRITERION

The maximum stress failure criterion, as it applies to the plane-stress case, can be stated as:

A fiber-reinforced composite material in a general state of stress will fail when: EITHER,

The maximum stress in the fiber direction equals the maximum stress in a uniaxial specimen of the same material loaded in the fiber direction when it fails;
OR,
The maximum stress perpendicular to the fiber direction equals the maximum stress in a uniaxial specimen of the same material loaded perpendicular to the fiber direction when it fails;
OR,
The maximum shear stress in the 1-2 plane equals the maximum shear stress in a specimen of the same material loaded in shear in the 1-2 plane when it fails.

Note the *either-or* nature of the criterion. Failure can occur for more than one reason. In addition, the first two portions of the criterion each involve tension and compression. Symbolically, the maximum stress criterion states that a fiber-reinforced material will not fail if at every point

$$\sigma_1^C < \sigma_1 < \sigma_1^T$$
$$\sigma_2^C < \sigma_2 < \sigma_2^T \tag{9.3}$$
$$|\tau_{12}| < \tau_{12}^S$$

While satisfaction of the inequalities of equation (9.3) guarantees, according to the criterion, no failure, it is the equalities associated with equation (9.3) that are important for determining failure loads. These equalities are

$$\sigma_1 = \sigma_1^C$$
$$\sigma_1 = \sigma_1^T$$
$$\sigma_2 = \sigma_2^C$$
$$\sigma_2 = \sigma_2^T \tag{9.4}$$
$$\tau_{12} = -\tau_{12}^F$$
$$\tau_{12} = \tau_{12}^F$$

Equation (9.4) defines the boundaries of the no-failure region in principal material coordinate system stress space σ_1-σ_2-τ_{12}. In this space each of the above equations defines a plane, and the totality of the planes defines a rectangular volume. Because of the differences in the tensile and compression failure loads, the geometric center of the volume does not coincide with the origin of the stress space. Figure 9.7 illustrates this rectangular volume, and it is important to note the proportions in the figure: the rectangular region is much longer in the 1 direction than in the other two directions. In fact, to scale, the rectangular box is much longer and narrower than is depicted in Figure 9.7. Even though we are dealing with a plane-stress problem, we must resort to a three-dimensional figure to describe the problem. Obviously, illustrating the case for a problem that involves all six components of stress becomes a challenge.

To illustrate the application of the maximum stress criterion, and to establish a procedure for using it in such a way that all possible failures in all layers are accounted for, three example failure problems will be studied in detail. These examples will be based on a fiber-reinforced tube, or cylinder, loaded in tension and in torsion. In the

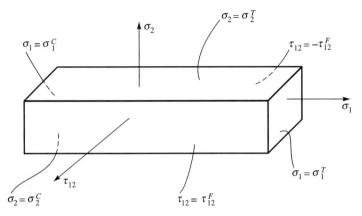

FIGURE 9.7
Maximum stress criterion in principal material system stress space

first example, the cylinder will be loaded with an axial load, and the question will be to determine the level of load required to produce failure. The second example is a combined-load problem: A fixed amount of axial load is applied, and the question is to determine the level of torsional load that can be applied before failure occurs. The third example is also a combined-load problem: The question is to determine what *combined* levels of torsional and axial load can be applied before failure occurs. This third example is a realistic design case; we determine the torsion-axial load envelope. This third example is quite involved but illustrates the complexity of a failure analysis for a fiber-reinforced composite material. The study of a tube is a deviation from our previous examples with the $[0/90]_S$ and $[\pm30/0]_S$ laminates discussed in Chapters 6, 7, and 8. However, the tube provides some variety to the examples and presents a problem of practical importance. We shall, however, return to our examples with the flat $[0/90]_S$ and $[\pm30/0]_S$ laminates and study failure in the context of those cases.

9.2
FAILURE EXAMPLE 1: TUBE WITH AXIAL LOAD — MAXIMUM STRESS CRITERION

Consider a tube with a mean radius of 25 mm made of graphite-reinforced composite with a 10-layer wall with a stacking sequence of $[\pm20/0_3]_S$. The tube is designed to resist an axial load but has the low-angle off-axis layers (the $\pm20°$ layers) to provide some circumferential and torsional stiffness, and to hold the load-carrying layers together. If we use the maximum stress failure criterion, what is the maximum allowable axial load? What layer or layers control failure? What is the mode of failure? This tube is illustrated in Figure 9.8, and the applied axial load is being denoted as P. The 25-mm mean radius is illustrated, as is the 1.50-mm wall thickness that results from using 10 layers, each of 0.150 mm thickness. We will assume that conditions within the tube wall do not vary with distance along the tube. The

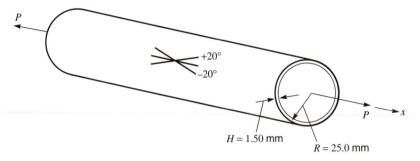

FIGURE 9.8
Tube with axial load P, Failure Example 1

procedure recommended throughout our analysis of failure will be to first determine the principal material systems stresses in each layer that are caused by a *unit* applied load, in this first example a load of $P = 1$ N. It will then assumed that the unit applied load is multiplied by a scale factor. Because we are dealing with a linear problem, the three components of stress in each layer will be multiplied by this same scale factor. These scaled stresses will then be used in the first equation of equation (9.4) to determine the value of the scale factor that causes failure in *compression* in the fiber direction in the first layer in the tube. The second equation of equation (9.4) will then be used to determine the value of the scale factor that causes *tensile* failure in the fiber direction in the first layer. Then the two equations representing failure perpendicular to the fibers will be checked, then the two shear equations. There will be six values of the scale factor for the first layer in the tube. We then repeat the analysis for the second layer, and six more scale factors result. For the third and subsequent layers, we will compute more scale factors. With the six values of the scale factors for all the layers in hand, the numerical value of the load to cause failure in the laminate, in this case the tube wall, is given by the value of the smallest scale factor. By keeping track of the scale factors, it is possible to determine which layer or layers are responsible for failure, and what limits the load capacity (fiber tension, tension perpendicular to the fibers, etc.). This approach, though tedious, can be programmed into a computer-based laminate analysis and the scale factors computed, sorted, listed in ascending order, and correlated to the specific layers and specific failure modes.

Though a tube should be analyzed in a polar coordinate system, let us use our rectangular notation and assign x to the axial direction, y to the circumferential direction, and z to the outward radial direction. To study the example problem, let us not be overly concerned with the method of transmitting the load P into the tube. We shall assume that P is distributed uniformly around the circumference of the ends of the tube, and thus, in keeping with the definition of N_x as being a load per unit length of laminate, N_x is the load P divided by the circumference of the end; that is:

$$N_x = \frac{P}{2\pi R} \tag{9.5}$$

This is the only load acting on the laminate constituting the tube wall and thus

$$N_x = \frac{P}{2\pi R}$$

$$N_y = 0$$

$$N_{xy} = 0 \qquad\qquad (9.6)$$

$$M_x = 0$$

$$M_y = 0$$

$$M_{xy} = 0$$

With this approach it is being assumed that the tube acts like a rolled-up flat laminate, and every point on the reference surface, in this case the mean radius of the tube, is subjected to the force resultants of equation (9.6). This approach leads to quite accurate answers as long as the ratio of the radius to wall thickness is greater than 10, and the details of the load introduction at the ends of the tube are not important. If we assume that $P = 1$ N, then, with $R = 25$ mm, equation (9.5) leads to

$$N_x = 6.37 \text{ N/m} \qquad\qquad (9.7)$$

Because the laminate is symmetric and balanced, with only force resultants applied, to compute the stresses in the various layers due to this the unit value of P we need only compute the elements of the A matrix. We can invert the A matrix and compute the reference surface strains, and knowing the reference surface strains, we can compute the strains throughout the wall thickness, and then the stresses. In particular, we can compute the stresses in the principal material system for each layer due to this load of $P = 1$ N. Details of those calculations will not be presented here because the methodology follows that of the examples in Chapters 6, 7, and 8, in particular CLT Examples 6 and 7. In addition, if Computer Exercise 1 in the Exercises for Section 7.2 has been satisfactorily completed, then computation of the stresses on a layer-by-layer basis is a matter of using that program.

Table 9.2 presents the stresses in the principal material system for a graphite-fiber reinforced $[\pm 20/0_3]_S$ laminate using the material properties of Table 2.1 and subjected to the stress resultant of equation (9.7). By the nature of the problem the stresses within each layer are independent of position within that layer.

If the axial load applied to the tube is p, then the stresses in each layer in the principal material system are as given in Table 9.3; these stresses are simply the stresses in Table 9.2 multiplied by p. The values of stresses in terms of p can be used in the maximum stress criterion to determine the value of p that will cause failure. In particular, in the $+20°$ layers, referring to the first equation of equation

TABLE 9.2
Principal material system stresses (Pa) in axially loaded tube for $P = 1$ N

Layer	σ_1	σ_2	τ_{12}
$+20°$	+3830	−112.3	−148.7
$-20°$	+3830	−112.3	+148.7
$0°$	+4770	−168.6	0

TABLE 9.3

Principal material system stresses (Pa) in axially loaded tube for $P = p$ (N)

Layer	σ_1	σ_2	τ_{12}
+20°	+3830p	−112.3p	−148.7p
−20°	+3830p	−112.3p	+148.7p
0°	+4770p	−168.6p	0

(9.3), we see that failure will not occur in the fiber direction if

$$\sigma_1^C < \sigma_1 < \sigma_1^T \tag{9.8}$$

or, substituting numerical values, if

$$-1250 \times 10^6 < 3830p < 1500 \times 10^6 \tag{9.9}$$

Using these results as in the first equation of equation (9.4), compression failure in the fiber direction is given by the condition

$$\sigma_1 = \sigma_1^C \tag{9.10}$$

or

$$3830p = -1250 \times 10^6 \tag{9.11}$$

This results in

$$p = -327\,000 \tag{9.12}$$

For failure of the fibers in tension

$$\sigma_1 = \sigma_1^T \tag{9.13}$$

or

$$3830p = 1500 \times 10^6 \tag{9.14}$$

This leads to

$$p = 392\,000 \tag{9.15}$$

Thus the $+20°$ layers fail in compression in the fiber direction when the applied load is $P = -0.327$ MN, and the layers fail in tension in the fiber direction when the applied load is $P = +0.392$ MN.

We now examine failure in the $+20°$ layers perpendicular to the fibers, that is, due to σ_2. By the second equation of equation (9.3) the layers are safe from failure due to σ_2 if

$$\sigma_2^C < \sigma_2 < \sigma_2^T \tag{9.16}$$

Numerically, the above becomes

$$-200 \times 10^6 < -112.3p < 50 \times 10^6 \tag{9.17}$$

Using the equalities from this equation to determine the values of p that cause failure, we find that failure of the layers due to compression in the 2 direction is given by the condition

$$\sigma_2 = \sigma_2^C \tag{9.18}$$

Numerically

$$-112.3p = -200 \times 10^6 \tag{9.19}$$

or

$$p = 1\,780\,000 \tag{9.20}$$

Failure due to a tensile failure of the material in the 2 direction is given by

$$\sigma_2 = \sigma_2^T \tag{9.21}$$

or
$$-122.3p = 50 \times 10^6 \tag{9.22}$$

which leads to
$$p = -445\ 000 \tag{9.23}$$

From these results, we can conclude that a *tensile* load of $P = +1.780$ MN on the tube causes the $+20°$ layers to fail in *compression* in the 2 direction, while a *compressive* load of $P = -0.445$ MN on the tube causes the $+20°$ layers to fail in *tension* in the 2 direction. It is important to keep track of the signs so that proper interpretation of the results is possible.

A shear failure in the $+20°$ layers is given by the condition from equation (9.3) of

$$-\tau_{12}^F < \tau_{12} < \tau_{12}^F \tag{9.24}$$

or, with numerical values,

$$-100 \times 10^6 < -148.7p < 100 \times 10^6 \tag{9.25}$$

From this it can be concluded that

$$p = 673\ 000 \tag{9.26}$$

will cause the $+20°$ layers to fail due to $-\tau_{12}$ and

$$p = -673\ 000 \tag{9.27}$$

will cause the $+20°$ layers to fail due to $+\tau_{12}$.

Table 9.4 summarizes the results from the analysis of the $+20°$ layers. From the table we can deduce that failure in the $+20°$ layers can be caused by a compressive load P of -0.327 MN due to a compressive stress in the fiber direction, or by a tensile load P of $+0.392$ MN, due to a tensile stress in the fiber direction.

Having determined the failure characteristics of the $+20°$ layers, we have completed one-third of the failure analysis. Failure analysis of the $-20°$ and $0°$ layers follows similar steps. For the $-20°$ layers the steps are as follows.

Compression failure in the 1 direction:

$$3830p = -1250 \times 10^6$$
$$p = -327\ 000 \tag{9.28}$$

Tension failure in the 1 direction:

$$3830p = 1500 \times 10^6$$
$$p = 392\ 000 \tag{9.29}$$

TABLE 9.4
Loads P (MN) to cause failure in $+20°$ layers: Maximum stress criterion

		Failure mode			
σ_1^C	σ_1^T	σ_2^C	σ_2^T	$-\tau_{12}^F$	$+\tau_{12}^F$
−0.327	+0.392	+1.780	−0.445	+0.673	−0.673

Compression failure in the 2 direction:

$$-112.3p = -200 \times 10^6$$

$$p = 1\,780\,000$$

(9.30)

Tension failure in the 2 direction:

$$-122.3p = 50 \times 10^6$$

$$p = -445\,000$$

(9.31)

Shear failure due to $-\tau_{12}$:

$$148.7p = -100 \times 10^6$$

$$p = -673\,000$$

(9.32)

Shear failure due to $+\tau_{12}$:

$$148.7p = 100 \times 10^6$$

$$p = 673\,000$$

(9.33)

Table 9.5 summarizes the results for the $-20°$ layers, and by comparing the results of this table with the results of Table 9.4, we see that the $+20°$ and $-20°$ layers have very similar failure characteristics; the only difference is in the sign of the load required to produce failure due to a $+\tau_{12}$ or $-\tau_{12}$ failure mode.

The failure analysis of the $0°$ layers is as follows:
Compression failure in the 1 direction:

$$4770p = -1250 \times 10^6$$

$$p = -262\,000$$

(9.34)

Tension failure in the 1 direction:

$$4770p = 1500 \times 10^6$$

$$p = 315\,000$$

(9.35)

Compression failure in the 2 direction:

$$-168.6p = -200 \times 10^6$$

$$p = 1\,186\,000$$

(9.36)

Tension failure in the 2 direction:

$$-168.6p = 50 \times 10^6$$

$$p = -297\,000$$

(9.37)

TABLE 9.5
Loads P (MN) to cause failure in $-20°$ layers: Maximum stress criterion

Failure mode					
σ_1^C	σ_1^T	σ_2^C	σ_2^T	$-\tau_{12}^F$	$+\tau_{12}^F$
-0.327	$+0.392$	$+1.780$	-0.445	-0.673	$+0.673$

Shear failure due to $-\tau_{12}$:

$$0p = -100 \times 10^6$$
$$p = -\infty$$

(9.38)

Shear failure due to $+\tau_{12}$:

$$0p = 100 \times 10^6$$
$$p = +\infty$$

(9.39)

The proper interpretation of the infinite values of applied load required to produce shear failure in the $0°$ layers is that shear failure cannot be produced in those layers with an applied axial load!

Table 9.6 summarizes the failure analysis for all the layers in the $[\pm 20/0_3]_S$ tube, and examination of the table shows that a tensile load of $P = +0.315$ MN causes the $0°$ layers to fail due to tensile stresses in the 1 direction. Discounting failure of the tube due to overall buckling, a compressive load of $P = -0.262$ MN causes the $0°$ layers to fail due to compressive stresses in the 1 direction, presumably due to fiber kinking and microbuckling. These are taken as the failure loads, and the accompanying failure modes, for the tube. As can be seen from this example, failure analysis based on the maximum stress criterion is quite straightforward and, as mentioned before, amenable to automation by computer programming when stresses are being computed.

This first failure example brings to light an important point regarding failure calculations. Considering the compressive failure to illustrate the point, and considering the load to start from zero and be increased in magnitude, we see from Table 9.6 that a load of $P = -0.297$ MN would cause the $0°$ layers to fail due to a tensile stress in the 2 direction, if the layers did not fail due to compression in the 1 direction due to a load $P = -0.262$ MN. We have mentioned the somewhat probabilistic nature of failure, so if only some of the fibers fail in compression at $P = -0.262$ MN, then the tube could sustain more compressive axial load than $P = -0.262$ MN. The value of $P = -0.297$ MN is only 13 percent larger in magnitude than the value $P = -0.262$ MN. Thus, it is possible, with only some of the fibers failing at $P = -0.262$ N, that the load could be increased by 13 percent, at which point failure in tension in the 2 direction would begin, say, due to matrix cracking parallel to the fibers. At this load level, it is highly likely that the tube would lose all load-carrying capacity. The value of $P = -0.262$ MN, then, could be interpreted as the value of the load at which failure first occurs, so-called first-ply failure. Using this value as

TABLE 9.6
Summary of loads P (MN) to cause failure in $[\pm 20/0_3]_S$ tube: Maximum stress criterion

Layer	σ_1^C	σ_1^T	σ_2^C	σ_2^T	$-\tau_{12}^F$	$+\tau_{12}^F$
			Failure mode			
$+20°$	-0.327	$+0.392$	$+1.780$	-0.445	$+0.673$	-0.673
$-20°$	-0.327	$+0.392$	$+1.780$	-0.445	-0.673	$+0.673$
$0°$	-0.262	$+0.315$	$+1.186$	-0.297	$-\infty$	$+\infty$

the load beyond which the tube is incapable of sustaining any more load, then, would be a conservative estimate of load capacity.

Exercises for Section 9.2

1. Suppose the off-axis layers in the tube of Failure Example 1 were at $\pm 30°$ instead of $\pm 20°$. (a) What would be the axial load capacity of the tube? (b) Would the failure mode and the layers that control failure be the same as when the fibers were at $\pm 20°$? To answer this question you essentially must redo the example problem, starting with the stresses due to $P = 1$ N (i.e., beginning at Table 9.2 and proceeding).

2. A $[\pm 45/0_2]_S$ graphite-reinforced plate is subjected to a biaxial loading such that the stress resultant in the y direction is opposite in sign to and one-half the magnitude of the stress resultant in the x direction. Call the stress resultant in the x direction N; the loading is given by

$$N_x = N$$
$$N_y = -0.5N$$
$$N_{xy} = 0$$

(a) Using the maximum stress criterion, compute the value of N to cause failure. (b) What layer or layers control failure? (c) What is the *mode* of failure? To answer these questions, compute the failure loads for each of the layers by using the procedure just discussed in connection with the axially loaded tube. Put the results in table form, as in Table 9.6, and answer the questions.

9.3
FAILURE EXAMPLE 2: TUBE IN TORSION — MAXIMUM STRESS CRITERION

This example of a failure analysis begins to address the question of a combined loading by looking at the case of tension and torsion on this same tube. Consider that the tube is designed to resist axial load but in a particular application the axial load is 0.225 MN tension and there is an unwanted amount of torsion, T. With the 0.225 MN tensile axial load acting on the tube, what is the maximum amount of torsion the tube can withstand before it fails? What layer or layers control failure and what is the mode of failure?

Figure 9.9 illustrates this case and, as with the previous example, we will not be overly concerned with the way the loads are transmitted to the ends. We will assume that both the axial and torsion loads are distributed uniformly over the ends. In particular,

$$N_x = \frac{P}{2\pi R} \qquad M_x = 0$$

$$N_y = 0 \qquad M_y = 0 \qquad (9.40)$$

$$N_{xy} = \frac{T}{2\pi R^2} \qquad M_{xy} = 0$$

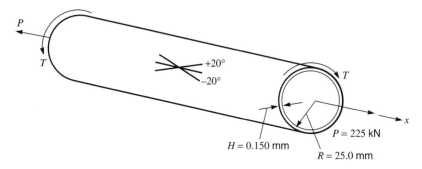

FIGURE 9.9
Tube with axial load $P = 0.225$ MN and torsion T, Failure Example 2

We shall follow the same procedure as before in that we shall examine the stresses in each layer due to a unit applied torsion and then use equation (9.4) to determine the level of torsion required to produce failure in the various modes in the various layers. The added feature in the problem is that the 0.225 MN axial tensile load is also acting on the tube.

Table 9.7 gives the stresses in each layer when 0.225 MN of axial load are applied. These are the stresses of Table 9.2 multiplied by 225 000. Table 9.8 presents the stresses in each layer when a torsion $T = 1$ N·m is applied to the tube (no axial load). These are calculated from the condition

$$N_{xy} = 255 \text{ N/m} \tag{9.41}$$

Note that in the $0°$ layers the applied torsion does not produce any stress in the fiber direction or perpendicular to the fiber direction. This is not the case for the $\pm 20°$ layers. When a torsional load of $T = t$ is applied to the tube, the stresses in each layer are t times the values in Table 9.8. When *both* a 0.225 MN tensile load *and* a torsional load of $T = t$ are applied to the tube, the stresses in each layer are those in Table 9.9. Superposition of the effects of the axial load alone and the torsional load alone can be used because this problem is strictly a linear problem. The goal of this failure analysis is to find the value or values of t that cause failure of the tube.

As in the previous example, the values of stress, in terms of t, from Table 9.9 can be used in the maximum stress criterion to determine the value of t that will cause failure. In the $+20°$ layers, referring to the first equation of equation (9.3), we find that failure will not occur in the fiber direction if

$$\sigma_1^C < \sigma_1 < \sigma_1^T \tag{9.42}$$

TABLE 9.7
Stresses (MPa) in axially loaded $[\pm 20/0_3]_S$ tube for $P = 0.225$ MN

Layer	σ_1	σ_2	τ_{12}
$+20°$	$+861$	-25.3	-33.5
$-20°$	$+861$	-25.3	$+33.5$
$0°$	$+1072$	-37.9	0

TABLE 9.8
Stresses (MPa) in torsionally loaded $[\pm 20/0_3]_S$ tube for $T = 1$ N·m

Layer	σ_1	σ_2	τ_{12}
+20°	+0.804	−0.0481	+0.0552
−20°	−0.804	+0.0481	+0.0552
0°	0	0	+0.0721

TABLE 9.9
Stresses (MPa) in $[\pm 20/0_3]_S$ tube loaded with tensile load $P = 0.225$ MN and torsional load $T = t$ (N·m)

Layer	σ_1	σ_2	τ_{12}
+20°	$+861 + 0.804t$	$-25.3 - 0.0481t$	$-33.5 + 0.0552t$
−20°	$+861 - 0.804t$	$-25.3 + 0.0481t$	$+33.5 + 0.0552t$
0°	$+1072 + 0t$	$-39.7 + 0t$	$0 + 0.0721t$

or, substituting from Table 9.9, if

$$-1250 < 861 + 0.0804t < 1500 \tag{9.43}$$

Using this as in the first equation of equation (9.4), we find that compression failure in the fiber direction is given by the condition

$$\sigma_1 = \sigma_1^C \tag{9.44}$$

or

$$861 + 0.804t = -1250 \tag{9.45}$$

This results in

$$t = -2620 \tag{9.46}$$

For failure of the fibers in tension

$$\sigma_1 = \sigma_1^T \tag{9.47}$$

or

$$861 + 0.804t = 1500 \tag{9.48}$$

This leads to

$$t = 795 \tag{9.49}$$

From these results we can say that the $+20°$ layers fail in compression in the fiber direction when, in addition to the applied axial load of $P = +0.225$ MN, the applied torque is $T = -2620$ N·m, and the layers fail in tension in the fiber direction when the additional applied torque is $T = +795$ N·m.

Turning to failure in the $+20°$ layers due to stresses perpendicular to the fibers, we find by the second equation of equation (9.3) that the layers are safe from failure due to σ_2 if

$$\sigma_2^C < \sigma_2 < \sigma_2^T \tag{9.50}$$

Numerically, the above becomes

$$-200 < -25.3 - 0.0481t < 50 \tag{9.51}$$

Using the equalities from this equation to determine the values of t that cause failure, we find that failure of the $+20°$ layers due to compression in the 2 direction is given by the condition

$$\sigma_2 = \sigma_2^C \tag{9.52}$$

Numerically
$$-25.3 - 0.0481t = -200 \tag{9.53}$$

or
$$t = 3630 \tag{9.54}$$

Failure due to a tensile failure of the material in the 2 direction is given by

$$\sigma_2 = \sigma_2^T \tag{9.55}$$

or
$$-25.3 - 0.0481t = 50 \tag{9.56}$$

which leads to

$$t = -1563 \tag{9.57}$$

From these results, we can see that a torsional load of $T = +3630$ N·m causes the $+20°$ layers to fail in compression in the 2 direction, while a torsional load of $T = -1563$ N·m causes the $+20°$ layers to fail in tension in the 2 direction.

A shear failure in the $+20°$ layers is given by the condition from equation (9.3) of

$$-\tau_{12}^F < \tau_{12} < \tau_{12}^F \tag{9.58}$$

or, if we use numerical values,

$$-100 < -33.5 + 0.0552t < 100 \tag{9.59}$$

From this, we can conclude that

$$t = -1205 \tag{9.60}$$

will cause the $+20°$ layers to fail due to $-\tau_{12}$ and that

$$t = 2420 \tag{9.61}$$

will cause the $+20°$ layers to fail due to $+\tau_{12}$. Unlike the case of a tensile load alone, Failure Example 1, the values of torsional load that cause shear failures due to $+\tau_{12}$ and $-\tau_{12}$ differ by more than a sign. The biasing effect of the $+0.225$ MN axial force applied to the tube causes this result.

The results from the analysis of the $+20°$ layers are summarized in Table 9.10, and we can see that in the presence of an axial force of $+0.225$ MN, failure in the $+20°$ layers can be caused by an applied torque T of -1205 N·m. Such a torque will cause the layers to fail due to a negative shear stress τ_{12}. Alternatively, an applied

TABLE 9.10
Torsions T (N·m) to cause failure in $+ 20°$ layers with $P = + 0.225$ MN: Maximum stress criterion

		Failure mode			
σ_1^C	σ_1^T	σ_2^C	σ_2^T	$-\tau_{12}^F$	$+\tau_{12}^F$
−2620	+795	+3630	−1563	−1205	+2420

torque T of $+795$ N·m causes the layers to fail due to tensile stresses in the 1 direction.

Failure of the $-20°$ and $0°$ layers follows similar steps. For the $-20°$ layers the steps are as follows.

Compression failure in the 1 direction:

$$861 - 0.804t = -1250$$
$$t = 2620$$

(9.62)

Tension failure in the 1 direction:

$$861 - 0.804t = 1500$$
$$t = -795$$

(9.63)

Compression failure in the 2 direction:

$$-25.3 + 0.0481t = -200$$
$$t = -3630$$

(9.64)

Tension failure in the 2 direction:

$$-25.3 + 0.0481t = 50$$
$$t = 1563$$

(9.65)

Shear failure due to $-\tau_{12}$:

$$33.5 + 0.0552t = -100$$
$$t = -2420$$

(9.66)

Shear failure due to $+\tau_{12}$:

$$33.5 + 0.0552t = 100$$
$$t = 1205$$

(9.67)

The failure analysis of the $0°$ layers is as follows:

Compression failure in the 1 direction:

$$1072 + 0t = -1250$$
$$t = -\infty$$

(9.68)

Tension failure in the 1 direction:

$$1072 + 0t = 1500$$
$$t = +\infty$$

(9.69)

From these last two statements it is clear that failure in the fiber direction in the $0°$ layers will not occur due to an applied torque. Continuing with the steps in the failure analysis, we find that the remaining steps for the $0°$ layers are as follows.

Compression failure in the 2 direction:

$$-39.7 + 0t = -200$$
$$t = -\infty$$

(9.70)

TABLE 9.11
Summary of torsions T (N·m) to cause failure in $[\pm 20/0_3]_S$ tube with
$P = 0.225$ MN: Maximum stress criterion

Layer	σ_1^C	σ_1^T	σ_2^C	σ_2^T	$-\tau_{12}^F$	$+\tau_{12}^F$
	\multicolumn{6}{c}{Failure mode}
+20°	−2620	+795	+3630	−1563	−1205	+2420
−20°	+2620	−795	−3630	+1563	−2420	+1205
0°	−∞	+∞	−∞	+∞	−1387	+1387

Tension failure in the 2 direction:

$$-39.7 + 0t = 50$$
$$t = +\infty \tag{9.71}$$

Like the equations for failure in the 0° direction, these last two statements lead to the conclusion that the 0° layers will not fail in the 2 direction due to the applied torque. Continuing, we find that shear failure due to $-\tau_{12}$ is determined with

$$0.0721t = -100$$
$$t = -1387 \tag{9.72}$$

and shear failure due to $+\tau_{12}$ with

$$0.0721t = 100$$
$$t = 1387 \tag{9.73}$$

Table 9.11 summarizes the failure analysis for all the layers, and here we see that a torque of $T = +795$ N·m causes the +20° layers to fail due to positive σ_1 and a torque of $T = -795$ N·m causes the −20° layers to fail also due to a positive σ_1. These torques are taken as the failure torques for the tube. Apparently the effects of the applied torsion added to the +0.225 MN axial load are enough to break the fibers in tension in the +20° layers or the −20° layers, depending on the direction of the applied torsion. Essentially, referring to Figure 9.7, we find that the initial axial load shifts the origin of the stress space relative to the failure surface.

Exercise for Section 9.3

A $[\pm 45/0_2]_S$ graphite-reinforced plate is subjected to a biaxial loading such that the stress resultant in the y direction is −0.200 MN/m and the stress resultant in the x direction is variable, that is:

$$N_x = N$$
$$N_y = -0.200 \text{ MN/m}$$
$$N_{xy} = 0$$

(a) Using the maximum stress criterion, compute the value of N required to cause failure. (b) What layer or layers control failure? (c) What is the mode of failure?

9.4
FAILURE EXAMPLE 3: TUBE WITH COMBINED LOAD —
MAXIMUM STRESS CRITERION

As a final example of the application of the maximum stress failure criterion, consider the following problem with the tube: In a particular application the tube is being used with an axial load P and a small torsional load T. According to the maximum stress criterion, what are the ranges of applied axial load and applied torsion the tube can withstand before it fails? What layer or layers control failure and what is the mode of failure? This is truly a combined-load problem, with unknown levels of both the axial and torsional loads, the goal being to find the bounds of P and T within which the tube is safe from failure. The approach here will be to revert to a P-T space and study the failure boundaries. Here P will be on the horizontal axis and T on the vertical axis. For multilayer laminates such as the tube, each layer will have an envelope in this space and the totality of envelopes represents the envelope for the laminate. Figure 9.10 illustrates the situation being studied.

We shall proceed exactly as before, using a scale factor to multiply the loads and then enforcing the equations representing various portions of the failure criterion to determine the scale factor. Here, however, both the scale factor for the axial load p and the scale factor for the torsional load t are unknown. We can, however, use the six equations of the failure criterion to determine relations between p and t. The stress components in each layer due to a unit axial load, $P = 1$ N, were given in Table 9.2, and the stress components in each layer due to a unit torque, $T = 1$ N·m, were given in Table 9.8. Accordingly, the stress components in each layer due to the combined effects of an axial load of p and a torsional load of t are given in Table 9.12. These stress components will now be used in the failure equations, equation (9.3), and the alternate form, equation (9.4).

For the $+20°$ layers, the first equation of equation (9.3), namely,

$$\sigma_1^C < \sigma_1 < \sigma_1^T \tag{9.74}$$

becomes $\qquad -1250 < 0.00383p + 0.804t < 1500 \qquad$ (9.75)

This inequality defines a region in the p-t coordinate system, that is, p-t space. The region defines the values of tension and torsion that can simultaneously be applied

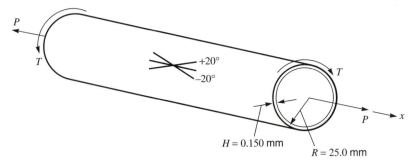

FIGURE 9.10
Tube with axial load P and torsion T, Failure Example 3

TABLE 9.12
Stresses (MPa) in $[\pm 20/0_3]_S$ tube loaded with $P = p$ (N) and $T = t$ (N·m)

Layer	σ_1	σ_2	τ_{12}
$+20°$	$+0.00383p + 0.804t$	$-0.0001123p - 0.0481t$	$-0.0001487p + 0.0552t$
$-20°$	$+0.00383p - 0.804t$	$-0.0001123p + 0.0481t$	$+0.0001487p + 0.0552t$
$0°$	$+0.00477p$	$-0.0001686p$	$+ 0.0721t$

to the tube and not cause the $+20°$ layers to fail. Of course, for no torsion equation (9.9) is recovered. As in previous discussions, it is the boundaries of the region that are important, and as in the previous examples, it is the equalities associated with the inequalities that define the boundaries. Specifically, in the present situation, the first equation of equation (9.4), namely, the compression side of the inequality,

$$\sigma_1 = \sigma_1^C \tag{9.76}$$

results in

$$0.00383p + 0.804t = -1250 \tag{9.77}$$

Unlike the past examples where enforcement of the equalities led to a specific value of p or t, here enforcement of the equality leads to an equation for a line in p-t space. This line, illustrated in Figure 9.11, is labeled σ_1^C, the notation identifying the failure mode represented by this line. The line divides p-t space into two regions,

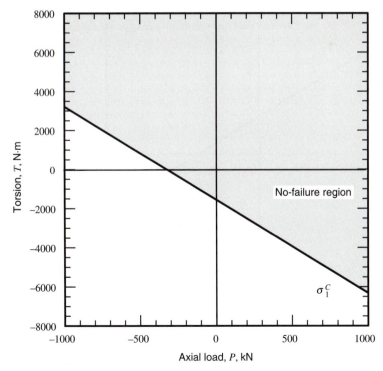

FIGURE 9.11
Failure boundary for compression failure in 1 direction in $+20°$ layers

where any combination of p and t on this line and below it represents a combination that will cause a compression failure in the fiber direction of the $+20°$ layers. A combination of p and t above this line represents values that will be safe from causing compression failure in the fiber direction of the $+20°$ layers.

The equation from the tension side of the inequality is

$$\sigma_1 = \sigma_1^T \tag{9.78}$$

or
$$0.00383p + 0.804t = 1500 \tag{9.79}$$

This represents another line in p-t space, illustrated in Figure 9.12, where the notation σ_1^T is assigned to the line represented by equation (9.79) to denote that the line represents failure due to tension in the fiber direction. The σ_1^T line again divides p-t space into two parts. In this case, the region on the line and above it represents the region where any combination of p and t will lead to failure of the $+20°$ layers because of tension in the fiber direction, and the region below the line represents combinations of loads that will not cause the $+20°$ layers to fail in tension in the fiber direction. Obviously the region between the σ_1^C line of Figure 9.11 and the σ_1^T line of Figure 9.12 represents combinations of p and t in p-t space that will not cause the $+20°$ layers to fail in the fiber direction. The two lines are shown in Figure 9.13; the region free from fiber-direction failure is shaded in and denoted as safe.

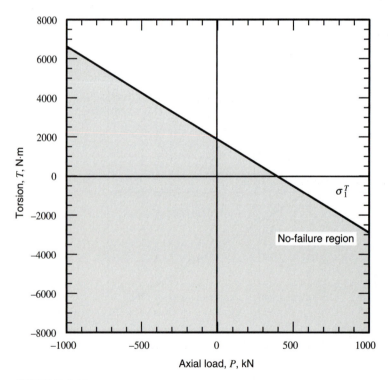

FIGURE 9.12
Failure boundary for tension failure in 1 direction in $+20°$ layers

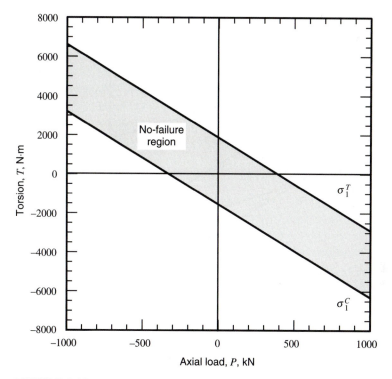

FIGURE 9.13
Failure envelope for failure in 1 direction in $+20°$ layers

From equation (9.3), the equation that represents failure in the 2 direction,

$$\sigma_2^C < \sigma_2 < \sigma_2^T \qquad (9.80)$$

leads to

$$-200 < -0.0001123p - 0.0481t < 50 \qquad (9.81)$$

This results in two equations for two more boundary lines in p-t space. From the compression portion,

$$-0.0001123p - 0.0481t = -200 \qquad (9.82)$$

and from the tension portion,

$$-0.0001123p - 0.0481t = 50 \qquad (9.83)$$

In Figure 9.14 the lines represented by these two equations are added to the lines of Figure 9.13 and the notation σ_2^C and σ_2^T is used to identify these lines. As we can see, the boundary associated with compression failure in the 2 direction is considerably removed from the previously established boundaries associated with failure in the fiber direction. This is interpreted to mean that failure due to compression in the 2 direction is not possible with any combination of p and t before failure occurs in the fiber direction due to tension. On the other hand, the line representing tension failure in the 2 direction intersects the previously established region for failure in

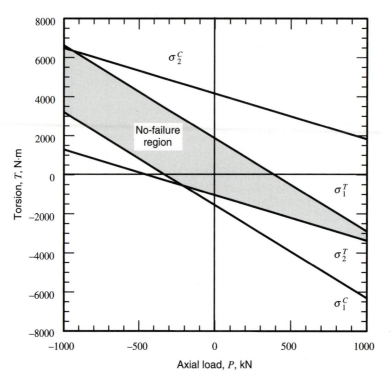

FIGURE 9.14
Failure envelope for failure in 1 and 2 directions in $+20°$ layers

the 1 direction. Thus, the safe region is impinged upon by failure in another mode and the size of the safe region is reduced.

Finally, failure in the $+20°$ layers due to shear stress τ_{12} is given by the third equation of equation (9.3), namely,

$$-\tau_1^F < \tau_{12} < \tau_{12}^F \tag{9.84}$$

which, for the case here, becomes

$$-100 < -0.0001487p + 0.0552t < 100 \tag{9.85}$$

This inequality defines yet a third region in p-t space, whose regional boundaries are given by

$$-0.0001487p + 0.0552t = -100 \tag{9.86}$$

and

$$-0.0001487p + 0.0552t = +100 \tag{9.87}$$

In Figure 9.15 the lines represented by these equations are added to Figure 9.14. Both of these new lines intersect the previously established safe region and further restrict its size. The irregular-shaped shaded region bounded by segments of the various lines corresponding to the six failure equalities represents the range of values of axial load and torsional load that can be applied to the tube without having the $+20°$ layers fail in any of the various modes. The shape of the region, the intersections of

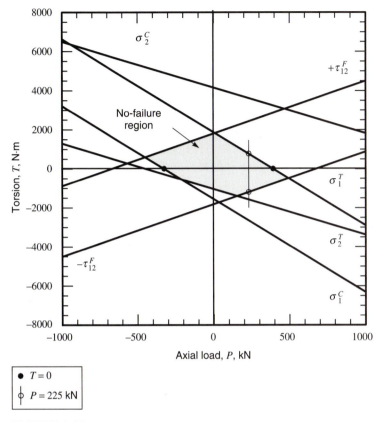

FIGURE 9.15
Complete failure envelope for $+20°$ layers

the lines with the coordinate axes, the intersections of the lines with each other, and other characteristics of the figure are a result of the material elastic properties, the fiber angles in the various layers, and the failure stresses of the material. Actually the tube radius is reflected in the characteristics of the region. Hence, Figure 9.15 summarizes a great deal of information regarding this particular problem. Also, on Figure 9.15 the points for $T = 0$, which correspond to the first example problem, are noted, as are the points for $P = 225$ kN, which correspond to the second example.

With the failure boundaries established for the $+20°$ layers, attention turns to the $-20°$ and $0°$ layers. The determination of the failure boundaries for these layers follows steps identical to those just completed for the $-20°$ layers. For the $-20°$ layers the analysis is as follows:

Compression failure in the 1 direction:

$$0.00383p - 0.804t = -1250 \tag{9.88}$$

Tension failure in the 1 direction:

$$0.00383p - 0.804t = 1500 \tag{9.89}$$

Compression failure in the 2 direction:

$$-0.0001123p + 0.0481t = -200 \tag{9.90}$$

Tension failure in the 2 direction:

$$-0.0001123p + 0.0481t = 50 \tag{9.91}$$

Shear failure:

$$0.0001487p + 0.0552t = -100 \tag{9.92}$$

and

$$0.0001487p + 0.0552t = 100 \tag{9.93}$$

Figure 9.16 illustrates the six lines associated with the failure boundaries and the region in p-t space free from failure for the $-20°$ layers. Note the differences and similarities between Figure 9.15 for the $+20°$ layers and Figure 9.16. Whereas the shaded region for the $+20°$ layers is skewed downward slightly on the right, the region for the $-20°$ layers is skewed upward slightly on the right. The points corresponding to the two previous failure examples are indicated.

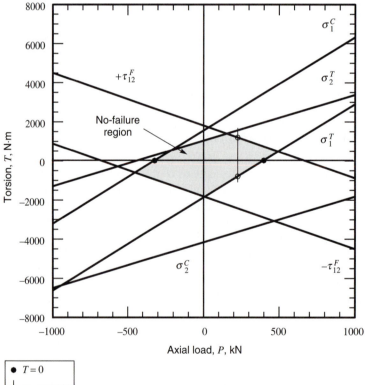

FIGURE 9.16
Complete failure envelope for $-20°$ layers

For the 0° layers:

Compression failure in the 1 direction:

$$0.00477p = -1250 \tag{9.94}$$

Tension failure in the 1 direction:

$$0.00477p = 1500 \tag{9.95}$$

Compression failure in the 2 direction:

$$-0.0001686p = -200 \tag{9.96}$$

Tension failure in the 2 direction:

$$-0.0001686p = 50 \tag{9.97}$$

Shear failure:

$$0.0721t = -100 \tag{9.98}$$

and

$$0.0721t = +100 \tag{9.99}$$

As shown in Figure 9.17, the lines representing failure of the 0° layers are all parallel to the coordinate axes in p-t space. The line representing compression failure

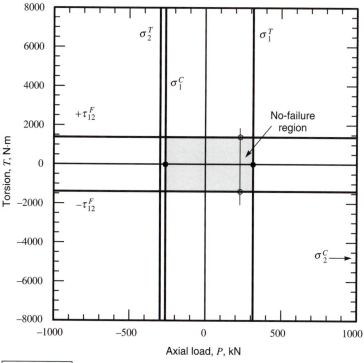

FIGURE 9.17

Complete failure envelope for 0° layers

in the 2 direction is not shown since it is far to the right. The region free from failure is a rectangular region bounded on the top and bottom by shear failure, and bounded on the left and right by failure in the fiber direction.

Finally, the superposition of the boundaries that control failure of the laminate is shown in Figure 9.18, where the boundaries are labeled as to the failure mode and the layer orientation that controls failure. In general, stress components σ_1 and σ_2 in the $\pm20°$ layers control failure. However, at extreme values of p, failure in the fiber direction in the 0° layers controls failure. Figure 9.18 is important because from it one can determine the level of torque that can be applied for a specific level of axial load.

We mentioned earlier that the analyses being presented were applicable to first-ply failure loads, and that there might be additional load capacity beyond the load at which the first failure occurs. This may be true particularly when the first failure does not involve fiber failure as when, for example, the first failure is due to tension in the 2 direction. For this case the matrix could develop cracks parallel to the fibers, but there would still be integrity to the fibers. To reflect the fact that matrix cracking

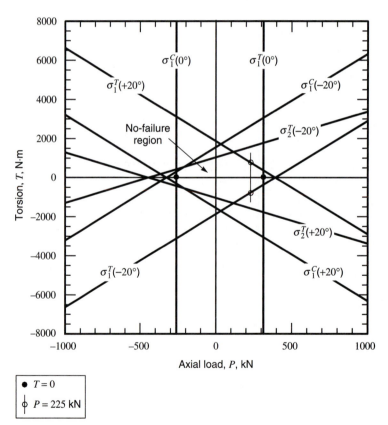

FIGURE 9.18
Superposition of failure envelopes for $+20°$, $-20°$, and $0°$ layers

has occurred in specific layers at a particular load level, E_2 and G_{12} in those layers could be reduced significantly or equated to zero, and the A matrix recomputed. A failure analysis could then be conducted on the altered laminate, and a new failure load could be predicted. This progressive recomputation of the A matrix (and the B and D matrices, if they are involved) can continue, and ultimate failure can be assumed to occur when stresses in the fiber direction in the layer most highly stressed in the fiber direction exceed the failure stress level.

Exercise for Section 9.4

A $[\pm 45/0_2]_S$ graphite-reinforced plate is subjected to a biaxial loading; that is:

$$N_x = N_x$$
$$N_y = N_y$$
$$N_{xy} = 0$$

Use the maximum stress failure criterion to determine the failure envelope of the plate in N_x-N_y space. Label the failure mode and layer or layers which control failure for each portion of the envelope. Note that the cases of Exercise 2 in the Exercises for Section 9.2 and the Exercise for Section 9.3 are included in this case.

   We shall now return to our previous examples with the flat $[0/90]_S$ and $[\pm 30/0]_S$ laminates and study failure for those cases. In particular, we shall study failure for the situation where the stress resultants, rather than the deformations, are specified. Because we have studied these examples in some depth, using them to study failure will provide us with additional insight into the response of fiber-reinforced laminated composites. Also, because we have studied these laminates with particular loadings, we have figures illustrating the stress distributions within the layers. With the stress distributions available, it might be tempting to focus only on the layer or layers with the highest stresses. Unfortunately, with the dramatically different failure strengths in the different directions in a layer, and in tension and compression, a focusing a priori on one layer or one failure mode in one layer can lead to the wrong conclusions. A failure mode may be overlooked, or the closeness of the loads for different failure modes in different layers may not be noted. Thus, we shall avoid the temptation to focus on one layer or one failure mode despite what may appear as overwhelming evidence that failure will occur in a particular way. Rather, we will use the same methodical approach to study failure that we have used in the tube examples. As this can all be automated by computer programming, there is every reason to take the safe and thorough approach to the study of failure.

9.5
FAILURE EXAMPLE 4: $[0/90]_S$ LAMINATE SUBJECTED TO N_X — MAXIMUM STRESS CRITERION

Consider the $[0/90]_S$ laminate subjected to a loading in the x direction, as in CLT Example 6 in Chapter 8. Figure 8.4 illustrated the stresses in the principal material system for that laminate with a loading of $N_x = 0.0504$ MN/m. From that figure it appears that when the loading is increased, if failure is going to be due to high stresses in the fiber direction, then it will occur in the $0°$ layers. If failure is going to be due to high stresses perpendicular to the fibers, then it will occur in the $90°$ layers. Shear failure is not an issue, and an analysis of failure in each layer and for each failure mode will either confirm that these are indeed the controlling modes, or it will reveal other failure modes. To begin investigating just what load N_x causes failure, let us determine the principal material system stresses due to $N_x = n_x$. These stresses are given in Table 9.13, where the numbers in this table are determined by scaling the stresses in Chapter 8 for the case of $N_x = +0.0504$ MN/m, namely, dividing the stresses given in equation (8.6) by 50 400. The key equations for determining failure, then, are the following.

For the $0°$ layers:

$$3090n_x = \sigma_1^C$$
$$3090n_x = \sigma_1^T$$
$$51.3n_x = \sigma_2^C$$
$$51.3n_x = \sigma_2^T \tag{9.100a}$$
$$0n_x = \tau_{12}^F$$
$$0n_x = -\tau_{12}^F$$

For the $90°$ layers:

$$-51.3n_x = \sigma_1^C$$
$$-51.3n_x = \sigma_1^T$$
$$240n_x = \sigma_2^C$$
$$240n_x = \sigma_2^T \tag{9.100b}$$
$$0n_x = \tau_{12}^T$$
$$0n_x = -\tau_{12}^F$$

where the solutions to these equations are summarized in Table 9.14.

TABLE 9.13
Principal material system stresses (Pa) in $[0/90]_S$ laminate due to $N_x = n_x$ (N/m)

Layer	σ_1	σ_2	τ_{12}
$0°$	$+3090n_x$	$+51.3n_x$	0
$90°$	$-51.3n_x$	$+240n_x$	0

TABLE 9.14
Summary of failure loads N_x (MN/m) for $[0/90]_S$ laminate: Maximum stress criterion

Layer	Failure mode					
	σ_1^C	σ_1^T	σ_2^C	σ_2^T	$-\tau_{12}^F$	$+\tau_{12}^F$
$0°$	-0.404	$+0.485$	-3.90	$+0.975$	$-\infty$	$+\infty$
$90°$	$+24.4$	-29.3	-0.835	$+0.209$	$-\infty$	$+\infty$

From these solutions it is evident that the following loads cause failure:

$$N_x = +0.209 \text{ MN/m}$$
$$N_x = -0.404 \text{ MN/m}$$

(9.101)

The physical interpretation is that a tensile load of $N_x = +0.209$ MN/m will cause the 90° layers to fail due to a tensile stress perpendicular to the fibers, while a compressive load of $N_x = -0.404$ MN/m will cause the 0° layers to fail due to a compressive stress in the fiber direction. In the first case, because the fibers remain intact in the 0° layers, the laminate still has the ability to support additional load. In the second case, the fibers fail in compression and the ability to support a compressive load much greater than $N_x = -0.404$ MN/m is unlikely. Hence, as the [0/90]$_S$ laminate is loaded in tension from zero load, cracks parallel to the fibers in the two 90° layers appear as the first evidence of failure. To fail the laminate, a considerable load beyond the load to cause this cracking is required, but just how much more load is difficult to say. These cracks in the 90° layers produce a stress concentration that increases the stresses in the 0°. Thus, the ultimate load capacity of the laminate is less than the load to cause fiber direction failure if the 90° layers were not part of the laminate, but more than the +0.209 MN/m. As the laminate is loaded in compression from zero load, the fibers in the 0° layers fail in compression and the entire laminate thus fails.

9.6
FAILURE EXAMPLE 5: [±30/0]$_S$ LAMINATE SUBJECTED TO N_X — MAXIMUM STRESS CRITERION

A failure analysis of the [±30/0]$_S$ laminate subjected to a loading in the x direction is similar to the analysis of the [0/90]$_S$ laminate above. Figure 8.8 shows the principal material system stresses in the laminate for the case of $N_x = 0.1024$ MN/m, and from that figure it can be seen that the leading candidates for failure are: fiber direction failure in the 0° layers, failure perpendicular to the fibers in the 0° layers, and shear failure in the ±30 layers. To determine which mechanism causes failure, the six components of the maximum stress failure criterion are studied for each layer. To do this, the principal material system stresses in each layer for $N_x = n_x$ are needed; these stresses are available from equation (8.14) by the simple scaling approach used with the [0/90]$_S$ laminate. The stresses in equation (8.14) were computed for the case of $N_x = 0.1024$ MN/m; thus, division of these stresses by 102 400 yields the required results, which are given in Table 9.15. Using these results in the maximum stress failure criterion leads to the following sets of equations.

TABLE 9.15
Principal material system stresses (Pa) in [±30/0]$_S$ laminate due to $N_x = n_x$ (N/m)

Layer	σ_1	σ_2	τ_{12}
+30°	$+866n_x$	$-77.3n_x$	$-102.8n_x$
−30°	$+866n_x$	$-77.3n_x$	$+102.8n_x$
0°	$+1894n_x$	$-138.9n_x$	0

For the $+30°$ layers:

$$866n_x = \sigma_1^C$$
$$866n_x = \sigma_1^T$$
$$-77.3n_x = \sigma_2^C$$
$$-77.3n_x = \sigma_2^T \qquad (9.102a)$$
$$-102.8n_x = \tau_{12}^F$$
$$-102.8n_x = -\tau_{12}^F$$

For the $-30°$ layers:

$$866n_x = \sigma_1^C$$
$$866n_x = \sigma_1^T$$
$$-77.3n_x = \sigma_2^C$$
$$-77.3n_x = \sigma_2^T \qquad (9.102b)$$
$$102.8n_x = \tau_{12}^T$$
$$102.8n_x = -\tau_{12}^F$$

For the $0°$ layers:

$$1894n_x = \sigma_1^C$$
$$1894n_x = \sigma_1^T$$
$$-138.9n_x = \sigma_2^C$$
$$-138.9n_x = \sigma_2^T \qquad (9.102c)$$
$$0n_x = \tau_{12}^T$$
$$0n_x = -\tau_{12}^F$$

Table 9.16 summarizes the resulting values of the failure loads. From these calculations we see that the $0°$ layers control failure. With a tensile load of $+0.791$ MN/m the $0°$ layers fail due to a tensile stress in the fiber direction, while with a compressive load of -0.360 MN/m they fail due to a tensile stress perpendicular to the fibers. The loads to cause failure in other layers and modes are not close to either of these loads. With a compressive load of -0.360 MN/m, cracks will appear

TABLE 9.16
Summary of failure loads N_x (MN/m) for $[\pm 30/0]_S$ laminate: Maximum stress criterion

Layer	σ_1^C	σ_1^T	σ_2^C	σ_2^T	$-\tau_{12}^F$	$+\tau_{12}^F$
			Failure mode			
$+30°$	-1.444	$+1.733$	$+2.59$	-0.647	$+0.973$	-0.973
$-30°$	-1.444	$+1.733$	$+2.59$	-0.647	-0.973	$+0.973$
$0°$	-0.659	$+0.791$	$+1.439$	-0.360	$-\infty$	$+\infty$

parallel to the fibers in the 0° layers. These cracks will not cause total failure of the laminate, and increased compressive loads can be tolerated. When the compressive load reaches −0.659 MN/m, compressive stresses in the fiber direction are predicted to cause failure. However, it is doubtful the −0.659 MN/m compressive load can be reached. The cracks that occur parallel to the fibers at −0.360 MN/m will lead to destabilization of the fibers because they are in compression and hence compressive loads much greater than −0.360 MN/m probably cannot be attained.

In many of the failure examples studied so far, the load limit of the laminate is determined not by fiber strength, but by the strength perpendicular to the fiber, or by shear strength. This is a characteristic of fiber-reinforced laminates, a characteristic that takes considerable effort to overcome. It makes little sense to develop strong and stiff fibers and then have the strength of the laminate be limited by the strength of the weak directions of a layer rather than the fiber direction. Yet this often happens!

9.7
FAILURE EXAMPLE 6: [±30/0]$_S$ LAMINATE SUBJECTED TO M_X — MAXIMUM STRESS CRITERION

Determining the load limits of a laminate when it is subjected to a bending stress resultant is somewhat more complicated than for inplane loading. The complication stems from the fact that the signs of stresses change from one side of the reference surface to the other, and the stresses vary linearly through the thickness of a layer. Fortunately, the stresses at the outer location of each layer are always larger in magnitude than the stresses at the inner location, and thus only the outermost location of each layer needs to be examined. Also, often the stresses at a negative z location within the laminate are related to the stresses at the identical positive z location by a simple sign change. The calculations done for positive z locations do double duty by simply changing signs. This characteristic of laminate bending is not, however, always the case, particularly for unsymmetric laminates. Figure 8.12 shows the distributions of principal material system stresses in the [±30/0]$_S$ laminate subjected to a bending stress resultant $M_x = 12.84$ N·m/m. With this figure available, it is possible to focus on the likely locations of failure. In particular, if the loading is increased, if failure is going to be due to high stresses in the fiber direction, then it will occur at the outer extremities of the −30° layers. If failure is going to be due to high stresses perpendicular to the fibers, it will also probably occur at the outer extremities of the −30° layers, though high stresses perpendicular to the fibers also occur at the outer extremities of the +30° layers. Without this figure, one would be tempted to assume that fiber direction stresses are highest at the outer extremities of the laminate. This would be an error. If failure is going to be due to high shear stresses, it will occur at the outer extremities of the laminate, in +30° layers. Table 9.17 uses the results of equation (8.27) and considers the outermost location of each layer on the positive z side of the reference surface to give the principal material system stresses due to a bending stress resultant of $M_x = m_x$. Accordingly, the key equations, using MPa, for studying failure at these $+z$ locations are:

TABLE 9.17
Principal material system stresses (MPa) in $[\pm30/0]_S$ laminate due to $M_x = m_x$ (N·m/m)*

Layer	σ_1	σ_2	τ_{12}
+30° ($z = 0.450$ mm)	$+5.79m_x$	$-0.850m_x$	$-1.415m_x$
−30° ($z = 0.300$ mm)	$+9.01m_x$	$-0.875m_x$	$+0.772m_x$
0° ($z = 0.150$ mm)	$+7.52m_x$	$-0.618m_x$	$-0.0858m_x$

* Positive z locations.

For the $+30°$ layer at $z = 0.450$ mm:

$$5.79m_x = \sigma_1^C$$
$$5.79m_x = \sigma_1^T$$
$$-0.850m_x = \sigma_2^C$$
$$-0.850m_x = \sigma_2^T \qquad (9.103a)$$
$$-1.415m_x = -\tau_{12}^F$$
$$-1.415m_x = \tau_{12}^F$$

For the $-30°$ layer at $z = 0.300$ mm:

$$9.01m_x = \sigma_1^C$$
$$9.01m_x = \sigma_1^T$$
$$-0.875m_x = \sigma_2^C$$
$$-0.875m_x = \sigma_2^T \qquad (9.103b)$$
$$0.772m_x = -\tau_{12}^T$$
$$0.772m_x = \tau_{12}^F$$

For the $0°$ layer at $z = 0.150$ mm

$$7.52m_x = \sigma_1^C$$
$$7.52m_x = \sigma_1^T$$
$$-0.618m_x = \sigma_2^C$$
$$-0.618m_x = \sigma_2^T \qquad (9.103c)$$
$$-0.0858m_x = -\tau_{12}^T$$
$$-0.0858m_x = \tau_{12}^F$$

The limiting bending moments determined from these equations are summarized in Table 9.18. It is important to note that another three sets of six equations can be written for each of the negative locations $z = -0.450$ mm, -0.300 mm, and -0.150 mm. These 18 equations would be identical to the 18 equations of equation (9.103) except the terms involving m_x would have a sign change. For example, the

TABLE 9.18
Summary of failure moments M_x (N·m/m) for $[\pm30/0]_S$ laminate: Maximum stress criterion[*]

Layer			Failure mode			
	σ_1^C	σ_1^T	σ_2^C	σ_2^T	$-\tau_{12}^F$	$+\tau_{12}^F$
+30°	−216	+259	+235	−58.8	+70.7	−70.7
−30°	−138.7	+166.4	+228	−57.1	−129.5	+129.5
0°	−166.3	+199.6	+324	−81.0	+1166	−1166

[*] Positive z locations.

first and last entries of equation (9.103a) would be

$$-5.79m_x = \sigma_1^C$$
$$1.415m_x = \tau_{12}^F \qquad (9.104)$$

If we constructed a table summarizing the results of these 18 equations, it would be identical to Table 9.18 except for a sign change with every entry. Thus, Table 9.18 can be thought of as having a double sign with every entry. From these results, then, it can be concluded that bending moment resultants of $M_x = \pm57.1$ N·m/m will cause the laminate to fail due to a tensile stress perpendicular to the fibers in the $-30°$ layers. At just a slightly greater bending moment, ±58.8 N·m/m, the $+30°$ layers will also fail due to a tensile stress perpendicular to the fibers. Thus, it appears that at $M_x \simeq 59$ N·m/m there would be excessive cracking in the $\pm30°$ layers on one side of the laminate, the side depending on the sign of M_x. Again we see the weak link in a laminate is a material strength that does not have the benefit of fiber reinforcing.

9.8
SUMMARY

As we can see, implementing the maximum stress failure criterion requires a number of calculations to determine the load level that causes failure. The different modes of failure must be checked, and each layer considered. If bending is present, positive and negative z locations must be checked, though the computation need only be done for positive z for many laminates. These steps can all be automated and really should be when applying the criterion routinely. For combined loading, graphical displays of the failure envelopes are very useful.

In the implementation of the maximum stress failure criterion, the individual failure modes were examined one at a time. When computing the load required to fail the fibers, for example, we found that there was no concern for the magnitude, or direction, of the stress perpendicular to the fibers, σ_2. That is, there was no concern that σ_2 might interact with σ_1, so the failure load predicted by accounting for the presence of σ_2 might be different from the failure load predicted by ignoring the presence of σ_2. The concept of considering more than one stress component at a time when studying failure is termed *stress interaction*. The next chapter examines a failure criterion that addresses this issue.

9.9
SUGGESTED READINGS

A good review of the failure and yielding theories for isotropic materials, particularly the maximum principal stress criterion, can be found in the following:

1. Dowling, N. E. *Mechanical Behavior of Materials*. Englewood Cliffs, NJ: Prentice Hall, 1993.

While fiber-reinforced composite materials may be thought of as contemporary materials, many of the issues related to the failure of composites have similiar counterparts in the development of failure theories for wood. The issues with wood date back to the 1920s and 1930s, and earlier. The following book presents a good discussion of the failure theories of wood, including some of the older references:

2. Bodig, J., and B. A. Jayne. *Mechanics of Wood and Wood Composites*. New York: Van Nostrand Reinhold, 1982.

Reviews of failure criteria up to the mid-1980s are given in:

3. Hashin, Z. "Analysis of Composite Materials—A Survey." *Transactions of the ASME, Journal of Applied Mechanics* 50, no. 3 (1983), pp. 481–505.
4. Nahas, M. N. "Survey of Failure and Post-Failure Theories of Laminated Fiber-Reinforced Composites." *Journal of Composites Technology and Research* 8, no. 4 (1986), pp. 138–53.

Much of the work in failure criteria since the mid-1980s has been directed at the study of compression failure. The following papers discuss the kinking mode of compression failure, which is quite similiar to the compression failure of wood:

5. Evans, A. G., and W. F. Adler. "Kinking as a Mode of Structural Degradation in Carbon Fiber Composites." *Acta Metallurgica* 26 (1978), pp. 725–38.
6. Budianski, B., and N. A. Fleck. "Compressive Failure of Fibre Composites." *Journal of Mechanics and Physics of Solids* 41 (1993), pp. 183–211.
7. Jelf, P. M., and N. A. Fleck. "Compression Failure Mechanisms in Unidirectional Composites." *Journal of Composite Materials* 26, no. 1 (1992), pp. 2706–26.
8. Fleck, N. A.; L. Deng; and B. Budiansky. "Prediction of Kink Width in Compressed Fiber Composites." *Transactions of the ASME, Journal of Applied Mechanics* 62, no. 2 (1995), pp. 329–37.

See the following article for a discussion of the effects of having the ratio of tube radius to wall thickness less than 10:

9. Hyer, M. W. "Hydrostatic Response of Thick Laminated Composite Cylinders." *Journal of Fiber Reinforced Plastics and Composites* 26, no. 7 (1988), pp. 852–57.

CHAPTER 10

Failure Theories for Fiber-Reinforced Materials: The Tsai-Wu Criterion

The von Mises criterion, introduced in strength-of-materials courses for studying yielding of metals, can be written as

$$\frac{1}{2}\left(\frac{1}{\sigma^Y}\right)^2 \left[(\sigma_1 - \sigma_2)^2 + (\sigma_1 - \sigma_3)^2 + (\sigma_2 - \sigma_3)^2\right] = 1 \qquad (10.1)$$

where σ^Y is the yield stress of the metal and σ_1, σ_2, and σ_3 are the principal stresses. This equation is of the form

$$F(\sigma_1, \sigma_2, \sigma_3) = 1 \qquad (10.2)$$

According to the von Mises criterion, if

$$F(\sigma_1, \sigma_2, \sigma_3) < 1 \qquad (10.3)$$

then the material has not yielded. Equation (10.1) represents the well-known von Mises ellipsoid and thus a surface in σ_1-σ_2-σ_3 principal stress space, and equation (10.3) represents the volume inside this surface. Because rolled metals have slightly different properties in the roll direction than in the other two perpendicular directions, Hill (see Suggested Readings) assumed that the yield criterion for these orthotropic metals was of the form

$$F(\sigma_1 - \sigma_2)^2 + G(\sigma_1 - \sigma_3)^2 + H(\sigma_2 - \sigma_3)^2 + 2L\tau_{12}^2 + 2M\tau_{13}^2 + 2N\tau_{23}^2 = 1 \qquad (10.4)$$

The constants F, G, H, and so forth, are related to the yield stresses in the different directions, like σ^Y in equation (10.1), and either the 1, 2, or 3 direction is aligned with the roll direction.

This view of a failure criterion can be extended to composite materials, which are, of course, orthotropic in the principal material coordinate system, by assuming an equation of the form

$$F(\sigma_1, \sigma_2, \sigma_3, \tau_{23}, \tau_{13}, \tau_{12}) = 1 \qquad (10.5)$$

can be used to represent the failure condition of a composite, while the condition of no failure is given by

$$F(\sigma_1, \sigma_2, \sigma_3, \tau_{23}, \tau_{13}, \tau_{12}) < 1 \qquad (10.6)$$

How do we determine the specific form of F for a composite? How do we know such a function can exist? Is the concept even valid? After all, composites are not like metals when it comes to failure. There are distinctly different failure mechanisms associated with failure of a composite: fiber kinking in compression, fiber fracture in tension, and failure at the fiber-matrix interface in shear or tension perpendicular to the fiber. Thus, why would extending the concepts from metals be valid for composites? In the strictest sense, it is not. However, if the generalization is viewed as a hypothesis for fitting empirical data, and if the fit is reasonable, then the hypothesis provides us with an indicator that can be used to study failure.

For a state of plane stress, if the power of the stress components is maintained at 2, as in equations (10.1) and (10.4), the most general form of F is

$$\begin{aligned} F(\sigma_1, \sigma_2, \tau_{12}) = F_1\sigma_1 + F_2\sigma_2 + F_6\tau_{12} + F_{11}\sigma_1^2 + F_{22}\sigma_2^2 + F_{66}\tau_{12}^2 \\ + 2F_{12}\sigma_1\sigma_2 + 2F_{16}\sigma_1\tau_{12} + 2F_{26}\sigma_2\tau_{12} \end{aligned} \qquad (10.7)$$

where in the above F_1, F_2, F_6, F_{11}, F_{22}, F_{66}, F_{12}, F_{16}, and F_{26} are constants. All stress components are represented to the first and second powers, and all products of the stresses are represented. The constants F_{12}, F_{16}, and F_{26} will be referred to as the *interaction constants*, and the magnitude of their value will dictate the degree of interaction among stress components. Interaction between the normal stresses σ_1 and σ_2 and the shear stress τ_{12} is included by virtue of constants F_{16} and F_{26}, and interaction between the normal stress components σ_1 and σ_2 is included with the F_{12} term.

With the above considerations, the failure criterion that we are seeking takes the form

$$\begin{aligned} F_1\sigma_1 + F_2\sigma_2 + F_6\tau_{12} + F_{11}\sigma_1^2 + F_{22}\sigma_2^2 + F_{66}\tau_{12}^2 \\ + 2F_{12}\sigma_1\sigma_2 + 2F_{16}\sigma_1\tau_{12} + 2F_{26}\sigma_2\tau_{12} = 1 \end{aligned} \qquad (10.8)$$

and the condition of no failure is given by the inequality

$$\begin{aligned} F_1\sigma_1 + F_2\sigma_2 + F_6\tau_{12} + F_{11}\sigma_1^2 + F_{22}\sigma_2^2 + F_{66}\tau_{12}^2 \\ + 2F_{12}\sigma_1\sigma_2 + 2F_{16}\sigma_1\tau_{12} + 2F_{26}\sigma_2\tau_{12} < 1 \end{aligned} \qquad (10.9)$$

The failure criterion represents a general second-order surface in the space with co-ordinates σ_1, σ_2, τ_{12}. Recall, the maximum stress criterion in Figure 9.7 represented a piecewise planar surface with sharp edges and corners. A second-order surface, on the other hand, is smooth, like an ellipsoid. If we know the values of constants $F_1, \ldots, F_{66}$, then we can construct the surface, if desired, and furthermore, the failure load of a laminate can be evaluated by using equation (10.8). Equation (10.8) is the plane-stress form of the failure criterion postulated by Tsai and Wu (see Suggested Readings). How do we determine the constants $F_1, \ldots, F_{66}$? The answer is simple! We evaluate them with the information we have regarding failure of an element of fiber-reinforced material, namely: we evaluate them in terms of $\sigma_1^T, \sigma_1^C, \sigma_2^T, \sigma_2^C$, and τ_{12}^F.

10.1
DETERMINATION OF THE CONSTANTS

We determine the constants $F_1, \ldots, F_{66}$ by referring to the results of simple failure tests with fiber-reinforced composite material. Consider an element of material subjected to a stress only in the fiber direction. For this situation

$$\sigma_1 \neq 0$$
$$\sigma_2 = 0 \qquad \qquad (10.10)$$
$$\tau_{12} = 0$$

The function $F(\sigma_1, \sigma_2, \tau_{12})$ of equation (10.7) reduces to the form

$$F(\sigma_1, \sigma_2, \tau_{12}) = F_1\sigma_1 + F_{11}\sigma_1^2 \qquad (10.11)$$

If, as in Figure 10.1, the stress σ_1 is tension, then failure occurs when $\sigma_1 = \sigma_1^T$. The failure criterion must be unity at this value of σ_1; that is:

$$F_1\sigma_1^T + F_{11}\left(\sigma_1^T\right)^2 = 1 \qquad (10.12)$$

If, on the other hand, the stress σ_1 is compression, as in Figure 10.2, then failure occurs when $\sigma_1 = \sigma_1^C$. The failure criterion must again be unity at this value of σ_1, namely,

$$F_1\sigma_1^C + F_{11}\left(\sigma_1^C\right)^2 = 1 \qquad (10.13)$$

Equations (10.12) and (10.13) resulting from the two tests for failure in the fiber direction provide enough information to solve for F_1 and F_{11}. These two equations result in

$$F_1 = \frac{1}{\sigma_1^T} + \frac{1}{\sigma_1^C} \qquad F_{11} = -\frac{1}{\sigma_1^T \sigma_1^C} \qquad (10.14)$$

$\sigma_1 \neq 0 \quad \sigma_2 = 0 \quad \tau_{12} = 0$
At failure $\sigma_1 = \sigma_1^T$
Failure criterion becomes
$F_1\sigma_1^T + F_{11}(\sigma_1^T)^2 = 1$

FIGURE 10.1
Tensile failure in 1 direction as it applies to the Tsai-Wu criterion

$\sigma_1 \neq 0 \quad \sigma_2 = 0 \quad \tau_{12} = 0$
At failure $\sigma_1 = \sigma_1^C$
Failure criterion becomes
$F_1\sigma_1^C + F_{11}(\sigma_1^C)^2 = 1$

FIGURE 10.2
Compression failure in 1 direction as it applies to the Tsai-Wu criterion

We can take a similar approach with tension and compression testing of an element of material in the 2 direction. For this situation

$$\sigma_1 = 0$$

$$\sigma_2 \neq 0 \qquad (10.15)$$

$$\tau_{12} = 0$$

The function $F(\sigma_1, \sigma_2, \tau_{12})$ of equation (10.7) reduces to the form

$$F(\sigma_1, \sigma_2, \tau_{12}) = F_2\sigma_2 + F_{22}\sigma_2^2 \qquad (10.16)$$

If the stress σ_2 is tension, then failure occurs when $\sigma_2 = \sigma_2^T$ and the failure criterion becomes

$$F_2\sigma_2^T + F_{22}\left(\sigma_2^T\right)^2 = 1 \qquad (10.17)$$

If the stress σ_2 is compression, then failure occurs when $\sigma_2 = \sigma_2^C$, resulting in

$$F_2\sigma_2^C + F_{22}\left(\sigma_2^C\right)^2 = 1 \qquad (10.18)$$

From equations (10.17) and (10.18),

$$F_2 = \frac{1}{\sigma_2^T} + \frac{1}{\sigma_2^C} \qquad F_{22} = -\frac{1}{\sigma_2^T \sigma_2^C} \qquad (10.19)$$

These two loading conditions with σ_2 are shown in Figures 10.3 and 10.4.

The results from testing to failure in shear an element of material can be used to determine two more constants in the criterion. If an element is loaded only in shear, then

$$\sigma_1 = 0$$

$$\sigma_2 = 0 \qquad (10.20)$$

$$\tau_{12} \neq 0$$

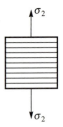

$\sigma_1 = 0 \quad \sigma_2 \neq 0 \quad \tau_{12} = 0$

At failure $\sigma_2 = \sigma_2^T$

Failure criterion becomes

$F_2\sigma_2^T + F_{22}(\sigma_1^T)^2 = 1$

Figure 10.3

Tension failure in 2 direction as it applies to the Tsai-Wu criterion

$\sigma_1 = 0 \quad \sigma_2 \neq 0 \quad \tau_{12} = 0$

At failure $\sigma_2 = \sigma_2^C$

Failure criterion becomes

$F_2\sigma_2^C + F_{22}(\sigma_2^C)^2 = 1$

Figure 10.4

Compression failure in 2 direction as it applies to the Tsai-Wu criterion

The function $F(\sigma_1, \sigma_2, \tau_{12})$ of equation (10.7) becomes

$$F(\sigma_1, \sigma_2, \tau_{12}) = F_6\tau_{12} + F_{66}\tau_{12}^2 \tag{10.21}$$

If the stress τ_{12} is positive, as shown in Figure 10.5, then failure occurs when $\tau_{12} = \tau_{12}^F$ and

$$F_6\tau_{12}^F + F_{66}\left(\tau_{12}^F\right)^2 = 1 \tag{10.22}$$

If the stress τ_{12} is reversed, as shown in Figure 10.6, then failure occurs when $\tau_{12} = -\tau_{12}^F$ and

$$-F_6\tau_{12}^F + F_{66}\left(-\tau_{12}^F\right)^2 = 1 \tag{10.23}$$

These two equations lead to

$$F_6 = 0 \qquad F_{66} = \left(\frac{1}{\tau_{12}^F}\right)^2 \tag{10.24}$$

Basically, F_6 is zero because in the principal material coordinate system failure is not sensitive to the sign of the shear stress. This certainly makes physical sense and extends to nonplanar stress situations as well. As a result, the failure criterion only involves the shear stress squared, reflecting the insensitivity to sign.

To this point, the Tsai-Wu failure criterion, equation (10.8), becomes

$$\left(\frac{1}{\sigma_1^T} + \frac{1}{\sigma_1^C}\right)\sigma_1 + \left(\frac{1}{\sigma_2^T} + \frac{1}{\sigma_2^C}\right)\sigma_2 + \left(-\frac{1}{\sigma_1^T\sigma_1^C}\right)\sigma_1^2 + \left(-\frac{1}{\sigma_2^T\sigma_2^C}\right)\sigma_2^2$$
$$+ \left(\frac{1}{\tau_{12}^F}\right)^2\tau_{12}^2 + 2F_{12}\sigma_1\sigma_2 + 2F_{16}\sigma_1\tau_{12} + 2F_{26}\sigma_2\tau_{12} = 1 \tag{10.25}$$

We need to evaluate coefficients that involve the *product* of two stress components. Whereas we could evaluate F_1 and F_{11} by looking at the results of testing to failure an element of material with a single stress component applied, namely, σ_1, to determine F_{12}, F_{16}, and F_{26} the failure of an element of material must be studied when stressed by more than one component. This form of testing is difficult and it can be expensive. However, two of the three remaining coefficients can be shown to be

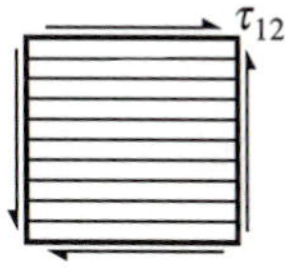

$\sigma_1 = 0 \quad \sigma_2 = 0 \quad \tau_{12} \neq 0$

At failure $\tau_{12} = \tau_{12}^F$

Failure criterion becomes

$F_6\tau_{12}^F + F_{66}(\tau_{12}^F)^2 = 1$

Figure 10.5
Failure due to positive τ_{12} as it applies to the Tsai-Wu criterion

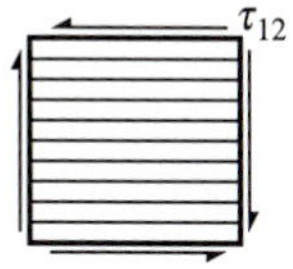

$\sigma_1 = 0 \quad \sigma_2 = 0 \quad \tau_{12} \neq 0$

At failure $\tau_{12} = -\tau_{12}^F$

Failure criterion becomes

$-F_6\tau_{12}^F + F_{66}(-\tau_{12}^F)^2 = 1$

Figure 10.6
Failure due to negative τ_{12} as it applies to the Tsai-Wu criterion

zero on physical grounds. The third coefficient can be estimated without resorting to actual experimental results.

To determine F_{16}, consider an element of fiber-reinforced material subjected to a tensile stress in the fiber direction. Assume, as in Figure 10.7(a), the tensile stress has a specific and known value, σ_1^*. Now suppose a shear stress τ_{12} is superposed on the element. Assume the shear stress is started from zero and increased until the element fails. Suppose, as Figure 10.7(b) shows, the value of τ_{12} that causes failure is τ_{12}^*. Because the element is stressed to failure, the failure criterion, equation (10.8), can be written as

$$F_1\sigma_1^* + F_{11}\left(\sigma_1^*\right)^2 + F_{66}\left(\tau_{12}^*\right)^2 + 2F_{16}\sigma_1^*\tau_{12}^* = 1 \qquad (10.26)$$

Recall that F_1, F_{11}, and F_{66} are known, and F_6 was shown to be zero. Now consider another element of material also loaded in the fiber direction by a tensile stress $\sigma_1 = \sigma_1^*$. A negative shear stress τ_{12} is applied to this element and increased in magnitude until the element fails. Because this experiment is being conducted in the principal material system, it is logical to assume, as in Figure 10.7(c), that the value of shear stress that causes failure is $\tau_{12} = -\tau_{12}^*$. If this is the case, then the failure criterion can be written as

$$F_1\sigma_1^* + F_{11}\left(\sigma_1^*\right)^2 + F_{66}\left(\tau_{12}^*\right)^2 - 2F_{16}\sigma_1^*\tau_{12}^* = 1 \qquad (10.27)$$

If equation (10.27) is subtracted from equation (10.26), the result is

$$4F_{16}\left(\tau_{12}^*\right)^2 = 0 \qquad (10.28)$$

from which we can conclude that

$$F_{16} = 0 \qquad (10.29)$$

A similar argument can be made regarding F_{26}, that is:

$$F_{26} = 0 \qquad (10.30)$$

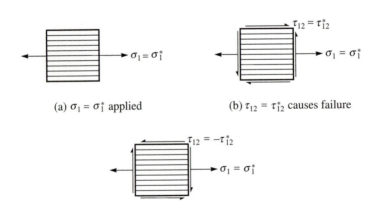

(a) $\sigma_1 = \sigma_1^*$ applied

(b) $\tau_{12} = \tau_{12}^*$ causes failure

(c) $\tau_{12} = -\tau_{12}^*$ also causes failure

FIGURE 10.7
Combined stresses σ_1 and τ_{12} as they apply to the Tsai-Wu criterion

The Tsai-Wu failure criterion thus simplifies one step further to become

$$F_1\sigma_1 + F_2\sigma_2 + F_{11}\sigma_1^2 + F_{22}\sigma_2^2 + F_{66}\tau_{12}^2 + 2F_{12}\sigma_1\sigma_2 = 1 \qquad (10.31)$$

We now turn to the evaluation of F_{12}.

The coefficient F_{12} involves both σ_1 and σ_2, and to determine F_{12} experimentally requires testing with nonzero values of both σ_1 and σ_2. There are several ways to accomplish this. One is to construct a loading device that applies both a σ_1 and a σ_2 simultaneously. The resulting biaxial state of stress would be that shown in Figure 10.8. Ideally the two components of stress would be controlled independently. Theoretically, a single pair of values of σ_1 and σ_2 is all that is needed to determine the value of F_{12}. In practice, a range of values of σ_1 and a range of values of σ_2, including both tensile and compressive values, should be studied and an average value computed for F_{12}. A second method of determining F_{12} experimentally is to use a uniaxial specimen with its fibers aligned at some known angle relative to the load direction. Then, as in Figure 10.9, the uniaxial stress is σ_x, and the components of stress in the principal material system, σ_1, σ_2, and τ_{12}, are given by the transformation equations, equation (5.10). If for a given off-axis angle $\theta = \theta^*$ failure occurs when $\sigma_x = \sigma_x^*$, then for this condition the Tsai-Wu failure criterion of equation (10.31) becomes

$$\left(F_1\cos^2(\theta^*) + F_2\sin^2(\theta^*)\right)\sigma_x^* + \left(F_{11}\cos^4(\theta^*) + F_{22}\sin^4(\theta^*)\right.$$
$$\left. + F_{66}\cos^2(\theta^*)\sin^2(\theta^*) + 2F_{12}\cos^2(\theta^*)\sin^2(\theta^*)\right)\sigma_x^{*2} = 1 \qquad (10.32)$$

From this, because the values of all quantities in the equation except F_{12} are known, the value of F_{12} can be determined. Obviously all that is needed is one test. However, several tests should be conducted at different angles to determine the consistency of the value of F_{12} determined in this manner. Finally, F_{12} could be determined by using a helically wound cylinder made from the material of interest. As Figure 10.10 shows, a cylinder can be internally or externally pressurized, loaded axially, loaded in torsion, or stressed in any combination of these to produce a variety of magnitudes and signs of σ_1, σ_2, and τ_{12} in the cylinder wall. The value of F_{12} can then be studied with such a specimen.

Another method of determining the value of F_{12} appeals to heuristic arguments and is as follows: For the case of plane stress, the von Mises criterion, equation (10.1), can be written

$$\left(\frac{1}{\sigma^Y}\right)^2\sigma_1^2 + \left(\frac{1}{\sigma^Y}\right)^2\sigma_2^2 - \left(\frac{1}{\sigma^Y}\right)^2\sigma_1\sigma_2 = 1 \qquad (10.33)$$

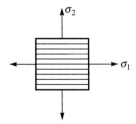

FIGURE 10.8
Biaxial loading for determining F_{12} in the Tsai-Wu criterion

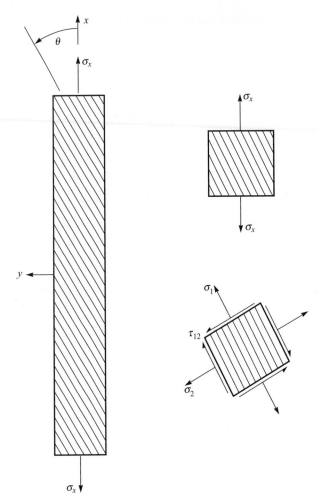

FIGURE 10.9
Off-axis tensile specimen for determining F_{12} in the Tsai-Wu criterion

If the Tsai-Wu criterion is applied to cases for which the von Mises criterion is valid, and the subscripts 1 and 2 in the Tsai-Wu criterion are identified with principal stress directions, then

$$\sigma_1^T = \sigma^Y \qquad \sigma_1^C = -\sigma^Y$$
$$\sigma_2^T = \sigma^Y \qquad \sigma_2^C = -\sigma^Y \qquad (10.34)$$

As a result, by their definitions in equations (10.14) and (10.19), F_1 and F_2 are zero. Because τ_{12} is zero when 1 and 2 are identified with principal stress directions, the Tsai-Wu criterion as given by equation (10.31) reduces to

$$\left(\frac{1}{\sigma^Y}\right)^2 \sigma_1^2 + \left(\frac{1}{\sigma^Y}\right)^2 \sigma_2^2 + 2F_{12}\sigma_1\sigma_2 = 1 \qquad (10.35)$$

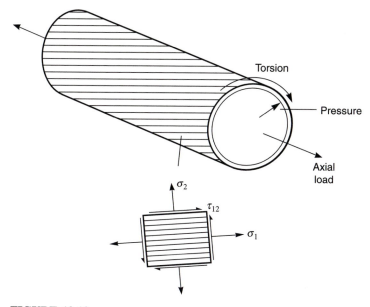

FIGURE 10.10
Cylindrical specimen for determining F_{12} in the Tsai-Wu criterion

For this reduced form of the Tsai-Wu criterion to yield the same result as the von Mises criterion, equation (10.33), it must be that

$$2F_{12} = -\left(\frac{1}{\sigma^Y}\right)^2 \tag{10.36}$$

This will be the case if in the Tsai-Wu criterion F_{12} is given by (see Suggested Readings)

$$F_{12} = -\frac{1}{2}\sqrt{F_{11}F_{22}} \tag{10.37}$$

Using this relation for composite materials, the Tsai-Wu criterion becomes

$$\boxed{F_1\sigma_1 + F_2\sigma_2 + F_{11}\sigma_1^2 + F_{22}\sigma_2^2 + F_{66}\tau_{12}^2 - \sqrt{F_{11}F_{22}}\sigma_1\sigma_2 = 1} \tag{10.38}$$

where

$$F_1 = \left(\frac{1}{\sigma_1^T} + \frac{1}{\sigma_1^C}\right) \qquad F_{11} = -\frac{1}{\sigma_1^T\sigma_1^C}$$

$$F_2 = \left(\frac{1}{\sigma_2^T} + \frac{1}{\sigma_2^C}\right) \qquad F_{22} = -\frac{1}{\sigma_2^T\sigma_2^C} \tag{10.39}$$

$$F_{66} = \left(\frac{1}{\tau_{12}^F}\right)^2$$

This is the form of the Tsai-Wu criterion we shall use. Table 10.1 gives the values of the failure stresses and the values of $F_1, \ldots, F_{66}$ for the graphite-reinforced material used throughout. Note the units of $F_1, \ldots, F_{66}$.

TABLE 10.1
Tsai-Wu failure parameters for graphite-reinforced composite

$\sigma_1^T = 1500$ MPa	$F_1 = 0.1333$ 1/GPa
$\sigma_1^C = -1250$ MPa	$F_2 = 15.00$ 1/GPa
$\sigma_2^T = 50$ MPa	$F_{11} = 0.533$ $(1/\text{GPa})^2$
$\sigma_2^C = -200$ MPa	$F_{22} = 100$ $(1/\text{GPa})^2$
$\tau_{12}^F = 100$ MPa	$F_{66} = 100$ $(1/\text{GPa})^2$

In the space formed by σ_1-σ_2-τ_{12} the Tsai-Wu criterion is an ellipsoid (see Figure 10.11). The ellipsoid is very long and slender, indicating the strong dependence on direction of the high strength of the fibers, and the weak strength of the matrix material. For the case of no shear in the principal material system, $\tau_{12} = 0$ and the Tsai-Wu criterion is an ellipse in σ_1-σ_2 space; Figure 10.12 shows the ellipse for the graphite-reinforced material considered here. The intersections of the ellipse with the coordinate axes are indicated by the letters A, B, C, and D, where these points represent the basic failure stresses σ_1^T, σ_2^T, σ_1^C, and σ_2^C, respectively. In the third quadrant the criterion predicts that compressive failure stresses in the fiber direction much more negative than σ_1^C are possible in the presence of compression in the 2 direction. This is a clear example of stress interaction; in this case the interaction predicts a significant strengthening influence. This characteristic of many interactive failure criteria causes concern.

Figure 10.13 shows several cross sections of the Tsai-Wu ellipsoid with the σ_2 axis expanded. The $\tau_{12} = 0$ case is a reproduction of Figure 10.12. The stress interaction is evident in all quadrants; the third quadrant again indicates that compressive stresses in the fiber direction exceeding σ_1^C in magnitude are possible with compressive σ_2. The other three quadrants predict that stress interaction effects can degrade strength. For example, with $\tau_{12} = 0$ and $\sigma_2 = -150$ MPa, failure in the fiber direction is predicted at just over $\sigma_1 = +700$ MPa. With no σ_2, failure in the fiber direction is predicted to occur, of course, at 1500 MPa. The cross sections

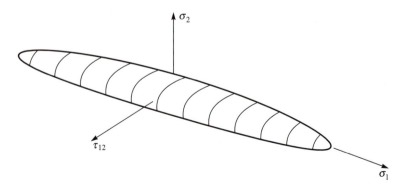

FIGURE 10.11
Tsai-Wu ellipsoid in principal material system stress space

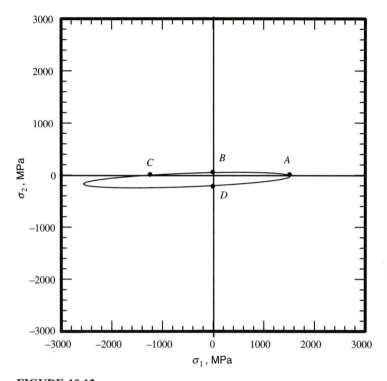

FIGURE 10.12
Cross section in σ_1-σ_2 plane ($\tau_{12} = 0$) of the Tsai-Wu ellipsoid for the graphite-reinforced composite

for nonzero values of τ_{12} indicate that the presence of shear stress τ_{12} even further degrades the levels of σ_1 and σ_2 that can be tolerated.

Before we turn to example problems to demonstrate the utility of the Tsai-Wu criterion, we should note three important characteristics of the criterion. First, in contrast to the six equations required to apply the maximum stress criterion, equation (9.4), the Tsai-Wu criterion involves only one equation, equation (10.38). This makes the application of the Tsai-Wu criterion simpler than the application of the maximum stress criterion. Second, because the Tsai-Wu criterion involves powers and products of the stresses, whenever the criterion is used to compute a failure load, the criterion will yield two answers, one positive and one negative. This will be apparent in the example problems to follow. However, this characteristic can be demonstrated with a rather simple example that, to some degree, represents a trivial application of the criterion: Consider a single layer of material subjected to a stress σ in the fiber direction. We ask what value of σ causes failure. For this case,

$$\sigma_1 = \sigma \qquad \sigma_2 = 0 \qquad \tau_{12} = 0 \tag{10.40}$$

and the Tsai-Wu criterion becomes

$$F_1\sigma + F_{11}\sigma^2 = 1 \tag{10.41}$$

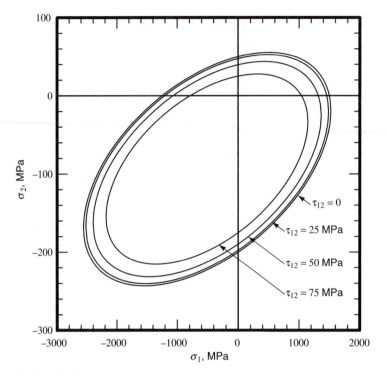

FIGURE 10.13
Several cross sections of the Tsai-Wu ellipsoid

This can be rearranged to read

$$F_{11}\sigma^2 + F_1\sigma - 1 = 0 \qquad (10.42)$$

This is a quadratic equation for σ and solution leads to

$$\sigma = \frac{-F_1 \pm \sqrt{F_1^2 + 4F_{11}}}{2F_{11}} \qquad (10.43)$$

Substituting for F_1 and F_{11} leads to

$$\sigma = \sigma_1^T \quad \text{and} \quad \sigma = \sigma_1^C \qquad (10.44)$$

In this case the stress to cause failure in tension and the stress to cause failure in compression, in the fiber direction, are the answers. This was expected for this problem but the results indicate the characteristic of the criterion to yield two answers. For this reason, the Tsai-Wu criterion is in a class of criteria called *quadratic* failure criteria. Finally, unlike the maximum stress criterion, the Tsai-Wu criterion does not directly indicate the mode of failure. The criterion predicts failure but does not indicate whether failure is due to fiber failure, shear failure, and so on. With additional calculations, however, some indication of the mode of failure is possible.

We will now turn to the application of the Tsai-Wu failure criterion to the previous example problems.

Exercises for Section 10.1

1. Verify the numerical values of the failure parameters in Table 10.1 and construct a similar table for glass-reinforced composite from the data in Table 9.1.

2. What would be the differences in Figure 10.13 if the ellipses from the Tsai-Wu criterion are drawn for the case of $\tau_{12} = -25$ MPa, -50 MPa, and -75 MPa?

10.2
FAILURE EXAMPLE 7: TUBE WITH AXIAL LOAD — TSAI-WU CRITERION

As with Failure Example 1, consider again a tube with a mean radius of 25 mm made of graphite-reinforced material that has a 10-layer wall with a stacking sequence of $[\pm 20/0_3]_S$. The tube is designed to resist axial load but has the low-angle off-axis layers (the $\pm 20°$ layers) to provide some circumferential and torsional stiffness, and to hold the load-carrying layers together. If we use the Tsai-Wu failure criterion, what is the maximum allowable axial load? What layer or layers control failure? What is the mode of failure? How do the predictions compare with those of the maximum stress criterion?

The tube and loading were illustrated in Figure 9.8, which was used in connection with studying this problem in the context of the maximum stress failure criterion. As with the procedure when using the maximum stress criterion, the stresses in the principal material system in each layer are computed for the case of a unit load, $P = 1$ N. The load is assumed to be multiplied by p, and so the stresses in each layer are also multiplied by p. The primary question is to determine the value or values of p that, according to the Tsai-Wu criterion, cause the tube to fail. These can be determined by substituting the stresses due to load p into the Tsai-Wu criterion and determining the value of p that causes the criterion to equal unity. Of course the stresses in each layer are different and so there is an equation, involving p, for each layer. The equation that leads to the lowest value of p indicates which layer controls failure, and the value of p from this equation is the failure load.

Table 9.2 presented the stresses in each layer due to a unit axial load, and Table 9.3 gave the stresses due to a load p. For the $+20°$ layers, then, the Tsai-Wu criterion, equation (10.38), predicts failure to occur when

$$F_1(3830p) + F_2(-112.3p) + F_{11}(3830p)^2$$

$$+ F_{22}(-112.3p)^2 + F_{66}(-148.7p)^2 \qquad (10.45)$$

$$- \sqrt{F_{11}F_{22}}(3830p)(-112.3p) = 1$$

which leads to a quadratic equation of the form

$$Ap^2 + Dp + F = 0 \qquad (10.46)$$

Using the values of the coefficients $F_1, \ldots, F_{66}$ from Table 10.1 results in

$$A = 0.1444 \times 10^{-10}$$
$$D = -22\,000 \times 10^{-10} \tag{10.47}$$
$$F = -1$$

and solving for p leads to

$$p = -198\,000 \quad \text{and} \quad p = +350\,000 \tag{10.48}$$

Thus the Tsai-Wu criterion predicts that the 20° layers will fail due either to a tensile load of $+350$ kN or to a compressive load of -198 kN. The simplicity of the Tsai-Wu approach can certainly be appreciated. One equation, equation (10.46), as compared to six, is much easier to deal with.

Using the stresses in the $-20°$ layers due to a load p, from Table 9.3, the Tsai-Wu criterion predicts failure to occur when

$$F_1(3830p) + F_2(-112.3p) + F_{11}(3830p)^2$$
$$+ F_{22}(-112.3p)^2 + F_{66}(148.7p)^2 \tag{10.49}$$
$$- \sqrt{F_{11}F_{22}}(3830p)(-112.3p) = 1$$

where this equation is, again, of the form

$$Ap^2 + Dp + F = 0 \tag{10.50}$$

with

$$A = 0.1444 \times 10^{-10}$$
$$D = -22\,000 \times 10^{-10} \tag{10.51}$$
$$F = -1$$

This equation is identical with the equation for the $+20°$ layers and the roots are thus

$$p = -198\,000 \quad \text{and} \quad p = +350\,000 \tag{10.52}$$

The load levels that cause the $+20°$ layers to fail also cause the $-20°$ layers to fail. Note that we have not discussed the mode of failure. Unlike the maximum stress criterion, the Tsai-Wu criterion does not explicitly address failure mode. However, shortly we shall address failure mode from the point of view of the Tsai-Wu criterion. Recall from Table 9.6 that the maximum stress criterion predicted the $+20°$ layers would fail at a tensile load of $P = +392$ kN due to tensile stresses in the fiber direction and at compressive load of $P = -327$ kN due to compressive stresses in the fiber direction. As Table 9.6 shows, these same tensile and compressive load levels and failure modes were predicted for the $-20°$ layers by the maximum stress criterion. These levels are larger in absolute value than the $+350$ kN and -198 kN predicted by the Tsai-Wu criterion. This is due to strength-reducing interactive effects in particular octants in σ_1-σ_2-τ_{12} space with the Tsai-Wu criterion.

The failure load for the $0°$ layers is determined in a similar manner. Using the stresses in the $0°$ layers due to load p in the Tsai-Wu criterion leads to

$$F_1(4770p) + F_2(-168.6p) + F_{11}(4770p)^2 + F_{22}(-168.6p)^2$$
$$+ F_{66}(0p)^2 - \sqrt{F_{11}F_{22}}(4770p)(-168.6p) = 1 \tag{10.53}$$

Again a quadratic of the form

$$Ap^2 + Dp + F = 0 \tag{10.54}$$

results, with

$$A = 0.208 \times 10^{-10}$$

$$D = -31\ 600 \times 10^{-10} \tag{10.55}$$

$$F = -1$$

leading to roots

$$p = -156\ 000 \quad \text{and} \quad p = +308\ 000 \tag{10.56}$$

For comparison, the maximum stress failure criterion predicts the values of $p = -262$ kN and $p = +315$ kN for the $0°$ layers. Table 10.2 summarizes the values of p for the three layer orientations, and the predictions of the maximum stress criterion are in parenthesis. Again, the primary reason for the difference between the two criteria is the stress interaction effects inherent in the Tsai-Wu criterion and absent from the maximum stress criterion. On an overall basis, the Tsai-Wu criterion predicts a tensile failure load of $+308$ kN, as opposed to $+315$ kN for the maximum stress criterion, and a compression failure load of -156 kN, as opposed to -262 kN. While the tensile load predictions are somewhat close, the compressive load predictions are significantly different. Which one is correct? Actually, neither is correct. Neither is incorrect. If, as discussed earlier, failure criteria are viewed as indicators rather than absolute predictors, then having two answers is not so disturbing.

To study the stress interaction effects, and the issue of failure mode as predicted by the Tsai-Wu criterion, consider the following: At the failure load level, the left-hand side of equation (10.38) sums to unity. Each of the six terms on the left-hand side contributes to, or subtracts from, the trend toward unity, depending on the values of the F_i and F_{ij}, and the values of σ_1, σ_2, and τ_{12} at failure. The contribution to unity of each of the six terms at the failure load level can easily be computed. For example, using the two values of p computed from equation (10.45) for the $+20°$ layers, we can construct a table to show the contribution of each term on the left-hand side of equation (10.38). Table 10.3 shows the results of these calculations for the two values of p. The values of σ_1, σ_2, and τ_{12} in the table are a result of

TABLE 10.2
Summary of loads P (kN) to cause failure in $[\pm 20/0_3]_S$ tube: Tsai-Wu criterion*

Layer	p (negative)	p (positive)
$+20°$	-198 (-327)	$+350$ $(+392)$
$-20°$	-198 (-327)	$+350$ $(+392)$
$0°$	-156 (-262)	$+308$ $(+315)$

* Maximum stress criterion prediction in parenthesis.

TABLE 10.3
Contribution of terms in Tsai-Wu criterion for $+20°$ layers in $[\pm 20/0_3]_S$ tube subject to tension

	$p = -198\ 000$ N	$p = +350\ 000$ N
σ_1, MPa	-758	1340
σ_2, MPa	22.2	-39.3
τ_{12}, MPa	29.4	-52.1
$F_1\sigma_1$	0.101	-0.179
$F_2\sigma_2$	0.334	-0.590
$F_{11}\sigma_1^2$	0.306	0.958
$F_{22}\sigma_2^2$	0.049	0.155
$F_{66}\tau_{12}^2$	0.087	0.271
$-\sqrt{F_{11}F_{22}}\sigma_1\sigma_2$	0.123	0.385

using the appropriate value of p and the entries of Table 9.3. For the condition of a tensile failure load, $P = +350$ kN, it appears that the term $F_{11}\sigma_1^2$ contributes significantly to the value of unity. However, the interaction term $-\sqrt{F_{11}F_{22}}\sigma_1\sigma_2$ also contributes, as does the shear term $F_{66}\tau_{12}^2$. The term $F_2\sigma_2$ tends to subtract. From these results, therefore, failure in the $+20°$ layers when the applied load is tensile appears to be dominated by tensile failure σ_1, but with some shear effects due to τ_{12} and interaction with compressive σ_2. Turning to Table 9.6, we can see that, according to the maximum stress criterion, in the $+20°$ layers fiber direction tension controls the level of applied tensile load. According to the Tsai-Wu criterion, this failure mode is of primary importance, but interaction lowers the tensile load level relative to the maximum stress criterion level, that is, $+392$ kN versus $+350$ kN.

For a compressive applied load, $P = -198$ kN, the $F_2\sigma_2$ and $F_{11}\sigma_1^2$ terms dominate, with the $F_1\sigma_1$ and $-\sqrt{F_{11}F_{22}}\sigma_1\sigma_2$ terms contributing. These four terms themselves nearly add to unity, signifying that interaction between these two components is important. The idea that interaction is an issue is further reinforced when the -198 kN level of allowable compressive load for the Tsai-Wu criterion is compared with the -327 kN applied load of the maximum stress criterion. The -327 kN level prediction is based on a fiber direction compression stress failure with no interaction assumed.

Table 10.4 summarizes the contributions of the terms in the Tsai-Wu failure criterion for each of the three-layer orientations in the $[\pm 20/0_3]_S$ tube at the failure load levels for each layer. For the $0°$ layers at a tensile load of $P = +308$ kN, the largest contribution to the unity value of the Tsai-Wu criterion is due to $F_{11}\sigma_1^2$. However, the interaction term $-\sqrt{F_{11}F_{22}}\sigma_1\sigma_2$ adds, while the $F_1\sigma_1$ and $F_2\sigma_2$ terms subtract. It could be stated that failure in the $0°$ layers due to an applied tensile load is due to fiber failure, with some interaction effects with σ_2. For a compressive axial load, the three largest terms are $F_2\sigma_2$, $F_{11}\sigma_1^2$, and the interaction term. Because compressive axial loads cause tensile σ_2 in the $0°$ layers, interaction between σ_1 and σ_2 might well be expected.

In summary, the results of Table 10.4 indicate the tube is limited in tension to $P = +308$ kN due to tensile failure in the fiber direction in the $0°$ layers. The tube

TABLE 10.4

Summary of loads P (kN) to cause failure in $[\pm 20/0_3]$ tube: Tsai-Wu criterion*[*]

	+20° layers		−20° layers		0° layers	
	$P = -198$	$P = +350$	$P = -198$	$P = +350$	$P = -156.0$	$P = +308$
σ_1, MPa	−758	1340	−758	1340	−743	1467
σ_2, MPa	22.2	−39.3	22.2	−39.3	26.3	−51.9
τ_{12}, MPa	29.4	−52.1	29.4	−52.1	0	0
$F_1\sigma_1$	0.101	−0.179	0.101	−0.179	0.099	−0.196
$F_2\sigma_2$	0.334	−0.590	0.334	−0.590	0.394	−0.779
$F_{11}\sigma_1^2$	0.306	0.958	0.306	0.958	0.295	1.149
$F_{22}\sigma_2^2$	0.049	0.155	0.049	0.155	0.069	0.269
$F_{66}\tau_{12}^2$	0.087	0.271	0.087	0.271	0	0
$-\sqrt{F_{11}F_{22}}\sigma_1\sigma_2$	0.123	0.385	0.123	0.385	0.143	0.556
Total	1.000	1.000	1.000	1.000	1.000	1.000

[*] The maximum stress failure criterion predicts failure with a positive axial load of $P = +315$ kN; failure is due to tensile stresses in the fiber direction in the 0° layers. Failure with a negative axial load occurs when $P = -262$ kN and is due to compression stresses in the fiber direction in the 0° layers.

is limited in compression to $P = -156$ kN due to an interaction between σ_1 and σ_2 in the 0° layers. Though the limiting number from the maximum stress criterion for a tensile value of P is similar (+315 kN), the compression value of P is quite different (−156 kN versus −262 kN). The 0° layers are predicted to be the critical layers for both criteria.

It should be mentioned that using the contributions in the Tsai-Wu criterion to indicate the failure mode is not as definitive as the failure mode prediction of the maximum stress criterion. However, using the contributions of the Tsai-Wu criterion to indicate possible failure modes is in keeping with the spirit of using failure criteria as indicators of failure rather than as absolute predictors.

Before continuing to the second example, it is logical to ask the significance of the negative contributions to the Tsai-Wu value of unity—that is, $F_1\sigma_1 = -0.196$ for the 0° layers with $P = +308$ kN. Figure 10.12 showed enhanced compressive strength in the fiber direction due to the presence of a compressive σ_2—that is, the elongation of the ellipse beyond $\sigma_1 = \sigma_1^C$ in the third quadrant. When such strengthening interaction is predicted, it appears as a subtraction from the sum toward unity of the terms on the left-hand side of the Tsai-Wu criterion.

Exercises for Section 10.2

1. Suppose the off-axis layers in the tube of Failure Example 1 were at $\pm 30°$ instead of $\pm 20°$.
   (a) Based on the Tsai-Wu criterion, what would be the axial load capacity of the tube?
   (b) Would the failure mode and the layers that control failure be the same as when the fibers were at $\pm 20°$? (To answer this question you essentially must redo the example problem, starting with the stresses due to $P = 1$ N, i.e., Table 9.2.) (c) Compare the results with the predictions of the maximum stress criterion, Exercise 1 in the Exercises for Section 9.2.

2. A $[\pm 45/0_2]_S$ graphite-reinforced plate is subjected to a biaxial loading such that the stress resultant in the y direction is opposite in sign to and one-half the magnitude of the stress resultant in the x direction. Call the stresses resultant in the x direction N; the loading is given by

$$N_x = N$$
$$N_y = -0.5N$$
$$N_{xy} = 0$$

(*a*) Use the Tsai-Wu criterion to compute the value of N to cause failure. (*b*) What layers control failure? (*c*) Use the contribution to unity of each term in the criterion to estimate the failure mode. To answer these questions, construct a table similar to Table 10.4. (*d*) Compare the results with the predictions of the maximum stress criterion, Exercise 2 in the Exercises for Section 9.2.

10.3
FAILURE EXAMPLE 8: TUBE IN TORSION — TSAI-WU CRITERION

Turn now to the second example solved previously with the maximum stress criterion, and consider that the $[\pm 20/0_3]_S$ tube is subjected to 225 kN tension and that there is an unwanted amount of torsion, T. According to the Tsai-Wu failure criterion, what is the maximum amount of torsion the tube can withstand before it fails? What layer or layers control failure and what is the mode of failure? How do the predictions compare with those of the maximum stress criterion?

The situation was illustrated in Figure 9.9, and as we were when considering the maximum stress failure criterion, we shall be interested in the stresses in each layer due to the 225 kN axial load, and an unknown to be solved for the amount of torsion, t. Table 9.9 provided us with the stresses in each layer for this situation. Referring to this table, then, for the $+20°$ layers, we find that the Tsai-Wu criterion, equation (10.38), becomes

$$F_1(861 + 0.804t) \times 10^6 + F_2(-25.3 - 0.048t) \times 10^6$$
$$+ F_{11}(861 + 0.804t)^2 \times 10^{12}$$
$$+ F_{22}(-25.3 - 0.048t)^2 \times 10^{12}$$
$$- \sqrt{F_{11}F_{22}}(861 + 0.804t)(-25.3 - 0.048t) \times 10^{12}$$
$$+ F_{66}(-33.5 + 0.0552t)^2 \times 10^{12} = 1$$

$$(10.57)$$

Substituting for the values of $F_1, \ldots, F_{66}$ results in a quadratic equation of the form

$$Bt^2 + Et + F = 0 \qquad (10.58)$$

Here F, rather than being -1, as in equation (10.46), involves the stresses due to the applied axial load and the F_i and F_{ij}. In this case

$$B = 1.163 \times 10^{-6}$$
$$E = 235 \times 10^{-6} \qquad (10.59)$$
$$F = -0.763$$

and the solution of equation (10.58) results in

$$t = -917 \quad \text{and} \quad t = +715 \tag{10.60}$$

For the $-20°$ layers, the Tsai-Wu criterion becomes

$$F_1(861 - 0.804t) \times 10^6 + F_2(-25.3 + 0.048t) \times 10^6$$
$$+ F_{11}(861 - 0.804t)^2 \times 10^{12}$$
$$+ F_{22}(-25.3 + 0.048t)^2 \times 10^{12}$$
$$- \sqrt{F_{11}F_{22}}(861 - 0.804t)(-25.3 + 0.0408t) \times 10^{12}$$
$$+ F_{66}(33.5 + 0.0552t)^2 \times 10^6 = 1$$

$$\tag{10.61}$$

resulting in $\qquad\qquad t = -715 \quad \text{and} \quad t = +917 \tag{10.62}$

Finally, following the above procedure for the $0°$ layers leads to

$$t = -1125 \quad \text{and} \quad t = +1125 \tag{10.63}$$

The results of this analysis are summarized in Table 10.5, and the contributions to unity for each layer and the six torsional load levels are included in the table.

In Table 10.5 we see that a torsional load of $T = \pm715$ N·m causes failure in the $\pm20°$ layers due to fiber tension. The stress component σ_1 is near its failure level of 1500 MPa, and the $F_{11}\sigma_1^2$ term is near unity. This failure load level is close to the prediction of ±795 N·m from the maximum stress criterion, Table 9.11, and the predicted failure mode is the same.

TABLE 10.5
Summary of torsions T (N·m) to cause failure in $[\pm20/0_3]$ tube with $P = 0.225$ MN: Tsai-Wu criterion*

	+20° layers		−20° layers		0° layers	
	$T = -917$	$T = +715$	$T = -715$	$T = +917$	$T = -1125$	$T = +1125$
σ_1, MPa	123.7	1436	1436	123.7	1072	1072
σ_2, MPa	18.86	59.7	−59.7	18.86	−37.9	−37.9
τ_{12}, MPa	−84.1	6.06	−6.06	84.1	−81.1	81.1
$F_1\sigma_1$	−0.016	−0.192	−0.192	−0.016	−0.143	−0.143
$F_2\sigma_2$	0.283	−0.896	−0.896	0.283	−0.569	−0.569
$F_{11}\sigma_1^2$	0.008	1.100	1.100	0.008	0.613	0.613
$F_{22}\sigma_2^2$	0.036	0.357	0.357	0.036	0.144	0.144
$F_{66}\tau_{12}^2$	0.707	0.004	0.004	0.707	0.658	0.658
$-\sqrt{F_{11}F_{22}}\sigma_1\sigma_2$	−0.017	0.627	0.626	−0.017	0.297	0.297
Total	1.000	1.000	1.000	1.000	1.000	1.000

* The maximum stress criterion predicts failure for a positive and negative torsion of $T = \pm795$ N·m due to tensile failure in the fiber direction in the $\pm20°$ layers.

Exercise for Section 10.3

A $[\pm 45/0_2]_s$ graphite reinforced plate is subjected to a biaxial loading such that the stress resultant in the y direction is -0.200 MN/m and the stress resultant in the x direction is variable, that is:

$$N_x = N$$
$$N_y = -0.200 MN/m$$
$$N_{xy} = 0$$

(a) Use the Tsai-Wu failure criterion to compute the value of N required to cause failure. (b) What layer or layers control failure? (c) What is the estimated mode of failure? (d) Compare your results with the predictions of the maximum stress criterion in the Exercise for Section 9.3.

10.4
FAILURE EXAMPLE 9: TUBE WITH COMBINED LOAD — TSAI-WU CRITERION

The third example demonstrates an advantage of the single-equation viewpoint of the Tsai-Wu failure criterion. The maximum stress criterion involves six equations per layer and, when interpreted graphically, leads to somewhat complicated figures (e.g., Figure 9.18). Consider the third example with the Tsai-Wu failure criterion applied rather than the maximum stress criterion. Specifically, in a particular application the tube is being used with an axial load P and a small torsional load T. According to the Tsai-Wu criterion, what are the ranges of applied axial load and applied torque the tube can withstand before it fails? What layer or layers control failure? What is the mode of failure? How do the results compare with those of the maximum stress criterion?

As stated at the time we studied this problem in the context of the maximum stress criterion, this is truly a combined-stress problem. The stresses in each layer due to an applied axial load p and a torsion t were given in Table 9.12. In this case, both p and t are unknown and we wish to establish a relationship between them based on the Tsai-Wu criterion. There will be a relation for each layer, and in p-t space this relation will describe an ellipse. Any point on the ellipse represents combinations of p and t that will produce failure in that layer, while any point inside the ellipse represents combinations of p and t that will not cause failure. Since each layer produces an ellipse, in p-t space there will be three ellipses: one ellipse represents the failure characteristics for the $+20°$ layers, the second ellipse represents the failure characteristics for the $-20°$ layers, and the third ellipse represents the failure characteristics for the $0°$ layers. The region interior to all three ellipses will represent the combinations of p and t that are "safe" for the tube as a whole. The calculations for these ellipses are as follows:

For the $+20°$ layers, the Tsai-Wu criterion becomes

$$F_1(0.00383p+0.804t)+F_2(-0.0001123p-0.0481t)+F_{11}(0.00383p+0.804t)^2$$
$$+ F_{22}(-0.0001123p - 0.0481t)^2 - \sqrt{F_{11}F_{22}}(0.00383p + 0.804t)$$
$$\times (-0.0001123p - 0.0481t) + F_{66}(-0.0001487p + 0.0552t)^2 = 1 \quad (10.64)$$

where substituting for the F_i and F_{ij} leads to an equation of the form

$$Ap^2 + Bt^2 + Cpt + Dp + Et + F = 0 \qquad (10.65)$$

In the above,

$$
\begin{aligned}
A &= 0.1444 \times 10^{-10} \\
B &= 11\,630 \times 10^{-10} \\
C &= 47.3 \times 10^{-10} \\
D &= -22\,000 \times 10^{-10} \\
E &= -82.9 \times 10^{-5} \\
F &= -1
\end{aligned}
\qquad (10.66)
$$

and in p-t space equation (10.65) represents an ellipse. For the case of $t = 0$, this equation reduces to the form of equation (10.46), and to the form of equation (10.58) when $P = +225$ kN. Figure 10.14 illustrates the ellipse represented by equation (10.65), and the figure is drawn on the same scale as Figure 9.15. Figure 10.14 is the Tsai-Wu criterion counterpart to Figure 9.15. The skewing of the shaded region

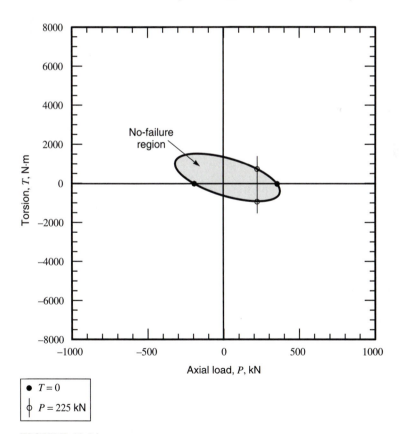

FIGURE 10.14
Tsai-Wu failure ellipse for $+20°$ layers

toward the fourth quadrant is evident in Figure 10.14, as it was in Figure 9.15. The values of p and t which represent load combinations that do not cause failure in the $+20°$ layers are given by all points inside the ellipse. Points on the ellipse represent values of p and t that result in failure of the $+20°$ layers.

On Figure 10.14 the points for $T = 0$ are noted. This corresponds to the first example problem and leads to the values of $P = -198.0\,\text{kN}$ and $P = +350\,\text{kN}$. (See equation [10.48] and the discussion leading up to it.) The points for $P = 225\,\text{kN}$, which correspond to the second example, are also shown, resulting in $T = -917\,\text{N·m}$ and $T = +715\,\text{N·m}$. (See equation [10.60] and the discussion leading up to it.) Of course, for the case of torsion only ($P = 0$), the intercepts of the ellipse with the vertical axis provide failure information.

Applying the Tsai-Wu criterion to the $-20°$ and $0°$ layers results in, respectively,

$$F_1(0.00383p - 0.804t) + F_2(-0.0001123p + 0.0481t)$$
$$+ F_{11}(0.00383p - 0.804t)^2$$
$$+ F_{22}(-0.0001123p + 0.0481t)^2$$
$$- \sqrt{F_{11}F_{22}}(0.00383p - 0.804t) \tag{10.67}$$
$$\times (-0.0001123p + 0.0481t)$$
$$+ F_{66}(0.0001487p + 0.0552t)^2 = 1$$

$$F_1(0.00477p) + F_2(-0.0001686p)$$
$$+ F_{11}(0.00477p)^2 + F_{22}(-0.0001686p)^2$$
$$- \sqrt{F_{11}F_{22}}(0.00477p)(-0.0001686p) \tag{10.68}$$
$$+ F_{66}(0.0721t)^2 = 1$$

These are both of the form of equation (10.65). For the $-20°$ layers

$$A = 0.1444 \times 10^{-10}$$
$$B = 11\,630 \times 10^{-10}$$
$$C = -47.3 \times 10^{-10}$$
$$D = -22\,000 \times 10^{-10} \tag{10.69}$$
$$E = 82.9 \times 10^{-5}$$
$$F = -1$$

while for the $0°$ layers

$$A = 0.208 \times 10^{-10}$$
$$B = 5200 \times 10^{-10}$$
$$C = 0$$
$$D = -31\,600 \times 10^{-10} \tag{10.70}$$
$$E = 0$$
$$F = -1$$

The ellipse represented by equation (10.67) (the −20° layers) is shown in Figure 10.15, the counterpart to Figure 9.16 for the maximum stress criterion, and the ellipse represented by equation (10.68) (the 0° layers) is shown in Figure 10.16, the counterpart to Figure 9.17. Note the interaction predicted by the Tsai-Wu criterion in all four quadrants for these three ellipses, interaction being identified by the smooth, as opposed to sharp, corners of the relations. In Figure 10.17 the ellipse for the +20° layers, the −20° layers, and the 0° layers are superimposed. This figure is the counterpart to Figure 9.18, and the area common to all three ellipses represents values of P and T that will not cause failure in the tube. The common area is bounded on the upper side by the ellipse for the −20° layers and on the lower side by the ellipse for the +20° layers. Except for low values of T combined with extreme values of P, failure in the tube is governed by the ±20° layers; this is somewhat similar to the conclusions reached by studying Figure 9.18. It is possible to associate various portions of the elliptical boundaries of the shaded region with various modes of failure, or the interaction of various modes, for the various layers. Alternatively, for specific values of P and T, a table similar to Table 10.4 or Table 10.5 can be constructed to determine the contributions of $F_1\sigma_1$, $F_{11}\sigma_1^2$, and the like. In either case, figures such

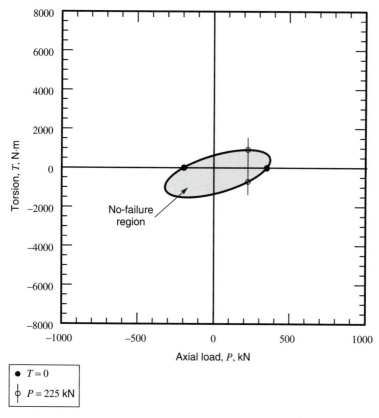

FIGURE 10.15
Tsai-Wu failure ellipse for −20° layers

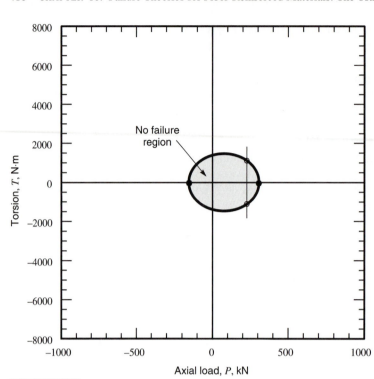

FIGURE 10.16
Tsai-Wu failure ellipse for 0° layers

as Figure 10.17, which show how the various layers contribute to the overall failure envelope, are very useful.

Before we leave the third example, it is interesting to compare the calculations from the Tsai-Wu criterion with those from the maximum stress criterion. Such a comparison is shown in Figure 10.18 (page 412) for the 20° layers, and in Figure 10.19 (page 413) for the 0° layers. In these figures the two criteria are superposed and the safe regions are indicated. For both layers the Tsai-Wu criterion is mostly within the maximum stress criterion. As stated earlier, the rounding off of the sharp corners, or cusps, of the maximum stress criterion by the Tsai-Wu criterion is due to interaction of the various stress components. The extensions of the Tsai-Wu criterion beyond the maximum stress criterion are also due to interaction effects, for example, negative σ_2 strengthening negative σ_1 in the third quadrant of Figure 10.12.

Finally, Figure 10.20 (page 414) combines the important portions of the maximum stress criterion, Figure 9.18, and the three ellipses from the Tsai-Wu criterion, Figure 10.17. The figure is rather complicated but it is clear that for this problem, the region within the three ellipses is within the region defined by the maximum stress

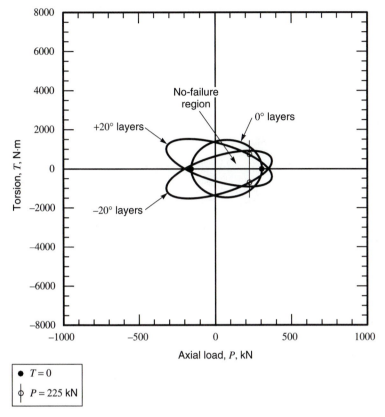

FIGURE 10.17
Superposition of the Tsai-Wu failure ellipses for $+20°$, $-20°$, and $0°$ layers

criterion. This, as seen from the figures of comparisons for specific layers, such as Figure 10.18, may not always be the case.

This completes our treatment of the three tube problems, which were studied using two different failure criteria. We shall now turn to solving the three familiar problems involving flat laminates, namely, the $[0/90]_S$ and $[\pm30/0]_S$ laminate subjected to force and moment resultants.

Exercise for Section 10.4

A $[\pm45/0_2]_S$ graphite-reinforced plate is subjected to a biaxial loading; that is:

$$N_x = N_x$$
$$N_y = N_y$$
$$N_{xy} = 0$$

(*a*) Use the Tsai-Wu failure criterion to determine the no-failure region of the plate in $N_x - N_y$ space. To do this, plot the ellipses for the various layers and indicate the safe region within the

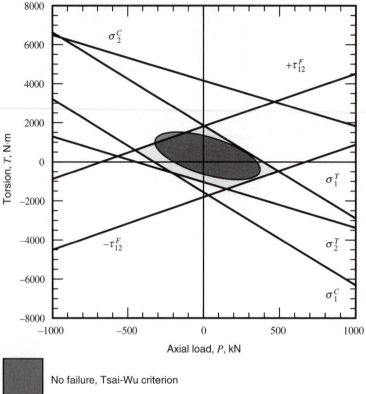

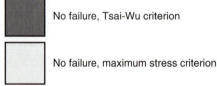

No failure, Tsai-Wu criterion

No failure, maximum stress criterion

FIGURE 10.18
Comparison between maximum stress criterion and Tsai-Wu criterion for $+20°$ layers
for the tube subjected to combined axial load P and torsion T

ellipses. Note that the cases of Exercise 2 in the Exercises for Section 10.2 and the Exercise
for Section 10.3 are included in this case. (*b*) Compare the results with the predictions of the
maximum stress criterion in the Exercise for Section 9.4.

10.5
FAILURE EXAMPLE 10: $[0/90]_S$ LAMINATE SUBJECTED TO N_x — TSAI-WU CRITERION

To continue to contrast the predictions of the maximum stress failure criterion, let us
examine the case of the flat $[0/90]_S$ laminate subjected to a force resultant N_x and
use the Tsai-Wu failure criterion to predict the value of load to produce failure. To
be thorough in our failure analysis, we will examine all layers. Because the Tsai-Wu

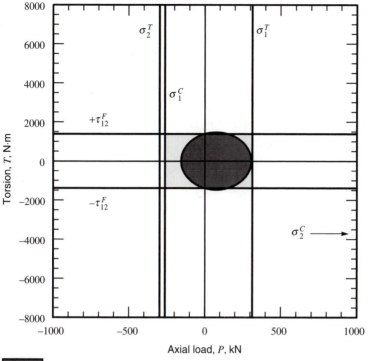

FIGURE 10.19

Comparison between maximum stress criterion and Tsai-Wu criterion for 0° layers for tube subjected to combined axial load P and torsion T

criterion involves all principal stress components, it is useful to refer to the tabulation of stresses in each layer that result from applying the force resultant $N_x = n_x$, as presented in Table 9.13. These stresses for each layer are substituted into the Tsai-Wu polynomial, resulting in a quadratic equation for n_x. For both the 0° and 90° layers the polynomial is of the form

$$An_x^2 + Dn_x + F = 0 \tag{10.71}$$

where for the 0° layers

$$A = 0.0421 \times 10^{-10}$$
$$D = 3560 \times 10^{-10} \tag{10.72a}$$
$$C = -1$$

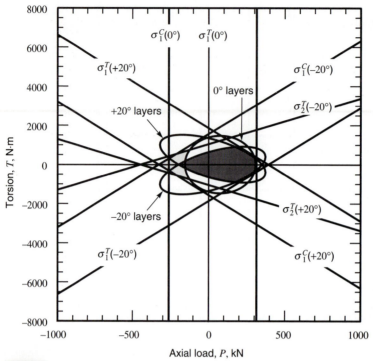

 No failure, Tsai-Wu criterion

 No failure, maximum stress criterion

FIGURE 10.20
Comparison between maximum stress criterion and Tsai-Wu criterion for all layers for tube subjected to combined axial load P and torsion T

while for the 90° layers

$$A = 0.0583 \times 10^{-10}$$

$$D = 36\,000 \times 10^{-10} \qquad (10.72b)$$

$$C = -1$$

The two values of n_x can be computed for each layer and the contributions of the various terms in the Tsai-Wu equation evaluated. Table 10.6 presents the results of these steps; the results indicate that the 90° layers control failure for positive, or tension, loading, and predict a failure load level of 0.208 MN/m. The 0° layers control failure for compression loading; the compressive failure load is −0.532 MN/m. The contributions of the various terms in the criterion indicate tension loading is limited by failure in the 90° layers due to tensile stresses perpendicular to the fibers, while failure in compression is limited by failure of the 0° layers in fiber direction

TABLE 10.6
Summary of failure loads N_x (MN/m) for $[0/90]_S$ laminate: Tsai-Wu criterion*[*]

	0° layers		90° layers	
	$N_x = -0.532$	$N_x = +0.447$	$N_x = -0.825$	$N_x = +0.208$
σ_1, MPa	-1644	1383	42.3	-10.65
σ_2, MPa	-27.2	22.9	-197.7	49.8
τ_{12}, MPa	0	0	0	0
$F_1\sigma_1$	0.219	-0.184	-0.006	0.001
$F_2\sigma_2$	-0.409	0.343	-2.97	0.747
$F_{11}\sigma_1^2$	1.442	1.020	0.001	0
$F_{22}\sigma_2^2$	0.074	0.052	3.91	0.248
$F_{66}\tau_{12}^2$	0	0	0	0
$-\sqrt{F_{11}F_{22}}\sigma_1\sigma_2$	-0.327	-0.231	0.0611	0.004
Total	1.000	1.000	1.000	1.000

[*] The maximum stress failure criterion predicts failure with a positive load of $N_x = +0.209$ MN/m, with failure due to tension perpendicular to the fibers in the 90° layers, and failure with a negative load of $N_x = -0.404$ MN/m, with failure due to compression failure in the fiber direction in the 0° layers.

compression. The failure load values predicted by the maximum stress criterion are included in the table. By comparing Tables 9.14 and 10.6, we see that for a tensile load the two failure criteria predict the same failure modes and nearly identical failure loads. The compressive failure load predicted by the Tsai-Wu criterion, however, is higher than that predicted by the maximum stress criterion, but the mode is the same, namely, fiber compression. A strengthening interaction between compressive σ_1 and compressive σ_2 in the 0° layers is responsible for the higher load prediction of the Tsai-Wu criterion, a strengthening that was illustrated in the third quadrant of Figure 10.12.

10.6
FAILURE EXAMPLE 11: $[\pm 30/0]_S$ LAMINATE SUBJECTED TO N_x — TSAI-WU CRITERION

To compute the failure loads for the $[\pm30/0]_S$ laminate subjected to a stress resultant N_x, we again use Table 9.15, constructed in Chapter 9 for the principal material system stresses for the condition $N_x = n_x$. The results from the table can be used directly to determine the Tsai-Wu criterion polynomials for n_x. The polynomials are all of the form of equation (10.71) and the values of the coefficients are:
   For the $\pm30°$ layers:

$$A = 0.0254 \times 10^{-10}$$

$$D = -12\,750 \times 10^{-10} \tag{10.73a}$$

$$F = -1$$

TABLE 10.7
Summary of failure loads N_x (MN/m) for $[\pm 30/0]_S$ laminate: Tsai-Wu criterion*

	+30° layers		−30° layers		0° layers	
	$N_x = -0.425$	$N_x = +0.926$	$N_x = -0.425$	$N_x = +0.926$	$N_x = -0.260$	$N_x = +0.665$
σ_1, MPa	−367	802	−367	802	−494	1261
σ_2, MPa	32.8	−71.6	32.8	−71.6	36.2	−92.5
τ_{12}, MPa	43.6	−95.2	−43.6	95.2	0	0
$F_1\sigma_1$	0.049	−0.106	0.049	−0.107	0.066	−0.168
$F_2\sigma_2$	0.492	−1.074	0.492	−1.074	0.542	−1.387
$F_{11}\sigma_1^2$	0.072	0.343	0.072	0.343	0.130	0.848
$F_{22}\sigma_2^2$	0.108	0.513	0.108	0.513	0.131	0.855
$F_{66}\tau_{12}^2$	0.190	0.906	0.190	0.906	0	0
$-\sqrt{F_{11}F_{22}}\sigma_1\sigma_2$	0.088	0.419	0.088	0.419	0.130	0.852
Total	1.000	1.000	1.000	1.000	1.000	1.000

*The maximum stress criterion predicts a maximum tensile load of +0.791 MN/m, with failure due to fiber direction tensile stress, and a maximum compressive load of −0.360 MN/m, with failure due to tensile stress perpendicular to the fibers.

For the 0° layers:

$$A = 0.0577 \times 10^{-10}$$
$$D = -23\,400 \times 10^{10} \tag{10.73b}$$
$$F = -1$$

Table 10.7 summarizes the results and illustrates the contributions of the various terms in the criterion. Examination of the table indicates that the laminate is limited in tension by a load of $N_x = +0.665$ MN/m, where it would appear that the mode of failure is due to a tensile stress in the fiber direction in the 0° layers. For compressive loading, the 0° layers limit the load to $N_x = -0.260$ MN/m due to what appears to be an interaction between tensile σ_2 and compressive σ_1.

From Table 9.16, the maximum stress criterion predicts a maximum tensile load of +0.791 MN/m, the limitation being fiber direction tension in the 0° layers. The two failure criteria are in agreement regarding this mode and load. For a compressive loading, the maximum stress criterion predicts a load of −0.360 MN/m, greater than the −0.260 MN/m compressive load predicted by the Tsai-Wu criterion; both criteria, however, predict this failure to be in the 0° layers. According to Table 10.7, it appears that for the 0° layers, interaction between tensile σ_2 and compressive σ_1 in the Tsai-Wu criterion leads to the lower predicted compressive load. The maximum stress criterion failure mode for a compressive load is due to tensile σ_2. The excessive tensile value of σ_2 is common to both criteria and hence must be an important aspect of failure.

10.7
FAILURE EXAMPLE 12: $[\pm 30/0]_S$ LAMINATE SUBJECTED TO M_x — TSAI-WU CRITERION

We now turn to the case of bending of the $[\pm 30/0]_S$ laminate. As with the maximum stress failure criterion, we must be concerned with examining the potential for failure at both positive z and negative z locations. The principal material system stresses at the outer location of each layer on the positive z side of the laminate for the case of $M_x = m_x$ were presented in Table 9.17. Substituting these principal material system stresses into the Tsai-Wu criterion results in an equation for all layers of the form

$$Am_x^2 + Dm_x + F = 0 \tag{10.74}$$

In particular, for the $+30°$ layer, at $z = 0.450$ mm,

$$A = 32.6 \times 10^{-5}$$
$$D = -1352 \times 10^{-5} \tag{10.75a}$$
$$F = -1$$

while for the $-30°$ layer, at $z = 0.300$ mm,

$$A = 23.7 \times 10^{-5}$$
$$D = -1433 \times 10^{-5} \tag{10.75b}$$
$$F = -1$$

and for the $0°$ layer, at $z = 0.150$ mm,

$$A = 10.29 \times 10^{-5}$$
$$D = -1027 \times 10^{-5} \tag{10.75c}$$
$$F = -1$$

The values of the two roots for each of the three polynomials and the contributions of the various terms in the Tsai-Wu criterion are summarized in Table 10.8. Care must be exercised in interpreting the results in this table. In particular, we must recall that the stresses on the positive z side of the laminate were used to compute the entries in this table. If the stresses on the negative z side of the laminate were used, the two roots to the three polynomials would be

$$m_x = -79.8 \quad \text{and} \quad m_x = 38.4$$
$$m_x = -101.8 \quad \text{and} \quad m_x = 41.4 \tag{10.76}$$
$$m_x = -160.3 \quad \text{and} \quad m_x = 60.6$$

Though the magnitude of the roots did not change, the signs of each pair switched. As a result, in a table like Table 10.8 the signs of the principal stresses would switch and the signs of some of the contributions would switch. It can thus be concluded that failure occurs when $M_x = \pm 38.4$ N·m/m. The failure is at the outer location of the $+30°$ layers and, according to the contributions to the criterion, is due to a combination of shear stress and tensile stress perpendicular to the fibers. Interestingly

TABLE 10.8

Summary of moments M_x (N·m/m) for $[\pm 30/0]_S$ laminate: Tsai-Wu criterion*

	Outer location of +30° layers		Outer location of −30° layers		Outer location of 0° layers	
	$M_x = -38.4$	$M_x = 79.8$	$M_x = -41.4$	$M_x = 101.8$	$M_x = -60.6$	$M_x = 160.3$
σ_1, MPa	−222	462	−373	918	−455	1205
σ_2, MPa	32.6	−67.8	36.2	−89.1	37.4	−99.0
τ_{12}, MPa	54.3	−113.0	−32.0	78.6	5.20	−13.75
$F_1\sigma_1$	0.030	−0.062	0.050	−0.122	0.061	−0.161
$F_2\sigma_2$	0.489	−1.017	0.544	−1.337	0.561	−1.486
$F_{11}\sigma_1^2$	0.026	0.1139	0.074	0.449	0.111	0.775
$F_{22}\sigma_2^2$	0.106	0.4600	0.131	0.794	0.140	0.981
$F_{66}\tau_{12}^2$	0.295	1.276	0.102	0.618	0.003	0.019
$-\sqrt{F_{11}F_{22}}\sigma_1\sigma_2$	0.053	0.229	0.099	0.597	0.124	0.872
Total	1.000	1.000	1.000	1.000	1.000	1.000

*Positive z locations. The maximum stress criterion predicts failure for a positive and negative moment at a load of $M_x = \pm 57.1$ N·m/m due to tensile stress perpendicular to the fibers in the −30° layers.

enough, for $M_x = \pm 41.4$ N·m/m, the outer location of the −30° layers fail due to the same combination of stresses. Thus, it can be concluded that near $M_x \simeq \pm 40$ N·m/m, the two outer layers experience cracking parallel with the fibers. This significantly deteriorates the laminate. However, the 0° layers are still intact and the laminate can continue to carry load past $M_x = \pm 40$ N·m/m.

In summary, for these last three examples the Tsai-Wu criterion generally results in lower failure loads than the maximum stress criterion. The exception to this was the higher load predicted for negative N_x in the $[0/90]_S$ laminate. It was mentioned that this was due to third-quadrant interaction between σ_1 and σ_2. When there is a difference between the maximum stress and Tsai-Wu criteria, examining the signs of the stresses and the magnitude of the various components in the Tsai-Wu criterion, and referring to information like that in Figures 10.12 and 10.13, can usually provide an explanation of the difference.

10.8
SUMMARY

This completes our chapters on failure. Two distinctly different failure criteria have been discussed and it has been shown in examples that they can lead to different results. Other failure criteria will, of course, lead to other results. The methodologies and procedures developed with the two criteria presented can be used with other criteria. In fact, besides presenting two criteria that are in use, the value of these past chapters is that they do point the way for performing a failure analysis. Stresses must be calculated, these stresses must be used in the equations representing the failure criteria, and the results interpreted. This approach would be followed with any stress-based criterion. With a strain-based criterion, strains in the principal material system would be used rather than stresses. The key issue with any one failure criterion is, "Does it accurately predict failure for your problem?" Generally, considerable testing is required to determine if this is the case. Unfortunately, there does not appear to be one universal criterion which works well for all situations and all materials. Material properties, lamination sequence, and type of loading all seem to influence which criterion works the best. For each particular class of problems and class of materials, a careful study of test data and predictions must be conducted before generalizations can be made. We suggest that more than one criterion be used and the results compared, as we have done here. Competing views are helpful!

10.9
SUGGESTED READINGS

A good review of the failure and yielding theories for isotropic materials, particularly the von Mises criterion, can be found in:

1. Dowling, N. E. *Mechanical Behavior of Materials*. Englewood Cliffs, NJ: Prentice Hall, 1993.

Yielding for orthotropic plasticity is discussed in:

2. Hill, R. *The Mathematical Theory of Plasticity*. New York: Oxford at the Clarendon Press, 1983.

One of the papers introducing the Tsai-Wu failure criterion is:

3. Tsai, S. W., and E. M. Wu. "A General Theory of Strength for Anisotropic Materials." *Journal of Composite Materials* 5 (1971), pp. 58–80.

Details of choosing $F_{12} = -\frac{1}{2}\sqrt{F_{11}F_{22}}$ can be found in:

4. Tsai, S. W., and M. T. Hahn. *Introduction to Composite Materials*. Lancaster, PA: Technomic Publishing Co., 1980, eq. 7.22.

An extension of the Tsai-Wu criterion is discussed in:

5. Tennyson, R. C.; D. MacDonald; and A. P. Nanyaro. "Evaluation of the Tensor Polynomial Failure Criterion for Composite Materials." *Journal of Composite Materials* 12 (1978), pp. 63–75.

A comparison of several interactive and noninteractive criteria, including experimental data, can be found in:

6. Sun, C. T., and B. J. Quinn. "Evaluation of Failure Criteria Using Off-Axis Laminate Specimens." In *Proceedings of the American Society for Composite Materials*. Lancaster, PA: Technomic Publishing Co., 1994. Eleventh Technical Conference.

CHAPTER 11

Environmentally Induced Stresses in Laminates

We have previously discussed how an element of fiber-reinforced material responds when its temperature is changed or it absorbs moisture. We discussed this topic in the context of having no stresses on the surface of the element, namely, free thermal- or moisture-induced strains, and in example problems in the context of specifying certain constraints on the element. However, the more important issue in the topic of free strains is determining their effect when the element of material is part of a laminate. In a $[0/90]_S$ laminate, the expansion of the 90° layers in their 2, or x, direction due to heating would be resisted by the contraction of the adjacent 0° layers in their 1 direction, also the x direction. The two adjacent layers would react in opposite manners when heated. The net result must be stresses in the layers. These stresses then add to stresses induced by any mechanical load, such as an N_x. The thermally induced stresses can be accounted for in the context of classical lamination theory. We do so in the following sections. The development parallels the development of classical lamination theory in Chapters 6 and 7; the primary difference is in the stress-strain relations used. Specifically, the effects of free thermal strains must be included in the stress-strain relations. We have looked at these relations in detail in previous chapters and thus are prepared to use them.

11.1
LAMINATE RESPONSE

11.1.1 Displacements and Strains

The key assumption when we include free thermal strains in the study of laminates is that, even with these strains present, the Kirchhoff hypothesis is valid. Because by the Kirchhoff hypothesis straight lines normal to the reference surface of the laminate remain straight and normal, it is implied that despite what can be radical dissimilarities in free thermal strains from one layer to the next, as in the case of a $[0/90]_S$ laminate, the internal stress state in the laminate adjusts so the lines remain straight. Determining just what this internal stress state actually is will be one outcome of the development of classical lamination theory that accounts for free thermal strains. Thus, we assume Figures 6.10 and 6.11 accurately represent the deformations of a laminate due to the actions of free thermal strains. The important difference between what we have learned so far and the case of accounting for free thermal strains is that now the displacements $u^o(x, y)$, $v^o(x, y)$, and $w^o(x, y)$ are the *total* displacements of a point on the reference surface. This is a subtle but extremely important point. Total displacements in the context of free thermal strains and laminate response do not mean displacements due to thermal effects added to displacements due to a mechanical load, such as N_x. Rather, by total displacements is meant the following: When an unrestrained element of material is heated, it deforms in accordance with the coefficients of thermal expansion of the material and the temperature change. When an element of material is part of a heated laminate, the element deforms, not in a free manner, but in a manner dictated by the coefficients of thermal expansion and the elastic properties of all the layers in the laminate, as well as by its own coefficients of thermal expansion and elastic properties. The element is no longer free; its displacement characteristics are dictated by the larger character of the laminate. These laminate displacements are considered the total displacements. It is the difference between the total displacements and the free thermal displacements in an element that cause stresses to develop in that element. Obviously, if the total displacements of the element are equal to the displacements due to the free thermal strains, then stresses would not develop. Generally, however, when the element is within a heated laminate, this is not the case. Stresses develop in the element due to differences between what the free element would like to do, and what it has to do because it is part of a laminate. Laminates can indeed be subjected to both a temperature change and a mechanical load, and for these situations there will be what could also be termed *total displacements*. For the moment, however, let us restrict our meaning of the word *total* to the former context.

With the understanding that the displacements represent the total displacements, and that the Kirchhoff hypothesis applies, equations (6.12) through (6.14) are then valid. They are repeated here for convenience as

$$
\begin{aligned}
u(x, y, z) &= u^o(x, y) - z\frac{\partial w^o(x, y)}{\partial x} \\
v(x, y, z) &= v^o(x, y) - z\frac{\partial w^o(x, y)}{\partial y} \\
w(x, y, z) &= w^o(x, y)
\end{aligned}
\tag{11.1}
$$

$$
\begin{aligned}
\varepsilon_x(x, y, z) &= \varepsilon_x^o(x, y) + z\kappa_x^o(x, y) \\
\varepsilon_y(x, y, z) &= \varepsilon_y^o(x, y) + z\kappa_y^o(x, y) \\
\gamma_{xy}(x, y, z) &= \gamma_{xy}^o(x, y) + z\kappa_{xy}^o(x, y)
\end{aligned}
\tag{11.2}
$$

with

$$
\varepsilon_x^o(x, y) = \frac{\partial u^o(x, y)}{\partial x} \quad \text{and} \quad \kappa_x^o(x, y) = -\frac{\partial^2 w^o(x, y)}{\partial x^2}
$$

$$
\varepsilon_y^o(x, y) = \frac{\partial v^o(x, y)}{\partial y} \quad \text{and} \quad \kappa_y^o(x, y) = -\frac{\partial^2 w^o(x, y)}{\partial y^2} \tag{11.3}
$$

$$
\gamma_{xy}^o(x, y) = \frac{\partial v^o(x, y)}{\partial x} + \frac{\partial u^o(x, y)}{\partial y} \quad \text{and} \quad \kappa_{xy}^o = -2\frac{\partial^2 w^o(x, y)}{\partial x \partial y}
$$

These equations define the total reference surface strains and curvatures, and hence the total strains as a function of thickness through the laminate.

11.1.2 Stresses

The second key assumption in the development of classical lamination theory to reflect free thermal strain effects is that the laminate is in a state of plane stress. From Chapter 5 the plane-stress stress-strain relations with free thermal strain effects included are understood and are available. If the reference surface strains and curvatures *and* the temperature at every point within the laminate are specified, then, using the stress-strain relations and the specification from equations (11.1)–(11.3) of how the strains vary through the thickness, the distribution of the stresses through the thickness of the laminate can be determined. The important point is that the stress-strain relations used must include free thermal strain effects. Specifically, with free thermal strain effects, the stress-strain relation for classical lamination theory follows from equation (5.163), with $\Delta M = 0$, as

$$
\left\{ \begin{array}{c} \sigma_x \\ \sigma_y \\ \tau_{xy} \end{array} \right\} = \left[\begin{array}{ccc} \bar{Q}_{11} & \bar{Q}_{12} & \bar{Q}_{16} \\ \bar{Q}_{12} & \bar{Q}_{22} & \bar{Q}_{26} \\ \bar{Q}_{16} & \bar{Q}_{26} & \bar{Q}_{66} \end{array} \right] \left\{ \begin{array}{c} \varepsilon_x^o + z\kappa_x^o - \alpha_x \Delta T \\ \varepsilon_y^o + z\kappa_x^o - \alpha_y \Delta T \\ \gamma_{xy}^o + z\kappa_{xy}^o - \alpha_{xy} \Delta T \end{array} \right\} \tag{11.4}
$$

The pertinent coefficients of thermal deformation in the off-axis system are given by equation (5.143) as

$$
\alpha_x = \alpha_1 \cos^2 \theta + \alpha_2 \sin^2 \theta
$$

$$
\alpha_y = \alpha_1 \sin^2 \theta + \alpha_2 \cos^2 \theta \tag{11.5}
$$

$$
\alpha_{xy} = 2(\alpha_1 - \alpha_2) \cos \theta \sin \theta
$$

Assuming the validity of the Kirchhoff hypothesis, having the concept of total displacements in place, and invoking the plane-stress assumption, we are ready to develop classical lamination theory to reflect the influence of free thermal strains. Recall that the original development was motivated by studying the strain and stress distributions in several examples, which began by stating that the reference surface strains of specific laminates were known. The strains and stresses through the thickness of the laminate were then computed. We will repeat these examples here, but they will include free thermal strain effects. As happened with the earlier examples, the graphical representation of the distribution of the stresses through the thickness of the laminate will provide motivation for defining force and moment resultants. This will lead to the definition of the ABD matrix for the laminate. While the definition of the ABD matrix is identical to the case for no thermal effects, the relationship between the ABD matrix and the force and moment resultants is not quite the same. Because stresses are the basis of the definition of the force and moment resultants, and because the stress-strain relations involve the free thermal strains, it is logical that this portion of the development would be different. That the definition of the ABD matrix is not altered with the introduction of free thermal strains is also logical. The ABD matrix involves nothing but laminate elastic properties and geometry, neither of which is altered by inclusion of free thermal strains. As we will see, the stress-strain relations and the relationship between the ABD matrix and the stress resultants are the only portions of the development of classical lamination theory that are altered by the inclusion of free thermal strains.

In the examples to follow, we shall discuss a temperature change that has physical significance. Polymer matrix composites generally are processed at an elevated temperature. Thermoset matrix materials require elevated temperatures to cure, and thermoplastic matrix materials require elevated temperatures for the melting required for consolidation of layers. After a period of time at the elevated temperature state the temperature of the composite is reduced, eventually reaching ambient temperature. Though the composite is soft at the elevated temperature, at some point in the cooling process the material hardens to the point that the elastic and thermal expansion properties of the composite are identifiable. Below this temperature, thermally induced stresses begin to develop in the laminate. The temperature at which the properties are identifiable is defined to be the stress-free temperature of the laminate. Using the difference in temperature between the stress-free state and the operating state of the structure provides an estimate of the processing stresses in the laminate at its operating state. The level of these processing stresses can influence the load-carrying capacity of the laminate. A 150°C temperature difference between the stress-free and operational temperatures of the structure would be considered representative of many composites. In the following examples, then, we shall use a ΔT of -150°C in the calculation of stresses.

11.2
EXAMPLES OF LAMINATE RESPONSE

11.2.1 CLT Example 9: [0/90]$_S$ Laminate Subjected to Known ε_x^o and Known ΔT

As the first example, consider the often-discussed [0/90]$_S$ laminate. Assume that at a particular point (x, y) on the reference surface

$$\varepsilon_x^o(x, y) = 1000 \times 10^{-6} \qquad \kappa_x^o(x, y) = 0$$
$$\varepsilon_y^o(x, y) = 0 \qquad\qquad \kappa_y^o(x, y) = 0 \qquad\qquad (11.6)$$
$$\gamma_{xy}^o(x, y) = 0 \qquad\qquad \kappa_{xy}^o(x, y) = 0$$

and the temperature change at every point within the laminate is $\Delta T = -150°C$. The distribution of the total strains through the thickness of the laminate at that point is given by

$$\varepsilon_x(x, y, z) = \varepsilon_x^o(x, y) + z\kappa_x^o(x, y) = 1000 \times 10^{-6}$$
$$\varepsilon_y(x, y, z) = \varepsilon_y^o(x, y) + z\kappa_y^o(x, y) = 0 \tag{11.7}$$
$$\gamma_{xy}(x, y, z) = \gamma_{xy}^o(x, y) + z\kappa_x^o(x, y) = 0$$

as shown in Figure 6.13. Because we are specifying the total strains, the picture of the strains is independent of how they are produced. Thus, this figure from Chapter 6, where thermal strains were not mentioned, is valid. The stresses in the x-y system that result from these strains are given by the stress-strain relations of equation (11.4). For the present situation, this equation becomes, for all layers,

$$\left\{ \begin{array}{c} \sigma_x \\ \sigma_y \\ \tau_{xy} \end{array} \right\} = \left[\begin{array}{ccc} \bar{Q}_{11} & \bar{Q}_{12} & \bar{Q}_{16} \\ \bar{Q}_{12} & \bar{Q}_{22} & \bar{Q}_{26} \\ \bar{Q}_{16} & \bar{Q}_{26} & \bar{Q}_{66} \end{array} \right] \left\{ \begin{array}{c} 1000 \times 10^{-6} - \alpha_x \Delta T \\ -\alpha_y \Delta T \\ -\alpha_{xy} \Delta T \end{array} \right\} \tag{11.8}$$

Even though there is no total extensional strain in the y direction nor total shear strain, there is a free thermal extensional strain in the y direction and, in general, a free thermal shear strain. These contribute to the mechanical strains and influence the stresses in the laminate, even though the total strain is the same as in the Chapter 6 example without thermally induced effects. For the $[0/90]_S$ laminate the coefficients of thermal deformation in x-y-z system are given by equation (11.5), for the $0°$ layers, as

$$\alpha_x = \alpha_1 = -0.018 \times 10^{-6}/°C$$
$$\alpha_y = \alpha_2 = 24.3 \times 10^{-6}/°C \tag{11.9a}$$
$$\alpha_{xy} = 0$$

while for the $90°$ layers

$$\alpha_x = \alpha_2 = 24.3 \times 10^{-6}/°C$$
$$\alpha_y = \alpha_1 = -0.018 \times 10^{-6}/°C \tag{11.9b}$$
$$\alpha_{xy} = 0$$

Numerically, then, for the $0°$ layers, from equations (11.8) and (11.9a), with $\Delta T = -150°C$,

$$\sigma_x = 166.3 \text{ MPa}$$
$$\sigma_y = 47.3 \text{ MPa} \tag{11.10a}$$
$$\tau_{xy} = 0$$

For the $90°$ layers, from equations (11.8) and (11.9b),

$$\sigma_x = 56.5 \text{ MPa}$$
$$\sigma_y = 13.59 \text{ MPa} \tag{11.10b}$$
$$\tau_{xy} = 0$$

The values of the reduced stiffnesses were given in Chapter 6 in connection with CLT Example 1. These stresses should be compared with the stresses for no thermal effects, equations (6.28) and (6.29). Figure 11.1 gives a graphical comparison of the $[0/90]_S$ laminate with and without thermal effects. The stress states represented by these two situations are different, the stresses being greater for the case of thermal effects, particularly in the y direction. This is not always the case and thus the findings of this example should not be generalized. Here the large free thermal strain in the 2 direction, due to the combination of the $-150°C$ temperature change and the large coefficient of thermal expansion in that direction, is responsible for the differences.

Each of the stresses of equation (11.10) can be thought of as being due to contributions from two parts. One part corresponds to the product of the reduced stiffness matrix and the total strain, and the second part to the product of the reduced stiffness matrix and the free thermal strains. Accordingly, equation (11.10a), which applies to the 0° layers, could be written to read

$$\sigma_x = 155.7 + 10.57 \text{ MPa}$$
$$\sigma_y = 3.02 + 44.3 \text{ MPa} \qquad (11.11a)$$
$$\tau_{xy} = 0$$

while for the 90° layers, equation (11.10b) could be written to read

$$\sigma_x = 12.16 + 44.3 \text{ MPa}$$
$$\sigma_y = 3.02 + 10.57 \text{ MPa} \qquad (11.11b)$$
$$\sigma_{xy} = 0$$

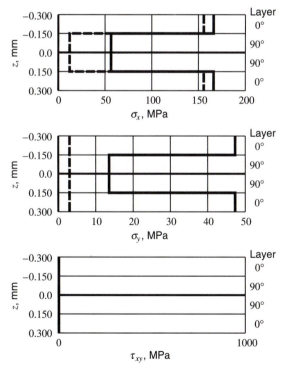

FIGURE 11.1
Stress distribution through the thickness of $[0/90]_S$ laminate subjected to $\varepsilon_x^o = 1000 \times 10^{-6}$, $\Delta T = -150°C$

—— Thermal effects   – – – No thermal effects

The second term in each sum is the stress due to free thermal strain effects, while the first term is identical to the stress for the case of no free thermal strain effects, equations (6.28) and (6.29). The material feels the sum and does not distinguish between the portion of the stress due to the total strains and the portion of the stress due to free thermal strains. However, for an illustration of physical effects, it is sometimes useful to examine the two portions.

By using the transformation relations, we find that the total strains in the principal material coordinate system are the same as for the case of no thermal effects in Chapter 6, equations (6.35) and (6.37), as illustrated in Figure 6.16. They are, for the 0° layers,

$$\varepsilon_1 = 1000 \times 10^{-6}$$
$$\varepsilon_2 = 0 \qquad\qquad (11.12a)$$
$$\gamma_{12} = 0$$

and for the 90° layers

$$\varepsilon_1 = 0$$
$$\varepsilon_2 = 1000 \times 10^{-6} \qquad\qquad (11.12b)$$
$$\gamma_{12} = 0$$

By transformation, the principal material system stresses are, for the 0° layers,

$$\sigma_1 = 166.3 \text{ MPa}$$
$$\sigma_2 = 47.3 \text{ MPa} \qquad\qquad (11.13a)$$
$$\tau_{12} = 0$$

and for the 90° layers

$$\sigma_1 = 13.59 \text{ MPa}$$
$$\sigma_2 = 56.5 \text{ MPa} \qquad\qquad (11.13b)$$
$$\tau_{12} = 0$$

It is important to note that σ_2 in the 90° layers is greater than the failure stress level of 50 MPa. Comparing this to the case of no thermal effects, equations (6.38) and (6.40), we can see that inclusion of the effects of cooldown from the processing temperature can be critical. Figure 11.2 compares the principal material system stresses for these two $[0/90]_S$ cases, and the large contribution to σ_2 from the free thermal strain is quite evident.

In equation (11.11), we studied the effect on the stresses of the free thermal strains in the x-y coordinate system by looking at the various contributions in the stress calculations. We can repeat this study for the 1-2 principal material coordinate system by looking at the total strains and the free thermal strains in that system. In fact, the separation of stresses into a portion due to total strain and a portion due to free thermal strain makes more sense in the principal material system. In that system it is possible to see directly how the portion of the stress attributable to free thermal strain influences failure. The key equation for each layer for examining the contributions, then, is equation (4.30), repeated here for convenience as

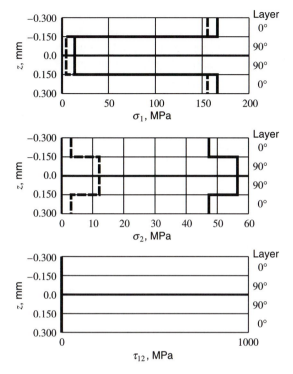

FIGURE 11.2
Principal material system stress distribution through the thickness of $[0/90]_S$ laminate subjected to $\varepsilon_x^o = 1000 \times 10^{-6}$, $\Delta T = -150°C$

$$
\left\{
\begin{array}{c}
\sigma_1 \\
\sigma_2 \\
\tau_{12}
\end{array}
\right\}
=
\left[
\begin{array}{ccc}
Q_{11} & Q_{12} & 0 \\
Q_{12} & Q_{22} & 0 \\
0 & 0 & Q_{66}
\end{array}
\right]
\left\{
\begin{array}{c}
\varepsilon_1 - \alpha_1 \Delta T \\
\varepsilon_2 - \alpha_2 \Delta T \\
\gamma_{12}
\end{array}
\right\}
\tag{11.14}
$$

In the principal material coordinate system the contributions of the free thermal strains to the stresses are the same for each layer, namely,

$$
\begin{aligned}
\sigma_1^T &= -(Q_{11}\alpha_1 + Q_{12}\alpha_2)\Delta T \\
\sigma_2^T &= -(Q_{12}\alpha_1 + Q_{22}\alpha_2)\Delta T
\end{aligned}
\tag{11.15}
$$

There is no influence of free thermal strains on the shear stresses in the principal material system. The notation σ_1^T and σ_2^T is introduced strictly for convenience, even though it risks the possibility of the physical interpretation that there are thermal stresses and nonthermal stresses in a problem. As mentioned earlier, this is not the case. There are stresses, period! The physical interpretation that can, however, be assigned to σ_1^T and σ_2^T is that these are the stresses in the principal material system that would result if a laminate, or a single layer for that matter, was restrained such that

$$
\varepsilon_x^o = \varepsilon_y^o = \gamma_{xy}^o = \kappa_x^o = \kappa_y^o = \kappa_{xy}^o = 0
\tag{11.16}
$$

One useful characteristic of the quantities σ_1^T and σ_2^T is that they are independent of the laminate, independent of how the laminate is being deformed, independent of the orientation of the layer, and depend only on principal material system properties and the temperature change. In fact, one can go one step further and introduce the quantities

$$\hat{\sigma}_1^T = Q_{11}\alpha_1 + Q_{12}\alpha_2$$
$$\hat{\sigma}_2^T = Q_{12}\alpha_1 + Q_{22}\alpha_2 \tag{11.17}$$

These are strictly material properties. For the problem here

$$\hat{\sigma}_1^T = 0.0705 \text{ MPa/}^\circ\text{C}$$
$$\hat{\sigma}_2^T = 0.295 \text{ MPa/}^\circ\text{C} \tag{11.18}$$

Using $\Delta T = -150^\circ\text{C}$, from equation (11.15),

$$\sigma_1^T = 10.57 \text{ MPa}$$
$$\sigma_2^T = 44.3 \text{ MPa} \tag{11.19}$$

and, hence, the stresses in the principal material system have the contributions, for the 0° layers,

$$\sigma_1 = 155.7 + 10.57 = 166.3 \text{ MPa}$$
$$\sigma_2 = 3.02 + 44.3 = 47.3 \text{ MPa} \tag{11.20a}$$

and for the 90° layers

$$\sigma_1 = 3.02 + 10.57 = 13.59 \text{ MPa}$$
$$\sigma_2 = 12.16 + 44.3 = 56.5 \text{ MPa} \tag{11.20b}$$

A close examination of Figure 11.2 reveals that in the principal material system, the difference between the stresses σ_1 for the thermal case and for the nonthermal case is σ_1^T for each layer. Likewise, the difference between the stresses σ_2 for the thermal case and for the nonthermal case is σ_2^T for each layer. Thermal effects simply shift horizontally the principal material system stresses in each layer; the shift is the same for each layer (see Figure 11.2). This same simple shift in each layer is not the situation with the stresses in the x-y global coordinate system (see Figure 11.1).

As mentioned in Chapter 2, the mechanical strains cannot be identified with any specific dimensional changes, as the free thermal strains and the total strains can, and thus they have no physical significance. However, it is useful to compare mechanical strains when studying problems with thermal effects. Recall, for the case of no thermal effects, the mechanical strains are the total strains. For the problem here, the total strains are given in the principal material system by equation (11.12). With thermal effects, from equation (4.31), the mechanical strains in the principal material system are, for the 0° layers,

$$\varepsilon_1^{mech} = 997 \times 10^{-6}$$
$$\varepsilon_2^{mech} = 3640 \times 10^{-6} \tag{11.21a}$$
$$\gamma_{12}^{mech} = 0$$

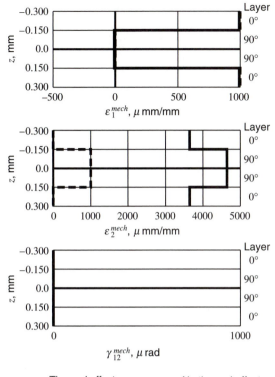

FIGURE 11.3
Principal material system
mechanical strain distribution
through the thickness of
$[0/90]_S$ laminate subjected to
$\varepsilon_x^o = 1000 \times 10^{-6}$, $\Delta T = -150°C$

and for the 90° layers,

$$\varepsilon_1^{mech} = -2.70 \times 10^{-6}$$
$$\varepsilon_2^{mech} = 4640 \times 10^{-6} \qquad (11.21b)$$
$$\gamma_{12}^{mech} = 0$$

Figure 11.3 presents the mechanical strains in the principal material system for both thermal effects and no thermal effects. If mechanical strains are viewed as the drivers for the stress-strain relations, from Figure 11.3 we can see that the case of including free thermal strains would lead to higher σ_2 stresses for this problem.

11.2.2 CLT Example 10: $[\pm 30/0]_S$ Laminate Subjected to Known κ_x^o and Known ΔT

As a second example, consider the $[\pm 30/0]_S$ laminate deformed such that at a point on the reference surface

$$\varepsilon_x^o(x, y) = 0 \qquad \kappa_x^o(x, y) = 2.22 \text{ m}^{-1}$$
$$\varepsilon_y^o(x, y) = 0 \qquad \kappa_y^o(x, y) = 0 \qquad (11.22)$$
$$\gamma_{xy}^o(x, y) = 0 \qquad \kappa_{xy}^o(x, y) = 0$$

with the temperature change being $\Delta T = -150°C$ at all locations. The distribution of total strains through the laminate is

$$\varepsilon_x(x, y, z) = \varepsilon_x^o(x, y) + z\kappa_x^o(x, y) = 2.22z$$
$$\varepsilon_y(x, y, z) = \varepsilon_y^o(x, y) + z\kappa_y^o(x, y) = 0 \quad\quad (11.23)$$
$$\gamma_{xy}(x, y, z) = \gamma_{xy}^o(x, y) + z\kappa_{xy}^o(x, y) = 0$$

These total strains are the same as those illustrated in Figure 6.33, and for this particular case equation (11.4) becomes

$$\left\{\begin{array}{c} \sigma_x \\ \sigma_y \\ \tau_{xy} \end{array}\right\} = \left[\begin{array}{ccc} \bar{Q}_{11} & \bar{Q}_{12} & \bar{Q}_{16} \\ \bar{Q}_{12} & \bar{Q}_{22} & \bar{Q}_{26} \\ \bar{Q}_{16} & \bar{Q}_{26} & \bar{Q}_{66} \end{array}\right] \left\{\begin{array}{c} 2.22z - \alpha_x \Delta T \\ -\alpha_y \Delta T \\ -\alpha_{xy} \Delta T \end{array}\right\} \quad (11.24)$$

The coefficients of thermal deformation are given by equation (11.5) as follows:
For the $+30°$ layers:

$$\alpha_x = 6.06 \times 10^{-6}/°C$$
$$\alpha_y = 18.22 \times 10^{-6}/°C \quad\quad (11.25a)$$
$$\alpha_{xy} = -21.1 \times 10^{-6}/°C$$

For the $-30°$ layers:

$$\alpha_x = 6.06 \times 10^{-6}/°C$$
$$\alpha_y = 18.22 \times 10^{-6}/°C \quad\quad (11.25b)$$
$$\alpha_{xy} = 21.1 \times 10^{-6}/°C$$

For the $0°$ layers:

$$\alpha_x = -0.018 \times 10^{-6}/°C$$
$$\alpha_y = 24.3 \times 10^{-6}/°C \quad\quad (11.25c)$$
$$\alpha_{xy} = 0$$

Using these in equation (11.24) for the various layers yields, we find that for the $+30°$ layers:

$$\sigma_x = 206\,000z + 19.00 \text{ MPa}$$
$$\sigma_y = 66\,800z + 35.9 \text{ MPa} \quad\quad (11.26a)$$
$$\tau_{xy} = 103\,800z - 14.61 \text{ MPa}$$

For the $-30°$ layers:

$$\sigma_x = 206\,000z + 19.00 \text{ MPa}$$
$$\sigma_y = 66\,800z + 35.9 \text{ MPa} \quad\quad (11.26b)$$
$$\tau_{xy} = -103\,800z + 14.61 \text{ MPa}$$

For the $0°$ layers:

$$\sigma_x = 346\,000z + 10.57 \text{ MPa}$$
$$\sigma_y = 6700z + 44.3 \text{ MPa} \quad\quad (11.26c)$$
$$\tau_{xy} = 0$$

These follow directly from the numerical values of the reduced stiffnesses, the co-efficients of thermal deformation, and the temperature change. Clearly the constant portion of each of the three stress components in the above equations is attributable to the free thermal strains, and the portions that vary linearly with z are identical to the calculations for this problem with no thermal effects, equation (6.118). Figure 11.4 shows the distribution of the three stresses through the thickness of the laminate. While the effects of free thermal strains on σ_x are not as great as their effects on σ_y or τ_{xy}, it is very important to note that with these particular free thermal strain effects, the stress distributions are not odd functions of z. For the stress σ_y, for example, the stresses are positive throughout the thickness. With no thermal effects, both positive and negative stresses occur. The constant stress portions in equation (11.26) prevent the stresses from being odd functions of z.

The total strains in the principal material system are as they were in Figure 6.36, where they were determined by transforming the total strains from the x-y system. For the record:

For the $+30°$ layers:

$$\varepsilon_1 = 1.667z$$

$$\varepsilon_2 = 0.556z \qquad (11.27a)$$

$$\gamma_{12} = -1.924z$$

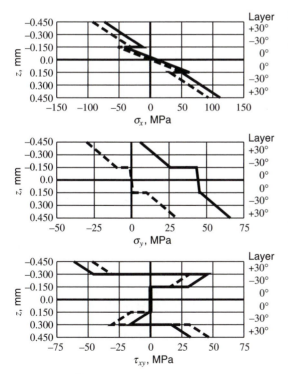

FIGURE 11.4
Stress distribution through the thickness of $[\pm 30/0]_S$ laminate subjected to $\kappa_x^o = 2.22$ m^{-1}, $\Delta T = -150°C$

For the $-30°$ layers:

$$\varepsilon_1 = 1.667z$$

$$\varepsilon_2 = 0.556z \qquad (11.27b)$$

$$\gamma_{12} = 1.924z$$

For the $0°$ layers:

$$\varepsilon_1 = 2.22z$$

$$\varepsilon_2 = \gamma_{12} = 0 \qquad (11.27c)$$

Stresses in the principal material system are determined by transforming the stresses from the x-y system, equation (11.26), or by using the total strains and the free thermal strains in the stress-strain relations in the principal material system. Though we have used the former method in the past, use of the latter method for this problem with free thermal strains affords an opportunity to see the contributions to the stress calculations in the spirit of equations (11.15) and (11.19). With this latter approach equation (4.30) is used directly for each layer orientation. With either approach, the principal material system stresses are:

For the $+30°$ layers:

$$\sigma_1 = 261\ 000z + 10.57 \text{ MPa}$$

$$\sigma_2 = 11\ 780z + 44.3 \text{ MPa} \qquad (11.28a)$$

$$\tau_{12} = -8470z \text{ MPa}$$

For the $-30°$ layers:

$$\sigma_1 = 261\ 000z + 10.57 \text{ MPa}$$

$$\sigma_2 = 11\ 780z + 44.3 \text{ MPa} \qquad (11.28b)$$

$$\tau_{12} = 8470z \text{ MPa}$$

For the $0°$ layers:

$$\sigma_1 = 346\ 000z + 10.57 \text{ MPa}$$

$$\sigma_2 = 6700z + 44.3 \text{ MPa} \qquad (11.28c)$$

$$\tau_{12} = 0$$

Again, in the principal material system, as seen in Figure 11.5, free thermal strain effects simply shift horizontally the stresses σ_1 and σ_2 relative to the no-thermal-effects case by the amount σ_1^T and σ_2^T, respectively. As in the $[0/90]_S$ laminate just studied, the stress component σ_2 is much larger in the presence of free thermal strain effects, while the stress component σ_1 is not influenced that much. In the principal material system, the shear stresses are unchanged. At $z = 0.450$ mm the influence of free thermal strains brings the stress σ_2 close to the failure value.

The mechanical strains are given by the following expressions:

For the $+30°$ layers:

$$\varepsilon_1^{mech} = 1.667z - 2.7 \times 10^{-6}$$
$$\varepsilon_2^{mech} = 0.556z + 3640 \times 10^{-6} \qquad (11.29a)$$
$$\gamma_{12}^{mech} = -1.924z$$

For the $-30°$ layers:

$$\varepsilon_1^{mech} = 1.667z - 2.7 \times 10^{-6}$$
$$\varepsilon_2^{mech} = 0.556z + 3640 \times 10^{-6} \qquad (11.29b)$$
$$\gamma_{12}^{mech} = 1.924z$$

For the $0°$ layers:

$$\varepsilon_1^{mech} = 2.22z - 2.7 \times 10^{-6}$$
$$\varepsilon_2^{mech} = 3640 \times 10^{-6} \qquad (11.29c)$$
$$\gamma_{12} = 0$$

The convenience of the principal material coordinate system is again obvious, with the same free thermal strains subtracting from the total strains in the same manner, independent of layer. In addition to convenience, stresses and strains have much more physical significance in the principal material system. Figure 11.6 illustrates the distributions of the mechanical strains.

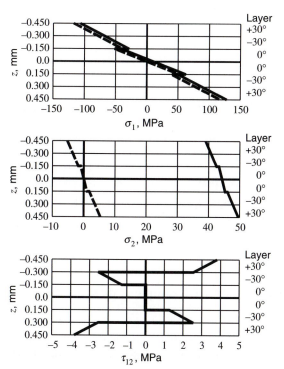

FIGURE 11.5
Principal material system stress distribution through the thickness of $[\pm30/0]_S$ laminate subjected to $\kappa_x^o = 2.22$ m^{-1}, $\Delta T = -150°C$

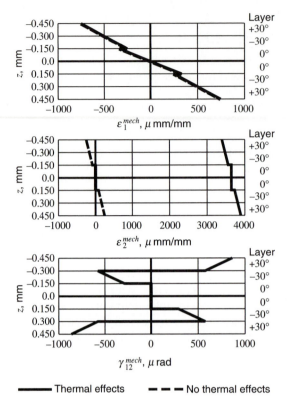

FIGURE 11.6
Principal material system mechanical strain distribution through the thickness of $[\pm30/0]_S$ laminate subjected to $\kappa_x^o = 2.22$ m^{-1}, $\Delta T = -150°C$

——— Thermal effects   — — — No thermal effects

11.3
FORCE AND MOMENT RESULTANTS IN EXAMPLES

By examining the distributions of the stresses through the thickness of the laminates, Figures 11.1 and 11.4, it is clear that stress resultants, in the form of force and moment resultants, can be defined. Furthermore, if we compare the case of thermal effects with the case of no thermal effects, it appears that the values of the force and moment resultants will not be the same. In this section we will integrate the stresses through the thickness and compute numerical values of the force and moment resultants. In the next section we shall formally define the resultants.

11.3.1 Stress Resultant Calculation for CLT Example 9: $[0/90]_S$ Laminate Subjected to Known ε_x^o and Known ΔT

From Chapter 6 for the $[0/90]_S$ case with no thermal effects, integration of the stresses through the thickness, equations (6.58)–(6.63), showed that the force resultants were

$$N_x = 50\ 400 \text{ N/m} \qquad M_x = 0$$
$$N_y = 1809 \text{ N/m} \qquad M_y = 0 \qquad (11.30)$$
$$N_{xy} = 0 \qquad M_{xy} = 0$$

From equation (11.10), integration of the stresses through the thickness of this laminate for the case of thermal effects leads to

$$N_x = [(166.3 + 56.5 + 56.5 + 166.3) \times 10^6](150 \times 10^{-6})$$
$$N_y = [(47.3 + 13.59 + 13.59 + 47.3) \times 10^6](150 \times 10^{-6}) \qquad (11.31)$$
$$N_{xy} = 0$$

Because in this case the stress distributions are even functions of z, there are no moment resultants. The above algebra results in

$$N_x = 66\ 800 \text{ N/m} \qquad M_x = 0$$
$$N_y = 18\ 270 \text{ N/m} \qquad M_y = 0 \qquad (11.32)$$
$$N_{xy} = 0 \qquad M_{xy} = 0$$

and the proper interpretation of this problem is that if a point on the reference surface of a [0/90]$_S$ laminate is to have the deformations given by equation (11.6) in the presence of a temperature decrease of 150°C relative to the stress-free state, then the resultants given by equation (11.32) are required at that point. If we compare the force resultants required for the case of no free thermal strain effects with the case of including free thermal strain effects, equation (11.30) versus equation (11.32), it becomes evident that to enforce the same state of deformation, the inclusion of free thermal strain effects requires that the stress resultants must be different. In particular, N_y must increase by a factor of 10. Why is this the case? We shall find that when heated or cooled, a laminate will deform. In particular, a [0/90]$_S$ laminate will contract when cooled. With the [0/90]$_S$ laminate we are specifying only a specific elongation strain in the x direction. To enforce this specified deformation requires greater force resultants to overcome the tendency of the laminate to contract in both the x and y directions when cooled 150°C. More will be said of the thermal deformations of laminates later. However, to emphasize that additional forces are indeed required, another interpretation of the results here is that if a [0/90]$_S$ laminate with length $L_x = 0.250$ m in the x direction and width $L_y = 0.125$ m in the y direction is to have the deformations of equation (11.6) valid everywhere on its 0.0312 m^2 of reference surface, then there must be a force

$$N_x \times L_y = 66\ 800 \text{ N/m} \times 0.125 \text{ m} = 8350 \text{ N} \qquad (11.33)$$

acting in the x direction and uniformly distributed along edge L_y, and a force

$$N_y \times L_x = 18\ 270 \text{ N/m} \times 0.250 \text{ m} = 4570 \text{ N} \qquad (11.34)$$

acting in the y direction and uniformly distributed along the edge L_x. For the case of no thermal effects, the forces required are 6300 and 452 N, respectively. The inclusion of free thermal strain effects requires real additional forces to produce a given state of deformation. The edge forces are illustrated in Figure 11.7; these results contrast the results of Figure 6.26, the case with no thermal effects.

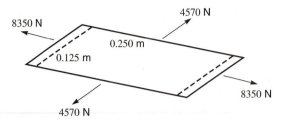

FIGURE 11.7
Forces required to produce state of deformation $\varepsilon_x^o = 1000 \times 10^{-6}$ in $[0/90]_S$ laminate with $\Delta T = -150°C$

11.3.2 Stress Resultant Calculation for CLT Example 10: $[\pm 30/0]_S$ Laminate Subjected to Known κ_x^o and Known ΔT

For the $[\pm 30/0]_S$ laminate with no thermal effects, in Chapter 6 we saw that no force resultants were required to enforce the curvature-only deformation of equation (11.22), but moment resultants were. Specifically, from equation (6.124), the force and moment resultants were

$$N_x = 0 \qquad M_x = 12.84 \text{ N·m/m}$$
$$N_y = 0 \qquad M_y = 3.92 \text{ N·m/m} \qquad (11.35)$$
$$N_{xy} = 0 \qquad M_{xy} = 2.80 \text{ N·m/m}$$

With thermal effects, from equation (11.26), the force resultant N_x required is

$$N_x = \left\{ \int_{-450\times10^{-6}}^{-300\times10^{-6}} (206\,000z + 19.00)dz + \int_{-300\times10^{-6}}^{-150\times10^{-6}} (206\,000z + 19.00)dz \right.$$

$$+ \int_{-150\times10^{-6}}^{0} (346\,000z + 10.57)dz + \int_{0}^{150\times10^{-6}} (346\,000z + 10.57)dz$$

$$+ \left. \int_{150\times10^{-6}}^{300\times10^{-6}} (206\,000z + 19.00)dz + \int_{300\times10^{-6}}^{450\times10^{-6}} (206\,000z + 19.00)dz \right\}$$

$$\times 10^6$$

$$(11.36)$$

These integrands consist of odd parts, those proportional to z, and even parts. The parts proportional to z in the first and sixth integrations sum to zero. There is a similar pairing and summing to zero for the odd parts of the other four integrals. The net result is

$$N_x = [(19.00 + 19.00 + 10.57 + 10.57 + 19.00 + 19.00) \times 10^6](150 \times 10^{-6})$$

or
$$N_x = 0.01457 \text{ MN/m} \qquad (11.37)$$

In a similar fashion, the odd parts summing to zero for the computations for N_y and N_{xy} results in

$$N_y = 0.0348 \text{ MN/m} \tag{11.38}$$

and

$$N_{xy} = 0 \tag{11.39}$$

Unlike the case with no thermal effects, force resultants are required if the curvature-only deformations of equation (11.22) are to be produced in the presence of free thermal strain effects. This is an interesting revelation!

The moment resultant M_x required is, from equation (11.26),

$$
\begin{aligned}
M_x = \Bigg\{ & \int_{-450\times10^{-6}}^{-300\times10^{-6}} (206\,000z + 19.00)z\,dz + \int_{-300\times10^{-6}}^{-150\times10^{-6}} (206\,000z + 19.00)z\,dz \\
& + \int_{-150\times10^{-6}}^{0} (346\,000z + 10.57)z\,dz + \int_{0}^{150\times10^{-6}} (346\,000z + 10.57)z\,dz \\
& + \int_{150\times10^{-6}}^{300\times10^{-6}} (206\,000z + 19.00)z\,dz + \int_{300\times10^{-6}}^{450\times10^{-6}} (206\,000z + 19.00)z\,dz \Bigg\} \\
& \times 10^6
\end{aligned}
\tag{11.40}
$$

The integrands again consist of odd parts, those proportional to z, and even parts, those proportional to z^2. The odd parts contribute nothing to the overall integration, whereas the even parts are identical to the integration with no thermal effects present, equation (6.120). The net result is that the moment is the same as the case with no thermal effects present, namely,

$$M_x = 12.84 \text{ N·m/m} \tag{11.41}$$

Likewise, the other two moment resultants are unaffected by the presence of thermal effects. So

$$M_y = 3.92 \text{ N·m/m} \tag{11.42}$$

and

$$M_{xy} = 2.80 \text{ N·m/m} \tag{11.43}$$

Thus, to produce the deformations of equation (11.22) at a point on the reference surface of a $[\pm30/0]_S$ laminate in the presence of a 150°C temperature decrease relative to the stress-free state requires the following resultants at that point:

$$
\begin{aligned}
N_x &= 0.01457 \text{ MN/m} & M_x &= 12.84 \text{ N·m/m} \\
N_y &= 0.0348 \text{ MN/m} & M_y &= 3.92 \text{ N·m/m} \\
N_{xy} &= 0 & M_{xy} &= 2.80 \text{ N·m/m}
\end{aligned}
\tag{11.44}
$$

If a laminate that is 0.250 m long by 0.125 m wide is to have these deformations at every point on its reference surface, then uniformly distributed bending moments of 1.605 N·m and uniformly distributed twisting moments of 0.350 N·m are required along the 0.125 m edges, and uniformly distributed bending moments of 0.981 N·m and uniformly distributed twisting moments of 0.701 N·m are required along the

0.250 m edges. In addition, along the 0.125 m edges uniformly distributed forces of 1822 N in the x direction are required, and along the 0.250 m edges uniformly distributed forces of 8700 N in the y direction are required. Figure 11.8 illustrates these forces, and this figure should be compared to Figure 6.35, the case with no thermal effects. The force resultants are required because of the natural tendency of this laminate to contract in both the x and y directions when cooled. These contractions would violate the requirement of no reference surface strains, that is, equation (11.22).

We will later make a general statement regarding symmetric laminates in the presence of temperature changes that are not functions of z, but the generalization is important and might as well be made at this time also. That general statement is: Temperature changes that are independent of the z coordinate have no effect on the moment resultants of a symmetric laminate. They do, however, have an effect on the force resultants.

As stated earlier, it has been assumed throughout the discussion that the temperature change ΔT is independent of z. This resulted in the use of $\Delta T = -150°C$ at all layer locations. It is certainly possible to solve problems where the temperature varies with z. The basic formulation used above would still be employed. The functional dependence of ΔT on z would be used in the stress-strain relations, equation (5.163), to determine how the stresses varied with z. If the temperature varied linearly with z in the problem with the $[0/90]_S$ laminate, the stresses would not be independent of z within a layer; rather, they would vary linearly with z. Then moment resultants would be required, in addition to force resultants, if the reference surface deformations of equation (11.6) were to remain valid. The exact moments required would be computed by integration of the stress distributions through the

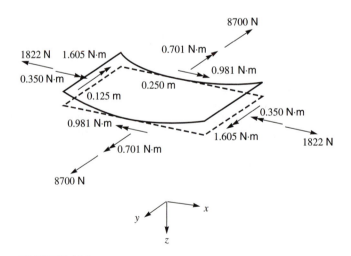

FIGURE 11.8
Forces and moments required to produce state of deformation $\kappa_x^o = 2.22$ m^{-1} in $[\pm30/0]_S$ laminate with $\Delta T = -150°C$

thickness, as in equation (11.40). Please note: The approaches being discussed are fundamental. They can be applied to problems more complicated than our example problems. Consequently, it is important—but sometimes not so easy—to comprehend all the subtleties of the simpler example problems before thinking of more advanced problems.

11.4
DEFINITION OF THERMAL FORCE AND MOMENT RESULTANTS

In the previous section we informally used the definitions of force and moment resultants introduced in Chapter 7, equations (7.1) and (7.3), to compute the numerical values of the resultants for the two thermal problems. We bring these definitions forward for convenience; they are

$$N_x \equiv \int_{-\frac{H}{2}}^{\frac{H}{2}} \sigma_x dz \qquad M_x \equiv \int_{-\frac{H}{2}}^{\frac{H}{2}} \sigma_x z dz$$

$$N_y \equiv \int_{-\frac{H}{2}}^{\frac{H}{2}} \sigma_y dz \qquad M_y \equiv \int_{-\frac{H}{2}}^{\frac{H}{2}} \sigma_y z dz \qquad (11.45)$$

$$N_{xy} \equiv \int_{-\frac{H}{2}}^{\frac{H}{2}} \tau_{xy} dz \qquad M_{xy} \equiv \int_{-\frac{H}{2}}^{\frac{H}{2}} \tau_{xy} z dz$$

Considering the differential elements of Figures 7.2 and 7.3, we find that the physical interpretation of these force and moment resultants is the same, independently of whether or not free thermal strain effects are being included. The stresses in the integrals are what the material "senses" and are what will cause the material to fail; they are physical. The integrals of these stresses, namely, the resultants, are the net effect of these stresses and are thus also physical. They can be measured with any force-measuring device.

In Chapter 7 we substituted the plane-stress stress-strain relations, in terms of the $\bar{Q}_{ij}$, into the integrands of the above equations and carried out the integration. The result was a relationship between the stress resultants and the reference surface strains and curvatures—the all-important ABD matrix relationship. As shown in Figure 7.4, this relationship completed the loop in the analysis of the stresses and strains of a laminate. For the case of including free thermal strains in the stress-strain relations, the relationship between the reference surface strains and curvatures and the stress resultants is different. This was indeed seen in the example problems. For the same states of strains and curvatures on the reference surface, the [0/90] and [±30/0]$_S$ laminates with free thermal strain effects and a temperature change of −150°C required different values of the force and moment resultants relative to the values required if free thermal strain effects were not present. To see this formally, let us examine N_x using the definition from equation (11.45) and the stress-strain relations with free thermal effects included, equation (11.4):

$$N_x = \int_{-\frac{H}{2}}^{\frac{H}{2}} \{\bar{Q}_{11}(\varepsilon_x^o + z\kappa_x^o - \alpha_x \Delta T) + \bar{Q}_{12}(\varepsilon_y^o + z\kappa_y^o - \alpha_y \Delta T)$$

$$+ \bar{Q}_{16}(\gamma_{xy}^o + z\kappa_{xy}^o - \alpha_{xy} \Delta T)\} \, dz \tag{11.46}$$

where expanding the integrand results in

$$N_x = \int_{-\frac{H}{2}}^{\frac{H}{2}} \{\bar{Q}_{11}\varepsilon_x^o + \bar{Q}_{11}z\kappa_x^o + \bar{Q}_{12}\varepsilon_y^o + \bar{Q}_{12}z\kappa_y^o + \bar{Q}_{16}\gamma_{xy}^o$$

$$+ \bar{Q}_{16}z\kappa_{xy}^o - (\bar{Q}_{11}\alpha_x + \bar{Q}_{12}\alpha_y + \bar{Q}_{16}\alpha_{xy})\Delta T\} \, dz \tag{11.47}$$

The integration can be distributed over the seven terms to give

$$N_x = \int_{-\frac{H}{2}}^{\frac{H}{2}} \bar{Q}_{11}\varepsilon_x^o dz + \int_{-\frac{H}{2}}^{\frac{H}{2}} \bar{Q}_{11}z\kappa_x^o dz + \int_{-\frac{H}{2}}^{\frac{H}{2}} \bar{Q}_{12}\varepsilon_y^o dz$$

$$+ \int_{-\frac{H}{2}}^{\frac{H}{2}} \bar{Q}_{12}z\kappa_y^o dz + \int_{-\frac{H}{2}}^{\frac{H}{2}} \bar{Q}_{16}\gamma_{xy}^o dz + \int_{-\frac{H}{2}}^{\frac{H}{2}} \bar{Q}_{16}z\kappa_{xy}^o dz \quad (11.48)$$

$$- \int_{-\frac{H}{2}}^{\frac{H}{2}} (\bar{Q}_{11}\alpha_x + \bar{Q}_{12}\alpha_y + \bar{Q}_{16}\alpha_{xy})\Delta T \, dz$$

The first six terms should be quite familiar, following the derivation in Chapter 7, while the seventh term is new. Integration with respect to z results in

$$N_x = A_{11}\varepsilon_x^o + A_{12}\varepsilon_y^o + A_{16}\gamma_{xy}^o + B_{11}\kappa_x^o + B_{12}\kappa_y^o + B_{16}\kappa_{xy}^o - N_x^T \tag{11.49}$$

where the seventh term,

$$N_x^T = \int_{-\frac{H}{2}}^{\frac{H}{2}} (\bar{Q}_{11}\alpha_x + \bar{Q}_{12}\alpha_y + \bar{Q}_{16}\alpha_{xy})\Delta T \, dz \tag{11.50}$$

has the units of the force resultants, namely, N/m, and is referred to as a *thermal force resultant*. In particular, it is the thermal force resultant in the x direction. Note, it involves material properties, the temperature change, and by virtue of the spatial integration with respect to z, the layer thicknesses and locations. The definitions of the other two force resultants follow similar steps, and the results are

$$N_y = A_{12}\varepsilon_x^o + A_{22}\varepsilon_y^o + A_{26}\gamma_{xy}^o + B_{12}\kappa_x^o + B_{22}\kappa_y^o + B_{26}\kappa_{xy}^o - N_y^T \tag{11.51}$$

and $\quad N_{xy} = A_{16}\varepsilon_x^o + A_{26}\varepsilon_y^o + A_{66}\gamma_{xy}^o + B_{16}\kappa_x^o + B_{26}\kappa_y^o + B_{66}\kappa_{xy}^o - N_{xy}^T \tag{11.52}$

where the thermal force resultants in the y direction and in shear are given by

$$N_y^T = \int_{-\frac{H}{2}}^{\frac{H}{2}} (\bar{Q}_{12}\alpha_x + \bar{Q}_{22}\alpha_y + \bar{Q}_{26}\alpha_{xy})\Delta T \, dz \tag{11.53}$$

and $\quad$ $$N_{xy}^T = \int_{-\frac{H}{2}}^{\frac{H}{2}} (\bar{Q}_{16}\alpha_x + \bar{Q}_{26}\alpha_y + \bar{Q}_{66}\alpha_{xy})\Delta T \, dz \tag{11.54}$$

The moment resultants can be studied by substituting into the definition of M_x, namely,

$$M_x = \int_{-\frac{H}{2}}^{\frac{H}{2}} \{ \bar{Q}_{11}(\varepsilon_x^o + z\kappa_x^o - \alpha_x \Delta T) + \bar{Q}_{12}(\varepsilon_y^o + z\kappa_y^o - \alpha_y \Delta T)$$

$$+ \bar{Q}_{16}(\gamma_{xy}^o + z\kappa_{xy}^o - \alpha_{xy}\Delta T) \} z \, dz \tag{11.55}$$

Expanding equation (11.55), we find that

$$M_x = \int_{-\frac{H}{2}}^{\frac{H}{2}} \{ \bar{Q}_{11}\varepsilon_x^o + \bar{Q}_{11}z\kappa_x^o + \bar{Q}_{12}\varepsilon_y^o + \bar{Q}_{12}z\kappa_y^o + \bar{Q}_{16}\gamma_{xy}^o$$

$$+ \bar{Q}_{16}z\kappa_{xy}^o - (\bar{Q}_{11}\alpha_x + \bar{Q}_{12}\alpha_y + \bar{Q}_{16}\alpha_{xy})\Delta T \} z \, dz \tag{11.56}$$

and the integration can be distributed over the seven terms to give

$$M_x = \int_{-\frac{H}{2}}^{\frac{H}{2}} \bar{Q}_{11}\varepsilon_x^o z \, dz + \int_{-\frac{H}{2}}^{\frac{H}{2}} \bar{Q}_{11}z\kappa_x^o z \, dz + \int_{-\frac{H}{2}}^{\frac{H}{2}} \bar{Q}_{12}\varepsilon_y^o z \, dz$$

$$+ \int_{-\frac{H}{2}}^{\frac{H}{2}} \bar{Q}_{12}z\kappa_y^o z \, dz + \int_{-\frac{H}{2}}^{\frac{H}{2}} \bar{Q}_{16}\gamma_{xy}^o z \, dz + \int_{-\frac{H}{2}}^{\frac{H}{2}} \bar{Q}_{16}z\kappa_{xy}^o z \, dz \tag{11.57}$$

$$- \int_{-\frac{H}{2}}^{\frac{H}{2}} (\bar{Q}_{11}\alpha_x + \bar{Q}_{12}\alpha_y + \bar{Q}_{16}\alpha_{xy})\Delta T z \, dz$$

The first six terms should again be familiar; the seventh term is new. Integration with respect to z results in

$$M_x = B_{11}\varepsilon_x^o + B_{12}\varepsilon_y^o + B_{16}\gamma_{xy}^o + D_{11}\kappa_x^o + D_{12}\kappa_y^o + D_{16}\kappa_{xy}^o - M_x^T \tag{11.58}$$

where the seventh term,

$$M_x^T = \int_{-\frac{H}{2}}^{\frac{H}{2}} (\bar{Q}_{11}\alpha_x + \bar{Q}_{12}\alpha_y + \bar{Q}_{16}\alpha_{xy})\Delta T z \, dz \tag{11.59}$$

has the units of the moment resultants, namely, N·m/m, and is referred to as a *thermal moment resultant*. It, like the thermal force resultants, involves material properties, the temperature change, and, by virtue of the spatial integration with respect to z, the layer thicknesses and locations. The definitions of the other two moment resultants follow similar steps, resulting in

$$M_y = B_{12}\varepsilon_x^o + B_{22}\varepsilon_y^o + B_{26}\gamma_{xy}^o + D_{12}\kappa_x^o + D_{22}\kappa_y^o + D_{26}\kappa_{xy}^o - M_y^T \tag{11.60}$$

and $M_{xy} = B_{16}\varepsilon_x^o + B_{26}\varepsilon_y^o + B_{66}\gamma_{xy}^o + D_{16}\kappa_x^o + D_{26}\kappa_y^o + D_{66}\kappa_{xy}^o - M_{xy}^T \quad (11.61)$

where the thermal moment resultants in the y direction and in shear are given by

$$M_y^T = \int_{-\frac{H}{2}}^{\frac{H}{2}} (\bar{Q}_{12}\alpha_x + \bar{Q}_{22}\alpha_y + \bar{Q}_{26}\alpha_{xy})\Delta T z \, dz \tag{11.62}$$

and

$$M_{xy}^T = \int_{-\frac{H}{2}}^{\frac{H}{2}} (\bar{Q}_{16}\alpha_x + \bar{Q}_{26}\alpha_y + \bar{Q}_{66}\alpha_{xy})\Delta T z \, dz \tag{11.63}$$

In summary, then, with a slight rearrangement, the relation between the stress resultants and the reference surface strains and curvatures is

$$
\left\{
\begin{array}{c}
N_x + N_x^T \\
N_y + N_y^T \\
N_{xy} + N_{xy}^T \\
M_x + M_x^T \\
M_y + M_y^T \\
M_{xy} + M_{xy}^T
\end{array}
\right\}
=
\left[
\begin{array}{cccccc}
A_{11} & A_{12} & A_{16} & B_{11} & B_{12} & B_{16} \\
A_{12} & A_{22} & A_{26} & B_{12} & B_{22} & B_{26} \\
A_{16} & A_{26} & A_{66} & B_{16} & B_{26} & B_{66} \\
B_{11} & B_{12} & B_{16} & D_{11} & D_{12} & D_{16} \\
B_{12} & B_{22} & B_{26} & D_{12} & D_{22} & D_{26} \\
B_{16} & B_{26} & B_{66} & D_{16} & D_{26} & D_{66}
\end{array}
\right]
\left\{
\begin{array}{c}
\varepsilon_x^o \\
\varepsilon_y^o \\
\gamma_{xy}^o \\
\kappa_x^o \\
\kappa_y^o \\
\kappa_{xy}^o
\end{array}
\right\}
\tag{11.64}
$$

This is a key equation for the thermo-mechanical analysis of laminates. In the above, gathering definitions, we find that

$$
N_x^T = \int_{-\frac{H}{2}}^{\frac{H}{2}} (\bar{Q}_{11}\alpha_x + \bar{Q}_{12}\alpha_y + \bar{Q}_{16}\alpha_{xy})\Delta T\, dz
$$

$$
N_y^T = \int_{-\frac{H}{2}}^{\frac{H}{2}} (\bar{Q}_{12}\alpha_x + \bar{Q}_{22}\alpha_y + \bar{Q}_{26}\alpha_{xy})\Delta T\, dz
$$

$$
N_{xy}^T = \int_{-\frac{H}{2}}^{\frac{H}{2}} (\bar{Q}_{16}\alpha_x + \bar{Q}_{26}\alpha_y + \bar{Q}_{66}\alpha_{xy})\Delta T\, dz
$$

$$
M_x^T = \int_{-\frac{H}{2}}^{\frac{H}{2}} (\bar{Q}_{11}\alpha_x + \bar{Q}_{12}\alpha_y + \bar{Q}_{16}\alpha_{xy})\Delta T z\, dz
\tag{11.65}
$$

$$
M_y^T = \int_{-\frac{H}{2}}^{\frac{H}{2}} (\bar{Q}_{12}\alpha_x + \bar{Q}_{22}\alpha_y + \bar{Q}_{26}\alpha_{xy})\Delta T z\, dz
$$

$$
M_{xy}^T = \int_{-\frac{H}{2}}^{\frac{H}{2}} (\bar{Q}_{16}\alpha_x + \bar{Q}_{26}\alpha_y + \bar{Q}_{66}\alpha_{xy})\Delta T z\, dz
$$

The inverse relation is

$$
\left\{
\begin{array}{c}
\varepsilon_x^o \\
\varepsilon_y^o \\
\gamma_{xy}^o \\
\kappa_x^o \\
\kappa_y^o \\
\kappa_{xy}^o
\end{array}
\right\}
=
\left[
\begin{array}{cccccc}
a_{11} & a_{12} & a_{16} & b_{11} & b_{12} & b_{16} \\
a_{12} & a_{22} & a_{26} & b_{21} & b_{22} & b_{26} \\
a_{16} & a_{26} & a_{66} & b_{61} & b_{62} & b_{66} \\
b_{11} & b_{21} & b_{61} & d_{11} & d_{12} & d_{16} \\
b_{12} & b_{22} & b_{62} & d_{12} & d_{22} & d_{26} \\
b_{16} & b_{26} & b_{66} & d_{16} & d_{26} & d_{66}
\end{array}
\right]
\left\{
\begin{array}{c}
N_x + N_x^T \\
N_y + N_y^T \\
N_{xy} + N_{xy}^T \\
M_x + M_x^T \\
M_y + M_y^T \\
M_{xy} + M_{xy}^T
\end{array}
\right\}
\tag{11.66}
$$

where the a_{ij}, b_{ij}, and d_{ij} are as defined in equation (7.87b). This is another key equation for the thermomechanical analysis of laminates.

From the form of equation (11.66), it is obvious that the thermal forces and moments have an important role in determining the deformations of a laminate. For the case of only a temperature change and no force or moment resultants, that is,

$$N_x = N_y = N_{xy} = M_x = M_y = M_{xy} = 0 \qquad (11.67)$$

equation (11.66) indicates there are deformations in the laminate caused strictly by the free thermal strains of each layer.

From the form of equation (11.64), it is quite evident why in our series of examples with the $[0/90]_S$ and $[\pm 30/0]_S$ laminates the enforcement of a specific state of deformation at a point on the reference surface required different values of $N_x, \ldots, M_{xy}$, depending on whether or not thermal effects were present. For a given state of reference surface deformation, the right-hand side of equation (11.64) is fixed. Thus, the values of $N_x, \ldots, M_{xy}$ depend on the values of $N_x^T, \ldots, M_{xy}^T$. With no thermal effects, one set of values is required, while with thermal effects, a different set of values is required.

The integral form of the definitions of the thermal force and moment resultants in equation (11.65) makes these definitions quite general. If the temperature change is a general but known function of z, the integration can be carried out. The values of the material properties, namely, the transformed reduced stiffnesses and coefficients of thermal expansion, are generally considered constant within a layer and, thus, integration with z shifts to how the temperature change varies with z. Heat transfer in thin laminates is generally quite rapid and, hence, thermal gradients are not often encountered. The temperature change ΔT would most likely, then, be independent of z, as it was in our examples. For thicker laminates, temperatures could vary with thickness, for example, linearly. If high temperatures are involved in these gradients, the dependence of elastic properties on temperature may have to be incorporated in the calculations. If the temperature changes with z, and if the material properties change with temperature, then the integrations can be complicated. They can, however, be approximated numerically and used directly in equations (11.64) or (11.66). In light of these considerations, a given problem will generally involve both thermal force resultants and thermal moment resultants, and the computation of these resultants may not be trivial. The concepts, however, are straightforward.

As mentioned at the start of this section, a parallel development can be used to address the influence of free moisture strains in laminates. Accordingly, the key point in such a parallel development would be the definition of moisture stress resultants. For the most general case of having no restrictions on the distribution of moisture with thickness, and no restrictions on the dependence of material properties on moisture content, the definitions of the moisture stress resultants would be

$$N_x^M = \int_{-\frac{H}{2}}^{\frac{H}{2}} (\bar{Q}_{11}\beta_x + \bar{Q}_{12}\beta_y + \bar{Q}_{16}\beta_{xy})\Delta M \, dz$$

$$N_y^M = \int_{-\frac{H}{2}}^{\frac{H}{2}} (\bar{Q}_{12}\beta_x + \bar{Q}_{22}\beta_y + \bar{Q}_{26}\beta_{xy})\Delta M \, dz$$

$$N_{xy}^M = \int_{-\frac{H}{2}}^{\frac{H}{2}} (\bar{Q}_{16}\beta_x + \bar{Q}_{26}\beta_y + \bar{Q}_{66}\beta_{xy})\Delta M dz$$

$$M_x^M = \int_{-\frac{H}{2}}^{\frac{H}{2}} (\bar{Q}_{11}\beta_x + \bar{Q}_{12}\beta_y + \bar{Q}_{16}\beta_{xy})\Delta M z dz$$

$$\text{(11.68)}$$

$$M_y^M = \int_{-\frac{H}{2}}^{\frac{H}{2}} (\bar{Q}_{12}\beta_x + \bar{Q}_{22}\beta_y + \bar{Q}_{26}\beta_{xy})\Delta M z dz$$

$$M_{xy}^M = \int_{-\frac{H}{2}}^{\frac{H}{2}} (\bar{Q}_{16}\beta_x + \bar{Q}_{26}\beta_y + \bar{Q}_{66}\beta_{xy})\Delta M z dz$$

11.5
DEFINITIONS OF UNIT THERMAL FORCE AND MOMENT RESULTANTS

If the temperature is independent of z, the temperature change ΔT can be removed from within the integrals, and the thermal force and moment resultants can be written as

$$\begin{aligned} N_x^T &= \hat{N}_x^T \Delta T & M_x^T &= \hat{M}_x^T \Delta T \\ N_y^T &= \hat{N}_y^T \Delta T & M_y^T &= \hat{M}_y^T \Delta T \\ N_{xy}^T &= \hat{N}_{xy}^T \Delta T & M_{xy}^T &= \hat{M}_{xy}^T \Delta T \end{aligned} \tag{11.69}$$

where the hatted quantities are strictly material properties and are given by

$$\hat{N}_x^T = \int_{-\frac{H}{2}}^{\frac{H}{2}} (\bar{Q}_{11}\alpha_x + \bar{Q}_{12}\alpha_y + \bar{Q}_{16}\alpha_{xy})dz$$

$$\hat{N}_y^T = \int_{-\frac{H}{2}}^{\frac{H}{2}} (\bar{Q}_{12}\alpha_x + \bar{Q}_{22}\alpha_y + \bar{Q}_{26}\alpha_{xy})dz$$

$$\hat{N}_{xy}^T = \int_{-\frac{H}{2}}^{\frac{H}{2}} (\bar{Q}_{16}\alpha_x + \bar{Q}_{26}\alpha_y + \bar{Q}_{66}\alpha_{xy})dz$$

$$\text{(11.70)}$$

$$\hat{M}_x^T = \int_{-\frac{H}{2}}^{\frac{H}{2}} (\bar{Q}_{11}\alpha_x + \bar{Q}_{12}\alpha_y + \bar{Q}_{16}\alpha_{xy})z dz$$

$$\hat{M}_y^T = \int_{-\frac{H}{2}}^{\frac{H}{2}} (\bar{Q}_{12}\alpha_x + \bar{Q}_{22}\alpha_y + \bar{Q}_{26}\alpha_{xy})z dz$$

$$\hat{M}_{xy}^T = \int_{-\frac{H}{2}}^{\frac{H}{2}} (\bar{Q}_{16}\alpha_x + \bar{Q}_{26}\alpha_y + \bar{Q}_{66}\alpha_{xy})z dz$$

If the material properties can be considered constant through the thickness of each layer—a highly likely situation if the temperature is independent of z—then the integrals become simple sums in the manner of the integrals defining the A_{ij}, B_{ij}, and D_{ij}. From integrating equation (11.70), we find that these sums are

$$\hat{N}_x^T = \sum_{k=1}^{N} (\bar{Q}_{11_k}\alpha_{x_k} + \bar{Q}_{12_k}\alpha_{y_k} + \bar{Q}_{16_k}\alpha_{xy_k})(z_k - z_{k-1})$$

$$\hat{N}_y^T = \sum_{k=1}^{N} (\bar{Q}_{12_k}\alpha_{x_k} + \bar{Q}_{22_k}\alpha_{y_k} + \bar{Q}_{26_k}\alpha_{xy_k})(z_k - z_{k-1})$$

$$\hat{N}_{xy}^T = \sum_{k=1}^{N} (\bar{Q}_{16_k}\alpha_{x_k} + \bar{Q}_{26_k}\alpha_{y_k} + \bar{Q}_{66_k}\alpha_{xy_k})(z_k - z_{k-1})$$

$$\hat{M}_x^T = \frac{1}{2}\sum_{k=1}^{N} (\bar{Q}_{11_k}\alpha_{x_k} + \bar{Q}_{12_k}\alpha_{y_k} + \bar{Q}_{16_k}\alpha_{xy_k})(z_k^2 - z_{k-1}^2)$$
(11.71)

$$\hat{M}_y^T = \frac{1}{2}\sum_{k=1}^{N} (\bar{Q}_{12_k}\alpha_{x_k} + \bar{Q}_{22_k}\alpha_{y_k} + \bar{Q}_{26_k}\alpha_{xy_k})(z_k^2 - z_{k-1}^2)$$

$$\hat{M}_{xy}^T = \frac{1}{2}\sum_{k=1}^{N} (\bar{Q}_{16_k}\alpha_{x_k} + \bar{Q}_{26_k}\alpha_{y_k} + \bar{Q}_{66_k}\alpha_{xy_k})(z_k^2 - z_{k-1}^2)$$

These quantities can be conveniently computed for a laminate in much the same manner as the A_{ij}, B_{ij}, and D_{ij} are computed. The quantities $\hat{N}_x^T, \ldots, \hat{M}_{xy}^T$ have the physical interpretation of thermal stress resultants per unit temperature change, and as a result, will be referred to as *unit thermal stress resultants*. The unit thermal stress resultants are material properties, just as the A_{ij}, B_{ij}, and D_{ij} are, and they yield the thermal stress resultants by simply multiplying them by the temperature change, ΔT. However, the concept of thermal stress resultants per unit temperature change as presented is not useful if the temperature change is a function of z. From this point forward, the discussions will consider only situations where the temperature is independent of z. We shall be thus able to effectively use $\hat{N}_x^T, \ldots, \hat{M}_{xy}^T$. For the $[0/90]_S$ laminate,

$$\hat{N}_x^T = 109.8 \text{ N/m/}^\circ\text{C} \qquad \hat{M}_x^T = 0$$
$$\hat{N}_y^T = 109.8 \text{ N/m/}^\circ\text{C} \qquad \hat{M}_y^T = 0 \qquad (11.72)$$
$$\hat{N}_{xy}^T = 0 \qquad \hat{M}_{xy}^T = 0$$

while for the $[\pm 30/0]_S$ laminate,

$$
\begin{aligned}
\hat{N}_x^T &= 97.2 \text{ N/m/}^\circ\text{C} & \hat{M}_x^T &= 0 \\
\hat{N}_y^T &= 232 \text{ N/m/}^\circ\text{C} & \hat{M}_y^T &= 0 \\
\hat{N}_{xy}^T &= 0 & \hat{M}_{xy}^T &= 0
\end{aligned}
\tag{11.73}
$$

The values of $\hat{N}_{xy}^T$, $\hat{M}_x^T$, $\hat{M}_y^T$, and $\hat{M}_{xy}^T$ are zero for both laminates. This is because both are symmetric balanced laminates. Like the values of the components of the ABD matrix, some components of the unit thermal stress resultants will be zero for certain types of laminates. This will be discussed in more detail shortly.

If the change in moisture, ΔM, is independent of z, then ΔM in equation (11.68) can be removed from within the integral and unit moisture force and moment resultants, in the spirit of equations (11.69), (11.70), and (11.71), can be defined. A word of caution, however: Whereas the diffusion of heat within a fiber-reinforced composite material is quite rapid, thus minimizing the number of situations where the temperature change ΔT is dependent on z, the diffusion of moisture within a fiber-reinforced material is quite slow. Diffusion of moisture to a condition of spatial uniformity can take months, even years, depending on the thickness of a laminate. Thus, within the time scale of a particular problem, ΔM may never be independent of z. The distribution must be known and used in the integral to evaluate the moisture force and moment resultants. Assuming they are being used in the proper context, and are based on the definitions of the unit thermal force and moment resultants, we find that the unit moisture force and moment resultants are defined as

$$
\begin{aligned}
\hat{N}_x^M &= \sum_{k=1}^{N} (\bar{Q}_{11_k}\beta_{x_k} + \bar{Q}_{12_k}\beta_{y_k} + \bar{Q}_{16_k}\beta_{xy_k})(z_k - z_{k-1}) \\[2mm]
\hat{N}_y^M &= \sum_{k=1}^{N} (\bar{Q}_{12_k}\beta_{x_k} + \bar{Q}_{22_k}\beta_{y_k} + \bar{Q}_{26_k}\beta_{xy_k})(z_k - z_{k-1}) \\[2mm]
\hat{N}_{xy}^M &= \sum_{k=1}^{N} (\bar{Q}_{16_k}\beta_{x_k} + \bar{Q}_{26_k}\beta_{y_k} + \bar{Q}_{66_k}\beta_{xy_k})(z_k - z_{k-1}) \\[2mm]
\hat{M}_x^M &= \frac{1}{2}\sum_{k=1}^{N} (\bar{Q}_{11_k}\beta_{x_k} + \bar{Q}_{12_k}\beta_{y_k} + \bar{Q}_{16_k}\beta_{xy_k})(z_k^2 - z_{k-1}^2) \\[2mm]
\hat{M}_y^M &= \frac{1}{2}\sum_{k=1}^{N} (\bar{Q}_{12_k}\beta_{x_k} + \bar{Q}_{22_k}\beta_{y_k} + \bar{Q}_{26_k}\beta_{xy_k})(z_k^2 - z_{k-1}^2) \\[2mm]
\hat{M}_{xy}^M &= \frac{1}{2}\sum_{k=1}^{N} (\bar{Q}_{16_k}\beta_{x_k} + \bar{Q}_{26_k}\beta_{y_k} + \bar{Q}_{66_k}\beta_{xy_k})(z_k^2 - z_{k-1}^2)
\end{aligned}
\tag{11.74}
$$

For the $[0/90]_S$ laminate

$$\hat{N}_x^M = 28\ 700 \text{ N/m/\%} \qquad \hat{M}_x^M = 0$$
$$\hat{N}_y^M = 28\ 700 \text{ N/m/\%} \qquad \hat{M}_y^M = 0 \qquad (11.75)$$
$$\hat{N}_{xy}^M = 0 \qquad\qquad\qquad \hat{M}_{xy}^M = 0$$

and for the $[\pm 30/0]_S$ laminate

$$\hat{N}_x^M = 36\ 600 \text{ N/m/\%} \qquad \hat{M}_x^M = 0$$
$$\hat{N}_y^M = 49\ 400 \text{ N/m/\%} \qquad \hat{M}_y^M = 0 \qquad (11.76)$$
$$\hat{N}_{xy}^M = 0 \qquad\qquad\qquad \hat{M}_{xy}^M = 0$$

where the coefficients of moisture expansion from Table 2.1 have been used.

11.6
THE EFFECT OF LAMINATE CLASSIFICATION ON THE UNIT THERMAL FORCE AND MOMENT RESULTANTS

Just as some components of the ABD matrix are identically zero, depending on whether a laminate is classified as symmetric, balanced, cross-plied, or some combination of these three characteristics, some of the thermal force and moment resultants may also be zero for these laminate classifications. For a temperature change that depends on z, no general statements can be made. The specific dependence of temperature on z, the specific fiber orientations, and the specific stacking sequence must be considered case by case. However, for temperature changes that are independent of z, some of the unit thermal force and moment resultants are identically zero for specific classes of laminates.

11.6.1 Symmetric Laminates

For symmetric laminates, all three components of the unit thermal moment resultants are zero. In equation (11.70), in the definitions of the unit thermal moment resultants, the material properties in each integrand are even functions of z, causing the entire integrand in each case to be an odd function of z. Each of the three moment integrals, and hence each of the three moment sums in equation (11.71), are zero. For symmetric laminates, then,

$$\hat{N}_x^T \neq 0 \qquad \hat{M}_x^T = 0$$
$$\hat{N}_y^T \neq 0 \qquad \hat{M}_y^T = 0 \qquad (11.77)$$
$$\hat{N}_{xy}^T \neq 0 \qquad \hat{M}_{xy}^T = 0$$

11.6.2 Balanced Laminates

For balanced laminates, the unit thermal moment resultant $\hat{N}_{xy}^T$ is zero. In equation (11.70), in the definition of $\hat{N}_{xy}^T$, the values of $\bar{Q}_{16}$, $\bar{Q}_{26}$, and α_{xy} in the layer oriented at $-\theta$ are opposite in sign to, but equal in magnitude to, respectively, the values of $\bar{Q}_{16}$, $\bar{Q}_{26}$, and α_{xy} for the $+\theta$ layer. Because in balanced laminates there is this layer pairing for all layers, the value of the integral for the entire laminate is zero, resulting in

$$
\begin{array}{ll}
\hat{N}_x^T \neq 0 & \hat{M}_x^T \neq 0 \\
\hat{N}_y^T \neq 0 & \hat{M}_y^T \neq 0 \\
\hat{N}_{xy}^T = 0 & \hat{M}_{xy}^T \neq 0
\end{array}
\tag{11.78}
$$

11.6.3 Symmetric Balanced Laminates

If the laminate is symmetric and balanced, then the combined effects of the previous two classifications apply, and the unit thermal force and moment resultants are

$$
\begin{array}{ll}
\hat{N}_x^T \neq 0 & \hat{M}_x^T = 0 \\
\hat{N}_y^T \neq 0 & \hat{M}_y^T = 0 \\
\hat{N}_{xy}^T = 0 & \hat{M}_{xy}^T = 0
\end{array}
\tag{11.79}
$$

There are only two nonzero unit resultants.

11.6.4 Cross-ply Laminates

If the laminate is a cross-ply, and thus consists only of layers at 0 and 90°, $\bar{Q}_{16}$, $\bar{Q}_{26}$, and α_{xy} are zero for every layer, and the unit resultants are

$$
\begin{array}{ll}
\hat{N}_x^T \neq 0 & \hat{M}_x^T \neq 0 \\
\hat{N}_y^T \neq 0 & \hat{M}_y^T \neq 0 \\
\hat{N}_{xy}^T = 0 & \hat{M}_{xy}^T = 0
\end{array}
\tag{11.80}
$$

11.6.5 Symmetric Cross-ply Laminates

For symmetric cross-ply laminates, the unit moments in equation (11.80) are zero. The unit force and moment resultants are as they are for the symmetric balanced case, equation (11.79), namely,

$$
\begin{array}{ll}
\hat{N}_x^T \neq 0 & \hat{M}_x^T = 0 \\
\hat{N}_y^T \neq 0 & \hat{M}_y^T = 0 \\
\hat{N}_{xy}^T = 0 & \hat{M}_{xy}^T = 0
\end{array}
\tag{11.81}
$$

11.6.6 Single Isotropic Layer

For a single isotropic layer

$$
\begin{array}{ll}
\hat{N}_x^T = \dfrac{EH\alpha}{1 - \nu^2} & \hat{M}_x^T = 0 \\[2ex]
\hat{N}_y^T = \hat{N}_x^T & \hat{M}_y^T = 0 \\[2ex]
\hat{N}_{xy}^T = 0 & \hat{M}_{xy}^T = 0
\end{array}
\tag{11.82}
$$

where α is the coefficient of thermal expansion of the isotropic material.

Exercise for Section 11.6

Compute the unit thermal stress resultants for the graphite-reinforced $[\pm30/0]_T$ of CLT Example 5.

Computer Exercise

Add to your program the capability to compute the unit thermal forces and moments. This will require you to compute the values of α_x, α_y, and α_{xy} for each layer. Print out the value of these material properties. Also add the capability to read in a temperature change that is independent of z. Check the values of equations (11.72) and (11.73) with your program, and the results from the Exercise above.

11.7
FREE THERMAL RESPONSE OF LAMINATES

We now have all the governing equations and accompanying formal definitions to study the stresses and strains within a laminate due to the inclusion of free thermal strain effects in the material. In Chapters 6, 7, and 8 we carefully studied many examples of laminate response due to loads. Studying the stresses and strains due to both loads and free thermal strain effects is now within our reach. We shall do that shortly and, as before, we shall compare the thermal and nonthermal cases. However, one class of problems is very important in studying thermal effects in laminated fiber-reinforced composite materials: the stresses and strains in laminates due only to a temperature change. For this class of problems the stress resultants are zero; hence, the terminology *free thermal response of laminates* is used. We begin our discussion with the presentation of some general theory, which will be followed with specific examples.

11.7.1 Equations Governing Free Thermal Response of Laminates

If the loads are zero, then the reference surface strains and curvatures are strictly a result of the thermal force and moment resultants. Equation (11.66), the inverse of the classical relationship involving the ABD matrix, is the preferred form for studying free thermal response. That equation is, for this case of no applied loads,

$$\begin{Bmatrix} \varepsilon_x^o \\ \varepsilon_y^o \\ \gamma_{xy}^o \\ \kappa_x^o \\ \kappa_y^o \\ \kappa_{xy}^o \end{Bmatrix} = \begin{bmatrix} a_{11} & a_{12} & a_{16} & b_{11} & b_{12} & b_{16} \\ a_{12} & a_{22} & a_{26} & b_{21} & b_{22} & b_{26} \\ a_{16} & a_{26} & a_{66} & b_{61} & b_{62} & b_{66} \\ b_{11} & b_{21} & b_{61} & d_{11} & d_{12} & d_{16} \\ b_{12} & b_{22} & b_{62} & d_{12} & d_{22} & d_{26} \\ b_{16} & b_{26} & b_{66} & d_{16} & d_{26} & d_{66} \end{bmatrix} \begin{Bmatrix} N_x^T \\ N_y^T \\ N_{xy}^T \\ M_x^T \\ M_y^T \\ M_{xy}^T \end{Bmatrix} \qquad (11.83)$$

Utilizing the definition of the unit thermal resultants, this equation becomes

$$\begin{Bmatrix} \varepsilon_x^o \\ \varepsilon_y^o \\ \gamma_{xy}^o \\ \kappa_x^o \\ \kappa_y^o \\ \kappa_{xy}^o \end{Bmatrix} = \begin{bmatrix} a_{11} & a_{12} & a_{16} & b_{11} & b_{12} & b_{16} \\ a_{12} & a_{22} & a_{26} & b_{21} & b_{22} & b_{26} \\ a_{16} & a_{26} & a_{66} & b_{61} & b_{62} & b_{66} \\ b_{11} & b_{21} & b_{61} & d_{11} & d_{12} & d_{16} \\ b_{12} & b_{22} & b_{62} & d_{12} & d_{22} & d_{26} \\ b_{16} & b_{26} & b_{66} & d_{16} & d_{26} & d_{66} \end{bmatrix} \begin{Bmatrix} \hat{N}_x^T \\ \hat{N}_y^T \\ \hat{N}_{xy}^T \\ \hat{M}_x^T \\ \hat{M}_y^T \\ \hat{M}_{xy}^T \end{Bmatrix} \Delta T \quad (11.84)$$

What is quite interesting is that except for ΔT, every quantity on the right side of the equation is a material property, or depends on laminate geometry. If $\Delta T = 1$, then the deformations of the reference surface, the left side of the above equation, are completely determined by layer material properties, layer orientation, layer location, and layer thickness. It can be concluded that the response of the laminate due to a spatially uniform unit temperature change is an inherent property of the laminate. Because for a general laminate each term in the inverse of the ABD matrix is nonzero, and because all the unit thermal resultants are nonzero, the reference surface deformation due to a temperature change involves all three strains and all three curvatures. Thus, if a general laminate is flat at its processing temperature, as the temperature is reduced to the ambient temperature, the laminate deforms in a complex fashion.

Whether or not the laminate is general, whether or not the temperature change being discussed is measured relative to the processing temperature, and whether or not the temperature change is unity, once the strains and curvatures of the reference surface are known, from equation (11.83), the strain distributions through the thickness of the laminate can be determined, as we have done many times. With the strains known, the stresses can be computed by using the stress-strain relations with the effects of free thermal strain included, that is, equation (11.4). If every point on the reference surface of a laminate has the same strains and curvatures, due to the fact that the temperature is not a function of x or y, then the change in dimensions, and the change in shape, of the laminate can be determined from the reference surface deformations. If the laminate is unsymmetric, reference surface curvatures develop, leading to out-of-plane displacements. In fact, if the reference surface strains and curvatures do not vary with x or y, then equation (11.3) can be integrated to give

$$u^o(x, y) = \varepsilon_x^o x + C_1 y + C_2$$
$$v^o(x, y) = \varepsilon_y^o y + (\gamma_{xy}^o - C_1)x + C_3 \tag{11.85}$$
$$w^o(x, y) = -\frac{1}{2}(\kappa_x^o x^2 + \kappa_y^o y^2 + \kappa_{xy}^o xy) + K_1 x + K_2 y + K_3$$

This integration was done in connection with equations (7.132) and (7.154), except here $\varepsilon_x^o, \ldots, \kappa_{xy}^o$ are due to thermal effects. As in previous chapters, the constants $C_1, C_2, \ldots, K_3$ represent rigid body displacements and rotations, and they can be arbitrarily equated to zero. Though we will not utilize the above equations much, they are quite valuable for determining deformations of a laminate due to temperature changes. If the laminate is symmetric, then equation (11.84) simplifies considerably to

$$\begin{Bmatrix} \varepsilon_x^o \\ \varepsilon_y^o \\ \gamma_{xy}^o \end{Bmatrix} = \begin{bmatrix} a_{11} & a_{12} & a_{16} \\ a_{12} & a_{22} & a_{26} \\ a_{16} & a_{26} & a_{66} \end{bmatrix} \begin{Bmatrix} \hat{N}_x^T \\ \hat{N}_y^T \\ \hat{N}_{xy}^T \end{Bmatrix} \Delta T \tag{11.86}$$

and
$$\kappa_x^o = \kappa_y^o = \kappa_{xy}^o = 0 \tag{11.87}$$

As a result of a temperature change, the laminate expands or contracts in the x and y directions, and the corner right angles, considering a rectangular laminate, change. The reference surface displacements are given by

$$u^o(x, y) = \varepsilon_x^o x$$
$$v^o(x, y) = \varepsilon_y^o y + \gamma_{xy}^o x \tag{11.88}$$
$$w^o(x, y) = 0$$

where the constants have been equated to zero.

If in addition to being symmetric, the laminate is balanced, then equations (11.86) and (11.87) become

$$\begin{Bmatrix} \varepsilon_x^o \\ \varepsilon_y^o \end{Bmatrix} = \begin{bmatrix} a_{11} & a_{12} \\ a_{12} & a_{22} \end{bmatrix} \begin{Bmatrix} \hat{N}_x^T \\ \hat{N}_y^T \end{Bmatrix} \Delta T \tag{11.89}$$

and
$$\gamma_{xy}^o = \kappa_x^o = \kappa_y^o = \kappa_{xy}^o = 0 \tag{11.90}$$

Equations (11.89) and (11.90) are also applicable if the laminate is of a symmetric cross-ply construction.

11.7.2 Laminate Coefficients of Thermal Expansion

Considering the special case of symmetric laminates with no stress resultants applied, and considering a unit temperature change, the reference surface strains are given by

$$\varepsilon_x^o = a_{11}\hat{N}_x^T + a_{12}\hat{N}_y^T + a_{16}\hat{N}_{xy}^T$$
$$\varepsilon_y^o = a_{12}\hat{N}_x^T + a_{22}\hat{N}_y^T + a_{26}\hat{N}_{xy}^T \qquad (11.91)$$
$$\gamma_{xy}^o = a_{16}\hat{N}_x^T + a_{26}\hat{N}_y^T + a_{66}\hat{N}_{xy}^T$$

Because there are no curvatures, and hence no out-of-plane displacements, these expressions quantify the strains of the laminate in the x and y directions, and the change in right angles, due simply to a unit temperature change. This is precisely like the definitions of the coefficient of thermal deformation in the first three expressions of equation (5.143), that is, the deformations due to a unit temperature change when an element of material is free in space. Though in a laminate there are stresses within the layers, overall, the integrals of these stresses (i.e., the stress resultants) are zero. Thus, in the global sense the laminate is free. Hence, we define the *laminate coefficients of thermal deformation* to be

$$\bar{\alpha}_x \equiv a_{11}\hat{N}_x^T + a_{12}\hat{N}_y^T + a_{16}\hat{N}_{xy}^T$$
$$\bar{\alpha}_y \equiv a_{12}\hat{N}_x^T + a_{22}\hat{N}_y^T + a_{26}\hat{N}_{xy}^T \qquad (11.92)$$
$$\bar{\alpha}_{xy} \equiv a_{16}\hat{N}_x^T + a_{26}\hat{N}_y^T + a_{66}\hat{N}_{xy}^T$$

These coefficients of thermal deformation are strictly material properties, and are thus an inherent property of the laminate. They, like the components of the ABD matrix, can be computed quite readily for a given laminate. We should emphasize that these coefficients of thermal deformation apply only to deformations in the x and y directions. Even though a temperature change does not cause any out-of-plane deformations of the reference surface if the laminate is symmetric, the thickness of a laminate changes as the temperature is increased or decreased. In the next chapter, we shall study the thickness deformations of a laminate due to a temperature change. If in addition to being symmetric, the laminate is balanced, or if the laminate is cross-plied, then A_{16}, A_{26}, and $\hat{N}_{xy}^T$ are identically zero and the remaining coefficients can be written in a somewhat more basic form as

$$\bar{\alpha}_x \equiv \frac{A_{22}\hat{N}_x^T - A_{12}\hat{N}_y^T}{A_{11}A_{22} - A_{12}^2}$$

$$\bar{\alpha}_y \equiv \frac{A_{11}\hat{N}_y^T - A_{12}\hat{N}_x^T}{A_{11}A_{22} - A_{12}^2} \qquad (11.93)$$

Because there is no right-angle change associated with this case, these coefficients are more commonly referred to as *laminate coefficients of thermal expansion*, a nomenclature more in keeping with traditional material properties. By examining the expressions for thermal expansion, we can see that one of the advantages of laminated fiber-reinforced composite materials is that the coefficient of thermal expansion in one direction can be made to be zero. By the proper choice of material properties and fiber angles, it is possible to have the numerator in one of the two expressions above be zero. It would be very rare if both coefficients could be made zero simultaneously. However, having even one coefficient zero leads to many applications, as having no dimensional change in spite of a temperature change is a useful property. As might be expected, the zero coefficient does not come without a price. In many cases, with the coefficient of thermal expansion in one direction being zero, the coefficient of expansion in the other direction may be quite large. In addition, knowing that heating individual layers causes them to expand in the 2 direction and contract in the 1 direction, we see that fabricating a laminate that has no expansion in one direction can lead to high stresses within the layers.

In analogy to the laminate coefficients of thermal deformation and coefficients of thermal expansion, the *laminate coefficients of moisture deformation* and *coefficients of moisture expansion* can be defined, as in equations (11.92) and (11.93), as follows:

$$\bar{\beta}_x \equiv a_{11}\hat{N}_x^M + a_{12}\hat{N}_y^M + a_{16}\hat{N}_{xy}^M$$

$$\bar{\beta}_y \equiv a_{12}\hat{N}_x^M + a_{22}\hat{N}_y^M + a_{26}\hat{N}_{xy}^M \qquad (11.94)$$

$$\bar{\beta}_{xy} \equiv a_{16}\hat{N}_x^M + a_{26}\hat{N}_y^M + a_{66}\hat{N}_{xy}^M$$

$$\bar{\beta}_x \equiv \frac{A_{22}\hat{N}_x^M - A_{12}\hat{N}_y^M}{A_{11}A_{22} - A_{12}^2}$$

$$\qquad (11.95)$$

$$\bar{\beta}_y \equiv \frac{A_{11}\hat{N}_y^M - A_{12}\hat{N}_x^M}{A_{11}A_{22} - A_{12}^2}$$

11.8
EXAMPLES OF LAMINATE FREE THERMAL RESPONSE

To gain further insight into the effects of free thermal strains on laminate response, let us examine the free thermal response of the familiar $[0/90]_S$ and $[\pm 30/0]_S$ laminates. In addition, because it represents an unsymmetric laminate, we shall also examine the free thermal response of the $[\pm 30/0]_T$ laminate. A temperature change of $-150°C$ will be considered; this, as mentioned before, represents the temperature difference between the processing temperature of the laminate, a condition where the stresses and deformations are assumed to be zero, and the operating temperature. The stresses that result from this cooldown from processing are called residual thermal stresses, or processing stresses. Here we will use the former nomenclature.

11.8.1 CLT Example 11: $[0/90]_S$ Laminate Subjected to a Known ΔT

It is useful to compute the coefficients of thermal expansion of a laminate to develop a feeling of how the various layers, each with its own specific orientations, interact to produce an overall expansion. Accordingly, we gave the values of the unit thermal stress resultants for the $[0/90]_S$ laminate in equation (11.72), and using these values in the definitions of the coefficient of thermal expansion, equation (11.93), and the values of A_{11}, A_{12}, and A_{22} from equation (7.88a), we compute the coefficients of thermal expansion for the $[0/90]_S$ graphite-reinforced laminate as

$$\bar{\alpha}_x = 2.10 \times 10^{-6}/°C$$
$$\bar{\alpha}_y = \bar{\alpha}_x \tag{11.96}$$

The coefficients of thermal expansion are the same in the x and y directions because for both directions there are two layers with fibers parallel to that direction and two layers with fibers perpendicular to that direction. That the coefficients are close to $2 \times 10^{-6}/°C$ demonstrates the strong influence of the fiber-direction stiffness. Table 2.1 indicates that this cross-ply has a thermal expansion coefficient 10 times smaller than aluminum and about 5 times smaller than steel, whose coefficient of thermal expansion is $10.80 \times 10^{-6}/°C$.

Turning now to the computation of residual thermal stresses, because the $[0/90]_S$ laminate is a cross-ply laminate, the distributions of the total strains through the thickness of the laminate, above and below some point (x, y) on the reference surface, are determined by direct application of equation (11.89). The numerical values of all the quantities required in those relations have been presented, and the total strains for the 150°C temperature decrease are

$$\varepsilon_x^o = \varepsilon_y^o = -316 \times 10^{-6} \tag{11.97}$$

Alternatively, the total strains in the x and y directions are the coefficients of thermal expansion for the laminate multiplied by $-150°C$. A $[0/90]_S$ laminate 0.250 m long by 0.125 m wide before the temperature change is 249.921 mm long by 124.960 mm wide after the temperature change, the laminate obviously having contracted when cooling. There are no thermally induced shear deformations for this problem. The distributions of the total strains through the thickness of the laminate are illustrated in Figure 11.9.

The residual stresses due to cooldown are calculated by using the stress-strain relations of equation (11.4). Because there are no curvatures and $\gamma_{xy}^o = 0$, the details of equation (11.4) are simplified. Using the total strains from equation (11.97), the values of α_x and α_y from equation (11.9), and the values of the reduced stiffnesses, the $\bar{Q}_{ij}$, from Chapter 6, we find that the residual stresses in the 0° layers are

$$\sigma_x = -39.5 \text{ MPa}$$
$$\sigma_y = 39.5 \text{ MPa} \tag{11.98a}$$
$$\tau_{xy} = 0$$

while the stresses in the 90° layers are

$$\sigma_x = 39.5 \text{ MPa}$$
$$\sigma_y = -39.5 \text{ MPa} \tag{11.98b}$$
$$\tau_{xy} = 0$$

where the equal and opposite nature of the stresses should be noted. The distributions of these stresses are illustrated in Figure 11.10, and the equal and opposite nature of the stresses must be the case if the integrals of the stresses, namely, the stress resultants, are to be zero (which they must be for laminate free thermal response). The sign of the stresses is interesting; cooling causes the fiber direction to

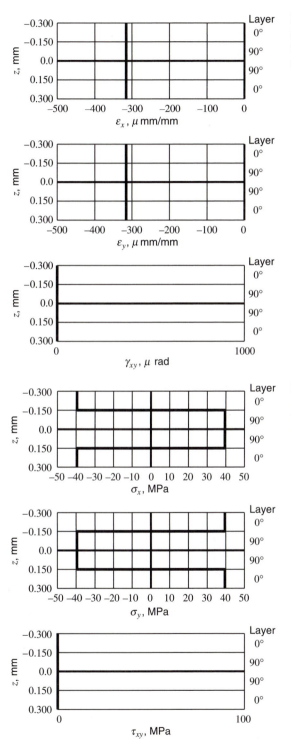

FIGURE 11.9
Strain distribution through the thickness of $[0/90]_S$ laminate subjected to $\Delta T = -150°C$

FIGURE 11.10
Stress distribution through the thickness of $[0/90]_S$ laminate subjected to $\Delta T = -150°C$

experience compression, and the direction perpendicular to the fibers to experience tension. The physics of the signs of these stresses is quite logical. Upon cooling, a free layer wants to expand in the fiber direction and contract perpendicular to the fiber direction. When in a laminate, however, the expansion in the fiber direction in one layer is resisted by contraction perpendicular to the fibers in the adjacent layer. Resistance to expansion is compression; hence, the fibers are in compression. Conversely, resistance to contraction, in the direction perpendicular to the fibers, is tension.

Figure 11.11 shows the principal material system residual stresses, and Figure 11.12 shows the principal material system strains. For this laminate the transformations of the stresses and strains from the x-y system to the 1-2 system are trivial; the principal material system stresses are as follows.

For the 0° layers:

$$\sigma_1 = -39.5 \text{ MPa}$$
$$\sigma_2 = 39.5 \text{ MPa} \tag{11.99a}$$
$$\tau_{12} = 0$$

For the 90° layers:

$$\sigma_1 = -39.5 \text{ MPa}$$
$$\sigma_2 = 39.5 \text{ MPa} \tag{11.99b}$$
$$\tau_{12} = 0$$

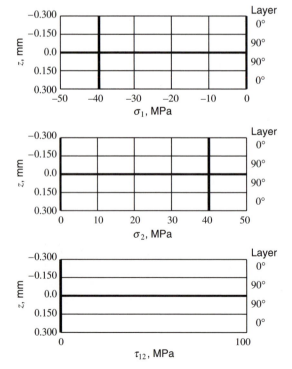

FIGURE 11.11
Principal material system stress distribution through the thickness of $[0/90]_S$ laminate subjected to $\Delta T = -150°C$

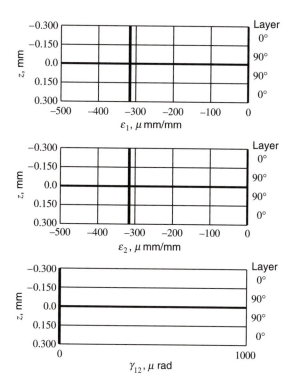

FIGURE 11.12
Principal material system strain distribution through the thickness of $[0/90]_S$ laminate subjected to $\Delta T = -150°C$

The 150°C cooldown results in tensile stresses σ_2 that are near the 50 MPa failure stress. We saw earlier in CLT Example 9 that not much load or specified strain would be required to increase these stresses to failure levels.

Figure 11.13 illustrates the thermally induced mechanical strains in the 1-2 system; these strains are computed directly from equation (4.31) for each layer and are as follows:

For the 0° layers:

$$\varepsilon_1^{mech} = -318 \times 10^{-6}$$
$$\varepsilon_2^{mech} = 3330 \times 10^{-6} \tag{11.100a}$$
$$\gamma_{12}^{mech} = 0$$

For the 90° layers:

$$\varepsilon_1^{mech} = -318 \times 10^{-6}$$
$$\varepsilon_2^{mech} = 3330 \times 10^{-6} \tag{11.100b}$$
$$\gamma_{12}^{mech} = 0$$

By the nature of the laminate construction, the mechanical strains are the same in each layer. Though the mechanical strains still have no physical meaning, for the free thermal response of a laminate these strains represent the difference between the free thermal response of the laminate and the free thermal response of the individual layers. The mechanical strains are thus clearly a measure of the mismatch of free thermal strain effects.

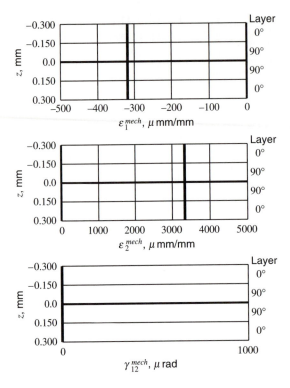

FIGURE 11.13

Principal material system mechanical strain distribution through the thickness of $[0/90]_S$ laminate subjected to $\Delta T = -150°C$

11.8.2 CLT Example 12: $[\pm 30/0]_S$ Laminate Subjected to a Known ΔT

The $[\pm 30/0]_S$ and the $[0/90]$ laminates are similar in that their free thermal responses are governed by equation (11.89). However, the similarity ends there. By the nature of the laminate, the response of the $[0/90]_S$ in the y direction is identical to its response in the x direction; not so for the $[\pm 30/0]_S$ laminate. In addition, the response in the x direction for the latter laminate is quite interesting.

The values of the unit thermal stress resultants for the $[\pm 30/0]_S$ were given in equation (11.73), and substituting these values, as well as the values of A_{11}, A_{12}, and A_{22} from equation (7.89a), into the expressions for the coefficients of thermal expansion, equation (11.93), results in

$$\bar{\alpha}_x = -2.16 \times 10^{-6}/°C$$
$$\bar{\alpha}_y = 16.80 \times 10^{-6}/°C \tag{11.101}$$

The coefficient in the x direction is negative! Like the fiber direction of a single layer, in the x direction the laminate contracts when heated and expands when cooled, while in the y direction, the laminate expands considerably when heated and contracts when cooled. According to the results from the $[0/90]_S$ laminate, it is clear that if the $\pm 30°$ layers were replaced with $90°$ layers, the coefficient of thermal expansion in the x direction would be positive. Thus, for some θ between $30°$ and $90°$, a $(\pm \theta/0)_S$ laminate will have $\bar{\alpha}_x = 0$.

The total strains due to the 150°C temperature drop can be determined from equation (11.89) or by multiplying the above coefficients of thermal expansion by $-150°C$. The result is

$$\varepsilon_x^o = 324 \times 10^{-6}$$
$$\varepsilon_y^o = -2520 \times 10^{-6}$$

(11.102)

These strains are illustrated in Figure 11.14. Calculation of the residual stresses is again based on equation (11.4); the values of the reduced stiffnesses come from Chapter 6 and the values of α_x, α_y, and α_{xy} come from equation (11.25). These residual stresses, illustrated in Figure 11.15, are:

For the $+30°$ layers:

$$\sigma_x = -26.7 \text{ MPa}$$
$$\sigma_y = -7.32 \text{ MPa}$$
$$\tau_{xy} = -38.5 \text{ MPa}$$

(11.103a)

For the $-30°$ layers:

$$\sigma_x = -26.7 \text{ MPa}$$
$$\sigma_y = -7.32 \text{ MPa}$$
$$\tau_{xy} = 38.5 \text{ MPa}$$

(11.103b)

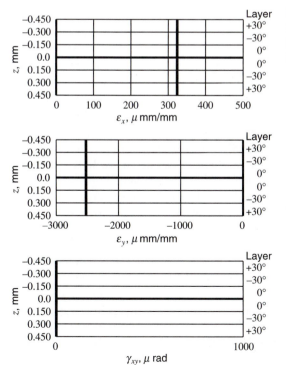

FIGURE 11.14
Strain distribution through the thickness of $[\pm 30/0]_S$ laminate subjected to $\Delta T = -150°C$

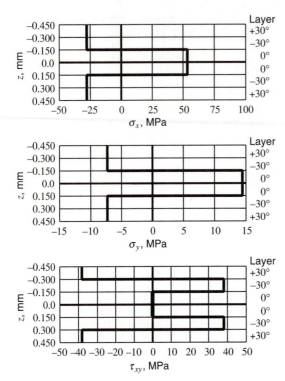

FIGURE 11.15
Stress distribution through the thickness of $[\pm 30/0]_S$ laminate subjected to $\Delta T = -150°C$

For the 0° layers:

$$\sigma_x = 53.4 \text{ MPa}$$
$$\sigma_y = 14.64 \text{ MPa} \qquad (11.103c)$$
$$\tau_{xy} = 0$$

Because there are only half as many 0° layers as $\pm 30°$ layers, because the values of σ_x and σ_y in the $\pm 30°$ layers are identical, and because the integrals of the stresses through the thickness must be zero, the residual stresses σ_x and σ_y in the 0° layers must be exactly twice those stresses in the other two layers, and of opposite sign. In fact, one really does not need to go through the calculations for finding the residual stresses in the 0° layers. Using the fact that N_x, N_y, and N_{xy} have to be zero for the free thermal response of a laminate, the stresses in the 0° layers could be determined once the stresses in the other two layer directions have been determined.

The residual stresses in the principal material system, shown in Figure 11.16, are found by using the transformations and result in:
For the +30° layers:

$$\sigma_1 = -55.2 \text{ MPa}$$
$$\sigma_2 = 21.1 \text{ MPa} \qquad (11.104a)$$
$$\tau_{12} = -10.84 \text{ MPa}$$

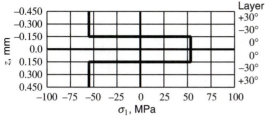

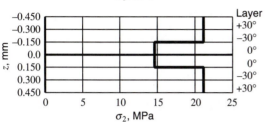

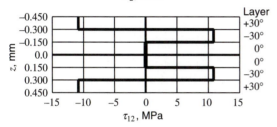

FIGURE 11.16
Principal material system stress distribution through the thickness of $[\pm30/0]_S$ laminate subjected to $\Delta T = -150°C$

For the $-30°$ layers:

$$\sigma_1 = -55.2 \text{ MPa}$$
$$\sigma_2 = 21.1 \text{ MPa} \tag{11.104b}$$
$$\tau_{12} = 10.84 \text{ MPa}$$

For the $0°$ layers:

$$\sigma_1 = 53.4 \text{ MPa}$$
$$\sigma_2 = 14.64 \text{ MPa} \tag{11.104c}$$
$$\tau_{12} = 0$$

The level of σ_2 due to the 150°C cooldown is generally lower than the level in the cross-ply laminate; the residual tensile stress σ_2 here is less than one-half the failure stress level. Recall that in the cross-ply case, Figure 11.11, the residual stress level of σ_2 was closer to the failure level. The lower stress level for the $[\pm30/0]_S$ laminate should not be interpreted as the superiority of that laminate as regards thermal stresses due to processing; rather, the comparison should be interpreted as a disadvantage, and an important one, of the cross-ply construction.

Figure 11.17 shows the distributions of the total strains in the principal material system, while Figure 11.18 shows the distributions of the principal material system mechanical strains. The total strains are as follows:

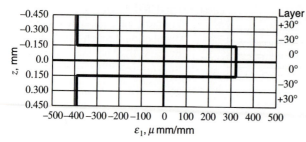

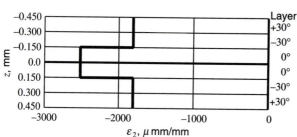

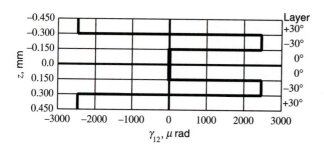

FIGURE 11.17
Principal material system strain distribution through the thickness of $[\pm 30/0]_S$ laminate subjected to $\Delta T = -150°C$

For the $+30°$ layers:

$$\varepsilon_1 = -387 \times 10^{-6}$$
$$\varepsilon_2 = -1809 \times 10^{-6} \qquad (11.105a)$$
$$\gamma_{12} = -2460 \times 10^{-6}$$

For the $-30°$ layers:

$$\varepsilon_1 = -387 \times 10^{-6}$$
$$\varepsilon_2 = -1809 \times 10^{-6} \qquad (11.105b)$$
$$\gamma_{12} = 2460 \times 10^{-6}$$

For the $0°$ layers:

$$\varepsilon_1 = 324 \times 10^{-6}$$
$$\varepsilon_2 = -2520 \times 10^{-6} \qquad (11.105c)$$
$$\gamma_{12} = 0$$

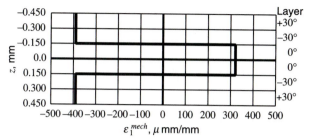

FIGURE 11.18
Principal material system
mechanical strain distribution through the thickness
of $[\pm 30/0]_S$ laminate
subjected to $\Delta T = -150°C$

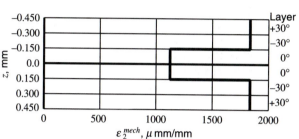

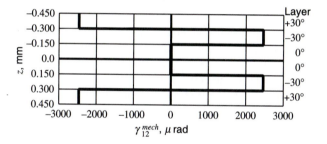

The mechanical strains are as follows:
For the $+30°$ layers:

$$\varepsilon_1^{mech} = -390 \times 10^6$$

$$\varepsilon_2^{mech} = 1836 \times 10^{-6} \tag{11.106a}$$

$$\gamma_{12}^{mech} = -2460 \times 10^{-6}$$

For the $-30°$ layers:

$$\varepsilon_1^{mech} = -390 \times 10^{-6}$$

$$\varepsilon_2^{mech} = 1836 \times 10^{-6} \tag{11.106b}$$

$$\gamma_{12}^{mech} = 2460 \times 10^{-6}$$

For the 0° layers:

$$\varepsilon_1^{mech} = 321 \times 10^{-6}$$
$$\varepsilon_2^{mech} = 1125 \times 10^{-6} \qquad (11.106c)$$
$$\gamma_{12}^{mech} = 0$$

Interpreting the mechanical strains as a measure of the mismatch of response of the laminate relative to the free thermal responses of the various layers, we find that the largest mismatch in the 2 direction is just under 2000×10^{-6}, whereas the mismatch in the $[0/90]_S$ laminate is over 3000×10^{-6}. Again, this attests to the disadvantage of the cross-ply construction in terms of residual thermal effects.

Using the total strains from equation (11.102), we find that the deformed length and width of a 0.250 m by 0.125 m laminate are 250.081 mm by 124.685 mm, with the length increasing with temperature decrease.

11.8.3 CLT Example 13: $[\pm 30/0]_T$ Laminate Subjected to a Known ΔT

The free thermal response of unsymmetric laminates is a very interesting topic, due mainly to the fact that with unsymmetric laminates reference surface curvatures, in addition to reference surface strains, develop when the temperature is changed. A laminate generally experiences changes in the bending curvature in both the x and y directions, as well as changes in the twisting curvature. Using the $-150°C$ temperature change and assuming the laminate is flat at the processing temperature (i.e., all curvatures are zero), we find that the curvatures that develop represent the shape of the unsymmetric laminate after it has cooled to the ambient temperature. We use equation (11.84) to determine the deformations. For the $[\pm 30/0]_T$ laminate, from equation (11.71) for the unit thermal forces and moments,

$$\hat{N}_x^T = 48.6 \text{ N/m/°C} \qquad \hat{M}_x^T = -12.65 \times 10^{-4} \text{N·m/m/°C}$$
$$\hat{N}_y^T = 116.1 \text{ N/m/°C} \qquad \hat{M}_y^T = 12.65 \times 10^{-4} \text{N·m/m/°C} \qquad (11.107)$$
$$\hat{N}_{xy}^T = 0 \qquad \hat{M}_{xy}^T = 21.9 \times 10^{-4} \text{N·m/m/°C}$$

Using $\Delta T = -150°C$ and the values of a_{ij}, b_{ij}, and d_{ij} for this laminate from Chapter 7, equation (7.93), results in

$$\varepsilon_x^o = 199.8 \times 10^{-6} \qquad \kappa_x^o = -0.919/\text{m}$$
$$\varepsilon_y^o = -2900 \times 10^{-6} \qquad \kappa_y^o = -2.32/\text{m} \qquad (11.108)$$
$$\gamma_{xy}^o = -742 \times 10^{-6} \qquad \kappa_{xy}^o = -9.40/\text{m}$$

The deformations in the y direction, both the reference surface strain and the reference surface curvature, are much larger than the reference surface deformations in the x direction. There is a significant twisting curvature κ_{xy}^o. If the reference surface strains and curvatures do not vary with x and y, then equation (11.85) can be used to determine the displacements; in particular, the third equation describes the out-of-plane displacements. The out-of-plane shape of a rectangular $[\pm 30/0]_T$ laminate is depicted in Figure 11.19; considerable warping from the originally flat condition

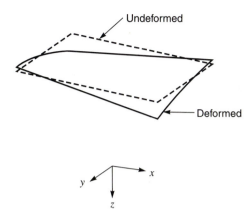

FIGURE 11.19
Shape of $[\pm 30/0]_T$ laminate subjected to $\Delta T = -150°C$ (shape assumed flat at $\Delta T = 0$)

is evident. The use of unsymmetric laminates has the potential for designing shape into a laminate. Of course, if the laminate is to be used in an environment with changing temperature, the shape of the laminate will change. For some applications this may be desirable, while for others it may not be desirable. If the displacements predicted by using equation (11.85) are many times the thickness of the laminate, then the predictions may not be accurate because the equation is not applicable. For large deflections, geometrically nonlinear effects must be included in the formulation governing the change in curvature. The inclusion of geometrically nonlinear effects results in significantly different predictions for $w^o(x, y)$ than are given by equations (11.84) and (11.85).

Figure 11.20 indicates the distributions of the total strains through the thickness of the $[\pm 30/0]_T$; the curvatures cause differences between the strains on the top of the laminate and the strains on the bottom. Figure 11.21 shows the distributions of the residual stresses, and Figure 11.22 illustrates the distribution of the stresses in the principal material system. It is evident that the stresses vary rapidly within a layer, and that the stress component σ_1 varies from positive to negative within the $+30°$ layer.

11.8.4 CLT Example 14: $[\pm 30/0]_S$ Laminate Subjected to a Known ΔM

Using the values of $\hat{N}_x^M$ and $\hat{N}_y^M$, as calculated from equation (11.76), in equation (11.95) leads to the coefficients of moisture expansion for the $[\pm 30/0]_S$ laminate, namely,

$$\bar{\beta}_x = -261 \times 10^{-6}/\%$$
$$\bar{\beta}_y = 3340 \times 10^{-6}/\% \qquad (11.109)$$

These numbers tacitly assume that the percent moisture absorbed by the laminate is uniform through the thickness. If that uniform level of moisture is 2%, for example, then the total strains due to this level of moisture absorption are

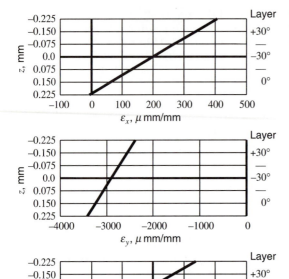

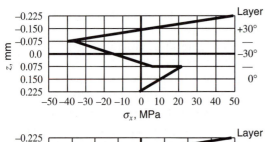

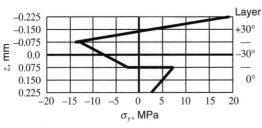

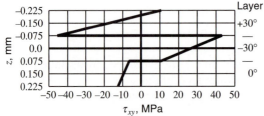

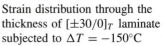

FIGURE 11.20
Strain distribution through the thickness of $[\pm 30/0]_T$ laminate subjected to $\Delta T = -150°C$

FIGURE 11.21
Stress distribution through the thickness of $[\pm 30/0]_T$ laminate subjected to $\Delta T = -150°C$

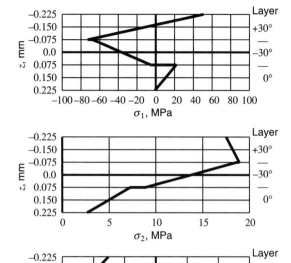

FIGURE 11.22
Principal material system stress distribution through the thickness of $[\pm 30/0]_T$ laminate subjected to $\Delta T = -150°C$

$$\varepsilon_x^o = -522 \times 10^{-6}$$

$$\varepsilon_y^o = 6690 \times 10^{-6} \qquad (11.110)$$

$$\gamma_{xy}^o = 0$$

These strains are determined by multiplying the coefficients of moisture expansion, equation (11.109), by $\Delta M = 2\%$, or alternatively, the analog to equation (11.89) for moisture expansion effects. It is implicit that these strains are measured relative to the state of the laminate before the 2% moisture was absorbed. This could be the completely dry state, for example.

Comparing these strains with the strains due to a $-150°C$ temperature drop in the $[\pm 30/0]_S$ laminate, equation (11.102), we see that the magnitudes of the moisture strains are greater, and the signs are opposite. The changes in the stresses due to this level of moisture absorption are given by using the moisture analog to equation (11.4).

For the $+30°$ layers:

$$\sigma_x = 67.7 \text{ MPa}$$

$$\sigma_y = 18.56 \text{ MPa} \qquad (11.111a)$$

$$\tau_{xy} = 97.5 \text{ MPa}$$

For the $-30°$ layers:
$$\sigma_x = 67.7 \text{ MPa}$$
$$\sigma_y = 18.56 \text{ MPa} \tag{11.111b}$$
$$\tau_{xy} = -97.5 \text{ MPa}$$

For the $0°$ layers:
$$\sigma_x = -135.4 \text{ MPa}$$
$$\sigma_y = -37.1 \text{ MPa} \tag{11.111c}$$
$$\tau_{xy} = 0$$

With free moisture absorption effects as with free thermal effects, the integrals of the stresses must be zero. Thus, for this case σ_x and σ_y in the $0°$ layers are twice the magnitude of these stress components in the $\pm30°$ layers, and of opposite sign.

The changes in the principal material system stresses for the 2 percent moisture increase are:

For the $+30°$ layers:
$$\sigma_1 = 139.9 \text{ MPa}$$
$$\sigma_2 = -53.6 \text{ MPa} \tag{11.112a}$$
$$\tau_{12} = 27.5 \text{ MPa}$$

For the $-30°$ layers:
$$\sigma_1 = 139.9 \text{ MPa}$$
$$\sigma_2 = -53.6 \text{ MPa} \tag{11.112b}$$
$$\tau_{12} = -27.4 \text{ MPa}$$

For the $0°$ layers:
$$\sigma_1 = -135.4 \text{ MPa}$$
$$\sigma_2 = -37.1 \text{ MPa} \tag{11.112c}$$
$$\tau_{12} = 0$$

Moisture absorption produces compression σ_2 components in each layer, whereas the cooldown residual stress components σ_2 are, from equation (11.104), tensile. As was mentioned, the tensile σ_2 due to cooldown can produce a stress that is a significant fraction of the failure stress. If we assume that after the cooldown the 2% moisture is absorbed, the net effect is *compressive* σ_2 in all layers. The principal material system stresses that result from a 2% moisture change, the stresses that result from a $-150°C$ temperature change, and the sum of the two effects are illustrated in Figure 11.23. We can conclude that for this case, because it relieves the residual tensile stress σ_2 that results from the cooldown from the processing temperature, it is advantageous to absorb moisture. Unfortunately, moisture degrades the strength of laminates. It is thus not really an advantage to utilize the stress-relieving tendencies of moisture absorption.

This completes the discussion of the free thermal response and free moisture response of laminates. This represents an important class of problems, particularly if the responses are considered as residual effects stemming from cooldown from the processing temperature. With our examples we found that the stress levels σ_2 resulting from the $-150°C$ temperature change may not be small when compared to the

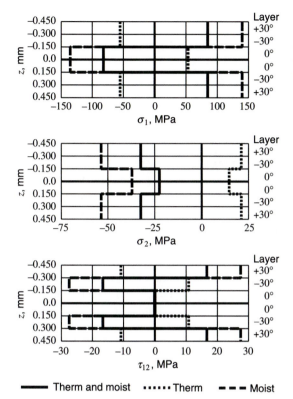

FIGURE 11.23

Principal material system stress distribution through the thickness of $[\pm30/0]_S$ laminate subjected to $\Delta T = -150°C$ alone, $\Delta M = 2\%$ alone, both $\Delta T = -150°C$ and $\Delta M = 2\%$

failure stress levels for that component of stress. In addition, we showed how laminate construction—for example, $[0/90]_S$ versus $[\pm30/0]_S$—influences thermally induced stresses. Though we have used the study of residual effects to illustrate free thermal response, any temperature change can be used. In fact, if the laminate is to operate at 50°C above ambient, then the difference in stress levels between the ambient condition and the 50°C condition can be determined by using $\Delta T = 50°C$ in equation (11.84) and proceeding as we did. Alternatively, the stresses relative to the stress-free condition can be determined by using $\Delta T = -100°C$.

The next section addresses one more aspect of including free thermal response in the analysis of composite laminates, specifically, cases where both the applied load and the temperature change will be specified. The reference surface strains and curvature will be dictated by the sum of these effects, and the resulting stresses and strains determined from these reference surface responses.

Exercises for Section 11.8

1. The ability to tailor a material so that the coefficient of thermal expansion in one direction is zero is an extreme advantage. There are a number of applications where this property would be useful. Consider a $[\pm\theta/0]_S$ laminate, $0 \le \theta \le \pi/2$. (a) Compute and plot $\bar\alpha_x$ and $\bar\alpha_y$ for this laminate as a function of θ. (b) Is there a value of θ that results in either $\bar\alpha_x$

or $\bar{\alpha}_y$ being zero? What value of θ is this? (c) Compute the residual thermal stresses for this case, assuming $\Delta T = -150°C$. (d) Are the magnitudes of the stresses in the principal material system significantly different for the value of θ that leads to $\bar{\alpha}_x$ or $\bar{\alpha}_y$ equal to zero than for the case $\theta = 30°$?

2. Using the values of $\hat{N}_x^M$ and $\hat{N}_y^M$ for the $[\pm 30/0]_S$ laminate, equation (11.76), verify equation (11.109).

11.9
RESPONSE DUE TO STRESS RESULTANTS
AND A TEMPERATURE CHANGE

As our final series of examples in this chapter, we shall revisit the problems of the $[0/90]_S$ and $[\pm 30/0]_S$ laminates subjected to a known normal force resultant N_x, and the $[\pm 30/0]_S$ laminate subjected to a known bending moment resultant M_x. These were CLT Examples 6, 7, and 8 in Chapter 8. Here we will consider the influence on laminate stresses and strains of a temperature change combined with the application of known stress resultants. The resultant levels will be the same as those used in Chapter 8, and in keeping with the examples in this chapter, and because it represents a temperature change of significant interest, we will assume the temperature change is the $-150°C$ temperature change associated with the cooldown from the processing temperature. Thus, we shall be able to evaluate the effects of residual thermal effects and an applied load on the stress levels in the laminates. In the final section of this chapter we shall examine how the residual thermal stress levels influence failure predictions. First, however, let us examine the character of the response when both residual effects and applied load are present.

The form in equation (11.66) is the most convenient for studying the combined effects of a temperature change and an applied load. For symmetric balanced laminates,

$$\left\{ \begin{array}{c} \varepsilon_x^o \\ \varepsilon_y^o \end{array} \right\} = \left[\begin{array}{cc} a_{11} & a_{12} \\ a_{12} & a_{22} \end{array} \right] \left\{ \begin{array}{c} N_x + N_x^T \\ N_y + N_y^T \end{array} \right\} \tag{11.113a}$$

with

$$\gamma_{xy}^o = a_{66}(N_{xy} + N_{xy}^T) \tag{11.113b}$$

and

$$\left\{ \begin{array}{c} \kappa_x^o \\ \kappa_y^o \\ \kappa_{xy}^o \end{array} \right\} = \left[\begin{array}{ccc} d_{11} & d_{12} & d_{16} \\ d_{12} & d_{22} & d_{26} \\ d_{16} & d_{26} & d_{66} \end{array} \right] \left\{ \begin{array}{c} M_x + M_x^T \\ M_y + M_y^T \\ M_{xy} + M_{xy}^T \end{array} \right\} \tag{11.113c}$$

We again use the concept of the unit thermal stress resultants, which were given in equations (11.72) and (11.73) for the $[0/90]_S$ and the $[\pm 30/0]_S$ laminates, respectively. The unit thermal moment resultants are zero for both of these laminates, and using $\Delta T = -150°C$, we find that the thermal force resultants for the $[0/90]_S$ laminate are

$$N_x^T = -0.016\,46 \text{ MN/m}$$
$$N_y^T = N_x^T \tag{11.114}$$
$$N_{xy} = 0$$

while for the $[\pm 30/0]_S$ laminate they are

$$N_x^T = -0.014\,57 \text{ MN/m}$$
$$N_y^T = -0.0348 \text{ MN/m} \tag{11.115}$$
$$N_{xy}^T = 0$$

If, in addition, moisture absorption effects are to be included, equation (11.113) takes the form

$$\left\{ \begin{array}{c} \varepsilon_x^o \\ \varepsilon_y^o \end{array} \right\} = \left[\begin{array}{cc} a_{11} & a_{12} \\ a_{12} & a_{22} \end{array} \right] \left\{ \begin{array}{c} N_x + N_x^T + N_x^M \\ N_y + N_y^T + N_y^M \end{array} \right\} \tag{11.116a}$$

with

$$\gamma_{xy}^o = a_{66}(N_{xy} + N_{xy}^T + N_{xy}^M) \tag{11.116b}$$

and

$$\left\{ \begin{array}{c} \kappa_x^o \\ \kappa_y^o \\ \kappa_{xy}^o \end{array} \right\} = \left[\begin{array}{ccc} d_{11} & d_{12} & d_{16} \\ d_{12} & d_{22} & d_{26} \\ d_{16} & d_{26} & d_{66} \end{array} \right] \left\{ \begin{array}{c} M_x + M_x^T + M_x^M \\ M_y + M_y^T + M_y^M \\ M_{xy} + M_{xy}^T + M_{xy}^M \end{array} \right\} \tag{11.116c}$$

Using a value of $\Delta M = 2\%$ and equations (11.75) and (11.76), we find that for the $[0/90]_S$ laminate

$$N_x^M = 0.0573 \text{ MN/m}$$
$$N_y^M = N_x^M \tag{11.117}$$
$$N_{xy}^M = 0$$

while for the $[\pm 30/0]_S$ laminate

$$N_x^M = 0.0732 \text{ MN/m}$$
$$N_y^M = 0.0988 \text{ MN/m} \tag{11.118}$$
$$N_{xy}^M = 0$$

11.9.1 CLT Example 15: $[0/90]_S$ Laminate Subjected to Known N_x and Known ΔT

To add physical dimensions to the developments that have been presented, let us cast a problem as follows: Consider a $[0/90]_S$ laminate that is cooled 150° from its stress-free processing temperature and cut to be 0.250 m long in the x direction by 0.125 m wide in the y direction. Forces of 6300 N, acting in the x direction and uniformly distributed along the 0.125-m end, are then applied. With this loading on the boundary, every point on the 0.0312 m² reference surface of this laminate is subjected to the stress resultants

$$N_x = 50\,400 \text{ N/m} \qquad M_x = 0$$
$$N_y = 0 \qquad\qquad M_y = 0 \tag{11.119}$$
$$N_{xy} = 0 \qquad\qquad M_{xy} = 0$$

This is exactly the situation studied in CLT Example 6 of Chapter 8, specifically equation (8.3), but with the addition of having thermal effects present. From equation (11.113), the total reference surface strains are given by

$$\varepsilon_x^o = a_{11}(N_x + N_x^T) + a_{12}N_y^T$$
$$\varepsilon_y^o = a_{12}(N_x + N_x^T) + a_{22}N_y^T \tag{11.120}$$
$$\gamma_{xy}^o = 0$$

or, if we use numerical values for the various quantities, equations (7.91a), (11.114), and (11.119),

$$\varepsilon_x^o = 686 \times 10^{-6}$$
$$\varepsilon_y^o = -351 \times 10^{-6} \tag{11.121}$$
$$\gamma_{xy}^o = 0$$

There are no curvatures for this problem and, as expected, the total strains for this problem are the sum of strains due to the free thermal response, equation (11.97),

$$\varepsilon_x^o = -316 \times 10^{-6}$$
$$\varepsilon_y^o = -316 \times 10^{-6} \tag{11.122}$$
$$\gamma_{xy}^o = 0$$

and the strains due to the stress resultant N_x, equation (8.4),

$$\varepsilon_x^o = 1001 \times 10^{-6}$$
$$\varepsilon_y^o = -36.0 \times 10^{-6} \tag{11.123}$$
$$\gamma_{xy} = 0$$

The free thermal response results in contraction in both the x and y directions, and the response due to N_x results in stretching in the x direction and, due to Poisson effects, further contraction in the y direction. With combined thermal effects and applied loads, the strains and the subsequent stresses in this problem are measured relative to the stress-free processing temperature of the laminate.

The distributions of the total strains through the thickness of the laminate with and without the effects of free thermal response are illustrated in Figure 11.24; the case of no thermal effects is repeated from Figure 8.2. The contractions in both the x and y directions are reflected in the differences in the strains with and without thermal effects. Equation (11.4) gives the stresses in each layer; all the quantities in that equation have been given previously. The stresses are:

For the 0° layers:

$$\sigma_x = 116.3 \text{ MPa}$$
$$\sigma_y = 42.1 \text{ MPa} \tag{11.124a}$$
$$\tau_{xy} = 0$$

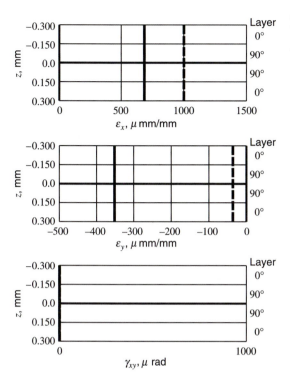

FIGURE 11.24
Strain distribution through the thickness of $[0/90]_S$ laminate subjected to $N_x = 50\,400$ N/m and $\Delta T = -150°C$

For the 90° layers:

$$\sigma_x = 51.6 \text{ MPa}$$
$$\sigma_y = -42.1 \text{ MPa} \qquad (11.124b)$$
$$\tau_{xy} = 0$$

Because these stresses represent the combined effects of the applied stress resultant N_x and the free thermal response of the laminate, the stresses are the sum of the stresses due to the two individual effects, namely, equation (8.5) and equation (11.98). For convenience, equation (8.5) is, for the 0° layers,

$$\sigma_x = 155.8 \text{ MPa}$$
$$\sigma_y = 2.58 \text{ MPa} \qquad (11.125a)$$
$$\tau_{xy} = 0$$

and for the 90° layers,

$$\sigma_x = 12.07 \text{ MPa}$$
$$\sigma_y = -2.58 \text{ MPa} \qquad (11.125b)$$
$$\tau_{xy} = 0$$

Figure 11.25 illustrates the distribution of the stresses through the thickness with and without the effects of free thermal response; the case of no thermal effects is

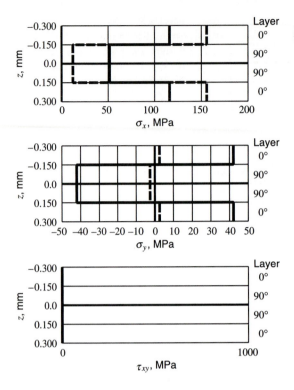

FIGURE 11.25
Stress distribution through the thickness of $[0/90]_S$ laminate subjected to $N_x = 50\,400$ N/m and $\Delta T = -150°C$

repeated from Figure 8.3. It is key to the understanding of thermal effects to realize that in Figure 11.25 the integral of the stress component σ_x with thermal effects present, the solid line, is $N_x = 50\,400$ N/m, and the integral of the stress component σ_x without thermal effects present, the dashed line, is also $N_x = 50\,400$ N/m! Likewise, the integral of the stress component σ_y with and without thermal effects is zero. Because the stress resultants have been specified for this problem in equation (11.119), independently of whether or not thermal effects are present, the integrals of the stress components must lead back to those resultants. This is different from the first example in this chapter, where the deformations of the reference surface were specified and inclusion or exclusion of thermal effects resulted in different values of the stress resultants. The distinction between specifying deformations and specifying resultants, though discussed in the past, is reemphasized here in the presence of thermal effects.

Figure 11.26 illustrates the distributions of the stresses in the principal material system; the inclusion of the temperature change causes the stress component σ_2 in the 90° layers to exceed the failure level of 50 MPa. The distributions of the total strains and the mechanical strains in the principal material system are shown in Figures 11.27 and 11.28, respectively. In all these figures, the difference between the solid lines and the dashed lines should be exactly the numbers from the free thermal response calculations, Section 11.8.1.

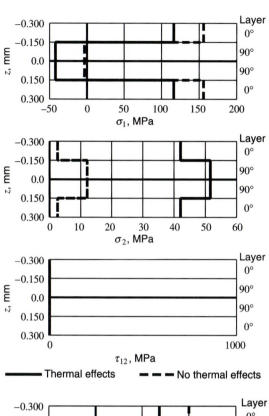

FIGURE 11.26
Principal material system stress distribution through the thickness of $[0/90]_S$ laminate subjected to $N_x = 50\,400$ N/m and $\Delta T = -150°C$

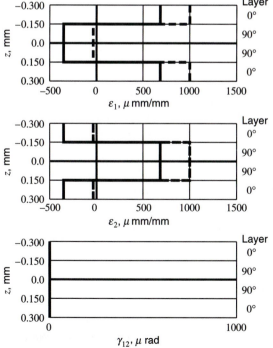

FIGURE 11.27
Principal material system strain distribution through the thickness of $[0/90]_S$ laminate subjected to $N_x = 50\,400$ N/m and $\Delta T = -150°C$

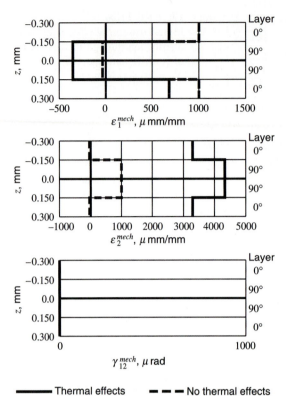

FIGURE 11.28
Principal material system mechanical strain distribution through the thickness of $[0/90]_S$ laminate subjected to $N_x = 50\ 400$ N/m and $\Delta T = -150°C$

11.9.2 CLT Example 16: $[\pm 30/0]_S$ Laminate Subjected to Known N_x and Known ΔT

For our second example of combined thermal effects and applied loads, consider a $[\pm 30/0]_S$ graphite-reinforced laminate that is cooled 150°, cut to be 0.250 m long by 0.125 m wide, and is then loaded with uniformly distributed loads of 0.0128 MN acting in the x direction on the 0.125-m ends. For this situation the stress resultants at every point on the reference surface are

$$N_x = 0.1024 \text{ MN/m} \qquad M_x = 0$$
$$N_y = 0 \qquad\qquad\qquad M_y = 0 \qquad\qquad (11.126)$$
$$N_{xy} = 0 \qquad\qquad\quad M_{xy} = 0$$

the situation of CLT Example 7 in Chapter 8, equation (8.11), but here with thermal effects included. Equation (11.113) is again applicable; the numerical values lead to

$$\varepsilon_x^o = 1599 \times 10^{-6}$$
$$\varepsilon_y^o = -4010 \times 10^{-6} \qquad\qquad (11.127)$$
$$\gamma_{xy}^o = 0$$

An examination of equation (8.12), the case of applied load only, and equation (11.102), the case of free thermal response, indicates the strains of equation (11.127) are again the sum of these two conditions. These strains are illustrated in Figure 11.29; the additional expansion in the x direction when thermal effects are included is due to the negative coefficient of thermal expansion of this laminate. The majority of the contraction in the y direction, again, is due to thermal effects; the large positive coefficient of thermal expansion in this direction for this laminate is responsible. The stresses, shown in Figure 11.30, are definitely influenced by the addition of thermal effects. The integral of the stress component σ_x both with and without thermal effects results in $N_x = 0.1024$ MN/m, and the integral of the stress component σ_y in both cases results in $N_y = 0$. The principal material system stresses, Figure 11.31, still the most effective view of the stress state in a laminate, show significant differences in the stress state, particularly the component σ_2. The component σ_2 is predicted to be everywhere tensile when accounting for thermal effects, while it is predicted to be everywhere compressive if thermal effects are not included. The shear stress component τ_{12} is more than doubled by the inclusion of thermal effects.

What must be realized in this and the other examples of the combined effects of a temperature change and an applied load is that the relative importance of each effect can change as a function of the particular application of a composite laminate. Using a composite in space applications may lead to an operating environment of $100°$ or so below $0°$C. For this case, then, a temperature change of $200–300°$C below the stress-free processing temperature must be used for computing stresses. If the

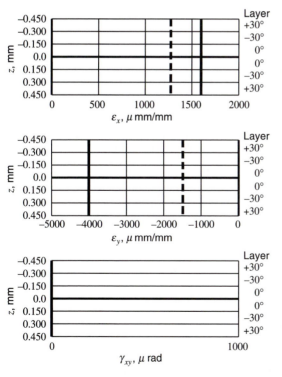

FIGURE 11.29
Strain distribution through the thickness of $[\pm30/0]_S$ laminate subjected to $N_x = 0.1024$ MN/m and $\Delta T = -150°$C

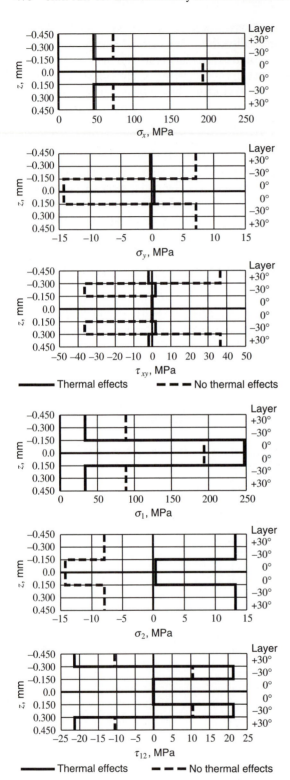

FIGURE 11.30

Stress distribution through the thickness of $[\pm 30/0]_S$ laminate subjected to $N_x = 0.1024$ MN/m and $\Delta T = -150°C$

FIGURE 11.31

Principal material system stress distribution through the thickness of $[\pm 30/0]_S$ laminate subjected to $N_x = 0.1024$ MN/m and $\Delta T = -150°C$

laminate is lightly loaded, as is often the case in space applications, thermal effects dominate. On the other hand, with the $-150°C$ temperature change, the applied load may be much larger than the loads that have been used in the examples. Thus, whether or not the inclusion or exclusion of thermal effects has a great influence on any one stress component can only be determined by a detailed study of the particular problem.

The distributions of the total strains and the mechanical strains in the principal material system are illustrated in Figures 11.32 and 11.33.

11.9.3 CLT Example 17: [±30/0]$_S$ Laminate Subjected to Known M_x and Known ΔT

As our last example of the combination of a temperature change and applied loads, consider again the [±30/0]$_S$ laminate. If the 0.125-m ends of the cooled laminate are subjected to uniformly distributed bending moments of 1.605 N·m, then the stress resultants at every point on the 0.0312 m² reference surface are given by

$$N_x = 0 \qquad M_x = 12.84 \text{ N·m/m}$$
$$N_y = 0 \qquad M_y = 0 \qquad\qquad (11.128)$$
$$N_{xy} = 0 \qquad M_{xy} = 0$$

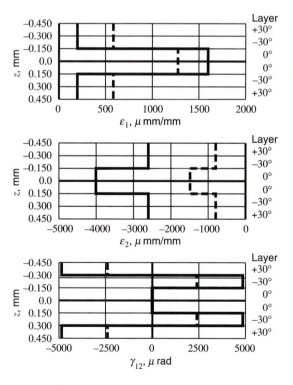

FIGURE 11.32
Principal material system strain distribution through the thickness of [±30/0]$_S$ laminate subjected to $N_x = 0.1024$ MN/m and $\Delta T = -150°C$

—— Thermal effects   --- No thermal effects

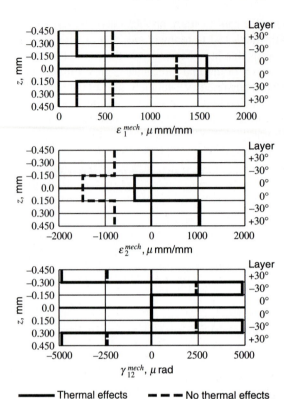

FIGURE 11.33
Principal material system mechanical strain distribution through the thickness of $[\pm30/0]_S$ laminate subjected to $N_x = 0.1024$ MN/m and $\Delta T = -150°$C

———— Thermal effects   — — — No thermal effects

To study the response of the laminate, we return again to equation (11.113). Though there are no applied force resultants, there are thermal force resultants that must be accounted for. For this problem, these governing equations reduce to

$$\varepsilon_x^o = a_{11}N_x^T + a_{12}N_y^T$$
$$\varepsilon_y^o = a_{12}N_x^T + a_{22}N_y^T \qquad (11.129a)$$
$$\gamma_{xy}^o = 0$$

and

$$\kappa_x^o = d_{11}M_x$$
$$\kappa_y^o = d_{12}M_x \qquad (11.129b)$$
$$\kappa_{xy}^o = d_{16}M_x$$

Using numerical values in these equations results in

$$\varepsilon_x^o = 324 \times 10^{-6} \qquad \kappa_x^o = 4.24 \text{ m}^{-1}$$
$$\varepsilon_y^o = -2520 \times 10^{-6} \qquad \kappa_y^o = -5.40 \text{ m}^{-1} \qquad (11.130)$$
$$\gamma_{xy}^o = 0 \qquad \kappa_{xy}^o = -1.669 \text{ m}^{-1}$$

As the applied bending moment results in no reference surface strains, they are due solely to free thermal response of the laminate and are identical to equation (11.102).

Likewise, as the temperature change results in no reference surface curvatures, they are due solely to the applied bending moment and are identical to equation (8.21). Figure 11.34 shows the distribution of the total strains with and without thermal effects. The linear variation of the moment-only strains, which are all zero at $z = 0$, is shifted uniformly by the free thermal strains of the laminate. Because of the negative coefficient of thermal expansion of this laminate in the x direction, the strains in the x direction are all more positive when free thermal strains are included. The strains in the y direction are all more negative, by a significant amount. As this is a balanced laminate, the shear strains are not influenced by thermal effects.

The stresses, illustrated in Figure 11.35, are strongly influenced by the inclusion of thermal effects, which prevent the stresses from being strictly odd functions of z. In addition, the integrals of all components of stress, both with and without thermal effects, are zero because there are no applied force resultants. On the other hand, the integral of the stress component σ_x times z yields $M_x = 12.84$ N·m/m. The integrals of the two other stress components, multiplied times z, are zero. Again we have this interesting point: Thermal effects will alter the stress state relative to the case with no thermal effects; however, the integrals of the stresses yield the same force and

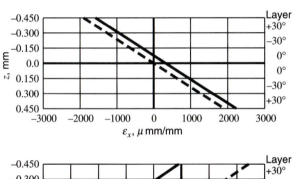

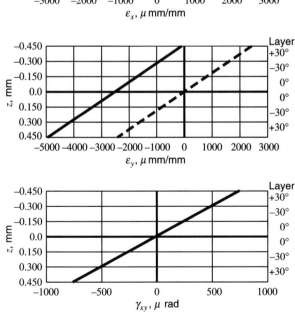

FIGURE 11.34
Strain distribution through the thickness of $[\pm30/0]_S$ laminate subjected to $M_x = 12.84$ N·m/m and $\Delta T = -150°C$

Thermal effects   — — — No thermal effects

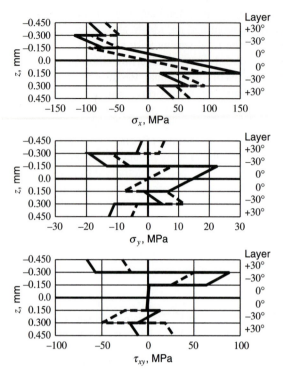

FIGURE 11.35
Stress distribution through the thickness of $[\pm 30/0]_S$ laminate subjected to $M_x = 12.84$ N·m/m and $\Delta T = -150°C$

————— Thermal effects   — — — No thermal effects

moment resultants as the nonthermal case. It is interesting to see how the stresses in the various layers adjust to compensate for the effect of free thermal strain in each layer, yet from the prospective of the entire laminate, they still integrate to the same value they had when thermal effects were not included. That this can all be predicted and explained is the essence of the study of the mechanics of composites with the effects of free thermal strains included.

Figure 11.36 shows the principal material system stresses; the stress component σ_2 is most influenced by thermal effects. With thermal effects, this stress component is everywhere tensile and much closer to failure levels than without thermal effects included. This is another example of the stresses due to the applied loads being quite small, and of opposite sign, relative to the stresses due to thermal effects. The shear stress component τ_{12} is also influenced. Thermal effects result in practically no shear stress in the upper $+30°$ layer, whereas the shear stress in the lower $+30°$ layer is nearly doubled. Thermal effects can alter the predicted location of failure. For example, if it is felt that high shear stresses should be avoided, then with no thermal effects, both $+30°$ layers have the same tendency to control the failure, in particular, at the outer location in each layer. However, with thermal effects, attention must be focused on the outer location in just the lower $+30°$ layer; the upper $+30°$ layer has no role due to its low level of shear stress. Interestingly, the maximum values of σ_1 and σ_2 do not occur in the outer layers. Also, if the sign of the applied

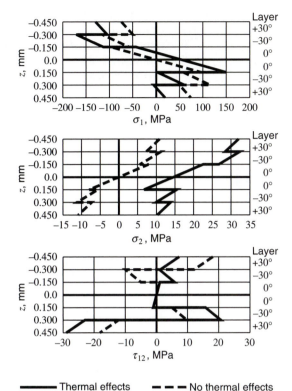

FIGURE 11.36
Principal material system stress distribution through the thickness of $[\pm30/0]_S$ laminate subjected to $M_x = 12.84$ N·m/m and $\Delta T = -150°C$

load changes (i.e., $M_x = -12.84$ N·m/m), then the stresses at a given point z do not simply change sign if thermal effects are present. The stresses at a given value of z consist of a portion that changes sign if the sign of the load changes, and a portion that does not change sign, that is, the portion due to ΔT. Thus, the location of failure will also depend on the sign of the applied load.

The principal material system strains and the principal material system mechanical strains are illustrated in Figures 11.37 and 11.38, respectively.

11.9.4 CLT Example 18: $[\pm30/0]_S$ Laminate Subjected to Known N_x, Known ΔT, and Known ΔM

Suppose that in addition to the applied load of $N_x = 0.1024$ MN/m and the temperature change of $\Delta T = -150°C$, the moisture change in the $[\pm30/0]_S$ laminate is $\Delta M = 2\%$. From equations (11.116a) and (11.116b), along with equation (11.118), the total reference surface strains are given by

$$\varepsilon_x^o = a_{11}(N_x + N_x^T + N_x^M) + a_{12}(N_y^T + N_y^M)$$
$$\varepsilon_y^o = a_{12}(N_x + N_x^T + N_x^M) + a_{22}(N_y^T + N_y^M) \qquad (11.131)$$
$$\gamma_{xy}^o = 0$$

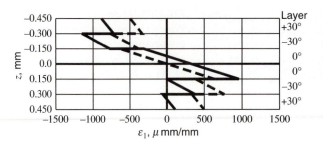

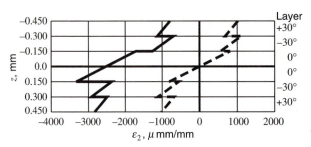

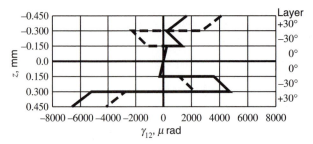

FIGURE 11.37
Principal material system total strain distribution through the thickness of $[\pm30/0]_S$ laminate subjected to $M_x = 12.84$ N·m/m and $\Delta T = -150°$C

——— Thermal effects   – – – No thermal effects

Using numerical values results in

$$\varepsilon_x^o = 1076 \times 10^{-6}$$

$$\varepsilon_y^o = 2680 \times 10^{-6} \tag{11.132}$$

$$\gamma_{xy}^o = 0$$

The principal material system stresses for the $[\pm30/0]_S$ laminate with load, residual thermal effects, and moisture absorption effects are shown in Figure 11.39. In addition, the stresses due to just the load, and the load plus thermal effects are illustrated. We see that for this example, the combination of load, temperature, and moisture tends to make σ_2 almost independent of z. This is a coincidence.

This concludes our examples of the combined response due to thermal effects and applied load. We examined only one set of material properties, namely, those

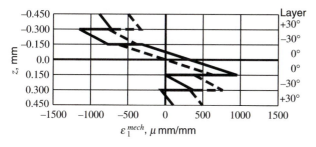

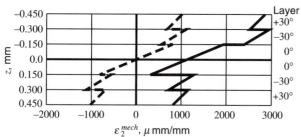

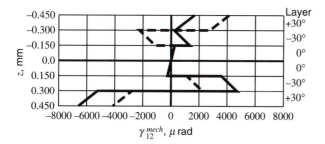

FIGURE 11.38
Principal material system mechanical strain distribution through the thickness of $[\pm30/0]_S$ laminate subjected to $M_x = 12.84$ N·m/m and $\Delta T = -150°C$

—— Thermal effects   – – – No thermal effects

of graphite-reinforced materials. Less-stiff materials, say, glass-reinforced materials, do not respond in the same way to a temperature change. Thermal effects may not be as pronounced. On the other hand, stiffer materials, say, higher modulus graphite-reinforced materials, may have much more pronounced thermal effects. Each particular problem must be examined in its own context. If loads are to be applied at elevated temperatures, the temperature dependence of the material properties may have to be included. The study of thermal effects is complex. What has been presented here represents a good foundation, but it is important to know that the findings have limitations. From the foundation, further topics can be pursued.

The final section in this chapter will examine failure in the context of combined thermal effects and applied loads. This also is a complex subject. However, we shall again lay the foundation for further study.

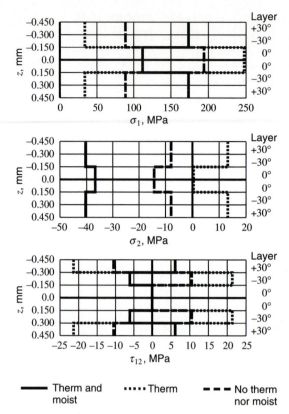

FIGURE 11.39

Principal material system stress distribution through the thickness of $[\pm 30/0]_S$ laminate subjected to $N_x = 0.1024$ MN/m, $\Delta T = -150°C$, and $\Delta M = 2\%$

———— Therm and moist   ••••• Therm   – – – No therm nor moist

11.10
INFLUENCE OF THERMAL EFFECTS ON FAILURE: THE MAXIMUM STRESS CRITERION

As we have mentioned several times, the inclusion of free thermal strain effects due to a temperature change of $-150°C$ causes the stress component σ_2 in the 90° layers of the $[0/90]_S$ laminate to approach and even exceed failure levels. In CLT Example 11, a $[0/90]_S$ laminate subjected to a ΔT of $-150°C$, the stress component σ_2, shown in Figure 11.11, was $+39.5$ MPa, or 79 percent of the failure level of $+50$ MPa. In CLT Example 15, the same laminate subjected to a ΔT of $-150°C$ and a loading of $N_x = 0.0504$ MN/m, the stress component σ_2 in Figure 11.26 exceeded the failure level of $+50$ MPa. Thus, properly accounting for thermal effects in the analysis of failure is important. The conscious or unconscious omission of thermal effects could lead to erroneous conclusions.

In the most basic form, accounting for thermal effects in the analysis of failure is exactly like the analysis of failure due to combined loads. In one of the examples involving failure of the tube, there was interest in knowing the level of torsion that was possible, given that there was a certain level of axial load present before the torsion was applied. Under the maximum stress criterion, we used the stresses due to the specific level of axial load P together with the stresses due to a unit torsion load to determine the levels of torsion that would cause failure. Under the Tsai-Wu criterion, we used these same stresses to find the roots of the quadratic equation for each layer; the roots represented the levels of torsion that caused failure. This same approach can be used to account for thermal effects, for example, residual thermal effects. The temperature change from the processing temperature causes stresses before any load is applied. The thermally induced stresses, together with the stresses due to a unit load, can then be used with either failure criterion to find failure load levels. In the tube examples, we developed failure envelopes by using a combination of stresses due to a unit tension and stresses due to a unit torsion. Both the maximum stress failure criterion and the Tsai-Wu criterion were used to find combinations of tension and torsion that would cause failure. In analogy to the tension-torsion problem, if temperature is to be varied, as well as load, temperature-load envelopes can be developed much like tension-torsion envelopes were. The key would be computing the stresses due to a unit temperature change, and combining these results with the stresses due to a unit load, whatever the loading may be. In Section 11.7, on the free thermal response of a laminate, we developed the procedure for computing the stresses due to a unit temperature change. In the chapters on failure we have computed the stresses due to a unit load. Hence, we really are in a position to study, in a comprehensive manner, a wide variety of problems involving failure in the presence of free thermal strain effects. We shall not present a general discourse on the topic here because it really is a matter of reinterpreting what has been covered in the examples dealing with combined load. Here we shall examine the problems in the last section with the $[0/90]_S$ and $[\pm30/0]_S$ laminates and determine what load level causes failure. The load levels that cause failure with residual thermal effects accounted for will be compared with the load levels that cause failure when residual thermal effects are not included. We will use both the maximum stress criterion and the Tsai-Wu criterion. The sections to follow will examine the problems first from the view of the maximum stress criterion, then from the view of the Tsai-Wu criterion. We have at our disposal the distributions of the stresses through the thickness. Thus, we might be tempted to again minimize the number of calculations and focus on specific stress components in specific layers for determining the failure characteristics. We shall again avoid this temptation and conduct a comprehensive analysis of possible failure conditions of all the layers. To do this we shall construct tables of the stresses in each layer due to a unit temperature change and use tables from Chapter 9 for the stresses due to a unit load. With the stresses due a unit load and unit temperature change in hand, failure analysis follows the previous procedures used for combined loads.

Recall from Chapter 9 that the boundaries of the maximum stress failure envelope for a layer are given by

$$\sigma_1 = \sigma_1^C$$
$$\sigma_1 = \sigma_1^T$$
$$\sigma_2 = \sigma_2^C$$
$$\sigma_2 = \sigma_2^T$$
$$\tau_{12} = \tau_{12}^F$$
$$\tau_{12} = -\tau_{12}^F$$

(11.133)

Examining each of these six equations in each layer makes it possible to determine the layer responsible for failure, the load level, and the mode of failure.

11.10.1 Failure Example 13: $[0/90]_S$ Laminate Subjected to N_x and Known ΔT

For the first example of the influence of thermal effects on failure predictions, consider the $[0/90]_S$ laminate that has experienced a temperature change of $-150°C$ and is loaded so the only nonzero stress resultant is N_x. To be determined is the value of N_x that causes failure. In Chapter 9, in Table 9.14, we found that, according to the maximum stress criterion, for no thermal effects, failure occurs with a tensile load of $N_x = +0.209$ MN/m due to failure of the $90°$ layers in tension perpendicular to the fibers. Failure also occurs with a compressive load of $N_x = -0.404$ MN/m due to failure of the $0°$ layers in compression in the fiber direction. To study failure with thermal effects, consider the stresses in the layers due to a temperature change ΔT, and the stresses in the layers due to a load n_x. The stresses in the principal material system are the important stresses for studying failure. The stresses due to load n_x were given in Chapter 9 in Table 9.13, and the stresses due to a temperature change ΔT are given in Table 11.1. These stresses, of course, are the stresses due to the free thermal response of the $[0/90]_S$ laminate, where the numerical values are determined by scaling the previously computed residual table thermal stresses. Specifically, the stresses in equation (11.99), the stresses due to a $-150°C$ temperature change, were divided by -150.

With both a load n_x and the temperature change ΔT, the stresses are given by: For the $0°$ layers:

$$\sigma_1 = 3090n_x + 263\,000\Delta T$$
$$\sigma_2 = 51.3n_x - 263\,000\Delta T \qquad (11.134a)$$
$$\tau_{12} = 0$$

For the $90°$ layers:

$$\sigma_1 = -51.3n_x + 263\,000\Delta T$$
$$\sigma_2 = 240n_x - 263\,000\Delta T \qquad (11.134b)$$
$$\tau_{12} = 0$$

With this particular situation $\Delta T = -150°C$, so:

TABLE 11.1

Principal material system stresses (MPa) in $[0/90]_S$ laminate due to a temperature change ΔT ($°C$)

Layer	σ_1	σ_2	τ_{12}
0°	+0.263ΔT	−0.263ΔT	0
90°	+0.263ΔT	−0.263ΔT	0

For the 0° layers:

$$\sigma_1 = 3090n_x - 39.5 \times 10^6$$
$$\sigma_2 = 51.3n_x + 39.5 \times 10^6 \qquad (11.135a)$$
$$\tau_{12} = 0$$

For the 90° layers:

$$\sigma_1 = -51.3n_x - 39.5 \times 10^6$$
$$\sigma_2 = 240n_x + 39.5 \times 10^6 \qquad (11.135b)$$
$$\tau_{12} = 0$$

To determine the layer responsible for failure, and the failure load and mode, the expressions for σ_1, σ_2, and τ_{12} are substituted into equation (11.133) and the resulting equations solved for n_x. The results of this substitution and solving for n_x are summarized in Table 11.2, where we can see that a load of $N_x = +0.0437$ MN/m will cause the 90° layers to fail in tension perpendicular to the fibers, and a load of $N_x = -0.391$ MN/m will cause the 0° layers to fail in compression parallel with the fibers. The results of this table should be compared with the results of Table 9.14, the summary of failure loads for this laminate but with no thermal effects accounted for. Thermal effects strongly influence the tensile capacity in this situation; the tensile capacity decreases from $N_x = +0.209$ to $+0.0437$ MN/m. The influence on compressive capacity is less pronounced. The comparison between the failure loads and failure characteristics with and without thermal effects is summarized in the upper portion of Table 11.3.

In light of the strong influence of residual thermal effects on failure, it is interesting to determine what temperature decrease alone, no load being applied, causes failure. The stresses in both the 0° and 90° layers with just thermal effects are given by, from equation (11.134),

$$\sigma_1 = 263\,000\Delta T$$
$$\sigma_2 = -263\,000\Delta T \qquad (11.136)$$
$$\tau_{12} = 0$$

Substituting these equations into the six of equation (11.133) results in the determination of the temperature change that causes failure in the various modes in the 0° and 90° layers; Table 11.4 summarizes these results. The table indicates that at 190.0°C below the processing temperature the layers will fail because stress component σ_2 reaches the tensile failure level. Based on these results, it appears that

TABLE 11.2
Summary of failure loads N_x (MN/m) for $[0/90]_S$ laminate, $\Delta T = -150°$C: Maximum stress criterion

Layer			Failure mode			
	σ_1^C	σ_1^T	σ_2^C	σ_2^T	$-\tau_{12}^F$	$+\tau_{12}^F$
0°	−0.391	+0.498	−4.67	+0.204	−∞	+∞
90°	+23.6	−30.0	−1.000	+0.0437	−∞	+∞

TABLE 11.3
Failure loads (MN/m) with and without thermal effects: Maximum stress criterion

		Thermal effects		No thermal effects	
Problem	Failure	< 0	> 0	< 0	> 0
$[0/90]_s$ with N_x	Load level Layer Mode	$N_x = -0.391$ 0° σ_1^C	$N_x = +0.0437$ 90° σ_2^T	$N_x = -0.404$ 0° σ_1^C	$N_x = +0.209$ 90° σ_2^T
$[\pm 30/0]_s$ with N_x	Load level Layer Mode	$N_x = -0.254$ 0° σ_2^T	$N_x = +0.763$ 0° σ_1^T	$N_x = -0.360$ 0° σ_2^T	$N_x = +0.791$ 0° σ_1^T
$[\pm 30/0]_s$ with M_x	Load level Layer Mode	$M_x = +33.0$ −30° σ_2^T	$M_x = -33.0$ −30° σ_2^T	$M_x = +57.1$ −30° σ_2^T	$M_x = -57.1$ −30° σ_2^T

TABLE 11.4
Temperature changes ΔT (°C) that cause failure in $[0/90]_S$ laminate: Maximum stress criterion

Layer	σ_1^C	σ_1^T	σ_2^C	σ_2^T	$-\tau_{12}^F$	$+\tau_{12}^F$
0° & 90°	−4740	+5690	+759	−190.0	−∞	+∞

the use of the $[0/90]_S$ laminate in cryogenic environments or in the cold of space would be unwise. Even at room temperature, its capacity in tension is inhibited by the residual stress state. This is a well-known characteristic of cross-ply laminates and one reason they see limited application.

11.10.2 Failure Example 14: $[\pm 30/0]_S$ Laminate Subjected to N_x and Known ΔT

In a manner similar to the approach for the $[0/90]_S$ laminate, we wish to evaluate the influence of residual thermal effects on the load capacity of a $[\pm 30/0]_S$ laminate subjected to a loading N_x. The stresses due to a loading n_x were given in Table 9.15 and the stresses due to a temperature change ΔT are given in Table 11.5. The entries in Table 11.5 are the stresses due to the free thermal response of the laminate and they were determined by scaling equation (11.104) by -150. For a temperature change of $-150°C$ the stresses in the layers are:

TABLE 11.5
Principal material system stresses (MPa) in $[\pm 30/0]_S$ laminate due to a temperature change ΔT (°C)

Layer	σ_1	σ_2	τ_{12}
+30°	$+0.368\Delta T$	$-0.1410\Delta T$	$+0.0723\Delta T$
−30°	$+0.368\Delta T$	$-0.1410\Delta T$	$-0.0723\Delta T$
0°	$-0.356\Delta T$	$-0.0976\Delta T$	0

For the $+30°$ layers:

$$\sigma_1 = 866n_x - 55.2 \times 10^6$$
$$\sigma_2 = -77.3n_x + 21.1 \times 10^6 \qquad (11.137a)$$
$$\tau_{12} = -102.8n_x - 10.84 \times 10^6$$

For the $-30°$ layers:

$$\sigma_1 = 866n_x - 55.2 \times 10^6$$
$$\sigma_2 = -77.3n_x + 21.1 \times 10^6 \qquad (11.137b)$$
$$\tau_{12} = 102.8n_x + 10.84 \times 10^6$$

For the $0°$ layers:

$$\sigma_1 = 1894n_x + 53.4 \times 10^6$$
$$\sigma_2 = -138.9n_x + 14.64 \times 10^6 \qquad (11.137c)$$
$$\tau_{12} = 0$$

Using these equations and equation (11.133) leads to a determination of the layer, load, and mode that causes failure. Table 11.6 summarizes these results; this table is the counterpart to Table 9.16, the case of no thermal effects. Table 11.6 indicates that a tensile load of $+0.763$ MN/m causes failure due to a tensile stress in the fiber direction in the $0°$ layer. For the case of no thermal load, Table 9.16, a tensile load of $+0.791$ MN/m causes failure for the same reason. Thus, due to thermal effects, tensile load capacity is reduced. Furthermore, Table 11.6 indicates a compressive load of -0.254 MN/m causes failure due to a tensile stress perpendicular to the fibers in the $0°$ layers. For the case of no thermal load, the compressive load required is, according to Table 9.16, -0.360 MN/m. Thermal effects dimish compressive capacity as well. Though it is not a critical issue, it is important to realize that with thermal effects, for a given layer the last two statements of equation (11.133) lead to loads of different magnitude, as well as different sign. With no thermal effects, the last two statements of equation (11.133) lead to loads of opposite sign but the same magnitude. This is a reflection of the fact that failure in shear is not influenced by the sign of the shear stress. However, the inclusion of thermal effects biases the shear stress in such a way that the shear stress produced by a positive loading is different in magnitude than the shear stress produced by a negative loading. The biasing can be strong enough to remove shear from contention as a possible failure mode. Only by systematically examining all facets of the maximum stress failure criterion, equation (11.133), can the failure mode and failure load be predicted for both positive and negative loads, and for inclusion or exclusion of thermal effects.

TABLE 11.6
Summary of failure loads N_x (MN/m) for $[\pm 30/0]_S$ laminate, $\Delta T = -150°$C: Maximum stress criterion

Layer			Failure mode			
	σ_1^C	σ_1^T	σ_2^C	σ_2^T	$-\tau_{12}^F$	$+\tau_{12}^F$
$+30°$	-1.380	$+1.796$	$+2.86$	-0.373	$+0.868$	-1.079
$-30°$	-1.380	$+1.796$	$+2.86$	-0.373	-1.079	$+0.868$
$0°$	-0.687	$+0.763$	$+1.545$	-0.254	$-\infty$	$+\infty$

The middle portion of Table 11.3 summarizes the comparison between the case of thermal effects and the case of no thermal effects.

11.10.3 Failure Example 15: $[\pm 30/0]_S$ Laminate Subjected to M_x and Known ΔT

With bending, the signs of the stresses depend not only on the sign of the applied load or the sign of the specified curvature, but also on whether one is considering positive or negative z locations within the laminate. To study the influence of thermally induced stresses on the level of bending moment M_x to cause failure in the $[\pm 30/0]_S$ laminate, we shall proceed as we did for the other two cases. The stresses produced at the outermost location in each layer as a result of bending moment $M_x = m_x$ are documented in Table 9.17. Recall that these stresses are for positive z locations. The stresses due to temperature change ΔT were given in Table 11.5. As a result, the stresses at the outermost locations of the layers for positive z are given by:

For the 30° layers:

$$\sigma_1 = (5.79m_x - 55.2) \times 10^6$$
$$\sigma_2 = (-0.850m_x + 21.1) \times 10^6 \qquad (11.138a)$$
$$\tau_{12} = (-1.415m_x - 10.84) \times 10^6$$

For the −30° layers:

$$\sigma_1 = (9.01m_x - 55.2) \times 10^6$$
$$\sigma_2 = (-0.875m_x + 21.1) \times 10^6 \qquad (11.138b)$$
$$\tau_{12} = (0.772m_x + 10.84) \times 10^6$$

For the 0° layers:

$$\sigma_1 = (7.52m_x + 53.4) \times 10^6$$
$$\sigma_2 = (-0.618m_x + 14.64) \times 10^6 \qquad (11.138c)$$
$$\tau_{12} = -0.0858m_x \times 10^6$$

Table 11.7 summarizes the results of substituting these equations into the six of equation (11.133) and of solving for the various values of m_x. Another table can be constructed for negative z locations. This table, which will not be shown, would be based on the fact that for negative z locations the stresses due to $M_x = m_x$ would be of opposite sign to those stresses in Table 9.17. As a result, the principal material system stresses at outer locations in the various layers would be as given by equation (11.138) except that the sign of the terms involving m_x would be changed. For example, from equation (11.138a), the stresses at the outer location of the 30° layer at the negative z location, would be

$$\sigma_1 = (-5.79m_x - 55.2) \times 10^6$$
$$\sigma_2 = (0.850m_x + 21.1) \times 10^6 \qquad (11.139)$$
$$\tau_{12} = (1.414m_x - 10.84) \times 10^6$$

Using these equations and the counterparts to equations (11.138b) and (c) in equation (11.133), we can find the values of m_x to cause failure at negative z locations and

TABLE 11.7
Summary of failure loads M_x (N·m/m) for $[\pm 30/0]_S$ laminate, $\Delta T = -150°C$: Maximum stress criterion*

Layer	σ_1^C	σ_1^T	σ_2^C	σ_2^T	$-\tau_{12}^F$	$+\tau_{12}^F$
			Failure mode			
+30°	−206	+269	+260	−34.0	+63.0	−78.3
−30°	−132.5	+172.5	+253	−33.0	−143.6	+115.5
0°	−173.4	+192.5	+348	−57.2	+1166	−1166

* Positive z locations.

can put them in tabular form. The table would be identical to Table 11.7 except that the signs of all the entries would be opposite. Thus, consider Table 11.7 to have ± signs with all the entries. From Table 11.7 we can then conclude that a bending moment of $M_x = -33.0$ N·m/m causes failure at the outer location of the −30° layer on the positive z side of the laminate due to a tensile stress perpendicular to the fibers in that layer, *and* that a bending moment of $M_x = +33.0$ N·m/m causes failure at the outer location of the −30° layer on the negative z side of the laminate, also due to a tensile stress perpendicular to the fibers in that layer. Table 11.7 is the counterpart to Table 9.18, which also has the implication of ± with each of the entries. A comparison of the thermal and nonthermal cases, which is given in the last entries of Table 11.3, indicates that for this example the inclusion of residual thermal effects, as with the previous examples, has a detrimental effect on laminate load capacity. In fact, an examination of all three problems in Table 11.3 indicates that residual thermal effects can generally be viewed as reducing the load capacity of a laminate. However, by comparing entries in Tables 9.14 versus 11.2, 9.16 versus 11.6, and 9.18 versus 11.7, we see that for some failure modes in some layers, inclusion of thermal effects actually leads to increased load capacity. It would be considered a special case, however, if, overall, thermal effects were beneficial.

Exercise for Section 11.10

Consider the $[\pm 30/0]_S$ graphite-reinforced laminate. (*a*) Use the maximum stress failure criterion to compute the value of ΔT below the processing temperature that causes failure in the laminate, assuming the stresses are zero at the processing temperature. (*b*) Discuss the failure mode associated with this temperature change. (*c*) Is this temperature higher or lower than the failure-producing temperature change for the $[0/90]_S$ laminate?

11.11
INFLUENCE OF THERMAL EFFECTS ON FAILURE:
THE TSAI-WU CRITERION

As with all failure examples, we shall use the Tsai-Wu criterion to study failure for the cases studied with the maximum stress criterion and compare the results. Recall from Chapter 10, equation (10.38), that the Tsai-Wu criterion is written as

$$F_1\sigma_1 + F_2\sigma_2 + F_{11}\sigma_1^2 + F_{22}\sigma_2^2 + F_{66}\tau_{12}^2 - \sqrt{F_{11}F_{22}}\sigma_1\sigma_2 = 1 \quad (11.140)$$

where the quantities $F_1, \ldots, F_{66}$ are related to the strengths of the material in the various directions and are given by Table 10.1. When solving for a load to produce failure, the Tsai-Wu criterion results in a quadratic equation, with one root being positive load and the other root being negative. In addition, the equation must be applied to each layer in the laminate, resulting in a number of failure loads for a given laminate. The lowest of these loads is considered the failure load. For this load, an examination of the contribution of each term in the criterion to the value of unity provides an indication of the failure mode.

11.11.1 Failure Example 16: $[0/90]_S$ Laminate Subjected to N_x and Known ΔT

The determination of the failure loads for the $[0/90]_S$ laminate subjected to a load N_x and a temperature change $\Delta T = -150°C$ can be determined by the direct substitution of previous results into equation (11.140). The methodical approach to the application of the maximum stress criterion has provided us with the necessary information. Specifically, equation (11.135) provides us with the stresses due to an applied load of $N_x = n_x$ and a temperature change $\Delta T = -150°C$ that are to be used in equation (11.140). Using these equations, we find that the Tsai-Wu criterion becomes:

For the 0° layers:

$$F_1(3090n_x - 39.5 \times 10^6) + F_2(51.3n_x + 39.5 \times 10^6)$$
$$+ F_{11}(3090n_x - 39.5 \times 10^6)^2 + F_{22}(51.3n_x + 39.5 \times 10^6)^2 \quad (11.141a)$$
$$- \sqrt{F_{11}F_{22}}(3090n_x - 39.5 \times 10^6)(51.3n_x + 39.5 \times 10^6) - 1 = 0$$

For the 90° layers:

$$F_1(-51.3n_x - 39.5 \times 10^6) + F_2(240n_x + 39.5 \times 10^6)$$
$$+ F_{11}(-51.3n_x - 39.5 \times 10^6)^2 + F_{22}(240n_x + 39.5 \times 10^6)^2 \quad (11.141b)$$
$$- \sqrt{F_{11}F_{22}}(-51.3n_x - 39.5 \times 10^6)(240n_x + 39.5 \times 10^6) - 1 = 0$$

Table 11.8 summarizes the results of solving these quadratic forms for n_x and gives the positive and negative root for each layer. Also tabulated are the principal material system stresses at the various failure load levels and the contributions of the terms in the failure criterion to unity. The predictions of the maximum stress failure criterion are indicated in parenthesis in the table.

Table 11.8 indicates that in the presence of residual thermal stresses, the $[0/90]_S$ can only sustain a tensile load $N_x = +0.0402$ MN/m and a compressive load of $N_x = -0.208$ MN/m. Failure due to tensile loading is predicted to be due to failure in the 90° layers due to tensile perpendicular to the fibers, while failure due to compressive loading appears to occur in the 0° layers due to a combination of fiber direction compression and tension perpendicular to the fiber. The results of this table

TABLE 11.8
Summary of failure loads N_x (MN/m) for $[0/90]_S$ laminate, $\Delta T = -150°C$: Tsai-Wu criterion*

	0° layers		90° layers	
	$N_x = -0.208$ (−0.391)	$N_x = +0.267$ (+0.204)	$N_x = -0.997$ (−1.000)	$N_x = +0.0402$ (+0.0437)
σ_1, MPa	−683	786	11.6	−41.6
σ_2, MPa	28.9	53.2	−199	49.1
τ_{12}, MPa	0	0	0	0
$F_1\sigma_1$	0.091	−0.105	−0.002	0.006
$F_2\sigma_2$	0.433	0.798	−2.99	0.737
$F_{11}\sigma_1^2$	0.249	0.329	0	0.001
$F_{22}\sigma_2^2$	0.083	0.283	3.98	0.242
$F_{66}\tau_{12}^2$	0	0	0	0
$-\sqrt{F_{11}F_{22}}\sigma_1\sigma_2$	0.144	−0.305	0.017	0.015
Total	1.000	1.000	1.000	1.000

* Maximum stress failure theory predictions in parenthesis.

can be compared with the results of Table 10.6, where the predictions are based on the Tsai-Wu criterion with no thermal stresses present. Table 10.6 indicated that if you do not account for residual thermal effects, the laminate can sustain a tensile load of $N_x = +0.208$ MN/m and a compressive load of $N_x = -0.532$ MN/m. Like the maximum stress criterion, the Tsai-Wu criterion predicts that residual thermal effects are detrimental, particularly in tension. In addition, the tensile failure load of $+0.0402$ MN/m compares well with the prediction of the maximum stress criterion, Table 11.2, of $+0.0437$ MN/m, while the compressive load prediction of -0.208 MN/m is somewhat lower than the maximum stress criterion prediction of -0.391 MN/m.

Again, it is interesting to determine the temperature change, with no load, that causes failure in the laminate. This can be determined by using equation (11.134) with $n_x = 0$ in the Tsai-Wu criterion. For this case and for this laminate the Tsai-Wu criterion is the same for each layer, namely,

$$(F_1 - F_2)263\,000\Delta T + (F_{11} + F_{22} - \sqrt{F_{11}F_{22}})(263\,000\Delta T)^2 - 1 = 0 \quad (11.142)$$

where the roots for ΔT are summarized in Table 11.9.

The results of the table indicate that at a temperature $-185.9°C$ below the stress-free processing temperature, both layers fail due to excessive stresses perpendicular to the fibers. This finding correlates well with the results of the maximum stress criterion.

Table 11.10 provides a good summary comparison of this problem as viewed with the two failure criteria with and without residual thermal effects. The failure load, the layer that fails, and the failure mode, as predicted by each criterion, are indicated. Regarding the tensile load to cause failure, both criteria agree on load level, the layer that causes failure, and the failure mode, not only for the case of including

TABLE 11.9
Temperature changes $\Delta T(^\circ C)$ that cause failure in $[0/90]_S$ laminate: Tsai-Wu criterion[*]

	0° and 90° layers	
	$\Delta T = -185.9$	$\Delta T = 719$
σ_1, MPa	−49.0	189.3
σ_2, MPa	49.0	−189.3
τ_{12}, MPa	0	0
$F_1\sigma_1$	0.007	−0.025
$F_2\sigma_2$	0.735	−2.84
$F_{11}\sigma_1^2$	0.001	0.019
$F_{22}\sigma_2^2$	0.240	3.58
$F_{66}\tau_{12}^2$	0	0
$-\sqrt{F_{11}F_{22}}\sigma_1\sigma_2$	0.018	0.272
Total	1.000	1.000

[*] Maximum stress theory predicts $\Delta T = -190.0$ and 759°C.

TABLE 11.10
Failure loads N_x (MN/m) for $[0/90]_S$ laminate with and without thermal effects: Tsai-Wu and maximum stress criteria

		Thermal effects		No thermal effects	
Criterion	Failure	< 0	> 0	< 0	> 0
Tsai-Wu	Load level	$N_x = -0.208$	$N_x = +0.0402$	$N_x = -0.532$	$N_x = +0.208$
	Layer	0°	90°	0°	90°
	Mode	σ_1^C & σ_2^T	σ_2^T	σ_1^C	σ_2^T
Maximum stress	Load level	$N_x = -0.391$	$N_x = +0.0437$	$N_x = -0.404$	$N_x = +0.209$
	Layer	0°	90°	0°	90°
	Mode	σ_1^C	σ_2^T	σ_1^C	σ_2^T

thermal effects, but also for the case of excluding thermal effects. The agreement on the compressive load to cause failure is not so good. The lack of agreement can be traced to interaction effects predicted by the Tsai-Wu criterion.

11.11.2 Failure Example 17: $[\pm 30/0]_S$ Laminate Subject to N_x and Known ΔT

According to the Tsai-Wu criterion, the influence of residual thermal stresses on the load capacity of the $[\pm 30/0]_S$ laminate can be determined in the same straightforward manner. Equation (11.137) provides us with the principal material system stresses for the case of $N_x = n_x$ and $\Delta T = -150°C$. These stresses can be directly substituted into the Tsai-Wu criterion to yield:

For the $+30°$ layers:

$$F_1(866n_x - 55.2 \times 10^6) + F_2(-77.3n_x + 21.1 \times 10^6)$$
$$+ F_{11}(866n_x - 55.2 \times 10^6)^2 + F_{22}(-77.3n_x + 21.1 \times 10^6)^2$$
$$+ F_{66}(-102.8n_x - 10.84 \times 10^6)^2 \tag{11.143a}$$
$$- \sqrt{F_{11}F_{22}}(866n_x - 55.2 \times 10^6)(-77.3n_x + 21.1 \times 10^6) - 1 = 0$$

For the $-30°$ layers:

$$F_1(866n_x - 55.2 \times 10^6) + F_2(-77.3n_x + 21.1 \times 10^6)$$
$$+ F_{11}(866n_x - 55.2 \times 10^6)^2 + F_{22}(-77.3n_x + 21.1 \times 10^6)^2$$
$$+ F_{66}(102.8n_x + 10.84 \times 10^6)^2 \tag{11.143b}$$
$$- \sqrt{F_{11}F_{22}}(866n_x - 55.2 \times 10^6)(-77.3n_x + 21.1 \times 10^6) - 1 = 0$$

For the $0°$ layers:

$$F_1(1894n_x + 53.4 \times 10^6) + F_2(-138.9n_x + 14.64 \times 10^6)$$
$$+ F_{11}(1894n_x + 53.4 \times 10^6)^2 + F_{22}(-138.9n_x + 14.64 \times 10^6)^2 \tag{11.143c}$$
$$- \sqrt{F_{11}F_{22}}(1894n_x + 53.4 \times 10^6)(-138.9n_x + 14.64 \times 10^6) - 1 = 0$$

Table 11.11 summarizes the results from solving these equations for n_x. Again, the values of the principal stresses at the failure load levels and the contributions to the Tsai-Wu criterion for the three layer directions are given. From the table it appears that a positive load is limited to $+0.679$ MN/m by the $0°$ layers and the mode of failure is a tensile stress in the fiber direction. A negative load is limited to -0.1966 MN/m by the $0°$ layers, and the layers fail due to the combined effect of tension perpendicular to the fibers and, to some degree, compression in the fiber direction. An examination of Table 10.7, the counterpart to this case but with no thermal effects, indicates that, overall, thermal effects are again detrimental and the positive and negative limiting loads cause similar failure characteristics. However, there is an anomaly for the $0°$ layers and positive N_x, the case of including thermal effects resulting in a higher value of N_x ($+0.679$ MN/m) than with no thermal effects ($+0.665$ MN/m). Table 11.12 summarizes the findings for the $[\pm30/0]_S$ laminate.

11.11.3 Failure Example 18: $[\pm30/0]_S$ Laminate Subjected to M_x and Known ΔT

The principal material system stresses at the outer location of each layer of the $[\pm30/0]_S$ laminate with $M_x = m_x$ and $\Delta T = -150°C$ are given by equation (11.138). Recall the dual role of these expressions for the stresses, where the stresses as written in equation (11.138) are for the positive z locations, and the stresses for the negative z locations are given by changing the sign of the term associated with m_x. The results of substituting equation (11.138) into the Tsai-Wu criterion are summarized in Table 11.13 in the usual fashion. If the negative z locations were considered and summarized in table form, then that table would be similar to Table

TABLE 11.11
Summary of failure loads N_x (MN/m) for $[\pm 30/0]_S$ laminate, $\Delta T = -150°C$: Tsai-Wu Criterion*

	+30° layers		−30° layers		0° layers	
	$N_x = -0.268$ (−1.079)	$N_x = +0.895$ (+0.868)	$N_x = -0.268$ (−1.079)	$N_x = +0.895$ (+0.868)	$N_x = -0.1966$ (−0.254)	$N_x = +0.679$ (+0.763)
σ_1, MPa	−287	720	−287	720	−319	1341
σ_2, MPa	41.8	−48.0	41.8	−48.0	42.0	−79.7
τ_{12}, MPa	16.66	−102.8	−16.66	102.8	0	0
$F_1\sigma_1$	0.038	−0.096	0.038	−0.096	0.043	−0.179
$F_2\sigma_2$	0.627	−0.721	0.627	−0.721	0.629	−1.196
$F_{11}\sigma_1^2$	0.044	0.276	0.044	0.276	0.054	0.959
$F_{22}\sigma_2^2$	0.175	0.231	0.175	0.231	0.176	0.635
$F_{66}\tau_{12}^2$	0.028	1.057	0.028	1.057	0	0
$-\sqrt{F_{11}F_{22}}\sigma_1\sigma_2$	0.088	0.253	0.088	0.253	0.098	0.780
Total	1.000	1.000	1.000	1.000	1.000	1.000

* Maximum stress criterion in parenthesis.

TABLE 11.12
Failure loads N_x (MN/m) for $[\pm 30/0]_S$ laminate with and without thermal effects: Tsai-Wu and maximum stress criteria

		Thermal effects		No thermal effects	
Criterion	Failure	< 0	> 0	< 0	> 0
Tsai-Wu	Load level	$N_x = -0.1966$	$N_x = +0.679$	$N_x = -0.260$	$N_x = +0.665$
	Layer	0°	0°	0°	0°
	Mode	σ_2^T	σ_1^T	$\sigma_2^T \& \sigma_1^C$	σ_1^T
Maximum stress	Load level	$N_x = -0.254$	$N_x = 0.763$	$N_x = -0.360$	$N_x = +0.791$
	Layer	0°	0°	0°	0°
	Mode	σ_2^T	σ_1^T	σ_2^T	σ_1^T

11.13 except that the signs of the failure loads would reverse, resulting in a switching of columns. The two roots to the three polynomials would be

$$M_x = -73.3 \qquad M_x = 25.5$$
$$M_x = -103.4 \qquad M_x = 24.8 \qquad (11.144)$$
$$M_x = -164.2 \qquad M_x = 45.6$$

for the +30°, −30°, and 0° layers, respectively. The signs of some of the entries in the columns below the failure loads would change. As a result, we can conclude that failure occurs when $M_x = \pm 24.8$ N·m/m, and that failure occurs at the outer location in the −30° layer due to a tensile stress perpendicular to the fibers. Failure for the case of no thermal effects occurs, from Table 10.8, when $M_x = \pm 38.4$ N·m/m, and failure occurs at the outer location in the +30° layer due to a combination of

TABLE 11.13

Summary of failure loads M_x (N·m/m) for $[\pm30/0]_S$ laminate, $\Delta T = -150°C$: Tsai-Wu criterion[*]

	Outer location of +30° layers		Outer location of −30° layers		Outer location of 0° layers	
	$M_x = -25.5$ (−34.0)	$M_x = +73.3$ (+63.0)	$M_x = -24.8$ (−33.0)	$M_x = +103.4$ (+115.5)	$M_x = -45.6$ (−57.2)	$M_x = +164.2$ (+192.5)
σ_1, MPa	−202	369	−279	877	−289	1288
σ_2, MPa	42.8	−41.2	42.9	−69.3	42.8	−86.8
τ_{12}, MPa	25.2	−114.6	−8.33	90.7	3.91	−14.08
$F_1\sigma_1$	0.027	−0.049	0.037	−0.117	0.039	−0.172
$F_2\sigma_2$	0.642	−0.617	0.643	−1.040	0.642	−1.302
$F_{11}\sigma_1^2$	0.022	0.073	0.042	0.410	0.044	0.884
$F_{22}\sigma_2^2$	0.183	0.169	0.184	0.481	0.183	0.753
$F_{66}\tau_{12}^2$	0.063	1.313	0.007	0.832	0.002	0.020
$-\sqrt{F_{11}F_{22}}\sigma_1\sigma_2$	0.063	0.111	0.087	0.444	0.090	0.816
Total	1.000	1.000	1.000	1.000	1.000	1.000

[*] Positive z locations. Maximum stress criterion in parenthesis.

TABLE 11.14

Failure loads M_x (N·m/m) for $[\pm30/0]_S$ laminate, with and without thermal effects: Tsai-Wu and maximum stress criteria

		Thermal effects		No thermal effects	
Criterion	Failure	< 0	> 0	< 0	> 0
Tsai-Wu	Load level Layer Mode	$M_x = -24.8$ −30° σ_2^T	$M_x = +24.8$ −30° σ_2^T	$M_x = -38.4$ +30° $\tau_{12}^F \& \sigma_2^T$	$M_x = +38.4$ +30° $\tau_{12}^F \& \sigma_2^T$
Maximum stress	Load level Layer Mode	$M_x = -33.0$ −30° σ_2^T	$M_x = +33.0$ −30° σ_2^T	$M_x = -57.1$ −30° σ_2^T	$M_x = +57.1$ −30° σ_2^T

a shear stress and a tensile stress perpendicular to the fibers. Table 11.14 provides an overview of this problem from several perspectives and it is evident that thermal effects are detrimental to this bending problem. The Tsai-Wu criterion predicts a change of failure mode and failure location due to thermal effects, whereas the maximum stress criterion predicts the same mode and same location, with merely a reduced bending moment capacity.

This completes our discussion of the influence of thermal effects on failure. We have examined it from the view of the maximum stress and the Tsai-Wu criteria. We have considered thermal effects due to cooldown from processing, a very realistic case. Obviously, other thermal conditions can be studied with the approaches outlined here. Care must be taken, however, to ensure that the material strengths used in the failure criterion reflect the strength at that temperature. Because the presence of moisture tends to weaken composite materials, we have not presented any examples of failure in the presence of moisture-induced stresses. We have seen that the moisture-induced stresses tend to relieve the residual stresses due to cooldown from processing, but a moisture-weakened material may not take advantage of this relief in stresses. If another failure criterion is more applicable to a certain situation, the work presented here lays the foundation for incorporating thermal effects into that criterion. To be sure, whatever criterion is being used, it is important to examine all aspects of its predictions by approaching each problem in a thorough manner, examining all the stresses in all the layers. It is easy to overlook an important feature of the problem if thoroughness is not used.

Exercise for Section 11.11

Use the Tsai-Wu failure criterion to compute the value of ΔT below the processing temperature that causes failure in the $[\pm30/0]_S$ laminate, assuming the stresses are zero at the processing temperature. Discuss the failure mode associated with this temperature change and compare the results with the results from the maximum stress criterion, the Exercise for Section 11.10.

11.12
SUMMARY

This brings to a close the chapter on the influence of thermal effects on laminates. In this chapter we started with the basic assumptions necessary to incorporate free thermal expansion strains into the analysis of laminates. Following the development for the case of no thermal effects, the definitions of the classical A, B, and D matrices remained intact, and equivalent thermal force and moment resultants evolved as natural definitions. These equivalent resultants could then be used with the ABD matrix to determine the reference surface strains and curvatures that resulted from a temperature change. The discussions were limited to the case of temperature-independent material properties and temperature changes that did not depend on z. Definitions of the coefficients of thermal deformation of the laminate followed naturally from the derivations. We discussed example problems that considered thermal effects due to the cooling of a laminate from its consolidation temperature, and we considered the influence of the resulting residual stresses on the stress distributions through the thickness of the laminate, and on failure. At certain key points analogies to the effects of free strains due to moisture absorption were made. In the next chapter, we will present further discussion of the effects of free thermal strains. However, the discussion will be not in the context of stresses, but rather in the context of through-thickness dimensional changes in a laminate.

11.13
SUGGESTED READINGS

The following are several of the original papers dealing with environmental effects in laminates:

1. Hahn, H. T., and N. J. Pagano. "Curing Stresses in Composite Laminates." *Journal of Composite Materials* 9 (1975), pp. 2–20.
2. Pipes, R. B.; J. R. Vinson; and T.-W. and Chou. "On the Hygrothermal Response of Laminated Composite Structures." *Journal of Composite Materials* 10 (1976), pp. 129–48.
3. Hahn, H. T., and R. Y. Kim. "Swelling of Composite Laminates." In *Advanced Composite Materials—Environmental Effects*, ASTM STP 658, ed. J. R. Vinson. Philadelphia: American Society for Testing and Materials, 1978.

Several classic references for understanding the moisture absorption characteristics of composite materials are the following:

4. Shen, C.-H., and G. S. Springer. "Moisture Absorption and Desorption of Composite Materials." *Journal of Composite Materials* 10 (1976), pp. 2–20.
5. Browning, C. E.; G. E. Husman; and J. M. Whitney. "Moisture Effects in Epoxy Matrix Composites." In *Composite Materials: Testing and Design: Fourth Conference*, ASTM STP 617, ed. S. W. Tsai. Philadelphia: American Society for Testing and Materials, 1977.
6. Vinson, J. R., ed. *Advanced Composite Materials—Environmental Effects*, ASTM STP 658. Philadelphia: American Society for Testing and Materials, 1978.
7. Loos, A. C., and G. S. Springer. "Moisture Absorption of Graphite/Epoxy Composites Immersed in Liquids and in Humid Air." *Journal of Composite Materials* 13 (1979), pp. 131–47.

In addition to the stresses within the laminate, other consequences of thermal effects in laminates are described in the following:

8. Springer, G. S., ed. *Environmental Effects on Composite Materials*. Vols. 1 and 2. Lancaster, PA: Technomic Publishing Co., 1981 and 1984.
9. Meyers, C. A., and M. W. Hyer. "Thermal Buckling and Postbuckling of Symmetrically Laminated Composite Plates." *Journal of Thermal Stress* 14 (1991), pp. 519–40.
10. Noor, A. K.; J. H. Starnes, Jr.; and J. M. Peters. "Thermomechanical Buckling and Postbuckling of Multilayer Composite Panels." *Composite Structures* 23 (1993), pp. 233–51.

Some unusual thermally induced effects in unsymmetric laminates, specifically when equation (11.85) is not valid, are discussed in:

11. Hyer, M. W. "Calculations of the Room-Temperature Shapes of Unsymmetric Laminates." *Journal of Composite Materials* 15 (1981), pp. 296–310.

CHAPTER 12

Through-Thickness Laminate Strains

The key assumption that resulted in a simplification of the stress analysis of composite materials, and that led to what we have called classical lamination theory, was the plane-stress assumption. This assumption stated that the stresses σ_3, τ_{23}, and τ_{13} in the principal material coordinate system are so small in comparison to σ_1, σ_2, and τ_{12} that they can be assumed to be zero. After introducing this assumption and examining plane-stress problems for a single layer, from time to time we reminded ourselves that a layer of material still experiences strains in the 3 direction. We computed $\varepsilon_3 (= \varepsilon_z)$ in several examples, and in Exercise 3 in the Exercises for Section 5.2 we computed ε_z and then the change in layer thickness, Δh, as a function of θ for the tensile loading of an off-axis layer. When studying the response of laminates, we must not forget that they also experience strains in the thickness direction when loaded, or when subjected to a temperature change, or when they absorb moisture. This effect is so important to having a complete understanding of laminate response within the context of classical lamination theory that we have devoted this separate chapter to the subject. Additionally, through-thickness strains are important when mechanical fasteners through the thickness of the laminate, such as bolts or rivets, are used to connect two laminates. A bolt and nut combination improperly torqued could loosen due to through-thickness contraction from inplane loading if the through-thickness strains of the laminate are not properly computed. Also, to ensure dimensional tolerances in the application of composite materials to optical devices that experience temperature changes, every aspect of the thermal expansion or contraction of the material must be known. By using a combination of σ_1, σ_2, and τ_{12} as computed from classical lamination theory and the 3-dimensional stress-strain relations, the through-thickness strains can be computed. In this chapter we shall compute the change in thickness of a laminate subjected to inplane loads, define two laminate Poisson's ratios in the thickness direction, and define a through-thickness thermal expansion coefficient. These relations will be exact within the context of classical lamination theory.

12.1
THICKNESS CHANGE OF A LAMINATE, NO FREE THERMAL OR MOISTURE STRAIN EFFECTS

According to equation (4.1), in the principal material coordinate system, the 3-dimensional stress-strain relations, with no free thermal or moisture absorption effects and written with the plane-stress assumption incorporated, are

$$
\begin{Bmatrix} \varepsilon_1 \\ \varepsilon_2 \\ \varepsilon_3 \\ \gamma_{23} \\ \gamma_{13} \\ \gamma_{12} \end{Bmatrix} = \begin{bmatrix} S_{11} & S_{12} & S_{13} & 0 & 0 & 0 \\ S_{12} & S_{22} & S_{23} & 0 & 0 & 0 \\ S_{13} & S_{23} & S_{33} & 0 & 0 & 0 \\ 0 & 0 & 0 & S_{44} & 0 & 0 \\ 0 & 0 & 0 & 0 & S_{55} & 0 \\ 0 & 0 & 0 & 0 & 0 & S_{66} \end{bmatrix} \begin{Bmatrix} \sigma_1 \\ \sigma_2 \\ 0 \\ 0 \\ 0 \\ \tau_{12} \end{Bmatrix} \tag{12.1}
$$

As indicated in equations (4.2) and (4.3), from the above equation the through-thickness strains ε_3, γ_{23}, and γ_{13} are given by

$$
\varepsilon_3 = S_{13}\sigma_1 + S_{23}\sigma_2 \qquad \gamma_{23} = 0 \qquad \gamma_{13} = 0 \tag{12.2}
$$

These equations remind us that by the plane-stress assumption the through-thickness shear strains are identically zero. The extensional strain is not zero and can be computed directly from the inplane stresses σ_1 and σ_2 and the compliances S_{13} and S_{23}. Because ε_3 is not zero, a layer in a state of plane stress experiences a change in thickness, Δh, given by

$$
\Delta h = \varepsilon_3 h \tag{12.3}
$$

h being the thickness of the layer.

The expressions of equations (12.2) and (12.3) for ε_3 and Δh are independent of whether the layer is within a laminate or is isolated by itself. In fact, to indicate that the above expressions for ε_3 and Δh are valid for any layer, specifically the kth layer, the expressions can be rewritten as

$$
\varepsilon_{3_k} = S_{13_k}\sigma_{1_k} + S_{23_k}\sigma_{2_k} \tag{12.4}
$$

and

$$
\Delta h_k = \varepsilon_{3_k} h_k \tag{12.5}
$$

The stress analysis of a laminate subjected to loads N_x, N_y, N_{xy}, M_x, M_y, M_{xy} leads to a layer-by-layer calculation of σ_1 and σ_2. Hence, the thickness change of each layer can be computed and the total change in laminate thickness determined by summing the various layer thickness changes. Using the above notation, we find that the total thickness change of an N-layer laminate, ΔH, is

$$
\Delta H = \sum_{k=1}^{N} \Delta h_k \tag{12.6}
$$

As an example of this calculation, consider the response of the graphite-reinforced $[\pm 30/0]_S$ laminate subjected to a load in the x direction of $N_x = 0.1024$ MN/m. Table 12.1 lists the principal material system stresses σ_1 and σ_2 in each layer based on the results of classical lamination theory as calculated in Chapter 8, equation (8.14).

TABLE 12.1
Thickness response of $[\pm 30/0]_S$ laminate to load of $N_x = 0.1024$ MN/m

Layer no.	Angle deg.	σ_1 MPa	σ_2 psi	ε_3 $\times 10^{-6}$	Δh $\times 10^{-9}$ m
1	+30	88.7	−7.92	157.8	23.7
2	−30	88.7	−7.92	157.8	23.7
3	0	194.1	−14.23	228	34.2
4	0	194.1	−14.23	228	34.2
5	−30	88.7	−7.92	157.8	23.7
6	+30	88.7	−7.92	157.8	23.7

$$\Delta H = 163.1 \times 10^{-9} \, \text{m}$$

From Chapter 2, equation (2.56),

$$S_{13} = -\frac{\nu_{13}}{E_1} = -1.600 \times 10^{-12}$$
$$\text{(12.7)}$$
$$S_{23} = -\frac{\nu_{23}}{E_2} = -37.9 \times 10^{-12}$$

Using equation (12.4), we find that for the $\pm 30°$ layers, layers 1, 2, 5, and 6,

$$\varepsilon_{3_1} = \varepsilon_{3_2} = \varepsilon_{3_5} = \varepsilon_{3_6} = (-1.600 \times 10^{-12})(88.7 \times 10^6)$$
$$+ (-37.9 \times 10^{-12})(-7.92 \times 10^6) \quad \text{(12.8)}$$
$$= 157.8 \times 10^{-6}$$

Since a layer is 150×10^{-6} m thick, the change in thickness of each of the $\pm 30°$ layers is as given by equation (12.5), namely,

$$\Delta h_1 = \Delta h_2 = \Delta h_5 = \Delta h_6 = \varepsilon_{3_1} h_1$$
$$= (157.8 \times 10^{-6})(150 \times 10^{-6}) = 23.7 \times 10^{-9} \, \text{m} \quad \text{(12.9)}$$

Each of the $\pm 30°$ layers expands in the thickness direction due to the compressive value of σ_2 in those layers. The thickness strain in the two $0°$ layers is, from equation (12.4),

$$\varepsilon_{3_3} = \varepsilon_{3_4} = (-1.600 \times 10^{-12})(194.1 \times 10^6)$$
$$+ (-37.9 \times 10^{-12})(-14.23 \times 10^6) \quad \text{(12.10)}$$
$$= 228 \times 10^{-6}$$

Each of the $0°$ layers also expands in the thickness direction. As a result, the thickness change of the $0°$ layers is, from equation (12.5),

$$\Delta h_3 = \Delta h_4 = 34.2 \times 10^{-9} \, \text{m} \quad \text{(12.11)}$$

Summing the change in thickness of all 6 layers, we find that the total change in thickness of the laminate, ΔH, is

$$\Delta H = \sum_{k=1}^{6} \Delta h_k = 163.1 \times 10^{-9} \text{ m} \tag{12.12}$$

These various numerical values are entered in Table 12.1.

An average through-thickness strain for an N layer laminate, $\bar{\varepsilon}_z$, can be defined as the sum of the change in the thicknesses of the individual layers divided by the original thickness of the laminate, H; that is:

$$\bar{\varepsilon}_z = \frac{\Delta H}{H} = \frac{1}{H} \sum_{k=1}^{N} (S_{13_k} \sigma_{1_k} + S_{23_k} \sigma_{2_k}) h_k \tag{12.13}$$

where

$$H = \sum_{k=1}^{N} h_k \tag{12.14}$$

Like all average quantities, no one layer necessarily experiences the thickness strain of the laminate. Continuing with our numerical example, we find, using equation (12.13), that the average thickness strain for the $[\pm 30/0]_S$ laminate is

$$\bar{\varepsilon}_z = \frac{\Delta H}{H} = \frac{163.1 \times 10^{-9}}{6(150 \times 10^{-6})} = 181.2 \times 10^{-6} \tag{12.15}$$

It should be realized that, in general, the average strain in the thickness direction is not the average of the individual layer strains. Rather, the entire thickness change due to individual layer strains must be computed, and the average strain computed from this. Here, however, with all layers the same thickness and made of the same material, the average of the six strains is identical to the average strain.

12.2
THROUGH-THICKNESS LAMINATE POISSON'S RATIOS

While the thickness change of a specific laminate in a particular loading situation may be of interest, the through-thickness Poisson's ratios of the laminate provide a general measure of the thickness response. We can define through-thickness Poisson's ratios much as we defined the three Poisson's ratios for a small element of material in Chapter 2 or the effective Poisson's ratio $\bar{\nu}_{xy}$ for a laminate in Chapter 7. A brief review of those chapters indicates that Poisson's ratios can be defined only in the context of the deformations that result from the application of a single uniaxial applied loading. Then, for a laminate the through-thickness Poisson's ratio $\bar{\nu}_{xz}$ is defined to be minus the ratio of the through-thickness strain of the laminate divided by the laminate strain in the x direction, given that the laminate is loaded only in the x direction. In such a situation the laminate strain in the x direction is given by ε_x^o, the reference surface strain in the x direction. With this,

$$\bar{\nu}_{xz} \equiv -\frac{\bar{\varepsilon}_z}{\varepsilon_x^o} \tag{12.16}$$

For a symmetric laminate loaded only by force resultant N_x, the strain ε_x^o is given by equation (7.102a) as

$$\varepsilon_x^o = a_{11}N_x \tag{12.17}$$

As a result, from equation (12.16),

$$\bar{v}_{xz} = -\frac{\bar{\varepsilon}_z}{\varepsilon_x^o} = -\frac{\bar{\varepsilon}_z}{a_{11}N_x} \tag{12.18}$$

For the $[\pm 30/0]_S$ laminate, using numerical results from equation (7.92a), with $N_x = 0.1024$ MN/m, and equation (12.15), we find that

$$\bar{v}_{xz} = -0.142 \tag{12.19}$$

A more formal expression for $\bar{v}_{xz}$ can be developed by using equations (12.4), (12.5), (12.13), and (12.18) to give

$$\bar{v}_{xz} = -\frac{\displaystyle\sum_{k=1}^{N}(S_{13_k}\sigma_{1_k} + S_{23_k}\sigma_{2_k})h_k}{N_x a_{11} H} \tag{12.20}$$

A unit value of N_x can be considered, so equation (12.20) becomes

$$\bar{v}_{xz} = -\frac{\displaystyle\sum_{k=1}^{N}(S_{13_k}\sigma_{1_k} + S_{23_k}\sigma_{2_k})h_k}{a_{11} H} \tag{12.21}$$

With a unit N_x, for a symmetric laminate, from equation (7.102a),

$$\varepsilon_x^o = a_{11} \qquad \varepsilon_y^o = a_{12} \qquad \gamma_{xy}^o = a_{16} \tag{12.22}$$

For the kth layer, then, the principal material system strains are given by equation (5.21) as

$$\varepsilon_{1_k} = m_k^2 a_{11} + n_k^2 a_{12} + m_k n_k a_{16}$$
$$\varepsilon_{2_k} = n_k^2 a_{11} + m_k^2 a_{12} - m_k n_k a_{16} \tag{12.23}$$

Here $\qquad\qquad m_k = \cos(\theta_k) \qquad n_k = \sin(\theta_k) \tag{12.24}$

and θ_k is the fiber orientation of the kth layer. Using the plane-stress stress-strain relation, equation (4.14), for the kth layer, we find that

$$\sigma_{1_k} = Q_{11_k}\varepsilon_{1_k} + Q_{12_k}\varepsilon_{2_k}$$
$$\sigma_{2_k} = Q_{12_k}\varepsilon_{1_k} + Q_{22_k}\varepsilon_{2_k} \tag{12.25}$$

Combining all these expressions, equation (12.21) becomes

$$\bar{v}_{xz} = -\frac{1}{a_{11}H}\sum_{k=1}^{N}[\{S_{13_k}(Q_{11_k}m_k^2 + Q_{12_k}n_k^2) + S_{23_k}(Q_{12_k}m_k^2 + Q_{22_k}n_k^2)\}a_{11}$$
$$+ \{S_{13_k}(Q_{11_k}n_k^2 + Q_{12_k}m_k^2) + S_{23_k}(Q_{12_k}n_k^2 + Q_{22_k}m_k^2)\}a_{12}$$
$$+ \{S_{13_k}(Q_{11_k} - Q_{12_k}) + S_{23_k}(Q_{12_k} - Q_{22_k})\}m_k n_k a_{16}]h_k$$

$$\tag{12.26}$$

For balanced laminates $a_{16} = 0$, and if in addition all layers are identical except for fiber orientation, then equation (12.26) simplifies to

$$\bar{\nu}_{xz} = -\frac{1}{Na_{11}} \sum_{k=1}^{N} [\{S_{13}(Q_{11}m_k^2 + Q_{12}n_k^2) + S_{23}(Q_{12}m_k^2 + Q_{22}n_k^2)\}a_{11}$$

$$+ \{S_{13}(Q_{11}n_k^2 + Q_{12}m_k^2) + S_{23}(Q_{12}n_k^2 + Q_{22}m_k^2)\}a_{12}] \tag{12.27}$$

The relative simplicity of equation (12.27) points to the importance of casting problems in terms of the principal material system.

All quantities in equations (12.26) or (12.27) are either a material or geometric property of the laminate and hence $\bar{\nu}_{xz}$ truly represents an inherent property of the laminate. Though equations (12.26) or (12.27) can be used to compute $\bar{\nu}_{xz}$, the approach taken in constructing Table 12.1, whereby Δh of each layer is computed and the results added, as in equation (12.6), provides physical insight on a layer-by-layer basis which can be valuable, particularly if some layers increase in thickness and some layers decrease. With equations (12.26) or (12.27) these important effects are hidden and the exact makeup of $\bar{\nu}_{xz}$ is obscured.

The Poisson's ratio $\bar{\nu}_{yz}$ based on the average through-thickness strain due to a load applied to the laminate in the y direction can be defined in a similar manner. Specifically, with only N_y acting,

$$\bar{\nu}_{yz} = -\frac{\bar{\varepsilon}_z}{\varepsilon_y^o} \tag{12.28}$$

where, for a unit value of N_y,

$$\varepsilon_x^o = a_{12} \qquad \varepsilon_y^o = a_{22} \qquad \gamma_{xy}^o = a_{26} \tag{12.29}$$

Then, like equation (12.21),

$$\bar{\nu}_{yz} = -\frac{\displaystyle\sum_{k=1}^{N}(S_{13_k}\sigma_{1_k} + S_{23_k}\sigma_{2_k})h_k}{a_{22}H} \tag{12.30}$$

Transforming the strains in the x-y system, equation (12.29), to the principal material system, as in equation (12.23), results in

$$\varepsilon_{1_k} = m_k^2 a_{12} + n_k^2 a_{22} + m_k n_k a_{26}$$
$$\varepsilon_{2_k} = n_k^2 a_{12} + m_k^2 a_{22} - m_k n_k a_{26} \tag{12.31}$$

Using the plane-stress stress-strain relations, equation (12.25), and equation (12.31) in equation (12.30) leads to the final expression for $\bar{\nu}_{yz}$, namely,

$$\bar{\nu}_{yz} = -\frac{1}{a_{22}H} \sum_{k=1}^{N} [\{S_{13_k}(Q_{11_k}m_k^2 + Q_{12_k}n_k^2) + S_{23_k}(Q_{12_k}m_k^2 + Q_{22_k}n_k^2)\}a_{12}$$

$$+ \{S_{13_k}(Q_{11_k}n_k^2 + Q_{12_k}m_k^2) + S_{23_k}(Q_{12_k}n_k^2 + Q_{22_k}m_k^2)\}a_{22}$$

$$+ \{S_{13_k}(Q_{11_k} - Q_{12_k}) + S_{23_k}(Q_{12_k} - Q_{22_k})\}m_k n_k a_{26}]h_k$$

$$\tag{12.32}$$

For the case of all layers being identical except for fiber orientation and the laminate being balanced, equation (12.32) simplifies to

$$\bar{\nu}_{yz} = -\frac{1}{a_{22}N} \sum_{k=1}^{N} [\{S_{13}(Q_{11}m_k^2 + Q_{12}n_k^2) + S_{23}(Q_{12}m_k^2 + Q_{22}n_k^2)\}a_{12}$$

$$+ \{S_{13}(Q_{11}n_k^2 + Q_{12}m_k^2) + S_{23}(Q_{12}n_k^2 + Q_{22}m_k^2)\}a_{22}]$$

(12.33)

Of course $\bar{\nu}_{yz}$ can be computed by constructing a table such as Table 12.1 for a given N_y, computing Δh of each layer, then ΔH, and directly computing

$$\bar{\varepsilon}_z = \frac{\Delta H}{H}$$

(12.34)

Then

$$\bar{\nu}_{yz} = -\frac{\bar{\varepsilon}_z}{a_{22}N_y}$$

(12.35)

This approach again provides insight into details of the change in thickness of the individual layers. We find, using either the tabular approach or equation (12.33) directly, that for the $[\pm30/0]_S$ laminate

$$\bar{\nu}_{yz} = 0.378$$

(12.36)

Figure 12.1 shows how the two through-thickness Poisson's ratios $\bar{\nu}_{xz}$ and $\bar{\nu}_{yz}$ for a $[\pm\theta/0]_S$ laminate vary with θ. As can be seen, the sign of $\bar{\nu}_{xz}$ depends strongly on θ, and there are two values of θ for which a loading in the x direction, $N_x \neq 0$, $N_y = N_{xy} = 0$, produces no thickness change; that is, $\bar{\nu}_{xz} = 0$. For other values of θ, the laminate will either expand or contract in the thickness direction. Of course for $\theta = 0°$, $\bar{\nu}_{xz} = \nu_{13}$ and $\bar{\nu}_{yz} = \nu_{23}$.

Exercise for Section 12.2

By constructing tables similar to Table 12.1, compute $\bar{\nu}_{xz}$ and $\bar{\nu}_{yz}$ for a $[0/90]_S$ laminate.

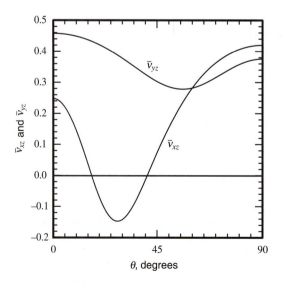

FIGURE 12.1
Variation of $\bar{\nu}_{xz}$ and $\bar{\nu}_{yz}$ with θ for a $[\pm\theta/0]_S$ laminate

12.3
THICKNESS CHANGE OF A LAMINATE DUE TO FREE THERMAL STRAIN EFFECTS

In Chapter 11, equations (11.92) and (11.93), we computed the laminate coefficients of thermal deformation $\bar{\alpha}_x, \bar{\alpha}_y$, and $\bar{\alpha}_{xy}$ by using classical lamination theory, incorporating the free thermal strain effects into the stress-strain relations, defining thermal force resultants, and properly interpreting the strain response of the laminate subjected to a temperature change. Nothing was said about the thickness expansion of the laminate, yet an isolated layer has a thermal expansion coefficient of α_3 in the thickness direction and changes thickness by the amount $\alpha_3 h \Delta T$ when the temperature is changed an amount ΔT. At first thought one may innocently assume that an entire laminate expands by an amount $\alpha_3 h \Delta T$. After all, the thickness direction of the laminate and the thickness direction of each individual layer coincide. However, nothing could be further from the truth. If such an assumption was made, then it would indicate that nothing had been learned about the interaction of layers when they are in a laminate. How *does* a laminate respond in the thickness direction when heated and cooled? Can we define a through-thickness expansion coefficient for the laminate? How does it depend on α_3?

To answer these questions, it is necessary to return to the case where free thermal strain effects are included in the three-dimensional stress-strain relations. Specifically, equation (4.28), with $\Delta M = 0$, is an expression for the through-thickness strain in the principal material system when free thermal strains must be accounted for, namely,

$$\varepsilon_3 = \alpha_3 \Delta T + S_{13}\sigma_1 + S_{23}\sigma_2 \tag{12.37}$$

We have seen throughout Chapter 11 that when a laminate is heated or cooled relative to some reference state, stresses σ_1 and σ_2 develop, and their values are changed with respect to that reference state. Equation (12.37) indicates that through the compliances S_{13} and S_{23} these stresses couple with the free thermal expansion $\alpha_3 \Delta T$ to produce a thickness strain in a layer that is different from simply $\alpha_3 \Delta T$. Only under very special circumstances is $S_{13}\sigma_1 + S_{23}\sigma_2 = 0$ and the thickness expansion given simply by $\alpha_3 \Delta T$. For the kth layer, then,

$$\varepsilon_{3_k} = \alpha_{3_k} \Delta T + S_{13_k}\sigma_{1_k} + S_{23_k}\sigma_{2_k} \tag{12.38}$$

and
$$\Delta h_k = \varepsilon_{3_k} h_k \tag{12.39}$$

We have implied, as in the past, that ΔT is not dependent on spatial location. For the entire N-layer laminate the total thickness change is

$$\Delta H = \sum_{k=1}^{N} \Delta h_k \tag{12.40}$$

As an example of a thermally induced thickness change, consider the $[\pm 30/0]_S$ laminate subjected to the cooldown from the processing temperature of $-150°C$. Table 12.2 lists the stresses in each layer that result from the free thermal deformation of the laminate. The results are taken directly from Chapter 11, equation (11.104). Using equation (12.38), for the $\pm 30°$ layers, layers 1, 2, 5, and 6, we find

TABLE 12.2
Thickness response of $[\pm 30/0]_S$ laminate due to $\Delta T = -150°C$

Layer no.	Angle deg.	σ_1 MPa	σ_2 MPa	ε_3 $\times 10^{-6}$	Δh $\times 10^{-9}$ m
1	+30	−55.2	21.1	−4360	−654
2	−30	−55.2	21.1	−4360	−654
3	0	53.4	14.64	−4280	−643
4	0	53.4	14.64	−4280	−643
5	−30	−55.2	21.1	−4360	−654
6	+30	−55.2	21.1	−4360	−654

$$\Delta H = -3900 \times 10^{-9}\,\text{m}$$

that

$$\varepsilon_{3_1} = \varepsilon_{3_2} = \varepsilon_{3_5} = \varepsilon_{3_6}$$

$$= (24.3 \times 10^{-6})(-150) + (-1.600 \times 10^{-12})(-55.2 \times 10^{6})$$
$$+ (-37.9 \times 10^{-12})(21.1 \times 10^{6}) \tag{12.41}$$
$$= -4360 \times 10^{-6}$$

with the value of α_3 being taken from Table 2.1. With a layer being 150×10^{-6} m thick, the thickness change for each of these four layers is given by equation (12.39), namely,

$$\Delta h_1 = \Delta h_2 = \Delta h_5 = \Delta h_6 = \varepsilon_{3_1} h_1$$
$$= (-4360 \times 10^{-6})(150 \times 10^{-6}) \tag{12.42}$$
$$= -654 \times 10^{-9}\,\text{m}$$

These layers contract in thickness due to the temperature decrease. Note that they decrease in thickness more than the value given by $\alpha_3 h \Delta T$; that value is

$$\alpha_3 h \Delta T = -547 \times 10^{-9}\,\text{m} \tag{12.43}$$

The through-thickness strain for the 0° layers is

$$\varepsilon_{3_3} = \varepsilon_{3_4} = (24.3 \times 10^{-6})(-150) + (-1.600 \times 10^{-12})(53.4 \times 10^{6})$$
$$+ (-37.9 \times 10^{-12})(14.64 \times 10^{6}) \tag{12.44}$$
$$= -4280 \times 10^{-6}$$

The change in thickness of each 0° layer is thus

$$\Delta h_3 = \Delta h_4 = \varepsilon_3 h_3 = -643 \times 10^{-9}\,\text{m} \tag{12.45}$$

and the total change in thickness of the laminate is, from equation (12.40),

$$\Delta H = -3900 \times 10^{-9} \tag{12.46}$$

The total thickness change due to the temperature change divided by the laminate thickness H is the average thermally induced strain in the thickness direction, $\bar{\varepsilon}_z^T$, the superscript T denoting thermal effects, that is,

$$\bar{\varepsilon}_z^T = \frac{\Delta H}{H} \tag{12.47}$$

Continuing with the numerical example for the $[\pm30/0]_S$ laminate, we find that

$$\bar{\varepsilon}_z^T = \frac{-3900 \times 10^{-9}}{6(150 \times 10^{-6})} = -4330 \times 10^{-6} \tag{12.48}$$

12.4
THROUGH-THICKNESS LAMINATE COEFFICIENT OF THERMAL EXPANSION

The average through-thickness coefficient of thermal expansion $\bar{\alpha}_z$ is defined as the average thermally induced strain in the thickness direction divided by the temperature change that caused it. Using this definition and the above expressions, we find that

$$\bar{\alpha}_z = \frac{\bar{\varepsilon}_z^T}{\Delta T} \tag{12.49}$$

or, using equation (12.48) for the $[\pm30/0]_S$ laminate, that

$$\bar{\alpha}_z = \frac{-4330 \times 10^{-6}}{-150} = 28.9 \times 10^{-6}/°C \tag{12.50}$$

This value of thickness expansion is different from the value of $\alpha_3 = 24.3 \times 10^{-6}/°C$, attesting to the fact that the layers interact to produce an overall thickness expansion that is a combination of free thermal expansion effects and through-thickness elastic properties of the individual layers. Specifically, it is the through-thickness Poisson's ratios that are responsible for the thickness expansion due to thermal effects being different from $\alpha_3 \Delta T$.

More formally, we can develop an expression for $\bar{\alpha}_z$ in a manner similar to the approach used find $\bar{\nu}_{xz}$, equation (12.26), and $\bar{\nu}_{yz}$, equation (12.32). Combining equations (12.38), (12.39), (12.40), (12.47), and (12.49) results in

$$\bar{\alpha}_z = \frac{1}{H \Delta T} \sum_{k=1}^{N} (\alpha_{3_k} \Delta T + S_{13_k} \sigma_{1_k} + S_{23_k} \sigma_{2_k}) h_k \tag{12.51}$$

For a unit temperature change, from equation (11.92) for a symmetric laminate, the strains are

$$\bar{\alpha}_x \equiv \varepsilon_x^o = a_{11}\hat{N}_x^T + a_{12}\hat{N}_y^T + a_{16}\hat{N}_{xy}^T$$

$$\bar{\alpha}_y \equiv \varepsilon_y^o = a_{12}\hat{N}_x^T + a_{22}\hat{N}_y^T + a_{26}\hat{N}_{xy}^T \tag{12.52}$$

$$\bar{\alpha}_{xy} \equiv \gamma_{xy}^o = a_{16}\hat{N}_x^T + a_{26}\hat{N}_y^T + a_{66}\hat{N}_{xy}^T$$

Transforming these strains to the principal material 1-2 system for the kth layer yields

$$\varepsilon_{1_k} = \bar{\alpha}_x m_k^2 + \bar{\alpha}_y n_k^2 + \bar{\alpha}_{xy} m_k n_k$$

$$\varepsilon_{2_k} = \bar{\alpha}_x n_k^2 + \bar{\alpha}_y m_k^2 - \bar{\alpha}_{xy} m_k n_k \tag{12.53}$$

and then the principal material system stresses become, for the kth layer,

$$\sigma_{1_k} = Q_{11_k}(\bar{\alpha}_x m_k^2 + \bar{\alpha}_y n_k^2 + \bar{\alpha}_{xy} m_k n_k - \alpha_{1_k})$$
$$+ Q_{12_k}(\bar{\alpha}_x n_k^2 + \bar{\alpha}_y m_k^2 - \bar{\alpha}_{xy} m_k n_k - \alpha_{2_k})$$
$$\sigma_{2_k} = Q_{12_k}(\bar{\alpha}_x m_k^2 + \bar{\alpha}_y n_k^2 + \bar{\alpha}_{xy} m_k n_k - \alpha_{1_k})$$
$$+ Q_{22_k}(\bar{\alpha}_x n_k^2 + \bar{\alpha}_y m_k^2 - \bar{\alpha}_{xy} m_k n_k - \alpha_{2_k})$$

(12.54)

Substituting equation (12.54) into equation (12.51) results in a rather cumbersome expression. However, if every layer is identical, except for fiber orientation, and the laminate is balanced as well as being symmetric, the expression for $\bar{\alpha}_z$ becomes

$$\bar{\alpha}_z = \frac{1}{N} \sum_{k=1}^{N} [\{S_{13}(Q_{11}m_k^2 + Q_{12}n_k^2) + S_{23}(Q_{12}m_k^2 + Q_{22}n_k^2)\}\bar{\alpha}_x$$
$$+ \{S_{13}(Q_{11}n_k^2 + Q_{12}m_k^2) + S_{23}(Q_{12}n_k^2 + Q_{22}m_k^2)\}\bar{\alpha}_y$$
$$+ \alpha_3 - \{S_{13}(Q_{11}\alpha_1 + Q_{12}\alpha_2) + S_{23}(Q_{12}\alpha_1 + Q_{22}\alpha_2)\}]$$

(12.55)

Figure 12.2 shows the variation with θ of $\bar{\alpha}_z$ for a $[\pm\theta/0]_S$ laminate. Of course, for $\theta = 0°$, $\bar{\alpha}_z = \alpha_3$ as the six layers act as one thicker unidirectional layer, and except for $\theta = 0°$, the value of $\bar{\alpha}_z$ is greater than α_3, being about 35 percent greater for $\theta > 60°$.

By direct analogy, expressions for the thickness response of a laminate due to moisture absorption can be developed. The moisture-induced strain and moisture-induced thickness change of the kth layer are given by

$$\varepsilon_{3_k} = \beta_{3_k}\Delta M + S_{13_k}\sigma_{1_k} + S_{23_k}\sigma_{2_k}$$

(12.56)

and

$$\Delta h_k = (\beta_{3_k}\Delta M + S_{13_k}\sigma_{1_k} + S_{23_k}\sigma_{2_k})h_k$$

(12.57)

The total change in thickness of the laminate due to moisture absorption is

$$\Delta H = \sum_{k=1}^{N} (\beta_{3_k}\Delta M + S_{13_k}\sigma_{1_k} + S_{23_k}\sigma_{2_k})h_k$$

(12.58)

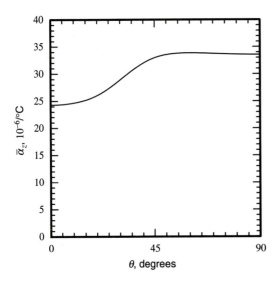

FIGURE 12.2
Variation of $\bar{\alpha}_z$ with θ for a $[\pm\theta/0]_S$ laminate

The average moisture-induced thickness strain is

$$\bar{\varepsilon}_z^M = \frac{\Delta H}{H} = \frac{1}{H} \sum_{k=1}^{N} (\beta_{3_k} \Delta M + S_{13_k} \sigma_{1_k} + S_{23_k} \sigma_{2_k}) h_k \qquad (12.59)$$

Finally, the average coefficient of moisture expansion in the thickness direction can be written as

$$\bar{\beta}_z = \frac{1}{H \Delta M} \sum_{k=1}^{N} (\beta_{3_k} \Delta M + S_{13_k} \sigma_{1_k} + S_{23_k} \sigma_{2_k}) h_k \qquad (12.60)$$

As long as it can validly be assumed that each layer is in a state of plane stress, the definitions just presented for through-thickness response are valid independently of laminate thickness.

Exercise for Section 12.4

By constructing a table similar to Table 12.2, compute $\bar{\alpha}_z$ for a $[0/90]_S$ laminate.

12.5
SUMMARY

Based on these simple examples, we can see that the through-thickness strain can be computed in a straightforward manner using fundamental principles. The through-thickness strain may or may not be important in a particular application, but it is important to remember that it is not zero and to assume so is wrong.

12.6
SUGGESTED READINGS

For further discussions of through-thickness Poisson's ratios, consult:

1. Herakovich, C. T. "Composite Laminates with Negative through-the-thickness Poisson's Ratios." *Journal of Composite Materials* 18 (1984), pp. 447–55.

Through-thickness thermal expansion is considered in:

2. Fahmy, A. A., and A. N. Ragai-Ellozy. "Thermal Expansion of Laminated Fiber Composites in the Thickness Direction." *Journal of Composite Materials* 8 (1974), pp. 90–92.

If you read the above citation, it is important to also read:

3. Pagano, N. J. "Thickness Expansion Coefficients of Composite Laminates." *Journal of Composite Materials* 8 (1974), pp. 310–12.

Another treatment of the same topic is given in:

4. Wetherhold, R. C., and C. S. Boss. "Transverse Thermal Expansion Coefficients for Composite Laminates." *Journal of Composite Materials* 22 (1988), pp. 812–17.

CHAPTER 13

Introduction to Fiber-Reinforced Laminated Plates

To this point the discussions have centered on the behavior of fiber-reinforced composites without regard to the structure they are part of. Chapters 2, 4, and 5 dealt with assumptions regarding the behavior of a fiber-reinforced material at a point, specifically the stress-strain relations, including the influence of free thermal strain. Stress-strain relations are algebraic relations that are defined at a point, so the results of those chapters are valid in any structure, independently of loading, shape, or size. Chapters 6, 7, and 8 focused on classical lamination theory, and rather than describing behavior at a point, described behavior on an entire line through the thickness of the laminate. Nonetheless, because the response on the line was directly tied to what was happening at a point on the reference surface, what we did in those chapters was still confined to a small region of a laminate. Ultimately, we want to be able to accurately describe the response at <u>all</u> points in a fiber-reinforced structure and examine the stresses, strains, and the issue of failure throughout the structure. We need to study the influence of the loading type, temperature change, laminate shape, laminate size, boundary conditions, as well as fiber orientation, material properties, and lamination sequence on the response. Maximum deflections, maximum stresses, loads to cause material failure, loads to cause buckling, and natural frequencies of vibration are some of the more important structural responses that depend on the details of the entire structure. To study these it is necessary to develop the tools for determining how laminate response varies with x and y. Within the context of classical lamination theory, we now know how laminate response varies with z, but we must expand our thinking to variations in the other two coordinate directions. The development of the tools depends to a large degree on the issue being studied (do we want to know maximum deflections, or buckling loads, or natural frequencies?) as well as the type of the structure (is it a thin plate, a thick plate, a cylindrical shell, a conical shell, etc.?). In this chapter we shall develop the tools necessary to study flat laminated plates and apply these tools to several problems that illustrate the unique response characteristics of fiber-reinforced structures in general, and plates in

514

particular. Because the study of laminated plates can lead to several books in itself, this chapter will be limited to the study of the linear response of laminated plates. Thus, for example, buckling will not be addressed. Furthermore, the discussions of this chapter will be limited to rectangular plates. Our primary purpose here is to establish the principles of classical laminated plate theory and illustrate these principles with simple examples. Advanced topics can then be pursued from these basic principles. The next section develops the equations governing the behavior of plates. Following that, the examples are presented.

13.1
EQUATIONS GOVERNING PLATE BEHAVIOR

There are several approaches to developing the equations that govern the behavior of plates. It must be clear from the onset that the equations that govern the behavior must include the proper specification of the conditions at the boundary, as well as the specification of the conditions that govern the behavior away from the boundary, in the interior of the plate. The latter conditions are generally referred to as the governing differential equations, whereas the former are, naturally, referred to as boundary conditions. The most consistent method of deriving the governing conditions is through energy and variational principles. With this approach, we obtain the governing differential equations and the boundary conditions. This is the preferred approach, but unfortunately not everybody feels comfortable with the principles of variational calculus. As a result, the alternative Newtonian approach, summing forces and moments, is often used to derive the governing differential equations. The disadvantage of this approach is that there is no direct information regarding the boundary conditions. Nonetheless, we will use the Newtonian approach in this chapter; the differential equations of equilibrium will be derived by summing forces and moments on a differential element of laminate, and the boundary conditions will simply be stated.

Consider a plate, as in Figure 13.1, with length a in the x direction, width b in the y direction, and thickness H. The plate is made of a number of layers of

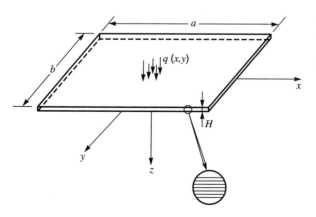

FIGURE 13.1
Geometry, nomenclature, and loading of a rectangular laminated plate

fiber-reinforced material, as the inset shows, and the reference surface of the plate will be taken as the reference surface of the laminate. The plate is subjected to a known distributed force $q(x, y)$ acting in the $+z$ direction. Plate theory cannot distinguish whether this distributed force is pushing on the top surface or whether it is pulling on the bottom surface. The fact that the distributed force can vary with spatial location (x, y) is indicated by the functional dependence of q on x and y. In addition to the distributed force, the boundaries of the plate can be subjected to a variety of forces and moments. Figure 13.2 illustrates the possible forces and moments that can be applied on the edge $x = +a/2$, namely, three forces and two moments. Note the nomenclature associated with these forces and moments. The superscript $+$ denotes that the application is at $x = +a/2$, and the functional dependence on y indicates that these forces and moments can vary with distance along the edge. Figure 13.3 illustrates such a variation for the normal force resultant N_x^+. In Figure 13.2 four of the five resultants are familiar. These are the normal force resultant $N_x^+(y)$, the shear force resultant $N_{xy}^+(y)$, the bending moment resultant $M_x^+(y)$, and the twisting moment resultant $M_{xy}^+(y)$. The transverse shear force resultant $Q_x^+(y)$ has not previously appeared. This resultant is defined by

$$Q_x^+(y) = \int_{z=-\frac{H}{2}}^{+\frac{H}{2}} \tau_{xz}\left(+\frac{a}{2}, y, z\right) dz \tag{13.1}$$

One can immediately raise the issue that because of the plane-stress assumption, $\tau_{xz} = 0$, and thus the transverse shear force resultant must be zero. Therefore we should not be able to say anything about defining this resultant. As will be seen shortly, there must be transverse shear force resultants Q_x, and, shortly to be introduced, Q_y to maintain equilibrium in the z direction. Thus, there is an inconsistency

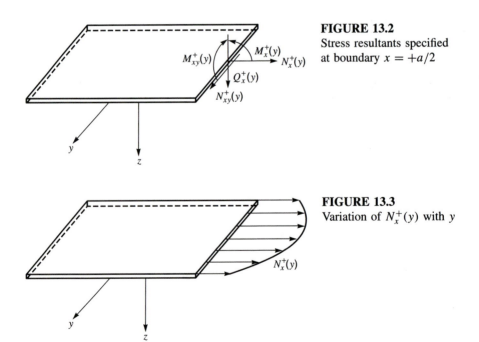

FIGURE 13.2
Stress resultants specified at boundary $x = +a/2$

FIGURE 13.3
Variation of $N_x^+(y)$ with y

in plate theory. The theory needs stresses τ_{xz} and τ_{yz} for equilibrium, yet the theory is based on the plane-stress assumption. The variational approach does not have to explicitly define Q_x and Q_y because it is not based on equilibrium principles.

Despite this inconsistency, the theory we are developing allows for the application of any or all of these five resultants on the edge $x = +a/2$. In particular, if the edge is free, as in a cantilevered plate, then all five resultants would be zero. On the edge $y = +b/2$, Figure 13.4, five other force resultants can be applied. Again, note the superscript $+$ signifying the location $y = +b/2$ and the possible variation of these resultants along the edge, in this case with the x coordinate. The subscripts yx instead of xy on N and M simply denote the difference between the edge $x = +a/2$ and the edge $y = +b/2$. The transverse shear force resultant $Q_y^+(x)$ is defined as

$$Q_y^+(x) = \int_{z=-\frac{H}{2}}^{+\frac{H}{2}} \tau_{yz}\left(x, +\frac{b}{2}, z\right) dz \qquad (13.2)$$

The forces and moments that can be applied on the other two boundaries are indicated in Figures 13.5 and 13.6. Note the superscript $-$ indicating the location of the edges, $x = -a/2$ and $y = -b/2$. Note also the sense of the applied resultants, the sense being consistent with past notation. The sense of the Q's is consistent with the sense of τ_{xz} and τ_{yz}.

To derive the governing equilibrium equations, we must consider force and moment equilibrium of a differential element of laminated plate. Because equilibrium considers the sums of forces and moments, not the sums of force and moment *resultants*, the force and moment resultants must be multiplied by an appropriate length. Recall that because the units of N_x, for example, are N/m, N_x must be multiplied by a length to have the units of force. To derive the equilibrium equations,

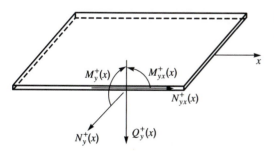

FIGURE 13.4
Stress resultants specified at boundary $y = +b/2$

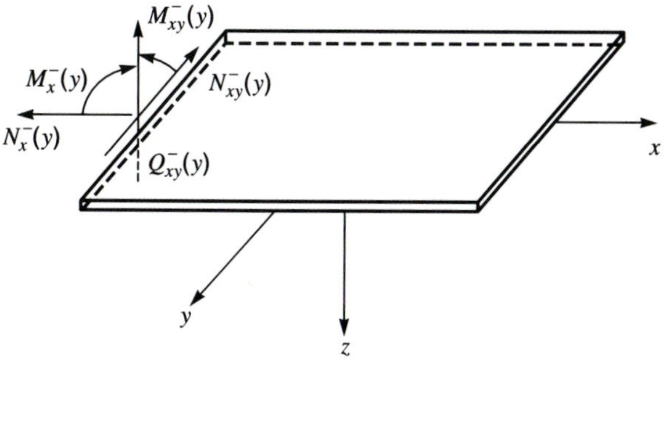

FIGURE 13.5
Stress resultants specified at boundary $x = -a/2$

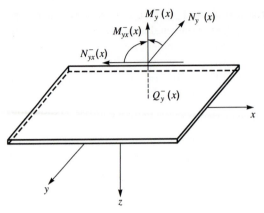

FIGURE 13.6
Stress resultants specified at
boundary $y = -b/2$

several figures of the differential element will be used, and each figure will depict a
specific set of force or moment resultants. In reality, all force and moment resultants
act simultaneously on the differential element, but to avoid cluttering the figure, sets
of forces are illustrated separately.

Six equilibrium conditions must be considered. The sum of the forces in each
of the three coordinate directions must be zero, and the sum of moments about each
of the three coordinate axes must be zero. To sum forces in the x direction, consider
Figure 13.7, which shows the differential element; the length in the x direction is
Δx and the length in the y direction is Δy. The element is centered about the point
(x, y). This differential element is cut from the interior of the plate, away from
the edges, and includes the entire thickness of the laminate. Because the force and
moment resultants vary with the x and y coordinates, and because the element is of
differential size, we can use a Taylor series expansion to represent the resultants on
the edges of the element in terms of the resultants at (x, y). In each case, only the first
term of the Taylor series is retained. In the figure the force resultant on each edge is

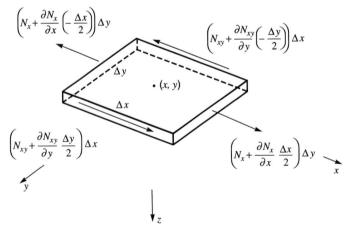

FIGURE 13.7
Summing forces in the x direction on a differential element of plate

multiplied by the length of the edge to properly define a force. For convenience, only the forces affecting equilibrium in the x direction are shown and the force is assumed to act at midlength along each edge. Summing forces in the $+x$ direction gives

$$\left(N_x + \frac{\partial N_x}{\partial x}\frac{\Delta x}{2}\right)\Delta y + \left(N_{xy} + \frac{\partial N_{xy}}{\partial y}\frac{\Delta y}{2}\right)\Delta x$$

$$- \left(N_x + \frac{\partial N_x}{\partial x}\left(-\frac{\Delta x}{2}\right)\right)\Delta y \qquad (13.3)$$

$$- \left(N_{xy} + \frac{\partial N_{xy}}{\partial y}\left(-\frac{\Delta y}{2}\right)\right)\Delta x = 0$$

The algebra leads to

$$\frac{\partial N_x}{\partial x} + \frac{\partial N_{xy}}{\partial y} = 0 \qquad (13.4)$$

This is one equilibrium equation. It is a partial differential equation, whose partial derivatives are gradients of the force resultants. The equation is an algebraic equation in the force resultant gradients, and the equation states that if N_x varies with x, then N_{xy} must vary with y.

The second equilibrium equation can be derived by summing forces in the y direction, where the force resultants involved in this summation are illustrated in Figure 13.8. Repeating the steps that led to equation (13.4), we find that the second equilibrium equation is

$$\frac{\partial N_{xy}}{\partial x} + \frac{\partial N_y}{\partial y} = 0 \qquad (13.5)$$

To derive the third equilibrium equation, we must define the transverse shear force resultants in the interior of the plate; these definitions are similar to the definitions of the transverse shear force resultants at the edges of the plate, equations

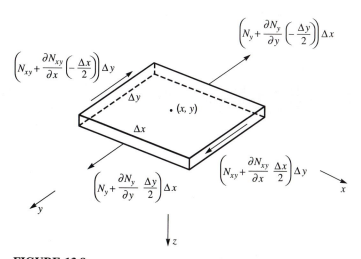

FIGURE 13.8
Summing forces in the y direction on a differential element of plate

(13.1) and (13.2). Accordingly, the transverse shear force resultant Q_x is defined as

$$Q_x = \int_{z=-\frac{H}{2}}^{+\frac{H}{2}} \tau_{xz} dz \qquad (13.6)$$

while the transverse shear force resultant Q_y is defined as

$$Q_y = \int_{z=-\frac{H}{2}}^{+\frac{H}{2}} \tau_{yz} dz \qquad (13.7)$$

both resultants, of course, varying with x and y.

Figure 13.9 shows the force resultants pertinent to summing forces in the z direction. The pertinent forces are these just-defined transverse shear force resultants, and the applied distributed force $q(x, y)$. To have the units of force, the distributed force must be multiplied by the differential area it acts on. The differential element is small enough that we can assume that the force due to the distributed load acts at point (x, y). Summing forces in the $+z$ (downward) direction results in

$$\left(Q_x + \frac{\partial Q_x}{\partial x} \frac{\Delta x}{2} \right) \Delta y + \left(Q_y + \frac{\partial Q_y}{\partial y} \frac{\Delta y}{2} \right) \Delta x$$
$$- \left(Q_x + \frac{\partial Q_x}{\partial x} \left(-\frac{\Delta x}{2} \right) \right) \Delta y$$
$$- \left(Q_y + \frac{\partial Q_y}{\partial y} \left(-\frac{\Delta y}{2} \right) \right) \Delta x \qquad (13.8)$$
$$+ q \Delta x \Delta y = 0$$

and the subsequent algebra leads to

$$\frac{\partial Q_x}{\partial x} + \frac{\partial Q_y}{\partial y} + q = 0 \qquad (13.9)$$

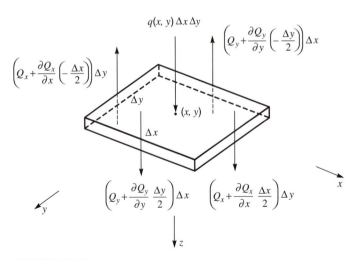

FIGURE 13.9
Summing forces in the z direction on a differential element of plate

We now must consider moment equilibrium; Figure 13.10 illustrates the force and moment resultants that produce a moment about the x axis. Summing moments about the point (x, y) results in

$$
\left(M_y + \frac{\partial M_y}{\partial y}\left(-\frac{\Delta y}{2}\right)\right)\Delta x + \left(M_{xy} + \frac{\partial M_{xy}}{\partial x}\left(-\frac{\Delta x}{2}\right)\right)\Delta y
$$

$$
+ \left(Q_y + \frac{\partial Q_y}{\partial y}\left(-\frac{\Delta y}{2}\right)\right)\Delta x \left(\frac{\Delta y}{2}\right)
$$

$$
+ \left(Q_y + \frac{\partial Q_y}{\partial y}\frac{\Delta y}{2}\right)\Delta x \left(\frac{\Delta y}{2}\right) \tag{13.10}
$$

$$
- \left(M_y + \frac{\partial M_y}{\partial y}\frac{\Delta y}{2}\right)\Delta x
$$

$$
- \left(M_{xy} + \frac{\partial M_{xy}}{\partial x}\frac{\Delta x}{2}\right)\Delta y
$$

Though the equivalent force of the distributed load, $q(x, y)\Delta x \Delta y$, acts vertically, it acts through point (x, y) and hence has no moment arm. Collecting terms, the above equation becomes

$$
Q_y = \frac{\partial M_y}{\partial y} + \frac{\partial M_{xy}}{\partial x} \tag{13.11}
$$

This equation, rather than equation (13.7), could be considered the definition of the shear force resultant Q_y. With this definition, it is unnecessary to appeal to using stresses that have been defined as zero, in this case τ_{yz}, to define a quantity that is

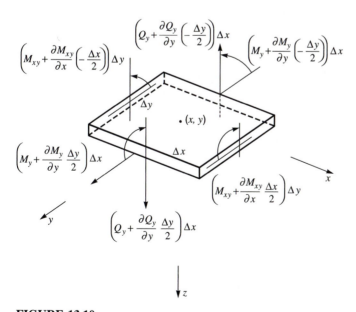

FIGURE 13.10
Summing moments about the x axis on a differential element of plate

absolutely necessary for equilibrium. Rather Q_y is defined to be equal to derivatives of moment resultants M_x and M_{xy}. These moment resultants have been an integral part of classical lamination theory and thus we are familiar with them. As we can see, if it were not for Q_x and Q_y, there would be nothing to react to the applied load $q(x, y)$. Plate theory based on the plane-stress assumption indeed presents a dilemma; however, the theory is quite accurate.

If we sum the moments about the y axis, Figure 13.11, an equation involving Q_x is derived:

$$Q_x = \frac{\partial M_x}{\partial x} + \frac{\partial M_{xy}}{\partial y} \tag{13.12}$$

Finally, moments must be summed about the z axis; Figure 13.12 illustrates a planform view of the important force resultants. Summing moments about point (x, y) results in

$$
\begin{aligned}
& -\left(N_{xy} + \frac{\partial N_{xy}}{\partial y}\left(-\frac{\Delta y}{2}\right)\right) \Delta x \left(\frac{\Delta y}{2}\right) \\
& -\left(N_{xy} + \frac{\partial N_{xy}}{\partial y}\frac{\Delta y}{2}\right) \Delta x \left(\frac{\Delta y}{2}\right) \\
& +\left(N_{xy} + \frac{\partial N_{xy}}{\partial x}\left(-\frac{\Delta x}{2}\right)\right) \Delta y \frac{\Delta x}{2} \\
& +\left(N_{xy} + \frac{\partial N_{xy}}{\partial x}\frac{\Delta x}{2}\right) \Delta y \frac{\Delta x}{2} = 0
\end{aligned}
\tag{13.13}
$$

Completing the algebra on the left-hand side leads to satisfaction of the equation identically. No new information is derived from this equation.

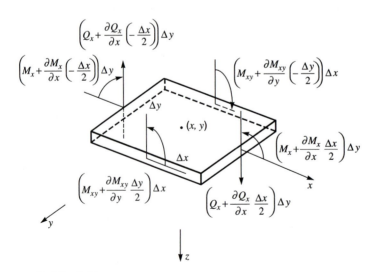

FIGURE 13.11
Summing moments about the y axis on a differential element of plate

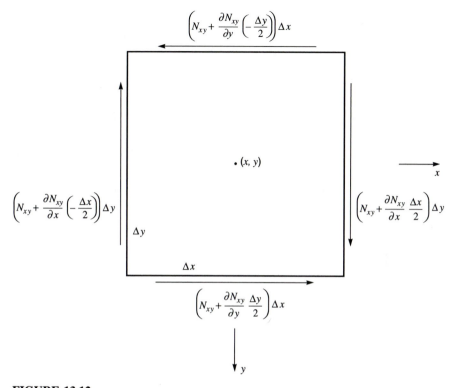

FIGURE 13.12
Summing moments about the z axis on a differential element of plate

Generally the equations defining Q_x and Q_y in terms of the moments, equations (13.11) and (13.12), are used to eliminate these variables from the problem. If this is done, equation (13.9) becomes

$$\frac{\partial^2 M_x}{\partial x^2} + 2\frac{\partial^2 M_{xy}}{\partial x \partial y} + \frac{\partial^2 M_y}{\partial y^2} + q = 0 \tag{13.14}$$

Thus, the equilibrium equations that govern the response of a laminated plate are

$$\boxed{\begin{aligned}
&\frac{\partial N_x}{\partial x} + \frac{\partial N_{xy}}{\partial y} = 0 \\[2mm]
&\frac{\partial N_{xy}}{\partial x} + \frac{\partial N_y}{\partial y} = 0 \\[2mm]
&\frac{\partial^2 M_x}{\partial x^2} + 2\frac{\partial^2 M_{xy}}{\partial x \partial y} + \frac{\partial^2 M_y}{\partial y^2} + q = 0
\end{aligned}} \tag{13.15}$$

Note that the issue of material properties has not entered the discussion. Material properties do not influence the equilibrium conditions; thus, the above equations are valid for any rectangular plate, independently of whether it is a single layer, a cross-ply laminate, a balanced laminate, an unsymmetric laminate, and so on.

The boundary conditions which must be satisfied along each edge are as follows:
Along edge $x = +a/2$:

(i)   Either $N_x = N_x^+$ or u^o must be specified

(ii) Either $N_{xy} = N_{xy}^+$ or v^o must be specified

(iii) Either $\dfrac{\partial M_x}{\partial x} + 2\dfrac{\partial M_{xy}}{\partial y} = Q_x^+ + \dfrac{\partial M_{xy}^+}{\partial y}$ or w^o must be specified

(iv) Either $M_x = M_x^+$ or $\dfrac{\partial w^o}{\partial x}$ must be specified   (13.16a)

Along edge $x = -a/2$:

(i)   Either $N_x = N_x^-$ or u^o must be specified

(ii) Either $N_{xy} = N_{xy}^-$ or v^o must be specified

(iii) Either $\dfrac{\partial M_x}{\partial x} + 2\dfrac{\partial M_{xy}}{\partial y} = Q_x^- + \dfrac{\partial M_{xy}^-}{\partial y}$ or w^o must be specified

(iv) Either $M_x = M_x^-$ or $\dfrac{\partial w^o}{\partial x}$ must be specified   (13.16b)

Along edge $y = +b/2$:

(i)   Either $N_y = N_y^+$ or v^o must be specified

(ii) Either $N_{xy} = N_{yx}^+$ or u^o must be specified

(iii) Either $\dfrac{\partial M_y}{\partial y} + 2\dfrac{\partial M_{xy}}{\partial x} = Q_y^+ + \dfrac{\partial M_{yx}^+}{\partial x}$ or w^o must be specified

(iv) Either $M_y = M_y^+$ or $\dfrac{\partial w^o}{\partial y}$ must be specified   (13.16c)

Along edge $y = -b/2$:

(i)   Either $N_y = N_y^-$ or v^o must be specified

(ii) Either $N_{xy} = N_{yx}^-$ or u^o must be specified

(iii) Either $\dfrac{\partial M_y}{\partial y} + 2\dfrac{\partial M_{xy}}{\partial x} = Q_y^- + \dfrac{\partial M_{yx}^-}{\partial x}$ or w^o must be specified

(iv) Either $M_y = M_y^-$ or $\dfrac{\partial w^o}{\partial y}$ must be specified   (13.16d)

These boundary conditions have important physical interpretations. At each edge four conditions must hold. The four conditions are determined by satisfying one option from each of four pairs of conditions, (i) through (iv). The first two conditions at each edge have rather simple interpretations. For example, along the edge at $x = +a/2$, condition (i) states that at that edge either the normal force resultant in the plate, N_x, must equal the resultant applied normal to that edge, N_x^+, or the displacement normal to the edge must be given. Condition (ii) states that either the shear force resultant in the plate, N_{xy}, must equal the resultant applied tangent to the edge, N_{xy}^+, or the displacement tangential to that edge must be given. In contrast to the simplicity of the first two conditions, the interpretation of the third condition at each edge is not so obvious; however, it is a well-known and important condition. The condition states that the combination of moment resultant derivatives must equal the sum of the applied shear force resultant and the derivative of the applied twisting moment resultant; otherwise the vertical displacement of the edge must be specified. In condition (iii), on each plate edge the derivative of the twisting moment resultant is taken with respect to the coordinate parallel with that edge. For example, at $x = +a/2$, the derivative of M_{xy}^+ is taken with respect to y, the coordinate parallel to that edge. Though this complicated boundary condition is a natural result of the variational process and is really nothing more than a result of the mathematical manipulation, the condition is not obvious when using the force equilibrium approach. Its physical interpretation is far-reaching. It basically implies that, for example, at $x = +a/2$, the edge of the plate does not respond to an applied transverse shear force resultant Q_x^+ and an applied twisting moment M_{xy}^+ separately if both are applied. Rather, the plate responds to the combination

$$Q_x^+ + \frac{\partial M_{xy}^+}{\partial y} \tag{13.17}$$

In fact, because the derivative in the above has the units of force per unit length, the above combination can be considered an applied effective transverse shear force resultant at the edge of the plate at $x = +a/2$; that is:

$$Q_x^+ + \frac{\partial M_{xy}^+}{\partial y} = Q_{x_{eff}}^+ \tag{13.18}$$

In the same vein, the quantity

$$\frac{\partial M_x}{\partial x} + 2\frac{\partial M_{xy}}{\partial y} \tag{13.19}$$

can also be considered as an effective transverse shear force resultant within the plate. Because the transverse shear force resultant within the plate is defined by equation (13.12) as

$$Q_x = \frac{\partial M_x}{\partial x} + \frac{\partial M_{xy}}{\partial y} \tag{13.20}$$

the expression in equation (13.19) can be written

$$Q_x + \frac{\partial M_{xy}}{\partial y} \tag{13.21}$$

or defining

$$Q_{x_{eff}} = Q_x + \frac{\partial M_{xy}}{\partial y} = \frac{\partial M_x}{\partial x} + 2\frac{\partial M_{xy}}{\partial y} \tag{13.22}$$

boundary condition (iii) for the edge $x = +a/2$ can be stated as

$$\text{either } Q_{x_{eff}} = Q_{x_{eff}}^+ \text{ or } w^o \text{ must be specified} \tag{13.23}$$

meaning that at the edge either the effective transverse shear force resultant in the plate must be equal to the applied effective transverse shear force resultant, or the displacement w^o must be specified. If the edge of the plate is free from forces or moments, then

$$Q_x^+ = M_{xy}^+ = 0 \tag{13.24}$$

but rather than applying each condition independently, the boundary condition becomes

$$\frac{\partial M_x}{\partial x} + 2\frac{\partial M_{xy}}{\partial y} = Q_x^+ + \frac{\partial M_{xy}^+}{\partial y} = 0 \tag{13.25}$$

This is the famous *Kirchhoff free-edge condition*. Equation (13.22) defines what is referred to as the *Kirchhoff shear force resultant*, namely $Q_{x_{eff}}$.

Condition (iv) is not so complicated. It states that either the bending moment resultant in the plate, M_x, must equal the bending moment applied to that edge, M_x^+, or the rotation (slope) of the edge must be given. For a clamped edge, for example, the slope would be specified to be zero and nothing could be stated regarding the moment. For a simply supported edge the applied bending moment resultant M_x^+ would be taken to be zero and nothing could be stated regarding the slope.

The boundary conditions on the other three edges follow similar interpretations. Within the plate the effective transverse shear force resultant $Q_{y_{eff}}$ can be defined as

$$Q_{y_{eff}} = Q_y + \frac{\partial M_{xy}}{\partial x} = \frac{\partial M_y}{\partial y} + 2\frac{\partial M_{xy}}{\partial x} \tag{13.26}$$

Likewise, the applied effective transverse shear force resultant at $y = +b/2$, for example, can be defined as

$$Q_{y_{eff}}^+ = Q_y^+ + \frac{\partial M_{xy}^+}{\partial x} \tag{13.27}$$

Thus, boundary condition (iii) at $y = +b/2$ can be alternatively stated as

$$\text{either } Q_{y_{eff}} = Q_{y_{eff}}^+ \text{ or } w^o \text{ must be specified} \tag{13.28}$$

The important conclusion from examining the boundary conditions is that *pairs of variables must be considered on each boundary*. Each pair always consists of what can be considered complementary response variables. One component of the pair involves a force or moment resultant, and the other component involves a displacement or rotation. In the study of plates, only the boundary conditions that conform to these conditions are legitimate. Imposing a boundary condition not covered by one of the four complementary pairs of variables results in an ill-posed problem. This will lead to erroneous conclusions if pursued, assuming, of course, you could even solve the problem.

In addition to conditions along each edge, certain conditions must be specified at each corner. Corner conditions, which result because the circumferential boundary of the plate is not smooth (i.e., it has corners) will not be discussed here. However, each of the corner conditions requires that either a particular twisting moment within the plate equal the applied twisting moment at the corner, or the displacement of the plate corner must be specified. If the displacements of all four edges of the plate are specified, as with simply supported or clamped edges, the corner conditions are automatically satisfied. If two adjacent edges are free, then the corner condition must be satisfied by the twisting moments being zero. It might be added that for a simply supported or clamped edge, boundary condition (iii) is satisfied by specification of the displacement. This is opposed to a free edge, where condition (iii) must be satisfied by the Kirchhoff free-edge condition (i.e., the effective transverse shear force resultant is zero). The required satisfaction of the force portion of boundary condition (iii) and the moment portion of the corner condition poses a larger challenge in finding solutions to the governing equations than enforcing displacement conditions. In practice, fortunately, the edges of a plate are usually simply supported or clamped, so the corner conditions and boundary condition (iii) can easily be satisfied.

Exercises for Section 13.1

1. Refer to Figure 13.8 to derive equation (13.5).

2. Fill in the details necessary for arriving at equation (13.12); refer to Figure 13.11.

13.2
GOVERNING CONDITIONS IN TERMS OF DISPLACEMENTS

Because the displacements are the basic variables in the problem, all other responses being derivable from the displacements, it is meaningful to express the equilibrium equations and boundary conditions in terms of the displacements. In addition, we usually have more of a physical feel for the displacement response of a structure and so it is useful to write the equations in terms of these quantities. In equation (7.85) the stress resultants were expressed in terms of the reference surface strains and curvatures. In equation (6.14) the reference surface strains and curvatures were defined in terms of the reference surface displacements. Accordingly, the stress resultants can be written in terms of the reference surface displacements as

$$N_x = A_{11}\frac{\partial u^o}{\partial x} + A_{12}\frac{\partial v^o}{\partial y} + A_{16}\left(\frac{\partial u^o}{\partial y} + \frac{\partial v^o}{\partial x}\right)$$
$$- B_{11}\frac{\partial^2 w^o}{\partial x^2} - B_{12}\frac{\partial^2 w^o}{\partial y^2} - 2B_{16}\frac{\partial^2 w^o}{\partial x \partial y}$$

$$N_y = A_{12}\frac{\partial u^o}{\partial x} + A_{22}\frac{\partial v^o}{\partial y} + A_{26}\left(\frac{\partial u^o}{\partial y} + \frac{\partial v^o}{\partial x}\right)$$
$$- B_{12}\frac{\partial^2 w^o}{\partial x^2} - B_{22}\frac{\partial^2 w^o}{\partial y^2} - 2B_{26}\frac{\partial^2 w^o}{\partial x \partial y}$$

$$N_{xy} = A_{16}\frac{\partial u^o}{\partial x} + A_{26}\frac{\partial v^o}{\partial y} + A_{66}\left(\frac{\partial u^o}{\partial y} + \frac{\partial v^o}{\partial x}\right)$$
$$- B_{16}\frac{\partial^2 w^o}{\partial x^2} - B_{26}\frac{\partial^2 w^o}{\partial y^2} - 2B_{66}\frac{\partial^2 w^o}{\partial x \partial y}$$

$$M_x = B_{11}\frac{\partial u^o}{\partial x} + B_{12}\frac{\partial v^o}{\partial y} + B_{16}\left(\frac{\partial u^o}{\partial y} + \frac{\partial v^o}{\partial x}\right)$$
$$- D_{11}\frac{\partial^2 w^o}{\partial x^2} - D_{12}\frac{\partial^2 w^o}{\partial y^2} - 2D_{16}\frac{\partial^2 w^o}{\partial x \partial y}$$

(13.29)

$$M_y = B_{12}\frac{\partial u^o}{\partial x} + B_{22}\frac{\partial v^o}{\partial y} + B_{26}\left(\frac{\partial u^o}{\partial y} + \frac{\partial v^o}{\partial x}\right)$$

$$- D_{12}\frac{\partial^2 w^o}{\partial x^2} - D_{22}\frac{\partial^2 w^o}{\partial y^2} - 2D_{26}\frac{\partial^2 w^o}{\partial x \partial y}$$

$$M_{xy} = B_{16}\frac{\partial u^o}{\partial x} + B_{26}\frac{\partial v^o}{\partial y} + B_{66}\left(\frac{\partial u^o}{\partial y} + \frac{\partial v^o}{\partial x}\right)$$

$$- D_{16}\frac{\partial^2 w^o}{\partial x^2} - D_{26}\frac{\partial^2 w^o}{\partial y^2} - 2D_{66}\frac{\partial^2 w^o}{\partial x \partial y}$$

Substituting these expressions into the equilibrium equations, equation (13.15), leads to three partial differential equations which govern the displacement response of a fiber-reinforced laminated plate, namely,

$$A_{11}\frac{\partial^2 u^o}{\partial x^2} + 2A_{16}\frac{\partial^2 u^o}{\partial x \partial y} + A_{66}\frac{\partial^2 u^o}{\partial y^2} + A_{16}\frac{\partial^2 v^o}{\partial x^2}$$

$$+ (A_{12} + A_{66})\frac{\partial^2 v^o}{\partial x \partial y} + A_{26}\frac{\partial^2 v^o}{\partial y^2} - B_{11}\frac{\partial^3 w^o}{\partial x^3} \qquad (13.30a)$$

$$- 3B_{16}\frac{\partial^3 w^o}{\partial x^2 \partial y} - (B_{12} + 2B_{66})\frac{\partial^3 w^o}{\partial x \partial y^2} - B_{26}\frac{\partial^3 w^o}{\partial y^3} = 0$$

$$A_{16}\frac{\partial^2 u^o}{\partial x^2} + (A_{12} + A_{66})\frac{\partial^2 u^o}{\partial x \partial y} + A_{26}\frac{\partial^2 u^o}{\partial y^2} + A_{66}\frac{\partial^2 v^o}{\partial x^2}$$

$$+ 2A_{26}\frac{\partial^2 v^o}{\partial x \partial y} + A_{22}\frac{\partial^2 v^o}{\partial y^2} - B_{16}\frac{\partial^3 w^o}{\partial x^3} \qquad (13.30b)$$

$$- (B_{12} + 2B_{66})\frac{\partial^3 w^o}{\partial x^2 \partial y} - 3B_{26}\frac{\partial^3 w^o}{\partial x \partial y^2} - B_{22}\frac{\partial^3 w^o}{\partial y^3} = 0$$

$$D_{11}\frac{\partial^4 w^o}{\partial x^4} + 4D_{16}\frac{\partial^4 w^o}{\partial x^3 \partial y} + 2(D_{12} + 2D_{66})\frac{\partial^4 w^o}{\partial x^2 \partial y^2} + 4D_{26}\frac{\partial^4 w^o}{\partial x \partial y^3}$$

$$+ D_{22}\frac{\partial^4 w^o}{\partial y^4} - B_{11}\frac{\partial^3 u^o}{\partial x^3} - 3B_{16}\frac{\partial^3 u^o}{\partial x^2 \partial y}$$

$$- (B_{12} + 2B_{66})\frac{\partial^3 u^o}{\partial x \partial y^2} - B_{26}\frac{\partial^3 u^o}{\partial y^3} - B_{16}\frac{\partial^3 v^o}{\partial x^3} \qquad (13.30c)$$

$$- (B_{12} + 2B_{66})\frac{\partial^3 v^o}{\partial x^2 \partial y} - 3B_{26}\frac{\partial^3 v^o}{\partial x \partial y^2} - B_{22}\frac{\partial^3 v^o}{\partial y^3} = q$$

The four boundary conditions along each of the four edges can also be expressed in terms of the displacements. Substituting for the stress resultants N_x, N_{xy}, and so on, in terms of the displacements in equation (13.16), we find that the boundary conditions become:

Along edge $x = +a/2$:

(i) Either $\left\{ A_{11} \dfrac{\partial u^o}{\partial x} + A_{12} \dfrac{\partial v^o}{\partial y} + A_{16} \left(\dfrac{\partial u^o}{\partial y} + \dfrac{\partial v^o}{\partial x} \right) - B_{11} \dfrac{\partial^2 w^o}{\partial x^2} \right.$

$$\left. - B_{12} \dfrac{\partial^2 w^o}{\partial y^2} - 2 B_{16} \dfrac{\partial^2 w^o}{\partial x \partial y} \right\} = N_x^+$$

or u^o must be specified

(ii) Either $\left\{ A_{16} \dfrac{\partial u^o}{\partial x} + A_{26} \dfrac{\partial v^o}{\partial y} + A_{66} \left(\dfrac{\partial u^o}{\partial y} + \dfrac{\partial v^o}{\partial x} \right) - B_{16} \dfrac{\partial^2 w^o}{\partial x^2} \right.$

$$\left. - B_{26} \dfrac{\partial^2 w^o}{\partial y^2} - 2 B_{66} \dfrac{\partial^2 w^o}{\partial x \partial y} \right\} = N_{xy}^+$$

or v^o must be specified

(iii) Either $\left\{ B_{11} \dfrac{\partial^2 u^o}{\partial x^2} + 2 B_{66} \dfrac{\partial^2 u^o}{\partial y^2} + 3 B_{16} \dfrac{\partial^2 u^o}{\partial x \partial y} + B_{16} \dfrac{\partial^2 v^o}{\partial^2 x^2} \right.$

$$+ 2 B_{26} \dfrac{\partial^2 v^o}{\partial y^2} + (B_{12} + 2 B_{66}) \dfrac{\partial^2 v^o}{\partial x \partial y}$$

$$- D_{11} \dfrac{\partial^3 w^o}{\partial x^3} - 2 D_{26} \dfrac{\partial^3 w^o}{\partial y^3} - (D_{12} + 4 D_{66}) \dfrac{\partial^3 w^o}{\partial x \partial y^2}$$

$$\left. - 4 D_{16} \dfrac{\partial^3 w^o}{\partial x^2 \partial y} \right\} = Q_x^+ + \dfrac{\partial M_{xy}^+}{\partial y}$$

or w^o must be specified

(iv) Either $\left\{ B_{11} \dfrac{\partial u^o}{\partial x} + B_{12} \dfrac{\partial v^o}{\partial y} + B_{16} \left(\dfrac{\partial u^o}{\partial y} + \dfrac{\partial v^o}{\partial x} \right) - D_{11} \dfrac{\partial^2 w^o}{\partial x^2} \right.$

$$\left. - D_{12} \dfrac{\partial^2 w^o}{\partial y^2} - 2 D_{16} \dfrac{\partial^2 w^o}{\partial x \partial y} \right\} = M_x^+$$

or $\dfrac{\partial w^o}{\partial x}$ must be specified   (13.31a)

Along edge $x = -a/2$:

(i) Either $\left\{ A_{11} \dfrac{\partial u^o}{\partial x} + A_{12} \dfrac{\partial v^o}{\partial y} + A_{16} \left(\dfrac{\partial u^o}{\partial y} + \dfrac{\partial v^o}{\partial x} \right) - B_{11} \dfrac{\partial^2 w^o}{\partial x^2} \right.$

$$\left. - B_{12} \dfrac{\partial^2 w^o}{\partial y^2} - 2 B_{16} \dfrac{\partial^2 w^o}{\partial x \partial y} \right\} = N_x^-$$

or u^o must be specified

(ii)   Either $\left\{ A_{16}\dfrac{\partial u^o}{\partial x} + A_{26}\dfrac{\partial v^o}{\partial y} + A_{66}\left(\dfrac{\partial u^o}{\partial y} + \dfrac{\partial v^o}{\partial x} \right) - B_{16}\dfrac{\partial^2 w^o}{\partial x^2} \right.$

$$\left. - B_{26}\dfrac{\partial^2 w^o}{\partial y^2} - 2B_{66}\dfrac{\partial^2 w^o}{\partial x \partial y} \right\} = N_{xy}^-$$

or v^o must be specified

(iii)   Either $\left\{ B_{11}\dfrac{\partial^2 u^o}{\partial x^2} + 2B_{66}\dfrac{\partial^2 u^o}{\partial y^2} + 3B_{16}\dfrac{\partial^2 u^o}{\partial x \partial y} + B_{16}\dfrac{\partial^2 v^o}{\partial x^2} \right.$

$$+ 2B_{26}\dfrac{\partial^2 v^o}{\partial y^2} + (B_{12} + 2B_{66})\dfrac{\partial^2 v^o}{\partial x \partial y} - D_{11}\dfrac{\partial^3 w^o}{\partial x^3}$$

$$- 2D_{26}\dfrac{\partial^3 w^o}{\partial y^3} - (D_{12} + 4D_{66})\dfrac{\partial^3 w^o}{\partial x \partial y^2}$$

$$\left. - 4D_{16}\dfrac{\partial^3 w^o}{\partial x^2 \partial y} \right\} = Q_x^- + \dfrac{\partial M_{xy}^-}{\partial y}$$

or w^o must be specified

(iv)   Either $\left\{ B_{11}\dfrac{\partial u^o}{\partial x} + B_{12}\dfrac{\partial v^o}{\partial y} + B_{16}\left(\dfrac{\partial u^o}{\partial y} + \dfrac{\partial v^o}{\partial x} \right) - D_{11}\dfrac{\partial^2 w^o}{\partial x^2} \right.$

$$\left. - D_{12}\dfrac{\partial^2 w^o}{\partial y^2} - 2D_{16}\dfrac{\partial^2 w^o}{\partial x \partial y} \right\} = M_x^-$$

or $\dfrac{\partial w^o}{\partial x}$ must be specified   (13.31b)

Along edge $y = +b/2$:

(i)   Either $\left\{ A_{12}\dfrac{\partial u^o}{\partial x} + A_{22}\dfrac{\partial v^o}{\partial y} + A_{26}\left(\dfrac{\partial u^o}{\partial y} + \dfrac{\partial v^o}{\partial x} \right) - B_{12}\dfrac{\partial^2 w^o}{\partial x^2} \right.$

$$\left. - B_{22}\dfrac{\partial^2 w^o}{\partial y^2} - 2B_{26}\dfrac{\partial^2 w^o}{\partial x \partial y} \right\} = N_y^+$$

or v^o must be specified

(ii)   Either $\left\{ A_{16}\dfrac{\partial u^o}{\partial x} + A_{26}\dfrac{\partial v^o}{\partial y} + A_{66}\left(\dfrac{\partial u^o}{\partial y} + \dfrac{\partial v^o}{\partial x} \right) - B_{16}\dfrac{\partial^2 w^o}{\partial x^2} \right.$

$$\left. - B_{26}\dfrac{\partial^2 w^o}{\partial y^2} - 2B_{66}\dfrac{\partial^2 w^o}{\partial x \partial y} \right\} = N_{yx}^+$$

or u^o must be specified

(iii) Either $\left\{ 2B_{16}\dfrac{\partial^2 u^o}{\partial x^2} + B_{26}\dfrac{\partial^2 u^o}{\partial y^2} + (B_{12}+2B_{66})\dfrac{\partial^2 u^o}{\partial x \partial y} \right.$

$$+ 2B_{66}\frac{\partial^2 v^o}{\partial x^2} + B_{22}\frac{\partial^2 v^o}{\partial y^2} + 3B_{26}\frac{\partial^2 v^o}{\partial x \partial y}$$

$$- 2D_{16}\frac{\partial^3 w^o}{\partial x^3} - D_{22}\frac{\partial^3 w^o}{\partial y^3} - 4D_{26}\frac{\partial^3 w^o}{\partial x \partial y^2}$$

$$\left. -(D_{12}+4D_{66})\frac{\partial^3 w^o}{\partial x^2 \partial y} \right\} = Q_y^+ + \frac{\partial M_{yx}^+}{\partial x}$$

or w^o must be specified

(iv) Either $\left\{ B_{12}\dfrac{\partial u^o}{\partial x} + B_{22}\dfrac{\partial v^o}{\partial y} + B_{26}\left(\dfrac{\partial u^o}{\partial y} + \dfrac{\partial v^o}{\partial x}\right) - D_{12}\dfrac{\partial^2 w^o}{\partial x^2} \right.$

$$\left. - D_{22}\frac{\partial^2 w^o}{\partial y^2} - 2D_{26}\frac{\partial^2 w^o}{\partial x \partial y} \right\} = M_y^+$$

or $\dfrac{\partial w^o}{\partial y}$ must be specified $(13.31c)$

Along edge $y = -b/2$:

(i) Either $\left\{ A_{12}\dfrac{\partial u^o}{\partial x} + A_{22}\dfrac{\partial v^o}{\partial y} + A_{26}\left(\dfrac{\partial u^o}{\partial y} + \dfrac{\partial v^o}{\partial x}\right) - B_{12}\dfrac{\partial^2 w^o}{\partial x^2} \right.$

$$\left. - B_{22}\frac{\partial^2 w^o}{\partial y^2} - 2B_{26}\frac{\partial^2 w^o}{\partial x \partial y} \right\} = N_y^-$$

or v^o must be specified

(ii) Either $\left\{ A_{16}\dfrac{\partial u^o}{\partial x} + A_{26}\dfrac{\partial v^o}{\partial y} + A_{66}\left(\dfrac{\partial u^o}{\partial y} + \dfrac{\partial v^o}{\partial x}\right) - B_{16}\dfrac{\partial^2 w^o}{\partial x^2} \right.$

$$\left. - B_{26}\frac{\partial^2 w^o}{\partial y^2} - 2B_{66}\frac{\partial^2 w^o}{\partial x \partial y} \right\} = N_{yx}^-$$

or u^o must be specified

(iii) Either $\left\{ 2B_{16}\dfrac{\partial^2 u^o}{\partial x^2} + B_{26}\dfrac{\partial^2 u^o}{\partial y^2} + (B_{12}+2B_{66})\dfrac{\partial^2 u^o}{\partial x \partial y} \right.$

$$+ 2B_{66}\frac{\partial^2 v^o}{\partial x^2} + B_{22}\frac{\partial^2 v^o}{\partial y^2} + 3B_{26}\frac{\partial^2 v^o}{\partial x \partial y}$$

$$-2D_{16}\frac{\partial^3 w^o}{\partial x^3} - D_{22}\frac{\partial^3 w^o}{\partial y^3} - 4D_{26}\frac{\partial^3 w^o}{\partial x \partial y^2}$$

$$\left. - (D_{12} + 4D_{66})\,\frac{\partial^3 w^o}{\partial x^2 \partial y} \right\} = Q_y^- + \frac{\partial M_{yx}^-}{\partial x}$$

or w^o must be specified

(iv) Either $\left\{ B_{12}\frac{\partial u^o}{\partial x} + B_{22}\frac{\partial v^o}{\partial y} + B_{26}\left(\frac{\partial u^o}{\partial y} + \frac{\partial v^o}{\partial x}\right) - D_{12}\frac{\partial^2 w^o}{\partial x^2} \right.$

$$\left. - D_{22}\,\frac{\partial^2 w^o}{\partial y^2} - 2D_{26}\frac{\partial^2 w^o}{\partial x \partial y} \right\} = M_y^-$$

or $\dfrac{\partial w^o}{\partial y}$ must be specified   (13.31d)

The rather awesome display of algebra in equations (13.30) and (13.31) is the proper formulation of the governing equilibrium equations and boundary conditions for the problem of determining the displacement response of a rectangular laminated fiber-reinforced plate subjected to loads on the plate edges, and/or normal loads acting on the top and/or bottom surfaces. All laminated plate problems that can be solved within the context of the assumptions of classical lamination theory are contained in the above formulation. Conversely, only within the above formulation can one solve for the response of a laminated plate that obeys the assumptions of classical lamination theory.

In spite of the overwhelming amount of algebra, several features of the response of laminated plates are evident in the above. First, each equilibrium equation, equations (13.30a), (13.30b), and (13.30c), involves all three components of displacement. This means that each component of displacement is coupled with the other two components of displacement. For a general unsymmetric laminate it is not possible to solve for the out-of-plane component of displacement, for example, without solving for the two inplane components. Second, the boundary conditions can involve all three components of displacement. This provides another coupling mechanism for the three components of displacement. Finally, the manner in which the material and geometric properties of the plate enter the problem, and hence influence the response, is seen explicitly. The elastic properties, thickness, location through the thickness, and fiber orientation of each layer influence the components of the A, B, and D matrices. Furthermore, because in equations (13.31a)–(d) the values of u^o, v^o, w^o and their derivatives are evaluated on the boundaries $x = \pm a/2$ and $y = \pm b/2$, the length and width of the plate enter the problem by way of the boundary conditions.

Operationally, the solutions for the displacements u^o, v^o, and w^o are determined from the governing set of partial differential equations. These solutions will be in terms of unknown constants of integration and known functions of x and y. The constants and functions resulting from the integration are determined by the application of the boundary conditions. For example, if the displacement portion of boundary condition (iii) of equation (13.31a) is being enforced because of a simple support condition, then setting $w^o(+a/2, y)$ to zero results in one equation from which to evaluate the constants of integration. If, on the other hand, the force portion of that boundary condition is being specified because of a free-edge condition, then the combination of derivatives of displacements in the bracketed term of condition (*iii*) of equation (13.31*a*) is set to zero to provide one equation from which to evaluate the unknown constants. Conditions exist on all four edges and when the problem is done properly, there are enough boundary conditions from which to solve for all of the unknown constants and functions. When one attempts to make up boundary conditions other than those specified by equations (13.31*a*)–(*d*), there may not be enough conditions from which to solve for all of the unknown constants, or there may be too may conditions. Not having exactly the same number of conditions and unknown constants of integration is an indication one may be trying to solve a physically impossible problem. Such a problem can be generated as a result of applying boundary conditions other than those allowed by the theory.

   In the next section we shall discuss simplifications to the above equations which result from considering special plates (e.g., symmetrically laminated plates). Though both the governing differential equations and the boundary conditions will simplify, the basic concepts of having three governing differential equations for the three components of displacement, and four boundary conditions along each of the four edges, still are valid. In the sections following the simplifications, we will solve specific problems and obtain numerical results. These problems will illustrate the steps necessary for solving the governing equations and enforcing boundary conditions, and will illustrate some important characteristics of the response of fiber-reinforced plates.

13.3
SIMPLIFICATIONS TO THE GOVERNING EQUATIONS

As we showed in Chapter 7, for specific classes of laminates some of the elastic coupling coefficients in the A, B, and D matrices vanish. The vanishing of these elastic coefficients can considerably simplify the differential equations and boundary conditions governing plate response. Some of the simplifications are minor, but some are quite significant. Three major simplifications are discussed next.

13.3.1 Symmetric Laminates

By far, the most dramatic simplification to the governing equations and boundary conditions for a plate occur when the plate is symmetrically laminated. For this situation, all B_{ij} terms are zero, and the equilibrium equations become

$$A_{11}\frac{\partial^2 u^o}{\partial x^2} + 2A_{16}\frac{\partial^2 u^o}{\partial x \partial y} + A_{66}\frac{\partial^2 u^o}{\partial y^2} + A_{16}\frac{\partial^2 v^o}{\partial x^2}$$

$$+ (A_{12} + A_{66})\frac{\partial^2 v^o}{\partial x \partial y} + A_{26}\frac{\partial^2 v^o}{\partial y^2} = 0$$

(13.32*a*)

$$A_{16}\frac{\partial^2 u^o}{\partial x^2} + (A_{12} + A_{66})\frac{\partial^2 u^o}{\partial x \partial y} + A_{26}\frac{\partial^2 u^o}{\partial y^2} + A_{66}\frac{\partial^2 v^o}{\partial x^2}$$

$$+ 2A_{26}\frac{\partial^2 v^o}{\partial x \partial y} + A_{22}\frac{\partial^2 v^o}{\partial y^2} = 0$$

(13.32b)

$$D_{11}\frac{\partial^4 w^o}{\partial x^4} + 4D_{16}\frac{\partial^4 w^o}{\partial x^3 \partial y} + 2(D_{12} + 2D_{66})\frac{\partial^4 w^o}{\partial x^2 \partial y^2}$$

$$+ 4D_{26}\frac{\partial^4 w^o}{\partial x \partial y^3} + D_{22}\frac{\partial^4 w^o}{\partial y^4} = q$$

(13.32c)

The boundary conditions reduce to:
  Along edge $x = +a/2$:

(i)   Either $\left\{ A_{11}\dfrac{\partial u^o}{\partial x} + A_{12}\dfrac{\partial v^o}{\partial y} + A_{16}\left(\dfrac{\partial u^o}{\partial y} + \dfrac{\partial v^o}{\partial x} \right) \right\} = N_x^+$

  or u^o must be specified

(ii)   Either $\left\{ A_{16}\dfrac{\partial u^o}{\partial x} + A_{26}\dfrac{\partial v^o}{\partial y} + A_{66}\left(\dfrac{\partial u^o}{\partial y} + \dfrac{\partial v^o}{\partial x} \right) \right\} = N_{xy}^+$

  or v^o must be specified

(iii)   Either $\left\{ -D_{11}\dfrac{\partial^3 w^o}{\partial x^3} - 2D_{26}\dfrac{\partial^3 w^o}{\partial y^3} - (D_{12} + 4D_{66})\dfrac{\partial^3 w^o}{\partial x \partial y^2} \right.$

$$\left. - 4D_{16}\frac{\partial^3 w^o}{\partial x^2 \partial y} \right\} = Q_x^+ + \frac{\partial M_{xy}^+}{\partial y}$$

  or w^o must be specified

(iv)   Either $\left\{ -D_{11}\dfrac{\partial^2 w^o}{\partial x^2} - D_{12}\dfrac{\partial^2 w^o}{\partial y^2} - 2D_{16}\dfrac{\partial^2 w^o}{\partial x \partial y} \right\} = M_x^+$

  or $\dfrac{\partial w^o}{\partial x}$ must be specified   (13.33a)

Along edge $x = -a/2$:

(i)   Either $\left\{ A_{11}\dfrac{\partial u^o}{\partial x} + A_{12}\dfrac{\partial v^o}{\partial y} + A_{16}\left(\dfrac{\partial u^o}{\partial y} + \dfrac{\partial v^o}{\partial x} \right) \right\} = N_x^-$

  or u^o must be specified

(ii)   Either $\left\{ A_{16}\dfrac{\partial u^o}{\partial x} + A_{26}\dfrac{\partial v^o}{\partial y} + A_{66}\left(\dfrac{\partial u^o}{\partial y} + \dfrac{\partial v^o}{\partial x} \right) \right\} = N_{xy}^-$

  or v^o must be specified

(*iii*)   Either $\left\{-D_{11}\dfrac{\partial^3 w^o}{\partial x^3} - 2D_{26}\dfrac{\partial^3 w^o}{\partial y^3} - (D_{12} + 4D_{66})\dfrac{\partial^3 w^o}{\partial x\partial y^2}\right.$

$$\left. - 4D_{16}\dfrac{\partial^3 w^o}{\partial x^2\partial y}\right\} = Q_x^- + \dfrac{\partial M_{xy}^-}{\partial y}$$

or w^o must be specified

(*iv*)   Either $\left\{-D_{11}\dfrac{\partial^2 w^o}{\partial x^2} - D_{12}\dfrac{\partial^2 w^o}{\partial y^2} - 2D_{16}\dfrac{\partial^2 w^o}{\partial x\partial y}\right\} = M_x^-$

or $\dfrac{\partial w^o}{\partial x}$ must be specified $\hspace{4cm}$ (13.33*b*)

Along edge $y = +b/2$:

(*i*)   Either $\left\{A_{12}\dfrac{\partial u^o}{\partial x} + A_{22}\dfrac{\partial v^o}{\partial y} + A_{26}\left(\dfrac{\partial u^o}{\partial y} + \dfrac{\partial v^o}{\partial x}\right)\right\} = N_y^+$

or v^o must be specified

(*ii*)   Either $\left\{A_{16}\dfrac{\partial u^o}{\partial x} + A_{26}\dfrac{\partial v^o}{\partial y} + A_{66}\left(\dfrac{\partial u^o}{\partial y} + \dfrac{\partial v^o}{\partial x}\right)\right\} = N_{yx}^+$

or u^o must be specified

(*iii*)   Either $\left\{-2D_{16}\dfrac{\partial^3 w^o}{\partial x^3} - D_{22}\dfrac{\partial^3 w^o}{\partial y^3} - 4D_{26}\dfrac{\partial^3 w^o}{\partial x\partial y^2}\right.$

$$\left. - (D_{12} + 4D_{66})\dfrac{\partial^3 w^o}{\partial x^2\partial y}\right\} = Q_y^+ + \dfrac{\partial M_{yx}^+}{\partial x}$$

or w^o must be specified

(*iv*)   Either $\left\{-D_{12}\dfrac{\partial^2 w^o}{\partial x^2} - D_{22}\dfrac{\partial^2 w^o}{\partial y^2} - 2D_{26}\dfrac{\partial^2 w^o}{\partial x\partial y}\right\} = M_y^+$

or $\dfrac{\partial w^o}{\partial y}$ must be specified $\hspace{4cm}$ (13.33*c*)

Along edge $y = -b/2$:

(*i*)   Either $\left\{A_{12}\dfrac{\partial u^o}{\partial x} + A_{22}\dfrac{\partial v^o}{\partial y} + A_{26}\left(\dfrac{\partial u^o}{\partial y} + \dfrac{\partial v^o}{\partial x}\right)\right\} = N_y^-$

or v^o must be specified

(*ii*)   Either $\left\{A_{16}\dfrac{\partial u^o}{\partial x} + A_{26}\dfrac{\partial v^o}{\partial y} + A_{66}\left(\dfrac{\partial u^o}{\partial y} + \dfrac{\partial v^o}{\partial x}\right)\right\} = N_{yx}^-$

or u^o must be specified

(iii) Either $\left\{-2D_{16}\dfrac{\partial^3 w^o}{\partial x^3} - D_{22}\dfrac{\partial^3 w^o}{\partial y^3} - 4D_{26}\dfrac{\partial^3 w^o}{\partial x \partial y^2}\right.$

$$\left. - (D_{12} + 4D_{66})\dfrac{\partial^3 w^o}{\partial x^2 \partial y}\right\} = Q_y^- + \dfrac{\partial M_{yx}^-}{\partial x}$$

or w^o must be specified

(iv) Either $\left\{-D_{12}\dfrac{\partial^2 w^o}{\partial x^2} - D_{22}\dfrac{\partial^2 w^o}{\partial y^2} - 2D_{26}\dfrac{\partial^2 w^o}{\partial x \partial y}\right\} = M_y^-$

or $\dfrac{\partial w^o}{\partial y}$ must be specified   (13.33d)

These simplifications are quite far-reaching. Note that in the governing equations and in the boundary conditions, the equations governing the inplane displacement components u^o and v^o separate from the equations governing the out-of-plane component w^o. The first two governing equilibrium equations, equations (13.32a) and (b), and boundary conditions (i) and (ii) in equations (13.33a)–(d) separate into a problem independent of the third equilibrium equation and boundary conditions (iii) and (iv). As a result, for a symmetrically laminated plate, the determination of the out-of-plane response, and the strains and stresses that accompany it, is totally independent of the determination of the inplane response. This is a major reduction in the level of effort if the out-of-plane response is the primary concern.

13.3.2 Symmetric Balanced Laminates

If, in addition to being symmetric, the laminate is also balanced, then both A_{16} and A_{26}, in addition to B_{ij}, are zero and the equations simplify further to become

$$A_{11}\frac{\partial^2 u^o}{\partial x^2} + A_{66}\frac{\partial^2 u^o}{\partial y^2} + (A_{12} + A_{66})\frac{\partial^2 v^o}{\partial x \partial y} = 0 \qquad (13.34a)$$

$$(A_{12} + A_{66})\frac{\partial^2 u^o}{\partial x \partial y} + A_{66}\frac{\partial^2 v^o}{\partial x^2} + A_{22}\frac{\partial^2 v^o}{\partial y^2} = 0 \qquad (13.34b)$$

$$D_{11}\frac{\partial^4 w^o}{\partial x^4} + 4D_{16}\frac{\partial^4 w^o}{\partial x^3 \partial y} + 2(D_{12} + 2D_{66})\frac{\partial^4 w^o}{\partial x^2 \partial y^2}$$

$$+ 4D_{26}\frac{\partial^4 w^o}{\partial x \partial y^3} + D_{22}\frac{\partial^4 w^o}{\partial y^4} = q \qquad (13.34c)$$

The boundary conditions reduce to:
Along edge $x = +a/2$:

(i) Either $\left\{A_{11}\dfrac{\partial u^o}{\partial x} + A_{12}\dfrac{\partial v^o}{\partial y}\right\} = N_x^+$

or u^o must be specified

(ii) Either $\left\{ A_{66} \left(\dfrac{\partial u^o}{\partial y} + \dfrac{\partial v^o}{\partial x} \right) \right\} = N_{xy}^+$

or v^o must be specified

(iii) Either $\left\{ -D_{11} \dfrac{\partial^3 w^o}{\partial x^3} - 2D_{26} \dfrac{\partial^3 w^o}{\partial y^3} - (D_{12} + 4D_{66}) \dfrac{\partial^3 w^o}{\partial x \partial y^2} \right.$

$$\left. -4D_{16} \dfrac{\partial^3 w^o}{\partial x^2 \partial y} \right\} = Q_x^+ + \dfrac{\partial M_{xy}^+}{\partial y}$$

or w^o must be specified

(iv) Either $\left\{ -D_{11} \dfrac{\partial^2 w^o}{\partial x^2} - D_{12} \dfrac{\partial^2 w^o}{\partial y^2} - 2D_{16} \dfrac{\partial^2 w^o}{\partial x \partial y} \right\} = M_x^+$

or $\dfrac{\partial w^o}{\partial x}$ must be specified   (13.35a)

Along edge $x = -a/2$:

(i) Either $\left\{ A_{11} \dfrac{\partial u^o}{\partial x} + A_{12} \dfrac{\partial v^o}{\partial y} \right\} = N_x^-$

or u^o must be specified

(ii) Either $\left\{ A_{66} \left(\dfrac{\partial u^o}{\partial y} + \dfrac{\partial v^o}{\partial x} \right) \right\} = N_{xy}^-$

or v^o must be specified

(iii) Either $\left\{ -D_{11} \dfrac{\partial^3 w^o}{\partial x^3} - 2D_{26} \dfrac{\partial^3 w^o}{\partial y^3} - (D_{12} + 4D_{66}) \dfrac{\partial^3 w^o}{\partial x \partial y^2} \right.$

$$\left. - 4D_{16} \dfrac{\partial^3 w^o}{\partial x^2 \partial y} \right\} = Q_x^- + \dfrac{\partial M_{xy}^-}{\partial y}$$

or w^o must be specified

(iv) Either $\left\{ -D_{11} \dfrac{\partial^2 w^o}{\partial x^2} - D_{12} \dfrac{\partial^2 w^o}{\partial y^2} - 2D_{16} \dfrac{\partial^2 w^o}{\partial x \partial y} \right\} = M_x^-$

or $\dfrac{\partial w^o}{\partial x}$ must be specified   (13.35b)

Along edge $y = +b/2$:

(i) Either $\left\{ A_{12} \dfrac{\partial u^o}{\partial x} + A_{22} \dfrac{\partial v^o}{\partial y} \right\} = N_y^+$

or v^o must be specified

(ii) Either $\left\{ A_{66} \left(\dfrac{\partial u^o}{\partial y} + \dfrac{\partial v^o}{\partial x} \right) \right\} = N_{yx}^+$

or u^o must be specified

(iii)   Either $\left\{ -2D_{16}\dfrac{\partial^3 w^o}{\partial x^3} - D_{22}\dfrac{\partial^3 w^o}{\partial y^3} - 4D_{26}\dfrac{\partial^3 w^o}{\partial x \partial y^2} \right.$

$$\left. - (D_{12} + 4D_{66})\dfrac{\partial^3 w^o}{\partial x^2 \partial y} \right\} = Q_y^+ + \dfrac{\partial M_{yx}^+}{\partial x}$$

or w^o must be specified

(iv)   Either $\left\{ -D_{12}\dfrac{\partial^2 w^o}{\partial x^2} - D_{22}\dfrac{\partial^2 w^o}{\partial y^2} - 2D_{26}\dfrac{\partial^2 w^o}{\partial x \partial y} \right\} = M_y^+$

or $\dfrac{\partial w^o}{\partial y}$ must be specified   (13.35c)

Along edge $y = -b/2$:

(i)   Either $\left\{ A_{12}\dfrac{\partial u^o}{\partial x} + A_{22}\dfrac{\partial v^o}{\partial y} \right\} = N_y^-$

or v^o must be specified

(ii)   Either $\left\{ A_{66}\left(\dfrac{\partial u^o}{\partial y} + \dfrac{\partial v^o}{\partial x} \right) \right\} = N_{yx}^-$

or u^o must be specified

(iii)   Either $\left\{ -2D_{16}\dfrac{\partial^3 w^o}{\partial x^3} - D_{22}\dfrac{\partial^3 w^o}{\partial y^3} - 4D_{26}\dfrac{\partial^3 w^o}{\partial x \partial y^2} \right.$

$$\left. - (D_{12} + 4D_{66})\dfrac{\partial^3 w^o}{\partial x^2 \partial y} \right\} = Q_y^- + \dfrac{\partial M_{yx}^-}{\partial x}$$

or w^o must be specified

(iv)   Either $\left\{ -D_{12}\dfrac{\partial^2 w^o}{\partial x^2} - D_{22}\dfrac{\partial^2 w^o}{\partial y^2} - 2D_{26}\dfrac{\partial^2 w^o}{\partial x \partial y} \right\} = M_y^-$

or $\dfrac{\partial w^o}{\partial y}$ must be specified   (13.35d)

The simplifications for this situation influence only the inplane portion of the problem. The inplane portion of the problem is strictly orthotropic, and, as will be seen, not too different from the form for isotropic materials.

13.3.3 Symmetric Cross-Ply Laminates

Finally, if the plate is a symmetric cross-ply lamination, then in addition to having zero values for A_{16}, A_{26}, and B_{ij}, D_{16} and D_{26} are also zero, and the equilibrium equations reduce to

$$A_{11}\dfrac{\partial^2 u^o}{\partial x^2} + A_{66}\dfrac{\partial^2 u^o}{\partial y^2} + (A_{12} + A_{66})\dfrac{\partial^2 v^o}{\partial x \partial y} = 0 \qquad (13.36a)$$

$$(A_{12} + A_{66})\frac{\partial^2 u^o}{\partial x \partial y} + A_{66}\frac{\partial^2 v^o}{\partial x^2} + A_{22}\frac{\partial^2 v^o}{\partial y^2} = 0 \qquad (13.36b)$$

$$D_{11}\frac{\partial^4 w^o}{\partial x^4} + 2(D_{12} + 2D_{66})\frac{\partial^4 w^o}{\partial x^2 \partial y^2} + D_{22}\frac{\partial^4 w^o}{\partial y^4} = q \qquad (13.36c)$$

The boundary conditions reduce to:
  Along edge $x = +a/2$:

(i) Either $\left\{ A_{11}\dfrac{\partial u^o}{\partial x} + A_{12}\dfrac{\partial v^o}{\partial y} \right\} = N_x^+$

or u^o must be specified

(ii) Either $\left\{ A_{66}\left(\dfrac{\partial u^o}{\partial y} + \dfrac{\partial v^o}{\partial x}\right) \right\} = N_{xy}^+$

or v^o must be specified

(iii) Either $\left\{ -D_{11}\dfrac{\partial^3 w^o}{\partial x^3} - (D_{12} + 4D_{66})\dfrac{\partial^3 w^o}{\partial x \partial y^2} \right\} = Q_x^+ + \dfrac{\partial M_{xy}^+}{\partial y}$

or w^o must be specified

(iv) Either $\left\{ -D_{11}\dfrac{\partial^2 w^o}{\partial x^2} - D_{12}\dfrac{\partial^2 w^o}{\partial y^2} \right\} = M_x^+$

or $\dfrac{\partial w^o}{\partial x}$ must be specified $\qquad (13.37a)$

Along edge $x = -a/2$:

(i) Either $\left\{ A_{11}\dfrac{\partial u^o}{\partial x} + A_{12}\dfrac{\partial v^o}{\partial y} \right\} = N_x^-$

or u^o must be specified

(ii) Either $\left\{ A_{66}\left(\dfrac{\partial u^o}{\partial y} + \dfrac{\partial v^o}{\partial x}\right) \right\} = N_{xy}^-$

or v^o must be specified

(iii) Either $\left\{ -D_{11}\dfrac{\partial^3 w^o}{\partial x^3} - (D_{12} + 4D_{66})\dfrac{\partial^3 w^o}{\partial x \partial y^2} \right\} = Q_x^- + \dfrac{\partial M_{xy}^-}{\partial y}$

or w^o must be specified

(iv) Either $\left\{ -D_{11}\dfrac{\partial^2 w^o}{\partial x^2} - D_{12}\dfrac{\partial^2 w^o}{\partial y^2} \right\} = M_x^-$

or $\dfrac{\partial w^o}{\partial x}$ must be specified $\qquad (13.37b)$

Along edge $y = +b/2$:

(i)   Either $\left\{ A_{12}\dfrac{\partial u^o}{\partial x} + A_{22}\dfrac{\partial v^o}{\partial y} \right\} = N_y^+$

or v^o must be specified

(ii)   Either $\left\{ A_{66}\left(\dfrac{\partial u^o}{\partial y} + \dfrac{\partial v^o}{\partial x} \right) \right\} = N_{yx}^+$

or u^o must be specified

(iii) Either $\left\{ -D_{22}\dfrac{\partial^3 w^o}{\partial y^3} - (D_{12} + 4D_{66})\dfrac{\partial^3 w^o}{\partial x^2 \partial y} \right\} = Q_y^+ + \dfrac{\partial M_{yx}^+}{\partial x}$

or w^o must be specified

(iv)   Either $\left\{ -D_{12}\dfrac{\partial^2 w^o}{\partial x^2} - D_{22}\dfrac{\partial^2 w^o}{\partial y^2} \right\} = M_y^+$

or $\dfrac{\partial w^o}{\partial y}$ must be specified   (13.37c)

Along edge $y = -b/2$:

(i)   Either $\left\{ A_{12}\dfrac{\partial u^o}{\partial x} + A_{22}\dfrac{\partial v^o}{\partial y} \right\} = N_y^-$

or v^o must be specified

(ii)   Either $\left\{ A_{66}\left(\dfrac{\partial u^o}{\partial y} + \dfrac{\partial v^o}{\partial x} \right) \right\} = N_{yx}^-$

or u^o must be specified

(iii) Either $\left\{ -D_{22}\dfrac{\partial^3 w^o}{\partial y^3} - (D_{12} + 4D_{66})\dfrac{\partial^3 w^o}{\partial x^2 \partial y} \right\} = Q_y^- + \dfrac{\partial M_{yx}^-}{\partial x}$

or w^o must be specified

(iv)   Either $\left\{ -D_{12}\dfrac{\partial^2 w^o}{\partial x^2} - D_{22}\dfrac{\partial^2 w^o}{\partial y^2} \right\} = M_y^-$

or $\dfrac{\partial w^o}{\partial y}$ must be specified   (13.37d)

In this case the simplifications occur to the out-of-plane portion of the problem. The vanishing of D_{16} and D_{26} eliminates odd derivatives with respect to x and y in the third equilibrium equation. This makes it easier to find exact solutions to the third equilibrium equation for certain situations. Vanishing of D_{16} and D_{26} also eliminates the twisting curvature κ_{xy}^o $(= -2\partial^2 w^o/\partial x \partial y)$ from the moment boundary condition, condition (iv) in equation (13.37).

13.3.4 Isotropic Plates

For completeness, the degenerate case of an isotropic plate is presented. Recall from Chapter 7, equation (7.111), that for a single isotropic layer of thickness H, Young's modulus E, and Poisson's ratio v,

$$A_{11} = A_{22} = \frac{EH}{1 - v^2} = A \qquad A_{12} = v\frac{EH}{1 - v^2} = vA$$

$$A_{66} = \frac{EH}{2(1 + v)} = \frac{1 - v}{2}A \qquad A_{16} = A_{26} = 0$$

$$D_{11} = D_{22} = \frac{EH^3}{12(1 - v^2)} = D \qquad D_{12} = v\frac{EH^3}{12(1 - v^2)} = vD$$

$$D_{66} = \frac{EH^3}{24(1 + v)} = \frac{1 - v}{2}D \qquad D_{16} = D_{26} = 0$$

(13.38)

Thus, the equilibrium equations reduce to

$$\frac{\partial^2 u^o}{\partial x^2} + \left(\frac{1 - v}{2}\right)\frac{\partial^2 u^o}{\partial y^2} + \left(\frac{1 + v}{2}\right)\frac{\partial^2 v^o}{\partial x \partial y} = 0 \qquad (13.39a)$$

$$\left(\frac{1 + v}{2}\right)\frac{\partial^2 u^o}{\partial x \partial y} + \left(\frac{1 - v}{2}\right)\frac{\partial^2 v^o}{\partial x^2} + \frac{\partial^2 v^o}{\partial y^2} = 0 \qquad (13.39b)$$

$$\frac{\partial^4 w^o}{\partial x^4} + 2\frac{\partial^4 w^o}{\partial x^2 \partial y^2} + \frac{\partial^4 w^o}{\partial y^4} = \frac{q}{D} \qquad (13.39c)$$

The boundary conditions reduce to:
 Along edge $x = +a/2$:

(i) Either $\left\{A\left(\dfrac{\partial u^o}{\partial x} + v\dfrac{\partial v^o}{\partial y}\right)\right\} = N_x^+$

 or u^o must be specified

(ii) Either $\left\{\dfrac{1 - v}{2}A\left(\dfrac{\partial u^o}{\partial y} + \dfrac{\partial v^o}{\partial x}\right)\right\} = N_{xy}^+$

 or v^o must be specified

(iii) Either $\left\{-D\left(\dfrac{\partial^3 w^o}{\partial x^3} + (2 - v)\dfrac{\partial^3 w^o}{\partial x \partial y^2}\right)\right\} = Q_x^+ + \dfrac{\partial M_{xy}^+}{\partial y}$

 or w^o must be specified

(iv) Either $\left\{-D\left(\dfrac{\partial^2 w^o}{\partial x^2} + v\dfrac{\partial^2 w^o}{\partial y^2}\right)\right\} = M_x^+$

 or $\dfrac{\partial w^o}{\partial x}$ must be specified

(13.40a)

Along edge $x = -a/2$:

(i)   Either $\left\{ A\left(\dfrac{\partial u^o}{\partial x} + v\dfrac{\partial v^o}{\partial y} \right) \right\} = N_x^-$

or u^o must be specified

(ii)   Either $\left\{ \dfrac{1-v}{2} A\left(\dfrac{\partial u^o}{\partial y} + \dfrac{\partial v^o}{\partial x} \right) \right\} = N_{xy}^-$

or v^o must be specified

(iii)   Either $\left\{ -D\left(\dfrac{\partial^3 w^o}{\partial x^3} + (2-v)\dfrac{\partial^3 w^o}{\partial x \partial y^2} \right) \right\} = Q_x^- + \dfrac{\partial M_{xy}^-}{\partial y}$

or w^o must be specified

(iv)   Either $\left\{ -D\left(\dfrac{\partial^2 w^o}{\partial x^2} + v\dfrac{\partial^2 w^o}{\partial y^2} \right) \right\} = M_x^-$

or $\dfrac{\partial w^o}{\partial x}$ must be specified   (13.40b)

Along edge $y = +b/2$:

(i)   Either $\left\{ A\left(v\dfrac{\partial u^o}{\partial x} + \dfrac{\partial v^o}{\partial y} \right) \right\} = N_y^+$

or v^o must be specified

(ii)   Either $\left\{ \dfrac{1-v}{2}\left(\dfrac{\partial u^o}{\partial y} + \dfrac{\partial v^o}{\partial x} \right) \right\} = N_{yx}^+$

or u^o must be specified

(iii)   Either $\left\{ -D\left(\dfrac{\partial^3 w^o}{\partial y^3} + (2-v)\dfrac{\partial^3 w^o}{\partial x^2 \partial y} \right) \right\} = Q_y^+ + \dfrac{\partial M_{yx}^+}{\partial x}$

or w^o must be specified

(iv)   Either $\left\{ -D\left(v\dfrac{\partial^2 w^o}{\partial x^2} + \dfrac{\partial^2 w^o}{\partial y^2} \right) \right\} = M_y^+$

or $\dfrac{\partial w^o}{\partial y}$ must be specified   (13.40c)

Along edge $y = -b/2$:

(i)   Either $\left\{ A\left(v\dfrac{\partial u^o}{\partial x} + \dfrac{\partial v^o}{\partial y} \right) \right\} = N_y^-$

or v^o must be specified

(ii)   Either $\left\{ \dfrac{1-v}{2} A\left(\dfrac{\partial u^o}{\partial y} + \dfrac{\partial v^o}{\partial x} \right) \right\} = N_{yx}^-$

or u^o must be specified

(iii) Either $\left\{ -D\left(\dfrac{\partial^3 w^o}{\partial y^3} + (2 - v)\dfrac{\partial^3 w^o}{\partial x^2 \partial y} \right) \right\} = Q_y^- + \dfrac{\partial M_{yx}^-}{\partial x}$

or w^o must be specified

(iv) Either $\left\{ -D\left(v\dfrac{\partial^2 w^o}{\partial x^2} + \dfrac{\partial^2 w^o}{\partial y^2} \right) \right\} = M_y^-$

or $\dfrac{\partial w^o}{\partial y}$ must be specified $\hspace{2cm}$ (13.40d)

The equations governing the response of isotropic plates are much simpler than the equations for the most general laminate or even a symmetric laminate. In fact, only the symmetric cross-ply plate has as simple a form as the isotropic case. For composite plates there are many situations where A_{16} and A_{26} are zero. There are very few situations where D_{16} and D_{26} are actually zero. However, assuming D_{16} and D_{26} are negligible and thus setting them to zero simplifies the equations for symmetric balanced laminates to the level of the cross-ply case. For this reason, often D_{16} and D_{26} are equated to zero even though they are not. Errors can be encountered in interpreting the results for these situations.

We will now focus on examples of the application of the plate equations to some very special cases. These cases are chosen as much to illustrate the steps necessary to obtain answers from the governing equations and boundary conditions as they are to illustrate some of the characteristics of the response of composite plates. For many cases of interest, it is not possible to find solutions to the governing equations. Thus, approximate methods must be employed that do not satisfy the governing equations or the boundary conditions. If one continually works with the approximate solutions, it is easy to loose track of the exact conditions required by the equations being discussed here.

13.4
PLATE EXAMPLE 1: A LONG $[0/90]_S$ PLATE

We will consider several plate examples. The first consists of a symmetrically laminated $[0/90]_S$ plate that is infinitely long in the y direction, loaded by a uniform load on its upper surface, and simply supported at $x = \pm a/2$. Because the plate is infinitely long in the y direction, the formulation is essentially one-dimensional.

The second example considers the same problem except that the laminate will be unsymmetric with a stacking sequence of $[0_2/90_2]_T$. We will study the coupling of the inplane and out-of-plane responses due to boundary conditions.

The third example focuses on an infinitely long uniformly loaded $[\pm 30/0]_S$ plate simply supported at $x = \pm a/2$. The influence of D_{16} is illustrated by this example.

The fourth example considers a simply supported, uniformly loaded rectangular plate. This example illustrates the effects at a finite geometry in both the x and y directions.

Let us consider a four-layer $[0/90]_S$ plate graphite-epoxy that has width a in the x direction and is infinitely long in the y direction. Because it is long, we can assume that the plate response is independent of the y coordinate. Also, for simplicity, assume the displacement in the y direction is restrained to be zero. Forces will be required to effect this restraint and they will be computed. Let us assume the plate is loaded by a uniformly distributed load of the form

$$q(x, y) = q_o \tag{13.41}$$

where q_o is a known constant. The situation is depicted in Figure 13.13, and interest will focus on the deflections as a function of x, and the stresses as a function of x and z. Using the boundary condition equations, we shall see what boundary conditions are allowed, and enforce conditions that correspond to simple supports.

13.4.1 Solution of Governing Differential Equations

Because the plate is constructed of a symmetric cross-ply laminate

$$A_{16} = A_{26} = D_{16} = D_{26} = 0 \tag{13.42a}$$

and $\qquad\qquad B_{ij} = 0 \qquad$ all i and j $\qquad\qquad$ (13.42b)

As a result of these material properties being zero, because the response is considered to be independent of y, and v^o is taken to be zero, the force and moment resultants, equation (13.29), reduce to

$$N_x = A_{11}\frac{du^o}{dx} \qquad N_y = A_{12}\frac{du^o}{dx} \qquad N_{xy} = 0$$

$$\tag{13.43}$$

$$M_x = -D_{11}\frac{d^2w^o}{dx^2} \qquad M_y = -D_{12}\frac{d^2w^o}{dx^2} \qquad M_{xy} = 0$$

Note that because the plate response is independent of y and v^o is taken to be zero,

$$\frac{\partial(\text{any quantity})}{\partial y} = 0 \tag{13.44}$$

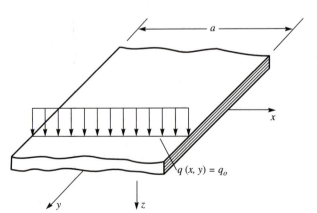

FIGURE 13.13
Semi-infinite plate loaded with a uniformly distributed load

$q(x, y) = q_o$

and the partial derivative on x has been replaced with the total derivative. Note also that there are force and moment resultants in the y direction even though the response is independent of y. These are the resultants which are required to have v^o be zero and w^o be independent of y.

The equilibrium equations in terms of force resultants, equation (13.15), reduce to

$$\frac{dN_x}{dx} = 0 \tag{13.45a}$$

$$0 = 0 \tag{13.45b}$$

$$\frac{d^2 M_x}{dx^2} + q_o = 0 \tag{13.45c}$$

The two nontrivial equations can be integrated directly, resulting in

$$N_x(x) = C_1 \tag{13.46a}$$

$$M_x(x) = -q_o \frac{x^2}{2} + C_2 x + C_3 \tag{13.46b}$$

where C_1, C_2, and C_3 are unknown but to-be-determined constants of integration.

Using equation (13.43), we find that in terms of the displacements the integrated results are

$$A_{11} \frac{du^o}{dx} = N_x = C_1 \tag{13.47a}$$

$$-D_{11} \frac{d^2 w^o}{dx^2} = M_x = -q_o \frac{x^2}{2} + C_2 x + C_3 \tag{13.47b}$$

Both of these equations can also be integrated directly to yield

$$u^o(x) = \frac{1}{A_{11}} \{C_1 x + C_4\} \tag{13.48a}$$

$$w^o(x) = \frac{1}{D_{11}} \left\{ q_o \frac{x^4}{24} - C_2 \frac{x^3}{6} - C_3 \frac{x^2}{2} - C_5 x - C_6 \right\} \tag{13.48b}$$

C_4, C_5, and C_6 being additional unknown constants of integration. All of the constants will be determined by examining the boundary conditions. Because the plate is infinitely long in the y direction, only the boundary conditions at $x = \pm a/2$ are examined. Because the plate is symmetrically laminated, the constants associated with u^o, namely, C_1 and C_4, do not appear in the equation for w^o. Conversely, the constants associated with w^o, namely, C_2, C_3, C_5, and C_6, do not appear in the equation for u^o. This will always be the case for symmetric laminates. For unsymmetric laminates, some of the same constants of integration will appear in both the equation for $u^o(x)$ and the equation $w^o(x)$. It should be noted well that because of the nature of classical lamination theory, specification of the boundary conditions means specification of conditions at the boundaries of the reference surface, not conditions at the boundaries of the top surface or bottom surface of the plate. If rotation is to be prevented at the boundary, then this means rotation of the reference surface is to be prevented at the boundary.

13.4.2 Application of Boundary Conditions

Equations (13.16a) and (13.16b), parts (i) and (ii), are the boundary conditions for u^o and v^o. Because v^o is specified to be zero everywhere, part (ii) is automatically satisfied. Thus, we only need to be concerned with part (i). If the plate is fixed by pinning it at $x = +a/2$ and at $x = -a/2$, as in Figure 13.14(a), then at both $x = +a/2$ and $x = -a/2$, u^o is specified to be zero, and from equation (13.48a)

$$u^o\left(+\frac{a}{2}\right) = \frac{1}{A_{11}}\left(C_1\left(+\frac{a}{2}\right) + C_4\right) = 0$$

$$u^o\left(-\frac{a}{2}\right) = \frac{1}{A_{11}}\left(C_1\left(-\frac{a}{2}\right) + C_4\right) = 0$$

(13.49)

which imply

$$C_1 = C_4 = 0$$

(13.50)

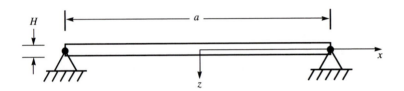

(a) Simple supports but no horizontal motion — fixed-fixed

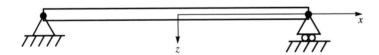

(b) Horizontal motion allowed at $x = + a/2$ — fixed-free

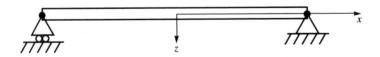

(c) Horizontal motion allowed at $x = - a/2$ — free-fixed

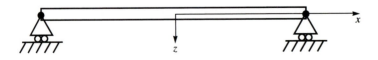

(d) Horizontal motion allowed at both ends — free-free

FIGURE 13.14
Various simple support boundary conditions at $x = \pm a/2$

If, as in Figure 13.14(b), the reference surface of the plate is fixed by pinning at $x = -a/2$ but is free to move horizontally because it is on rollers at $x = +a/2$, then u^o is not specified at $x = +a/2$, but rather N_x^+ is zero there, so

$$N_x\left(+\frac{a}{2}, y\right) = N_x^+ = 0 \tag{13.51}$$

But by equation (13.46a)

$$N_x = C_1 \tag{13.52}$$

so it must be concluded that

$$C_1 = 0 \tag{13.53}$$

At $x = -a/2$, u^o is specified to be zero; that is, from equation (13.48a)

$$u^o\left(-\frac{a}{2}\right) = 0 = \frac{1}{A_{11}}\left(0 \times \left(-\frac{a}{2}\right) + C_4\right) \tag{13.54}$$

which implies

$$C_4 = 0 \tag{13.55}$$

For the case of Figure 13.14(c), namely, the free-fixed case, it is also true that

$$C_1 = C_4 = 0 \tag{13.56}$$

The case of Figure 13.14(d), the free-free case, is quite interesting. Both edges are on rollers and so u^o is not specified at either edge. However, $N_x^+ = N_x^- = 0$ and boundary conditions of equation (13.16a) and (13.16b), part (i), become

$$N_x\left(+\frac{a}{2}, y\right) = N_x^+ = 0 \tag{13.57a}$$

and

$$N_x\left(-\frac{a}{2}, y\right) = N_x^- = 0 \tag{13.57b}$$

Both conditions can be satisfied by choosing, from equation (13.46a),

$$C_1 = 0 \tag{13.58}$$

If this is the case, then, from equation (13.48a),

$$u^o(x) = \frac{C_4}{A_{11}} \tag{13.59}$$

That is, $u^o(x)$ is everywhere the same and is a constant. This corresponds to the plate experiencing rigid body translation in the x direction. This makes sense, as both edges are on rollers and nothing prevents translation in the x direction. Translation can be suppressed by setting C_4 to zero. Doing this has no influence on the problem, so it will be done.

Independently of the whether the ends are fixed or free, as long as there are no resultant forces acting in the x direction at either edge of the plate,

$$C_1 = C_4 = 0 \tag{13.60}$$

This implies that for this loading and this laminate, there are no boundary conditions which will lead to values of $u^o(x)$ other than zero! This is an important finding. For unsymmetric laminates, as we shall see, the findings will be different.

To satisfy the boundary conditions of equations (13.16a) and (b), parts (iii) and (iv), the support conditions of no vertical deflection and no applied moment will be enforced. This implies

$$w^o\left(+\frac{a}{2}, y\right) = 0 \quad \text{and} \quad M_x\left(+\frac{a}{2}, y\right) = M_x^+ = 0 \tag{13.61a}$$

and $\qquad w^o\left(-\frac{a}{2}, y\right) = 0 \quad \text{and} \quad M_x\left(-\frac{a}{2}, y\right) = M_x^- = 0 \tag{13.61b}$

The second expressions of each of equations (13.61a) and (b), along with equation (13.47b), lead to

$$-q_o\frac{1}{2}\left(+\frac{a}{2}\right)^2 + C_2\left(+\frac{a}{2}\right) + C_3 = 0$$
$$-q_o\frac{1}{2}\left(-\frac{a}{2}\right)^2 + C_2\left(-\frac{a}{2}\right) + C_3 = 0 \tag{13.62}$$

or $\qquad\qquad C_2 = 0 \qquad C_3 = \dfrac{q_o a^2}{8} \tag{13.63}$

The first expressions of each of equations (13.61a) and (b), along with equation (13.48b), lead to

$$\frac{1}{D_{11}}\left\{\frac{q_o}{24}\left(+\frac{a}{2}\right)^4 - \frac{q_o a^2}{8}\frac{1}{2}\left(+\frac{a}{2}\right)^2 - C_5\left(+\frac{a}{2}\right) - C_6\right\} = 0$$
$$\frac{1}{D_{11}}\left\{\frac{q_o}{24}\left(-\frac{a}{2}\right)^4 - \frac{q_o a^2}{8}\frac{1}{2}\left(-\frac{a}{2}\right)^2 - C_5\left(-\frac{a}{2}\right) - C_6\right\} = 0 \tag{13.64}$$

or $\qquad\qquad C_5 = 0 \qquad C_6 = -\dfrac{5q_o a^4}{384} \tag{13.65}$

As a result of applying the boundary conditions, then, the complete solution for the displacements is

$$u^o(x) = 0$$
$$v^o(x) = 0 \tag{13.66}$$
$$w^o(x) = \frac{q_o a^4}{384 D_{11}}\left\{16\left(\frac{x}{a}\right)^4 - 24\left(\frac{x}{a}\right)^2 + 5\right\}$$

The support conditions at each end are classic simple-support conditions. However, here we have had to differentiate as to whether the simple support allows motion in the x direction (free), or whether it prevents motion in the x direction (fixed). Though the distinction was not important for this symmetric laminate, it will be quite important for the unsymmetric laminate studied in the next section.

The force and moment resultants across the width of the plate are given by equations (13.43), (13.46), and (13.66) as

$$N_x(x) = N_y(x) = N_{xy}(x) = 0$$
$$M_x(x) = \frac{q_o a^2}{8}\left\{1 - 4\left(\frac{x}{a}\right)^2\right\}$$
$$M_y(x) = \frac{D_{12}}{D_{11}}\frac{q_o a^2}{8}\left\{1 - 4\left(\frac{x}{a}\right)^2\right\} \tag{13.67}$$
$$M_{xy}(x) = 0$$

From the expressions for the displacements, equation (13.66), the strains, and hence stresses, at any point in the plate can be determined. The stresses and strains will, of course, be independent of y. The reference surface strains are zero, as

$$\varepsilon_x^o = \frac{\partial u^o}{\partial x} = 0 \qquad \varepsilon_y^o = \frac{\partial v^o}{\partial y} = 0 \qquad \gamma_{xy}^o = \frac{\partial u^o}{\partial y} + \frac{\partial v^o}{\partial x} = 0 \quad (13.68)$$

and the reference surface curvatures are

$$\kappa_x^o = -\frac{\partial^2 w^o}{\partial x^2} = \frac{q_o a^2}{8 D_{11}} \left\{ 1 - 4 \left(\frac{x}{a} \right)^2 \right\}$$

$$\kappa_y^o = -\frac{\partial^2 w^o}{\partial y^2} = 0 \qquad \kappa_{xy}^o = -2 \frac{\partial^2 w^o}{\partial y \partial x} = 0 \tag{13.69}$$

As a result, the strains at any location through the thickness, z, are given by

$$\varepsilon_x = \varepsilon_x^o + z \kappa_x^o = z \frac{q_o a^2}{8 D_{11}} \left\{ 1 - 4 \left(\frac{x}{a} \right)^2 \right\} \tag{13.70}$$

$$\varepsilon_y = 0 \qquad \gamma_{xy} = 0$$

In the two 0° layers

$$\varepsilon_1 = \varepsilon_x = z \frac{q_o a^2}{8 D_{11}} \left\{ 1 - 4 \left(\frac{x}{a} \right)^2 \right\} \tag{13.71a}$$

$$\varepsilon_2 = \gamma_{12} = 0$$

and in the two 90° layers

$$\varepsilon_2 = \varepsilon_x = z \frac{q_o a^2}{8 D_{11}} \left\{ 1 - 4 \left(\frac{x}{a} \right)^2 \right\} \tag{13.71b}$$

$$\varepsilon_1 = \gamma_{12} = 0$$

where z is in a range appropriate to each layer. The stresses in each layer are:
For the 0° layers:

$$\sigma_x = \sigma_1 = z \frac{Q_{11}}{D_{11}} \frac{q_o a^2}{8} \left\{ 1 - 4 \left(\frac{x}{a} \right)^2 \right\}$$

$$\sigma_y = \sigma_2 = z \frac{Q_{12}}{D_{11}} \frac{q_o a^2}{8} \left\{ 1 - 4 \left(\frac{x}{a} \right)^2 \right\} \tag{13.72a}$$

$$\tau_{xy} = \tau_{12} = 0$$

For the 90° layers:

$$\sigma_y = \sigma_1 = z \frac{Q_{12}}{D_{11}} \frac{q_o a^2}{8} \left\{ 1 - 4 \left(\frac{x}{a} \right)^2 \right\}$$

$$\sigma_x = \sigma_2 = z \frac{Q_{22}}{D_{11}} \frac{q_o a^2}{8} \left\{ 1 - 4 \left(\frac{x}{a} \right)^2 \right\} \tag{13.72b}$$

$$\tau_{xy} = \tau_{12} = 0$$

Numerical results for the deflections and stresses for a [0/90]$_s$ plate will be presented after the solution for Plate Example 2 is discussed. However, at this point it is important to note that nothing that has been done in solving the equations has restricted the solution to the $[0/90]_S$ case. The results obtained apply to any simply supported symmetrically laminated cross-ply plate. It is the value of D_{11} that makes the results unique to any particular cross-ply plate. In fact, the solution procedure, and indeed, the actual form of the solution for $w^o(x, y)$, equation (13.66), applies to the case of any symmetrically laminated plate. Because for the general symmetric laminate $D_{16} \neq 0$, there will be an M_{xy} associated with the response, as opposed to $M_{xy} = 0$ for the cross-ply case. A nonzero M_{xy} will be the case for the $[\pm 30/0]_S$ plate to be discussed shortly.

Exercise for Section 13.4

Consider the semi-infinite plate of Figure 13.13. Assume that the laminate is symmetric and the ends at $x = \pm a/2$ are prevented from vertical displacement, rotational displacement, and horizontal displacement; that is:

$$u^o\left(-\frac{a}{2}, y\right) = 0 \qquad u^o\left(+\frac{a}{2}, y\right) = 0$$

$$w^o\left(-\frac{a}{2}, y\right) = 0 \qquad w^o\left(+\frac{a}{2}, y\right) = 0$$

$$\frac{dw^o}{dx}\left(-\frac{a}{2}, y\right) \qquad \frac{dw^o}{dx}\left(+\frac{a}{2}, y\right) = 0$$

These boundary conditions correspond to the case of clamped edges. Develop expressions for $u^o(x)$, $w^o(x)$, $N_x(x)$, ..., $M_{xy}(x)$ for this case. Note that the expressions of equations (13.45) through (13.48) are valid for this case. Thus, apply the above boundary conditions to these expressions and find the constants C_1 through C_6. Comment on the location of the maximum magnitude of $M_x(x)$ for this clamped boundary condition case compared to the simple support boundary condition case of Figure 13.14(a), equation (13.67).

13.5
PLATE EXAMPLE 2: A LONG $[0_2/90_2]_T$ PLATE

Let us now consider the plate in the previous example to be unsymmetric, and in particular, a $[0_2/90_2]_T$ laminate. This laminate has the same number of layers in both the $0°$ and $90°$ directions as the first example. Because of the cross-ply nature of this unsymmetric laminate,

$$A_{16} = A_{26} = D_{16} = D_{26} = B_{16} = B_{26} = B_{12} = B_{66} = 0 \qquad (13.73)$$

13.5.1 Solution of Governing Differential Equations

As a result of the above properties being zero, and the conditions of this response being independent of y, and there being no displacement in the y direction, the force and moment resultants of equation (13.29) reduce to

$$N_x = A_{11}\frac{du^o}{dx} - B_{11}\frac{d^2w^o}{dx^2} \qquad N_y = A_{12}\frac{du^o}{dx} \qquad N_{xy} = 0$$

$$M_x = B_{11}\frac{du^o}{dx} - D_{11}\frac{d^2w^o}{dx^2} \qquad M_y = -D_{12}\frac{d^2w^o}{dx^2} \qquad M_{xy} = 0$$

(13.74)

The equilibrium equations in terms of the force resultants, as with the symmetric laminate, are, from equation (13.15),

$$\frac{dN_x}{dx} = 0 \qquad\qquad (13.75a)$$

$$0 = 0 \qquad\qquad (13.75b)$$

$$\frac{d^2M_x}{dx^2} + q_o = 0 \qquad\qquad (13.75c)$$

The solutions to these two equations are as before, equation (13.46a) and (b),

$$N_x(x) = C_1 \qquad\qquad (13.76a)$$

$$M_x(x) = -q_o\frac{x^2}{2} + C_2x + C_3 \qquad\qquad (13.76b)$$

In terms of the displacements, by equation (13.74), equation (13.76) becomes

$$A_{11}\frac{du^o}{dx} - B_{11}\frac{d^2w^o}{dx^2} = N_x = C_1 \qquad\qquad (13.77a)$$

$$B_{11}\frac{du^o}{dx} - D_{11}\frac{d^2w^o}{dx^2} = M_x = -q_o\frac{x^2}{2} + C_2x + C_3 \qquad\qquad (13.77b)$$

Unlike equations (13.47a) and (b) for the symmetric laminate, for this case the equations governing u^o and w^o are coupled and each equation cannot be solved directly. To solve for u^o and w^o, the first equation can be substituted into the second equation to yield

$$\left(\frac{B_{11}^2}{A_{11}} - D_{11}\right)\frac{d^2w^o}{dx^2} = -q_o\frac{x^2}{2} + C_2x + C_3 - \frac{B_{11}}{A_{11}}C_1 \qquad\qquad (13.78)$$

This equation can now be integrated directly to result in

$$w^o(x) = \left(\frac{A_{11}}{A_{11}D_{11} - B_{11}^2}\right)$$

$$\times \left\{q_o\frac{x^4}{24} - C_2\frac{x^3}{6} - \left(C_3 - \frac{B_{11}}{A_{11}}C_1\right)\frac{x^2}{2} - C_5x - C_6\right\}$$

(13.79)

Note the differences between this expression and the counterpart for the symmetrically laminated plate, equation (13.48b). Note also, with $B_{11} = 0$ in equation (13.79), equation (13.48b) is recovered.

Back-substituting and using the expression for $w^o(x)$ in the first expression of equation (13.77), we can write an equation for determining $u^o(x)$, namely

$$A_{11}\frac{du^o}{dx} = \left(\frac{A_{11}D_{11}}{A_{11}D_{11} - B_{11}^2}\right)C_1$$

$$+ \left(\frac{A_{11}B_{11}}{A_{11}D_{11} - B_{11}^2}\right)\left(q_o\frac{x^2}{2} - C_2x - C_3\right)$$

(13.80)

Integration results in

$$u^o(x) = \frac{D_{11}}{A_{11}D_{11} - B_{11}^2}C_1x + \left(\frac{B_{11}}{A_{11}D_{11} - B_{11}^2}\right)$$

$$\times \left(q_o\frac{x^3}{6} - C_2\frac{x^2}{2} - C_3x\right) + \frac{C_4}{A_{11}}$$

(13.81)

When $B_{11} = 0$, this reduces to equation (13.48a). As with the first example, six unknown constants of integration, $C_1 - C_6$, are involved. However, unlike the first example, some constants appear in both the expression for $u^o(x)$ and in the expression for $w^o(x)$, indicating the coupling of the inplane and out-of-plane response due to the unsymmetric laminate.

13.5.2 Application of Boundary Conditions

Let us consider the fixed-fixed simple supported case, Figure 13.14(a). For this situation, examining all three boundary conditions at each end, we find that

$$u^o\left(+\frac{a}{2}, y\right) = 0 \qquad w^o\left(+\frac{a}{2}, y\right) = 0 \qquad M_x\left(+\frac{a}{2}, y\right) = M_x^+ = 0 \quad (13.82a)$$

and

$$u^o\left(-\frac{a}{2}, y\right) = 0 \qquad w^o\left(-\frac{a}{2}, y\right) = 0 \qquad M_x\left(-\frac{a}{2}, y\right) = M_x^- = 0 \quad (13.82b)$$

If we use equation (13.76b), the third expressions of equations (13.82a) and (b) lead to the conclusion that

$$C_2 = 0 \qquad C_3 = \frac{q_oa^2}{8}$$

(13.83)

The first expressions of equations (13.82a) and (b) result in, from equation (13.81),

$$D_{11}C_1\left(+\frac{a}{2}\right) - B_{11}\frac{q_oa^3}{24} + \left(\frac{A_{11}D_{11} - B_{11}^2}{A_{11}}\right)C_4 = 0$$

$$D_{11}C_1\left(-\frac{a}{2}\right) + B_{11}\frac{q_oa^3}{24} + \left(\frac{A_{11}D_{11} - B_{11}^2}{A_{11}}\right)C_4 = 0$$

(13.84)

From these relations it follows that

$$C_4 = 0 \qquad C_1 = \frac{B_{11}}{D_{11}}\frac{q_oa^2}{12}$$

(13.85)

Finally, the second expressions of equations (13.82a) and (b) and equation (13.79) lead to

$$\frac{q_o}{24}\left(+\frac{a}{2}\right)^4 - q_o a^2 \left(\frac{1}{8} - \frac{B_{11}^2}{12 A_{11} D_{11}}\right)\frac{1}{2}\left(+\frac{a}{2}\right)^2 - C_5\left(+\frac{a}{2}\right) - C_6 = 0$$

$$\frac{q_o}{24}\left(-\frac{a}{2}\right)^4 - q_o a^2 \left(\frac{1}{8} - \frac{B_{11}^2}{12 A_{11} D_{11}}\right)\frac{1}{2}\left(-\frac{a}{2}\right)^2 - C_5\left(-\frac{a}{2}\right) - C_6 = 0$$

(13.86)

or $\qquad C_5 = 0 \qquad C_6 = -q_o a^4 \left(\frac{5}{384} - \frac{B_{11}^2}{96 A_{11} D_{11}}\right)$ (13.87)

As a result of applying the simple support boundary conditions of Figure 13.14(a), then, we find that

$$u^o(x) = \frac{B_{11}}{D_{11}} \frac{q_o a^3}{24 A_{11}^R}\left\{4\left(\frac{x}{a}\right)^3 - \left(\frac{x}{a}\right)\right\}$$

$$v^o(x) = 0$$

$$w^o(x) = \frac{q_o a^4}{384 D_{11}^R}\left\{16\left(\frac{x}{a}\right)^4 - 8\left(3 - 2\frac{B_{11}^2}{A_{11} D_{11}}\right)\left(\frac{x}{a}\right)^2 \right.$$
(13.88)
$$\left. + \left(5 - 4\frac{B_{11}^2}{A_{11} D_{11}}\right)\right\}$$

where $\qquad A_{11}^R = A_{11}\left(1 - \frac{B_{11}^2}{A_{11} D_{11}}\right)$ (13.89a)

and $\qquad D_{11}^R = D_{11}\left(1 - \frac{B_{11}^2}{A_{11} D_{11}}\right)$ (13.89b)

are referred to as the *reduced extensional stiffness* and the *reduced bending stiffness*, respectively. These displacements should be compared with the results in equation (13.66) for the symmetric laminate. For the unsymmetric laminate there is a displacement in the x direction, even though it is zero at the edges of the plate, and the expression for $w^o(x)$ is more complicated. With $B_{11} = 0$, equation (13.66) is recovered from equation (13.88).

The force and moment resultants across the width of the plate are given by equations (13.74) and (13.76) as

$$N_x(x) = \frac{B_{11}}{D_{11}}\frac{q_o a^2}{12}$$

$$N_y(x) = \frac{B_{11}}{D_{11}}\frac{A_{12}}{A_{11}^R}\frac{q_o a^2}{12}\left\{6\left(\frac{x}{a}\right)^2 - \frac{1}{2}\right\}$$

$$N_{xy}(x) = 0$$
(13.90)

$$M_x(x) = \frac{q_o a^2}{8}\left\{1 - 4\left(\frac{x}{a}\right)^2\right\}$$

$$M_y(x) = \frac{D_{12}}{D_{11}^R}\frac{q_o a^2}{8}\left\{\left(1 - \frac{2}{3}\frac{B_{11}^2}{A_{11} D_{11}}\right) - 4\left(\frac{x}{a}\right)^2\right\}$$

$$M_{xy}(x) = 0$$

We see that because of the coupling due to B_{11}, the applied loading causes inplane force resultants in both the x and y directions. Whether these two force resultants are tensile or compressive depends on the sign of B_{11}. The expressions here have been developed with a $[0_2/90_2]_T$ laminate in mind; however, the analysis is equally valid for a $[90_2/0_2]_T$ laminate and these two laminates have B_{11}'s of opposite sign, and hence cause $N_x(x)$ and $N_y(x)$ to have opposite signs. While $N_x(x)$ is not a function of plate width, $N_y(x)$ is. The nonzero moment resultants are both functions of plate width but are not sensitive to the sign of B_{11}.

The reference surface strains for the unsymmetric laminate are

$$\varepsilon_x^o = \frac{\partial u^o}{\partial x} = \frac{B_{11}}{D_{11}} \frac{q_o a^2}{24 A_{11}^R} \left\{ 12 \left(\frac{x}{a}\right)^2 - 1 \right\}$$

$$\varepsilon_y^o = \frac{\partial v^o}{\partial y} = 0 \qquad \gamma_{xy}^o = \frac{\partial u^o}{\partial y} + \frac{\partial v^o}{\partial x} = 0$$

(13.91)

and the reference surface curvatures are

$$\kappa_x^o = -\frac{\partial^2 w^o}{\partial x^2} = \frac{q_o a^2}{8 D_{11}^R} \left\{ \left(1 - \frac{2}{3} \frac{B_{11}^2}{A_{11} D_{11}} \right) - 4 \left(\frac{x}{a}\right)^2 \right\}$$

$$\kappa_y^o = -\frac{\partial^2 w^o}{\partial y^2} = 0 \qquad \kappa_{xy}^o = -2 \frac{\partial^2 w^o}{\partial y \partial x} = 0$$

(13.92)

In a manner similar to the first example, the strains, and hence stresses, at any location through the thickness and across the width of the plate can be determined.

Let us now consider the boundary conditions of Figure 13.14(b), the fixed-free case. With these boundary conditions, the left edge of the plate is fixed against displacement in the x direction, while the right edge is free to move. Thus, the three conditions to be enforced on the edges are

$$N_x \left(+\frac{a}{2}, y \right) = N_x^+ = 0 \quad w^o \left(+\frac{a}{2}, y \right) = 0 \quad M_x \left(+\frac{a}{2}, y \right) = M_x^+ = 0 \quad (13.93a)$$

and

$$u^o \left(-\frac{a}{2}, y \right) = 0 \qquad w^o \left(-\frac{a}{2}, y \right) = 0 \qquad M_x \left(-\frac{a}{2}, y \right) = M_x^- = 0 \quad (13.93b)$$

Obviously the solutions to the governing differential equations given by equations (13.76), (13.79), and (13.81) are valid for this case because boundary conditions had not been applied to obtain those general solutions. Thus, from the first expressions of equations (13.93a) and (13.76a),

$$N_x(x) = C_1 = 0$$

(13.94)

From the third expressions of equations (13.93a) and (b), and (13.76b), as before,

$$C_2 = 0 \qquad C_3 = \frac{q_o a^2}{8}$$

(13.95)

Using these results and the first expression of equation (13.93b), we find from (13.81) that

$$C_4 = -\frac{A_{11} B_{11} q_o a^3}{24(A_{11} D_{11} - B_{11}^2)}$$

(13.96)

Finally, using the second expression of equations (13.93a) and (b) and equation (13.79) we find that

$$\frac{q_o}{24}\left(+\frac{a}{2}\right)^4 - \frac{q_oa^2}{8}\frac{1}{2}\left(+\frac{a}{2}\right)^2 - C_5\left(+\frac{a}{2}\right) - C_6 = 0$$

$$\frac{q_o}{24}\left(-\frac{a}{2}\right)^4 - \frac{q_oa^2}{8}\frac{1}{2}\left(-\frac{a}{2}\right)^2 - C_5\left(-\frac{a}{2}\right) - C_6 = 0$$

(13.97)

or

$$C_5 = 0 \qquad C_6 = -\frac{5q_oa^4}{384}$$

(13.98)

As a result

$$u^o(x) = \frac{B_{11}}{D_{11}}\frac{q_oa^3}{24A_{11}^R}\left\{4\left(\frac{x}{a}\right)^3 - 3\left(\frac{x}{a}\right) - 1\right\}$$

$$v^o(x) = 0$$

(13.99)

$$w^o(x) = \frac{q_oa^4}{384D_{11}^R}\left\{16\left(\frac{x}{a}\right)^4 - 24\left(\frac{x}{a}\right)^2 + 5\right\}$$

It is important to compare the functional forms of $u^o(x)$ and $w^o(x)$ for the unsymmetric laminate with the two types of boundary conditions, Figure 13.14(a) and (b). The response for the fixed-fixed boundary conditions of Figure 13.14(a) is given by equation (13.88), while the response for the fixed-free case of Figure 13.14(b) is given by the equations just derived. In fact, comparison of the fixed-free unsymmetric laminate just derived and the symmetric laminate, equation (13.66), is interesting. The forms of $w^o(x)$ for the two cases are identical except for magnitude. For any laminate the relation

$$\frac{1}{D_{11}^R} = \frac{1}{D_{11}\left(1 - \frac{B_{11}^2}{A_{11}D_{11}}\right)} > \frac{1}{D_{11}}$$

(13.100)

is valid. Thus, the $w^o(x)$ deflection of the unsymmetric laminate with the fixed-free boundary conditions will be greater than the deflection of the symmetric laminate.

The force and moment resultants for the fixed-free conditions are given by

$$N_x(x) = 0 = N_{xy}(x)$$

$$N_y(x) = \frac{B_{11}}{D_{11}}\frac{A_{12}}{A_{11}^R}\frac{q_oa^2}{12}\left\{6\left(\frac{x}{a}\right)^2 - \frac{3}{2}\right\}$$

$$M_x(x) = \left(\frac{q_oa^2}{8}\right)\left\{1 - 4\left(\frac{x}{a}\right)^2\right\}$$

(13.101)

$$M_y(x) = \frac{D_{12}}{D_{11}^R}\frac{q_oa^2}{8}\left\{1 - 4\left(\frac{x}{a}\right)^2\right\}$$

$$M_{xy}(x) = 0$$

Note there is no N_x because the free end prevents N_x from developing. Interestingly enough, there is an N_y. The reference surface strains and curvatures for the fixed-free case are

$$\varepsilon_x^o(x) = \frac{B_{11}}{D_{11}} \frac{q_o a^2}{12 A_{11}^R} \left\{ 6\left(\frac{x}{a}\right)^2 - \frac{3}{2} \right\}$$

$$\varepsilon_y^o(x) = \gamma_{xy}^o(x) = 0$$

$$\kappa_x^o = \frac{q_o a^2}{8 D_{11}^R} \left\{ 1 - 4\left(\frac{x}{a}\right)^2 \right\}$$

$$\kappa_y^o = \kappa_{xy}^o = 0$$

(13.102)

The strains as a function of thickness and length, and hence the stresses, can be computed from the above.

For the edge support conditions studied here, the plate is statically determinant. Thus, the moment expression, for example, is independent of material properties. If the edges of the plate were clamped, or if one edge was clamped, the moment M_x would not be known at that edge. The problem would not be statically determinant and the evaluation of C_2 and C_3 would be coupled with the evaluation of the other constants.

The response of the unsymmetric laminate is dependent on the details of the boundary conditions on u^o and N_x at the edges of the plate. We saw that the response of the symmetric laminate was independent of these boundary conditions. The impact of this dependence on boundary conditions for the unsymmetric laminate will be seen shortly when we look at numerical results.

Exercise for Section 13.5

Consider the unsymmetric $[0_2/90_2]_T$ laminate discussed in Plate Example 2. In the discussion we considered the boundary conditions of Figure 13.14(a) and (b). Show that for the $[0_2/90_2]_T$ laminate, the boundary condition of Figure 13.14(c) leads to practically the same results for the displacements as a function of x as the boundary condition in Figure 13.14(b) does, the difference being one term in $u^o(x)$.

13.6
PLATE EXAMPLE 3: A LONG [± 30/0]$_S$ PLATE

As a final example of a long plate, consider a uniformly loaded $[\pm 30/0]_S$ plate simply supported at the edges $x = \pm a/2$. With this laminate

$$A_{16} = A_{26} = 0 \tag{13.103a}$$

and

$$B_{ij} = 0 \qquad \text{all } i \text{ and } j \tag{13.103b}$$

but neither D_{16} nor D_{26} are zero for this laminate. Assume also that v^o is zero. The force and moment resultants, equation (13.29), reduce to

$$N_x = A_{11} \frac{du^o}{dx} \qquad N_y = A_{12} \frac{du^o}{dx} \qquad N_{xy} = 0$$

$$M_x = -D_{11} \frac{d^2 w^o}{dx^2} \qquad M_y = -D_{12} \frac{d^2 w^o}{dx^2} \qquad M_{xy} = -D_{16} \frac{d^2 w^o}{dx^2}$$

(13.104)

The equilibrium equations in terms of the force resultants, equation (13.15), reduce to

$$\frac{dN_x}{dx} = 0 \tag{13.105a}$$

$$0 = 0 \tag{13.105b}$$

$$\frac{d^2 M_x}{dx^2} + q_o = 0 \tag{13.105c}$$

Integration of the two nontrivial equations above results in

$$N_x(x) = C_1 \tag{13.106a}$$

$$M_x(x) = -q_o \frac{x^2}{2} + C_2 x + C_3 \tag{13.106b}$$

In terms of the displacements,

$$A_{11} \frac{du^o}{dx} = N_x = C_1 \tag{13.107a}$$

$$-D_{11} \frac{d^2 w^o(x)}{dx^2} = M_x = -q_o \frac{x^2}{2} + C_2 x + C_3 \tag{13.107b}$$

These equations can be integrated to yield

$$u^o(x) = \frac{1}{A_{11}} \{C_1 x + C_4\} \tag{13.108a}$$

$$w^o(x) = \frac{1}{D_{11}} \left\{ q_o \frac{x^4}{24} - C_2 \frac{x^3}{6} - C_3 \frac{x^2}{2} - C_5 x - C_6 \right\} \tag{13.108b}$$

Because the laminate is symmetric, the constants of integration, C_1 and C_4, associated with the inplane displacement, $u^o(x)$, do not appear in the equation for the out-of-plane displacement, $w^o(x)$, nor do the constants of integration associated with the out-of-plane displacements, C_2, C_3, C_4, and C_6, appear in the equation for the inplane displacement. Thus, the results for all four boundary conditions depicted in Figure 13.14 will be identical:

$$u^o(x) = 0$$

$$v^o(x) = 0 \tag{13.109}$$

$$w^o(x) = \frac{q_o a^4}{384 D_{11}} \left\{ 16 \left(\frac{x}{a}\right)^4 - 24 \left(\frac{x}{a}\right)^2 + 5 \right\}$$

and

$$N_x(x) = N_y(x) = N_{xy}(x) = 0$$

$$M_x(x) = \frac{q_o a^2}{8} \left\{ 1 - 4 \left(\frac{x}{a}\right)^2 \right\}$$

$$M_y(x) = \frac{D_{12}}{D_{11}} \frac{q_o a^2}{8} \left\{ 1 - 4 \left(\frac{x}{a}\right)^2 \right\} \tag{13.110}$$

$$M_{xy}(x) = \frac{D_{16}}{D_{11}} \frac{q_o a^2}{8} \left\{ 1 - 4 \left(\frac{x}{a}\right)^2 \right\}$$

Except for the presence of a nonzero value of $M_{xy}(x)$, the development for this $[\pm30/0]_S$ laminate parallels exactly the development for the $[0/90]_S$ laminate. As with $M_y(x)$ and the value of D_{12} relative to D_{11}, it is the value of D_{16} relative to D_{11} that is important. Though it is not the case for the $[\pm30/0]_S$ laminate, D_{16} can be larger than D_{12}. For these cases, the twisting moment M_{xy} will be larger than M_y.

From the expressions for the displacements, the reference surface and curvatures are given by

$$\varepsilon_x^o = \frac{\partial u^o}{\partial x} = 0 \qquad \varepsilon_y^o = \frac{\partial v^o}{\partial y} = 0 \qquad \gamma_{xy}^o = \frac{\partial u^o}{\partial y} + \frac{\partial v^o}{\partial x} = 0$$

$$\kappa_x^o = -\frac{\partial^2 w^o}{\partial x^2} = \frac{q_o a^2}{8 D_{11}} \left\{ 1 - 4\left(\frac{x}{a}\right)^2 \right\} \tag{13.111}$$

$$\kappa_y^o = -\frac{\partial^2 w^o}{\partial y^2} = 0 \qquad \kappa_{xy}^o = -2\frac{\partial^2 w^o}{\partial y \partial x} = 0$$

The strains at any location through the thickness are given by

$$\varepsilon_x = \varepsilon_x^o + z\kappa_x^o = z\frac{q_o a^2}{8 D_{11}} \left\{ 1 - 4\left(\frac{x}{a}\right)^2 \right\} \tag{13.112}$$

$$\varepsilon_y = 0 \qquad \gamma_{xy} = 0$$

and the stresses are given by

$$\sigma_x = z\frac{\bar{Q}_{11}}{D_{11}}\frac{q_o a^2}{8} \left\{ 1 - 4\left(\frac{x}{a}\right)^2 \right\}$$

$$\sigma_y = z\frac{\bar{Q}_{12}}{D_{11}}\frac{q_o a^2}{8} \left\{ 1 - 4\left(\frac{x}{a}\right)^2 \right\} \tag{13.113}$$

$$\tau_{xy} = z\frac{\bar{Q}_{16}}{D_{11}}\frac{q_o a^2}{8} \left\{ 1 - 4\left(\frac{x}{a}\right)^2 \right\}$$

The values of $\bar{Q}_{11}$, $\bar{Q}_{12}$, and $\bar{Q}_{16}$ are a function of whether z is within a $+30°$ layer, a $-30°$ layer, or a $0°$ layer. The value of $\bar{Q}_{16}$ in the $+30°$ layers is opposite in sign to the value of $\bar{Q}_{16}$ in the $-30°$ layers, and in the $0°$ layers $\bar{Q}_{16} = 0$.

Next, numerical results for the plates just discussed will be presented.

13.7
NUMERICAL RESULTS FOR PLATE EXAMPLES 1, 2, AND 3

It is now appropriate to turn to numerical examples to illustrate the differences between the $[0/90]_S$, $[0_2/90_2]_T$, and the $[\pm30/0]_S$ plates, and the differences between the various boundary conditions applied to the edges of the $[0_2/90_2]_T$ plate. The material properties for a graphite-reinforced material are used. For the various plates, the pertinent laminate properties are as follows:

For the $[0/90]_S$ plate:

$$D_{11} = 2.48 \text{ N·m} \qquad D_{12} = 0.0543 \text{ N·m} \tag{13.114a}$$

For the $[0_2/90_2]_T$ plate:

$$A_{11} = 50.4 \text{ MN/m} \qquad A_{12} = 1.809 \text{ MN/m} \qquad B_{11} = -6\,460 \text{ N}$$

$$D_{11} = 1.511 \text{ N·m} \qquad D_{12} = 0.0543 \text{ N·m}$$
(13.114b)

For the $[\pm30/0]_S$ plate:

$$D_{11} = 5.78 \text{ N·m} \qquad D_{12} = 1.766 \text{ N·m} \qquad D_{16} = 1.261 \text{ N·m} \qquad (13.114c)$$

A unit pressure load $q_o = 1 \text{ N/m}^2$ and a plate span of $a = 0.5$ m are considered for purpose of the examples.

13.7.1 Cross-Ply Plates

Figure 13.15 illustrates the out-of-plane deflection, $w^o(x)$, as a function of plate width for the three cross-ply cases. In this and subsequent figures, nondimensionalized responses are shown as a function of normalized distance across the width. The nondimensionalizing factor is

$$\left(\frac{384\,^S D_{11}}{5q_o a^4} \right)$$
(13.115)

where $^S D_{11}$ is the value of D_{11} for the symmetric cross-ply laminate, namely,

$$^S D_{11} = 2.48 \text{ N·m}$$
(13.116)

This factor was chosen so the out-of-plane deflections of the three cross-ply cases could be easily compared.

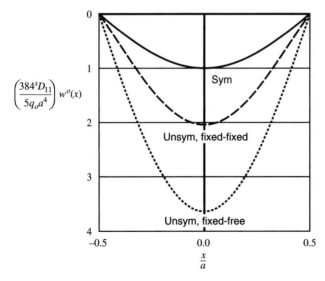

FIGURE 13.15
Out-of-plane deflection response $w^o(x)$ of semi-infinite, uniformly loaded $[0/90]_S$ and $[0_2/90_2]_T$ plates

As we can see from Figure 13.15, the symmetric $[0/90]_S$ plate deflects out-of-plane the least. This is because the symmetric laminate has a larger bending stiffness D_{11} than the unsymmetric laminate, and D_{11} is an important material property for resisting bending deflection. Having the two $0°$ layers on opposite sides of the laminate, as opposed to having them both on the same side, results in a larger D_{11}. The unsymmetric $[0_2/90_2]_T$ laminate with the fixed-free ends deflects out of plane the most; the smaller value of D_{11} is largely responsible for this increased deflection. However, for the unsymmetric laminate, D_{11} alone is not responsible for controlling the out-of-plane deflections. The fixed-fixed unsymmetric $[0_2/90_2]_T$ laminate deflects much less than the fixed-free one (see Figure 13.15). The coupling of $u^o(x)$ and $w^o(x)$ is responsible for this; the restraint on $u^o(x)$ at both edges causes the reduced level of out-of-plane deflection for the fixed-fixed case.

Figure 13.16 shows the inplane displacement $u^o(x)$. The displacements are nondimensionalized by the same parameter used with the out-of-plane displacements, but using a scale factor of 100. The magnitudes of the inplane displacements for the unsymmetric laminates are considerably smaller than the magnitude of the out-of-plane displacements. In addition, among the three cross-ply cases, there is a significant difference in the magnitudes of this displacement component. The symmetric laminate simply has zero displacement in the x direction, independently of the boundary conditions. This was emphasized in the discussion relating to equation (13.60). The fixed-free unsymmetric laminate, on the other hand, has its maximum

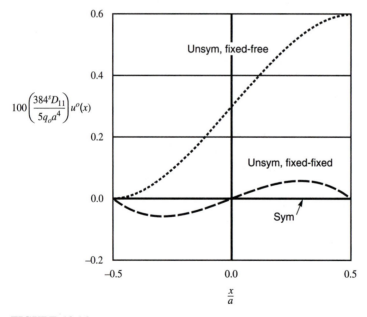

FIGURE 13.16
Inplane deflection response $u^o(x)$ of semi-infinite, uniformly loaded $[0/90]_S$ and $[0_2/90_2]_T$ plates

inplane displacement on the free end. Furthermore, the sign of the displacement indicates that every point across the width moves in the same direction. For the fixed-fixed unsymmetric laminate, the $u^o(x)$ displacements are dramatically less than for the fixed-free case. Also, for the fixed-fixed boundary conditions the sign of $u^o(x)$ depends on location across the width. Please note that the sign of B_{11} controls the sign of $u^o(x)$ for the two unsymmetric cases.

For the symmetric case and the unsymmetric fixed-free case

$$N_x = 0 \tag{13.117a}$$

while for the unsymmetric fixed-fixed case

$$\left(\frac{D_{11}}{B_{11}}\right)\left(\frac{12}{q_o a^2}\right) N_x = 1 \tag{13.117b}$$

The variation of the force resultant N_y across the plate is illustrated in Figure 13.17; the resultant is nondimensionalized by the quantity

$$\left(\frac{D_{11}}{B_{11}}\right)\left(\frac{12}{q_o a^2}\right) \tag{13.118}$$

This force resultant is responsible for enforcing the condition $v^o(x) = 0$. The symmetric laminate needs no N_y to enforce $v^o(x) = 0$, whereas the unsymmetric ones do. Also, for the fixed-free case, even though there is no N_x due to the loading, there is an N_y. From the figure we can see that the primary difference in the force resultants for the two unsymmetric cases amounts to a vertical shift in the values.

The variation of M_x with width is the same for all three cases; however, the variation of M_y with width is a function of laminate and boundary conditions. This

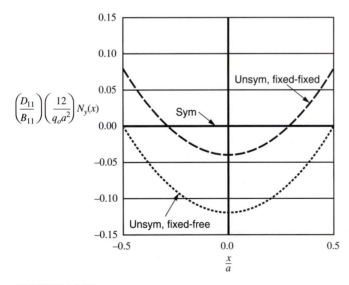

FIGURE 13.17
Force resultant $N_y(x)$ in semi-infinite, uniformly loaded $[0/90]_S$ and $[0_2/90]_T$ plates

point is illustrated in Figures 13.18 and 13.19, where the moments have been nondimensionalized by the parameter

$$\left(\frac{8}{q_o a^2}\right) \tag{13.119}$$

Compared to M_x, the magnitude of M_y for all three cases is quite small. Figure 13.19 is a more detailed illustration of M_y, and we see that the magnitude of M_y is greatest for the unsymmetric fixed-free unsymmetric laminate. The condition of being fixed against u^o displacement at both ends causes a slight reversal of the sign of M_y for the unsymmetric fixed-fixed case.

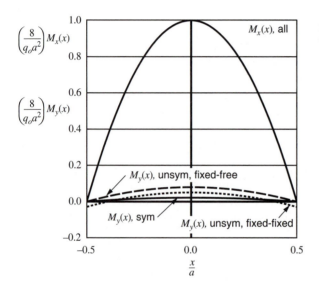

FIGURE 13.18
Moment resultant $M_x(x)$ in semi-infinite, uniformly loaded $[0/90]_S$ and $[0_2/90_2]_T$ plates

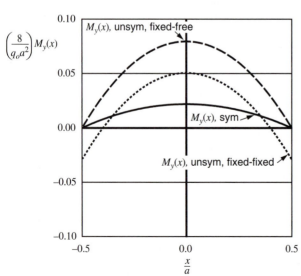

FIGURE 13.19
Moment resultant $M_y(x)$ in semi-infinite, uniformly loaded $[0/90]_S$ and $[0_2/90_2]_T$ plates

Finally, the all-important stress response is illustrated for the three cases in Figures 13.20–13.22; the stresses are shown as a function of location through the thickness at midspan; that is, $x = 0$. Keeping in mind the fiber orientations for these cross-ply configurations, we see that the symmetric laminate, Figure 13.20, clearly is subjected to the least stresses; the stresses in the fiber direction are ± 0.589 MPa, and the stresses perpendicular to the fibers are ± 0.0230 MPa. The fixed-free laminate experiences the largest stresses; the fiber direction stress σ_1 reaches -1.22 MPa and σ_2 reaches $+0.238$ MPa.

13.7.2 $[\pm 30/0]_S$ Plate

The response of the $[\pm 30/0]_S$ plate is quite similar to the response of the symmetric $[0/90]_S$ plate. Because the bending stiffness D_{16} is involved, there is a nonzero distribution of $M_{xy}(x)$ across the span of the plate, and in the $+30°$ and $-30°$ layers τ_{xy} is nonzero.

The distribution of the out-of-plane deflection across the width of the plate is shown in Figure 13.23, where again, nondimensionalization has been used. The distributions of $M_x(x)$, $M_y(x)$, and $M_{xy}(x)$ across the width of the plate are illustrated in Figure 13.24. Comparing the distributions of the $[\pm 30/0]_S$ plate with those of the cross-ply plates shown in Figure 13.18, we see that relative to the magnitude of $M_x(x)$, $M_y(x)$ for the $[\pm 30/0]_S$ plate is larger, and $M_{xy}(x)$ is of comparable

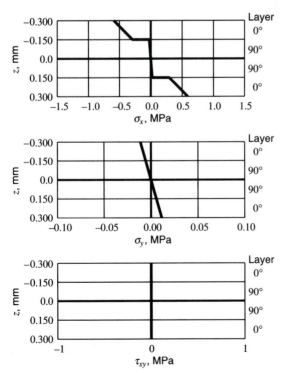

FIGURE 13.20

Stresses at midspan in a semi-infinite, uniformly loaded $[0/90]_S$ plate, $q_o = 1$, $a = 0.5$ m

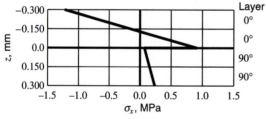

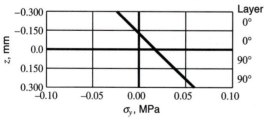

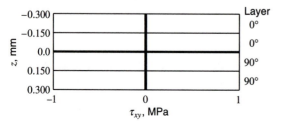

FIGURE 13.21
Stresses at midspan in a semi-infinite, uniformly loaded $[0_2/90_2]_T$ fixed-fixed plate, $q_o = 1$, $a = 0.5$ m

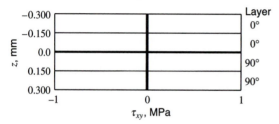

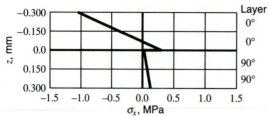

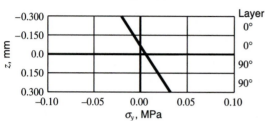

FIGURE 13.22
Stresses at midspan in a semi-infinite, uniformly loaded $[0_2/90_2]_T$ fixed-free plate, $q_o = 1$, $a = 0.5$ m

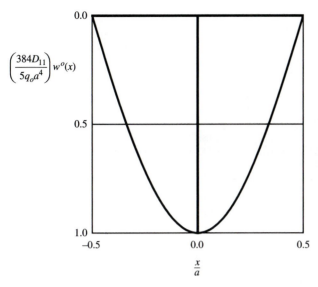

FIGURE 13.23
Out-of-plane deflection response $w^o(x)$ of a semi-infinite, uniformly loaded $[\pm30/0]_S$ plate

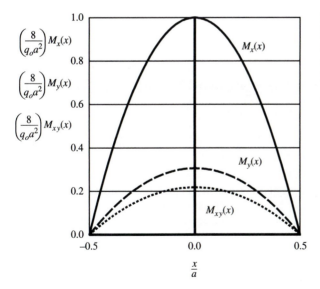

FIGURE 13.24
Moment resultants $M_x(x)$, $M_y(x)$, and $M_{xy}(x)$ of a semi-infinite, uniformly loaded $[\pm30/0]_S$ plate

magnitude. The stresses through the thickness of the $[\pm30/0]_S$ plate at midspan ($x = 0$) are shown in Figure 13.25. These are not the principal material system stresses, but we see that the magnitude of the shear stress τ_{xy} is greater than the magnitude of σ_y.

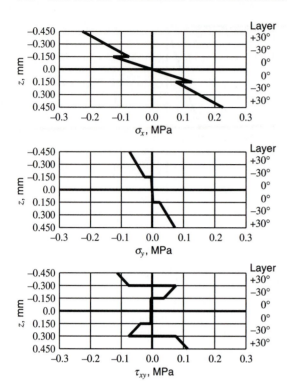

FIGURE 13.25

Stresses at midspan in a semi-infinite, uniformly loaded $[\pm 30/0]_S$ plate, $q_o = 1$, $a = 0.5$ m

Exercise for Section 13.7

Consider the semi-infinite plates just discussed in considerable detail. Instead of studying the response to the uniform load, consider a harmonic load, that is, $q(x, y) = q_n \cos\left(\frac{n\pi x}{a}\right)$. The counterpart to equation (13.45) is

$$\frac{dN_x}{dx} = 0$$

$$\frac{d^2 M_x}{dx^2} + q_n \cos\left(\frac{n\pi x}{a}\right) = 0$$

Develop equations for the displacements $u^o(x)$ and $w^o(x)$, and the six stress resultants for the boundary conditions in Figure 13.14(a) for a symmetric laminate. These equations provide the solution for a typical term in a Fourier series representation of a load that is distributed symmetrically about $x = 0$. Fourier series representations will be discussed shortly. Note, this solution is also valid for the boundary conditions of Figure 13.14(b), (c), and (d) because the laminate is symmetric.

13.8
PLATE EXAMPLE 4: A SQUARE, UNIFORMLY LOADED, CROSS-PLY PLATE

As a final example, consider a uniformly loaded, rectangular, symmetric, cross-ply plate that is free from any inplane forces and moments on its edges and, in fact, is free to move in the x and y direction on all the edges, but is prevented from out-of-plane motions at the edges. Assume that the origin of the x-y-z coordinate system is at the geometric center of the plate, as in Figure 13.1. Since the plate is symmetric and of cross-ply construction

$$A_{16} = A_{26} = D_{16} = D_{26} = 0$$
$$B_{ij} = 0 \quad \text{all } i \text{ and } j \tag{13.120}$$

As a result, the governing equations, from equation (13.36), are

$$A_{11}\frac{\partial^2 u^o}{\partial x^2} + A_{66}\frac{\partial^2 u^o}{\partial y^2} + (A_{12} + A_{66})\frac{\partial^2 v^o}{\partial x \partial y} = 0$$

$$(A_{12} + A_{66})\frac{\partial^2 u^o}{\partial x \partial y} + A_{66}\frac{\partial^2 v^o}{\partial x^2} + A_{22}\frac{\partial^2 v^o}{\partial y^2} = 0 \tag{13.121}$$

$$D_{11}\frac{\partial^4 w^o}{\partial x^4} + 2(D_{12} + 2D_{66})\frac{\partial^4 w^o}{\partial x^2 \partial y^2} + D_{22}\frac{\partial^4 w^o}{\partial y^4} = q_o$$

with q_o being the magnitude of the uniform load. Formally, the boundary conditions, since N_x^+, N_x^-, N_{xy}^+, N_{xy}^-, N_{yx}^+, N_{yx}^-, M_x^+, M_x^-, M_y^+, and M_y^- are zero, are as follows:
At $x = \pm a/2$:

(i) $\quad N_x = A_{11}\dfrac{\partial u^o}{\partial x} + A_{12}\dfrac{\partial v^o}{\partial y} = 0$

(ii) $\quad N_{xy} = A_{66}\left(\dfrac{\partial u^o}{\partial y} + \dfrac{\partial v^o}{\partial x}\right) = 0$

(iii) $\quad w^o = 0$ $\tag{13.122a}$

(iv) $\quad M_x = -\left\{D_{11}\dfrac{\partial^2 w^o}{\partial x^2} + D_{12}\dfrac{\partial^2 w^o}{\partial y^2}\right\} = 0$

At $y = \pm b/2$:

(i) $\quad N_y = A_{12}\dfrac{\partial u^o}{\partial x} + A_{22}\dfrac{\partial v^o}{\partial y} = 0$

(ii) $\quad N_{xy} = A_{66}\left(\dfrac{\partial u^o}{\partial y} + \dfrac{\partial v^o}{\partial x}\right) = 0$

(iii) $\quad w^o = 0$ $\tag{13.122b}$

(iv) $\quad M_y = -\left\{D_{12}\dfrac{\partial^2 w^o}{\partial x^2} + D_{22}\dfrac{\partial^2 w^o}{\partial y^2}\right\} = 0$

As is sometimes the case, for a particular problem the form of the solution can be determined by inspection. For this problem a solution of the form

$$u^o(x, y) = 0$$
$$v^o(x, y) = 0$$
$$w^o(x, y) = \sum_{m=1,3}^{\infty} \sum_{n=1,3}^{\infty} W_{mn} \cos\left(\frac{m\pi x}{a}\right) \cos\left(\frac{n\pi x}{b}\right) \tag{13.123}$$

satisfies the first two governing differential equations in equation (13.121), and *all* the boundary conditions at each edge, equation (13.122a) and (b). The W_{mn} can be determined by substituting the form for $w^o(x, y)$ into the third governing differential equation, expanding the uniform load in the same double cosine series, and equating coefficients. The form given by equation (13.123) happens to be the exact solution for this problem. If the plate was not of cross-ply construction, then the terms involving D_{16} and D_{26} would most likely appear in the third governing differential equation and the boundary conditions (iv) at each edge, and the form given by equation (13.123) would not be the solution form. However, for any symmetric cross-ply laminate with these same boundary conditions, the following procedure is valid if the load can be expanded in terms of a Fourier series.

The Fourier series representation of the uniform load is

$$q_o = \sum_{m=1,3}^{\infty} \sum_{n=1,3}^{\infty} Q_{mn} \cos\left(\frac{m\pi x}{a}\right) \cos\left(\frac{n\pi y}{b}\right) \tag{13.124a}$$

with

$$Q_{mn} = \frac{16 q_o}{mn\pi^2}(-1)^{\frac{m+n}{2}-1} \quad m, n = 1, 3, \ldots \tag{13.124b}$$

Substituting the series expressions for $w^o(x, y)$ and q_o into the governing equation, the third expression of equation (13.121), and equating like terms in m and n lead to

$$\left\{D_{11}\left(\frac{m\pi}{a}\right)^4 + 2(D_{12} + 2D_{66})\left(\frac{m\pi}{a}\right)^2\left(\frac{n\pi}{b}\right)^2 + D_{22}\left(\frac{n\pi}{b}\right)^4\right\} W_{mn} = Q_{mn}$$

$$\tag{13.125}$$

This results in

$$W_{mn} = \frac{Q_{mn}}{D_{11}\left(\frac{m\pi}{a}\right)^4 + 2(D_{12} + 2D_{66})\left(\frac{m\pi}{a}\right)^2\left(\frac{n\pi}{b}\right)^2 + D_{22}\left(\frac{n\pi}{b}\right)^4} \tag{13.126}$$

or

$$W_{mn} = 16 q_o \frac{(-1)^{\frac{m+n}{2}-1}}{mn\pi^6\left\{D_{11}\left(\frac{m}{a}\right)^4 + 2(D_{12} + 2D_{66})\left(\frac{m}{a}\right)^2\left(\frac{n}{b}\right)^2 + D_{22}\left(\frac{n}{b}\right)^4\right\}}$$

$$\tag{13.127}$$

With the above solution for W_{mn}, expressions for the curvatures, moments, strains, and stresses can be obtained in series form. There are no inplane force resultants or inplane strains for this problem. With the W_{mn} known, the curvatures are

$$\kappa_x^o = -\frac{\partial^2 w^o}{\partial x^2} = \sum_{m=1,3}^{\infty} \sum_{n=1,3}^{\infty} W_{mn}\left(\frac{m\pi}{a}\right)^2 \cos\left(\frac{m\pi x}{a}\right) \cos\left(\frac{n\pi y}{b}\right)$$

$$\kappa_y^o = -\frac{\partial^2 w^o}{\partial y^2} = \sum_{m=1,3}^{\infty} \sum_{n=1,3}^{\infty} W_{mn}\left(\frac{n\pi}{b}\right)^2 \cos\left(\frac{m\pi x}{a}\right) \cos\left(\frac{n\pi y}{b}\right) \tag{13.128}$$

$$\kappa_{xy}^o = -2\frac{\partial^2 w^o}{\partial x \partial y} = -2 \sum_{m=1,3}^{\infty} \sum_{n=1,3}^{\infty} W_{mn}\left(\frac{m\pi}{a}\right)\left(\frac{n\pi}{b}\right) \sin\left(\frac{m\pi x}{a}\right) \sin\left(\frac{n\pi y}{b}\right)$$

and the moment resultants are

$$M_x = D_{11}\kappa_x^o + D_{12}\kappa_y^o = \sum_{m=1,3}^{\infty} \sum_{n=1,3}^{\infty} \left(D_{11} \left(\frac{m\pi}{a}\right)^2 + D_{12} \left(\frac{n\pi}{b}\right)^2 \right)$$
$$\times W_{mn} \cos\left(\frac{m\pi x}{a}\right) \cos\left(\frac{n\pi y}{b}\right)$$

$$M_y = D_{12}\kappa_x^o + D_{22}\kappa_y^o = \sum_{m=1,3}^{\infty} \sum_{n=1,3}^{\infty} \left(D_{12} \left(\frac{m\pi}{a}\right)^2 + D_{22} \left(\frac{n\pi}{b}\right)^2 \right) \qquad (13.129)$$
$$\times W_{mn} \cos\left(\frac{m\pi x}{a}\right) \cos\left(\frac{n\pi y}{b}\right)$$

$$M_{xy} = D_{66}\kappa_{xy}^o = -2D_{66} \sum_{m=1,3}^{\infty} \sum_{n=1,3}^{\infty} W_{mn} \left(\frac{m\pi}{a}\right)\left(\frac{n\pi}{b}\right) \sin\left(\frac{m\pi x}{a}\right) \sin\left(\frac{n\pi y}{b}\right)$$

The transverse shear force resultants are

$$Q_x = \frac{\partial M_x}{\partial x} + \frac{\partial M_{xy}}{\partial y} = -\sum_{m=1,3}^{\infty} \sum_{n=1,3}^{\infty} W_{mn} \left\{ D_{11} \left(\frac{m\pi}{a}\right)^3 + D_{12} \left(\frac{n\pi}{b}\right)^2 \left(\frac{m\pi}{a}\right) \right.$$
$$\left. + 2D_{66} \left(\frac{m\pi}{a}\right)\left(\frac{n\pi}{b}\right)^2 \right\} \sin\left(\frac{m\pi x}{a}\right) \cos\left(\frac{n\pi y}{b}\right)$$

$$Q_y = \frac{\partial M_y}{\partial y} + \frac{\partial M_{xy}}{\partial x} = -\sum_{m=1,3}^{\infty} \sum_{n=1,3}^{\infty} W_{mn} \left\{ D_{22} \left(\frac{n\pi}{b}\right)^3 + D_{12} \left(\frac{m\pi}{a}\right)^2 \left(\frac{n\pi}{b}\right) \right.$$
$$\left. + 2D_{66} \left(\frac{m\pi}{a}\right)^2 \left(\frac{n\pi}{b}\right) \right\} \cos\left(\frac{m\pi x}{a}\right) \sin\left(\frac{n\pi y}{b}\right)$$

$$(13.130)$$

Note, these are not the effective, or Kirchhoff, transverse shear force resultants discussed earlier. The strains at any point x, y, or z within the laminate are

$$\varepsilon_x = z\kappa_x^o = z \sum_{m=1,3}^{\infty} \sum_{n=1,3}^{\infty} W_{mn} \left(\frac{m\pi}{a}\right)^2 \cos\left(\frac{m\pi x}{a}\right) \cos\left(\frac{n\pi y}{b}\right)$$

$$\varepsilon_y = z\kappa_y^o = z \sum_{m=1,3}^{\infty} \sum_{n=1,3}^{\infty} W_{mn} \left(\frac{n\pi}{b}\right)^2 \cos\left(\frac{m\pi x}{a}\right) \cos\left(\frac{n\pi y}{b}\right) \qquad (13.131)$$

$$\gamma_{xy} = z\kappa_{xy}^o = -2z \sum_{m=1,3}^{\infty} \sum_{n=1,3}^{\infty} W_{mn} \left(\frac{m\pi}{a}\right)\left(\frac{n\pi}{b}\right) \sin\left(\frac{m\pi x}{a}\right) \sin\left(\frac{n\pi y}{b}\right)$$

The stresses σ_x, σ_y, and τ_{xy} can be determined on a layer-by-layer basis. Of course, the strains, and stresses, in the principal material system depend on the particular laminate.

As a specific case, consider a $[0/90]_S$ laminate, as in the previous example, and assume the laminate has an aspect ratio $a/b = 0.5$, namely, the plate is twice as long as it is wide. The deflection of the center of the plate is given, from equation

(13.123), by

$$w^o(0,0) = \sum_{m=1,3}^{\infty} \sum_{n=1,3}^{\infty} W_{mn} \tag{13.132}$$

Obviously the sums cannot be made infinite and so it is of value to determine the predicted deflection as a function of the upper limits of the summation M and N, where now

$$w^o(0,0) = \sum_{m=1,3}^{M} \sum_{n=1,3}^{N} W_{mn} \tag{13.133}$$

Table 13.1 shows the convergence characteristics of the series for the midplate deflection. The midplate deflection has been normalized as before for the sake of comparison. Recall that $^{S}D_{11}$ is the value of D_{11} for a $[0/90]_S$ laminate, equation (13.116). About four terms ($M = N = 7$) are all that are required to provide the converged answer for the deflection at the center of the plate. With this number of terms the nondimensional deflection at the center for the plate is just slightly greater than 1. This is interesting in that for the semi-infinite $[0/90]_S$ plate the nondimensionalized midspan deflection was 1.0 (see Figure 13.15). Obviously as far as the deflection in the middle of the plate is concerned, the plate being only twice as long as it is wide, that is, $a/b = 0.5$, is equivalent to the plate being infinitely long, that is, $b \to \infty$, or $a/b \to 0$.

TABLE 13.1
Convergence characteristics for $[0/90]_S$ plate

	$a/b = 0.5$		
M, N	$\left(\dfrac{384^{s}D_{11}}{5q_o a^4}\right) w^o(0,0)$	$\left(\dfrac{8}{q_o a^2}\right) M_x(0,0)$	$\left(\dfrac{8}{q_o a^2}\right) M_y(0,0)$
1	1.209758	1.250537	0.095158
3	1.035223	1.033101	0.004721
5	1.058815	1.062241	0.037980
7	1.054079	1.055854	0.025005
9	1.055400	1.057856	0.030969
.	.	.	.
.	.	.	.
.	.	.	.
15	1.055032	1.057028	0.028457
17	1.055084	1.057319	0.029301
.	.	.	.
.	.	.	.
.	.	.	.
35	1.055065	1.057179	0.028907
37	1.055066	1.057207	0.028989
.	.	.	.
.	.	.	.
.	.	.	.
45	1.055066	1.057202	0.028973
47	1.055066	1.057189	0.028933

Also of interest are the moment resultants M_x and M_y at the center of the plate ($M_{xy} = 0$ there). From equation (13.129) these are

$$M_x(0,0) = \sum_{m=1,3}^{M} \sum_{n=1,3}^{N} \left(D_{11} \left(\frac{m\pi}{a} \right)^2 + D_{12} \left(\frac{n\pi}{b} \right)^2 \right) W_{mn}$$

$$M_y(0,0) = \sum_{m=1,3}^{M} \sum_{n=1,3}^{N} \left(D_{12} \left(\frac{m\pi}{a} \right)^2 + D_{22} \left(\frac{n\pi}{b} \right)^2 \right) W_{mn} \qquad (13.134)$$

$$M_{xy}(0,0) = 0$$

The convergence characteristics of these moments are also shown in Table 13.1; convergence is slightly less rapid than for the deflection of the center, and the moments are nondimensionalized as before, simply for comparison. As with the deflections, the values of the nondimensionalized moments M_x and M_y at the center of the plate are similar in magnitude to the values of these moments at the midspan of the semi-infinite $[0/90]_S$ plate. We can see this by comparing the converged values of the nondimensionalized values of M_x and M_y at the center of the plate ($M = N = 47$) with the values from the plots of Figures 13.18 and 13.19.

The stresses σ_x, σ_y, and τ_{xy} at the center of the plate for a unit load and with $a = 0.5$ m are illustrated in Figure 13.26. Comparing these stresses with the stresses at midspan in a semi-infinite $[0/90]_S$ plate, Figure 13.20, again reveals a similarity between the semi-infinite plate and the rectangular plate.

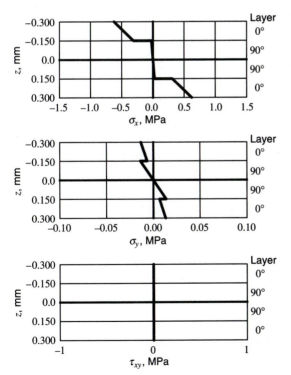

FIGURE 13.26
Stresses at center of pressure-loaded, simply supported, rectangular $[0/90]_S$ plate, $q_o = 1$, $a = 0.5$ m, $b = 1.0$ m

The results of Table 13.1 and the stresses in Figure 13.26 can be contrasted with the results of Table 13.2 and the stresses of Figure 13.27, where the midplate deflections and moments for a $[90/0]_S$ plate are illustrated in this second table and it is evident a simple switching of layer orientations makes a large change in the plate response. The deflection is four times larger than for the $[0/90]_S$ plate. In addition, overall, the moment and stress levels have increased for the following reason: For most plate problems, the bending stiffness in the direction of the shorter span controls the plate response. In this problem the bending stiffness in the short span direction is D_{11}. The $[0/90]_S$ laminate has a larger D_{11} than the $[90/0]_S$ laminate; therefore, the $[90/0]_S$ laminate deflects more, and as a result, has larger stresses. Viewed another way, the $0°$ layers are closer together in the $[90/0]_S$ laminate than in the $[0/90]_S$ laminate, yet they must support the same load across span a; as they are closer together in the $[90/0]_S$ laminate and are the primary load-carrying layers for span a, they must be more highly stressed than in the $[0/90]_S$ laminate to produce the same moment to support the load.

Finally, it is worth commenting on the convergence in the representation of the uniform load q_o. By equation (13.124) the load too is being represented by a double cosine series. This series is truncated at $m = M$ and $n = N$, as are the series for $w^o(x, y)$, $M_x(x, y)$, and the like. The load is represented as a function of the number of terms in the series in Figure 13.28. Here the load, normalized by the value of the intended uniform load q_o, as a function of location along the plate centerline $y = 0$

TABLE 13.2
Convergence characteristics for $[90/0]_S$ plate

M, N	$\left(\dfrac{384^s D_{11}}{5q_o a^4}\right) w^o(0,0)$	$\left(\dfrac{8}{q_o a^2}\right) M_x(0,0)$	$\left(\dfrac{8}{q_o a^2}\right) M_y(0,0)$
		$a/b = 0.5$	
1	3.946309	0.908713	1.103067
3	3.853063	0.851638	0.930295
5	3.858660	0.859876	0.964304
7	3.858419	0.856518	0.951863
9	3.858766	0.858028	0.957629
·	·	·	·
·	·	·	·
15	3.858665	0.857344	0.955100
17	3.858679	0.857572	0.955960
·	·	·	·
·	·	·	·
35	3.858674	0.857463	0.955556
37	3.858674	0.857485	0.955639
·	·	·	·
·	·	·	·
45	3.858674	0.857480	0.955622
47	3.858674	0.857470	0.955814

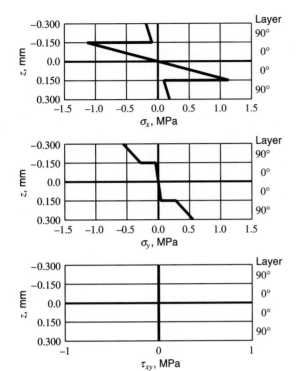

FIGURE 13.27

Stresses at center of pressure-loaded, simply supported, rectangular $[90/0]_S$ plate, $q_o = 1$, $a = 0.5$ m, $b = 1.0$ m

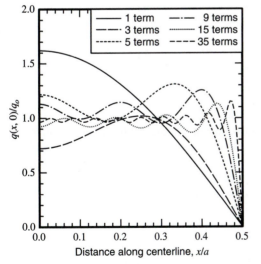

FIGURE 13.28

Variation of load along centerline $y = 0$ as a function of the number of terms in series

varies as the number of terms in the series increases. Obviously $M = N = 1$ does not represent a uniform load well and seriously biases the load toward the center of the plate. Thus, convergence of the various series for $w^o(x, y)$, $M_x(x, y)$, and so on, also involves how well the intended uniform load is represented.

Exercises for Section 13.8

1. As we saw from Tables 13.1 and 13.2, the moments converged less rapidly with increasing M and N than did the deflection. This is because the coefficients in the sum, W_{mn}, are inversely proportional to the fourth power of m and n. The larger m and n are, the less the contribution of W_{mn} to the summation. Hence, the converged displacement solution is contained in the $(1, 1)$, $(3, 3)$, $(5, 5)$, and $(7, 7)$ terms. The moments, however, are inversely proportional to the second power of m and n. Hence, the coefficients in the sums for the moments make contributions until $m = n = 15$. (a) To what power of m and n are the transverse shear stress resultants proportional? (b) Add two columns to Tables 13.1 and 13.2 and compute the value of $Q_x(a/2, 0)$ and $Q_y(0, b/2)$ as a function of M and N.

2. It is instructional to plot the variation of the plate response as a function of spatial location, and as a function of the number of terms in the series, M and N. For the $[0/90]_S$ laminate with $a/b = 0.5$, plot the variation of $w^o(x, 0)$, $M_x(x, 0)$, and $Q_x(x, 0)$ versus x, $0 \leq x \leq a/2$, as a function of M and N. Do a separate plot for each variable and let $M = N$. Note that besides requiring a certain number of terms for the response to converge at a particular point (e.g., $x = 0$, $y = 0$), a certain number of terms is required to have the functional relationships converge also.

3. We spent considerable time studying the semi-infinite plate. Is there any value to this? To answer this question, let's compare the response of a semi-infinite plate with the response of plates with $a/b = 1$, and less. For variety, consider a special laminate, specifically one with a stacking sequence of $[+\theta/-\theta/-\theta/+\theta/-\theta/+\theta/+\theta/-\theta]_S$, $\theta = 30°$. This laminate has no D_{16} or D_{26} terms, and hence the analysis for this laminate is identical to the analysis for a cross-ply laminate. For this laminate plot $w^o(x)$ versus x as determined by a semi-infinite plate analysis and $w^o(x, b/2)$ versus x as determined by the double series rectangular plate solution. Do this for $a/b = 1, 0.75, 0.50$, and so on. Comment on the value of a/b for which, as far as out-of-plane deflection is concerned, the rectangular plate behaves as an infinitely long plate.

13.9
SUMMARY

This completes our brief introduction to plates. Though limited, the examples illustrate the fundamentals of studying plate response. For general plate problems convenient exact solutions are not always readily available. However, approximate solutions based on variational methods, such as the Rayleigh-Ritz or Galerkin techniques, can give good results. Finite-element analysis, which is based on variational methods, is also very useful for difficult problems. Important results are often revealed by the proper use of approximate methods. One example of this is the out-of-plane deflection characteristics of a $[\pm30/0]_S$ simply supported plate. Figure 13.29(a) shows a contour plot of the out-of-plane deflections of the $[0/90]_S$ cross-ply plate. This contour plot is obtained directly from equations (13.123) and (13.127). Figure 13.29(b) shows the out-of-plane deflection contours of a $[\pm30/0]_S$ plate; this plot was obtained by a Rayleigh-Ritz solution to the governing equation for a $[\pm30/0]_S$ laminate, namely, equations (13.34) and (13.35). With a careful consideration, we can see there is a slight skewing of the contour lines of the $[\pm30/0]_S$ laminate when compared to the contour lines of the $[0/90]_S$ laminate. This skewing is due to the nonzero values of D_{16} and D_{26} for the $[\pm 30/0]_S$ laminate. The contour plot for an eight-layer $[+45_2/-45_2]_S$ laminate of Figure 13.29(c) exemplifies this skewing even more. This can be contrasted to the skewing in the opposite direction for the $[-45_2/+45_2]_S$ laminate of Figure 13.29(d). The clustering of the $+45$ and -45 layers (i.e., having two adjacent layers with the same fiber orientation), results in large values for D_{16} and D_{26}, and hence more skewing. Furthermore, the direction of skewing is dependent on the signs of D_{16} and D_{26}. The signs of D_{16} and D_{26} for the $[-45_2/+45_2]_S$ laminate are opposite the signs of D_{16} and D_{26} for the $[+45_2/-45_2]_S$ laminate. Skewing is an important effect and can only be studied by using approximate techniques.

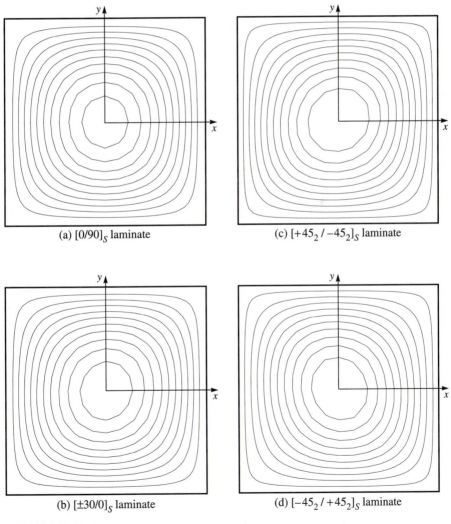

(a) $[0/90]_S$ laminate

(c) $[+45_2/-45_2]_S$ laminate

(b) $[\pm 30/0]_S$ laminate

(d) $[-45_2/+45_2]_S$ laminate

FIGURE 13.29
Out-of-plane deflection characteristics of various laminated plates

A concluding comment is in order regarding the approach taken to study the rectangular cross-ply plate versus the approach taken to study the semi-infinite plate. With the semi-infinite plate, the solution to the governing equations and boundary conditions was obtained by a methodical step-by-step integration of equations and determination of the constants of integration. By contrast, the solution to the rectangular plate more or less appeared out of thin air. The approach taken with the semi-infinite plate is the preferred approach. Unfortunately, for many problems even the first step in obtaining exact solutions is not possible; hence, the analysis is stopped early. However, often it is possible to envision, guess, or assume what the deflected shape might look like, and then to translate this shape into known functions of x and y. These known functions are then used in the equations, and often it is possible to move forward and obtain answers in the context of these functions. This is the basis for many approximate methods, including the Rayleigh-Ritz technique. More often than not, this is the approach that must be taken. The key issue is this: How good are the assumed functions? The answer to this question is not really known unless the assumed-function approach is used for a situation where there is actually an exact solution and the answers are compared. For the rectangular cross-ply plates considered here, the assumed double series of harmonic functions is the exact solution. Though this was not proven, it was stated when the double series was introduced. This assumed-function approach to the rectangular-plate example was taken because for laminates other than cross-ply, it is the only approach that will yield numerical results. However, studying problems that can be solved by integrating the governing differential equations and applying the boundary conditions always provides good insight into a problem. These problems also serve as a benchmark for approximate-solution techniques. With the exact solution, there is never a concern as to whether an unusual response is a physical reality or is the result of not assuming quite the right functions. With the exact solution, the physics of the problem are never in doubt, nor are they ever masked or biased by approximations.

13.10
SUGGESTED READINGS

For additional derivations, solutions to other plate problems (such as vibrations and thermal effects, through-thickness shear deformations), and extensions of the concepts developed in this chapter to other structural forms (e.g., cylindrical shells), consult these works:

1. Vinson, J. R. *The Behavior of Shells Composed of Isotropic and Composite Materials.* Hingham, MA: Kluwer Academic Publishers, 1993.
2. Vinson, J. R., and R. L. Sierakowski. *The Behavior of Structures Composed of Composite Materials.* Hingham, MA: Kluwer Academic Publishers, 1987.
3. Whitney, J. M. *Structural Analysis of Laminated Anisotropic Plates.* Lancaster, PA: Technomic Publishing Co., 1987.

For insights into more advanced topics, consult the following:

4. Turvey, G. J., and I. H. Marshall, eds. *Buckling and Postbuckling of Composite Plates.* New York: Chapman & Hall, 1994.
5. Palazotto, A. N., and S. T. Dennis. *Nonlinear Analysis of Shell Structures.* Washington, DC: American Institute of Aeronautics and Astronautics, 1992.

6. Vasiliev, V. V. *Mechanics of Composite Structures*. Bristol, PA: Taylor and Francis, 1993.

The definitive text for inplane loading of laminates that contain holes—and which is based on analysis using the theory of complex variables—is the following:

7. Lekhnitskii, S. G. *Anisotropic Plates*. New York: Gordon and Breach Publishers, 1968.

Several other classic texts in the field address structural level problems:

8. Jones, R. M. *Mechanics of Composite Materials*. New York: McGraw-Hill, 1975.
9. Calcote, L. R. *The Analysis of Laminated Composite Structures*. New York: Van Nostrand Reinhold, 1969.

CHAPTER 14

Appendix: Manufacturing Composite Laminates

14.1
BACKGROUND AND OVERVIEW

The multitude of tasks involved in the manufacturing of composite laminates can be categorized into two phases: (1) *fabrication* and (2) *processing*. In the fabrication phase the fiber reinforcement and accompanying matrix material are placed or shaped into a structural form such as a flat or curved plate, a cylinder or other body of revolution, and the like. The fiber and matrix may be in preimpregnated form, or the fiber and matrix material may be combined for the first time during this step of developing the structural form. During the processing phase, heat and pressure are used to densify and consolidate the structure. For thermoset matrices the chemical cross-linking reaction (i.e., curing) solidifies the structure, whereas thermoplastic matrices become hard after cooling from their melting temperature.

Fabrication techniques for composites are not dependent on the type of matrix material. In fact, some metal forming techniques have been adapted to composites fabrication (e.g., matched-metal die molding). However, processing conditions are entirely dependent on the type of matrix material used. For instance, thermosets require long processing times, whereas thermoplastics require relatively high pressures and temperatures.

In this appendix we present a brief introduction to the manufacturing of composites by addressing three important areas. First, we discuss fabrication techniques. Next, processing issues are presented. We conclude with a short discussion of manufacturing defects. Overall, we approach the topic of manufacturing from a general perspective. However, to keep the discussion in the context of this book, we begin by focusing on the manufacture of structural components from layers, or plies, of preimpregnated material, called prepreg. Specifically, layers of material, with the fibers in each layer aligned in a specific direction, are used to form a laminate. We will assume that the laminate is fabricated by hand, and we will describe the necessary steps of fabrication. We will further assume that this hand-fabricated laminate will be processed in an autoclave, which is a pressurized oven that provides the proper levels of heat and pressure to solidify and consolidate the structure. In the early years of the development of fiber-reinforced materials, structural components were fabricated by hand. Even today, in prototype development, hand fabrication is common, and this is also the case for specialty manufacturing and in many university laboratories. However, as labor costs and the need for consistency have increased, engineers have been charged with designing low-cost automated manufacturing techniques. Now, automated techniques like robotic tow and tape placement methods, injection molding, and pultrusion have dramatically reduced the cost of manufacturing some composite structures. Later in the appendix we will consider these various other approaches to fabricating and processing a laminate.

Common to all manufacturing methods is the use of a die, mold, or mandrel. They provide the structural shape for the composite material, and in this discussion they will be referred to generally as the tool. Tools are usually an inverse, or female, replica of the desired structural shape. The design of the tool is a critical and expensive process. The cost of the tool often far exceeds the material and labor costs to produce a composite structure. Also common to all manufacturing methods is, as mentioned, the need to apply temperature and pressure to the structural component after the fiber and matrix are brought together into the desired structural form. The pressure takes two forms: actual pressure, ideally hydrostatic, to consolidate the tows and layers; a vacuum to remove air entrapped between the layers and to reduce the amount of unwanted gases given off by the resin as it cures. The application of pressure can be in the form of closing both halves of the tool or, as with a flat structural component, pressing the laminate in a hot press. More commonly, however, and as will be assumed here, pressure is applied by putting the uncured structural component into an autoclave. Finally, the vacuum requirement is met by enclosing the structural component in a vacuum-tight bag and drawing a vacuum.

14.2
FABRICATION

14.2.1 Tooling and Specialty Materials

Tooling

As all fabrication methods require tools to provide the shape of the composite structure during processing, the design and construction of the tool are critical components of the manufacturing process. Because the tool is heated and pressurized, especially critical is the choice of tooling material. Factors which must be considered in tool material selection are dimensional stability and compatibility, cost, surface finish, and durability. Table A.1 lists the coefficients of thermal expansion for several tool materials. Of the metals, tool steel most closely matches that of the composites. Steel tools are also highly durable and have good thermal conductivity;

TABLE A.1
Thermal expansion of tooling materials

Material system	Coefficient of thermal expansion ($\times 10^{-6}$/°C)
Polymer matrix composites (fiber dir., α_1)	
• Aramid or graphite-reinforced	−1.5 to 1.1
• Boron fiber-reinforced	2.3 to 3.0
• Glass fiber-reinforced	6.3 to 8.4
Slip cast and fired ceramic	0.83
Tool steel	11.1
Electroformed iron	11.7
Electroformed nickel	12.8
Plaster	13.9
High-temperature epoxy	19.4
Aluminum	23.3

however, they are extremely heavy and they take substantial time to heat and cool. Ceramic tools have the lowest thermal expansion, so their dimensional stability is the best, and they also have a thermal conductivity close to that of tool steels. However, they are brittle and must be protected from chipping and cracking. Sometimes ceramic inserts are used in steel tools to combine the best characteristics of both materials.

Aluminum tools are easily machined and less expensive than steel or ceramic. They are lighter than steel tools and they heat and cool faster than steel; however, they are not as durable as steel tools and their thermal expansion is excessive. Plaster tools are sometimes used when durability is not required. They can be made easily by pouring the uncured plaster around a model. Once the plaster has been cast, it is cured and then hardened by coating with a varnish. Actually, graphite- or glass-reinforced composite materials can be used to fabricate a tool. If this is the case the thermal expansion of the tool can be exactly matched to that of the composite structure. Composite tools are durable, their surface finish is excellent, and they are less expensive than steel tools. However, they usually require that a plaster casting be made of the structure first.

Specialty materials

Many secondary or specialty materials are used in composites manufacturing. Before we discuss the various types of specialty materials it is helpful to examine a typical lay-up of a composite structure prepared for autoclave processing. Figure A.1 shows a cross section of an autoclave specimen. In addition to the actual composite laminate, there are release coatings, peel plies, release film, bleeder plies, breather plies, vacuum bags, sealant tape, and damming material. Each of these materials serves a specific function. For instance, release agents are used to prevent the composite material from bonding to the tool. The illustration in Figure A.1 is based on processing a flat laminate, and thus the portion

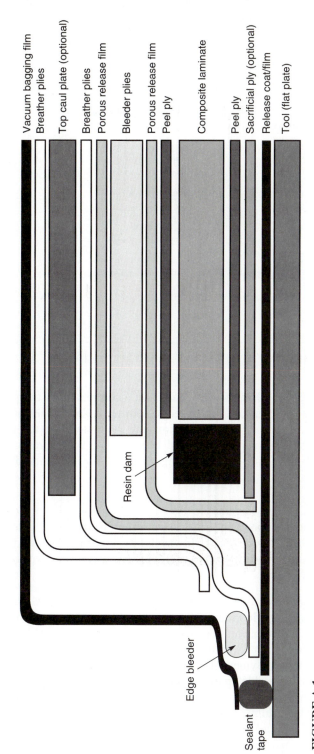

FIGURE A.1

Typical autoclave lay-up

Vacuum bagging film
Breather plies
Top caul plate (optional)
Breather plies
Porous release film
Bleeder plies
Porous release film
Peel ply
Composite laminate
Peel ply
Sacrificial ply (optional)
Release coat/film
Tool (flat plate)

Resin dam

Edge bleeder

Sealant tape

581

TABLE A.2
Mold release agents

Type	Form	Examples
Fluorocarbons	Films or dispersions	Teflon (tetrafluoroethylene), Tedlar (polyvinylfluoride)
Polymer films	Coated paper, extruded film	Polyvinyl alcohol (PVA), polyamines, polyethylene, cellophane
Silicones	Liquids, resins, greases	Silicone polymers
Waxes	Paste	Parafin, carnuba, microcrystalline waxes
Metal salts	Liquids or particles (external and internal release agents)	Stearic acids (calcium, zinc, lead, aluminum, magnesium salts)
Inorganic compounds	Powders	Talcum, mica

of the tool at the bottom of the figure is actually a flat plate and the portion of the tool toward the top is also a flat plate, sometimes called a caul plate when used in this fashion. However, it is easy to imagine the bottom and top tool components having curvature, or only a curved bottom plate being present and the peel plies, release film, and so on, being draped over the curved uncured laminate.

Release agents.   Release agents are used to coat the tool so that the composite structure is prevented from bonding to the surface. They are usually a paste or liquid that is coated onto the tool surface and allowed to dry. Films are also used, but they are limited to flat or single-curvature surfaces. Table A.2 lists several types of release agents that are used.

1. *Fluorocarbons.* Fluorocarbon polymers are used extensively in autoclave molding. Tedlar®, a polyvinyl fluoride, is a fluorocarbon film manufactured by DuPont. FEP (fluoroethylene propylene) forms a continuous film on the mold surface and is used for composites cured up to 177°C. Above this temperature the fluorine disassociates from the polymer. PTFE (polytetrafluoroethylene), a polymeric dispersion of Teflon®, is stable up to 260°C.
2. *Polymer films.* These polymers are insoluble in most solvents, so they are applied as extruded or blown films. PVA (polyvinyl alcohol), cellophane, polyamines, and polyethylene have all been used as release agents.
3. *Silicones.* The commercial silicones are cured polymers with high melting points and low volatility. They are applied in liquid form or as a grease. Some special

silicone release agents are stable up to 480°C; however, most are limited to about 200°C.

4. *Waxes*. Carnuba paste wax is cheap and easily applied. It is sometimes polished before the composite is laid onto the tool surface. Carnuba waxes are excellent mold release agents for composites cured below 250°F. Above this temperature the wax begins to degrade.

5. *Metal salts*. Stearic acid, a fatty acid, is used widely for mold release. It has a sharply defined melting point at 71.5°C and has good wetting properties. The main derivatives of stearic acid (such as calcium, zinc, and lead salts) are also used as release agents. The choice of metallic stearate to use for a specific application depends primarily on the type of polymer matrix.

6. *Inorganic compounds*. These are probably the oldest known release agents. Because they are insoluble, they are applied as powders. Talcum and mica are the most common compounds used. In some cases they are mixed with metal stearates to improve their release ability.

Peel plies and release films and fabrics.   Surfaces are protected from contamination by peel plies, and they are normally removed from the composite structure just before bonding or secondary coating operations. The most common peel plies are heat-cleaned-and-scoured nylon, heat-cleaned lightweight fiberglass, or polyester fabrics. User preference of the surface texture after the removal of the peel ply dictates the choice of a specific type.

Release films and fabrics serve many different purposes. Sometimes they are used as separators between successive layers of preimpregnated material, and they are also used to separate the bleeder or breather materials from the composite laminate. They are most commonly a Teflon-coated fiberglass fabric, and in some cases the release film is porous so that resin can flow through the film. For example, porous release films are used to separate the bleeder plies from the composite laminate. This allows the resin to flow from the laminate through the release film and into the bleeder plies.

Bleeder and breather plies.   Bleeder and breather plies are porous, high-temperature fabrics which are used to absorb excess resin during processing. Most preimpregnated materials are supplied with excess resin, which is subsequently removed during processing. The resin is removed to increase the fiber volume fraction and flush voids from the laminate. The excess resin is trapped and absorbed by the porous bleeder plies. Fiberglass, cellulose, and polyester fabrics are all used as bleeder plies.

Breather plies are used to provide a vacuum pathway into the composite laminate, and they also act as a conduit for the removal of volatiles during cure. They must remain porous at high temperatures and pressures. Most bleeder materials can also be used as breather plies. In addition, perforated fluorocarbon or nylon films and Teflon-coated fiberglass are sometimes used.

Bagging films.   Bagging films form a barrier between the composite laminate and the oven or autoclave environment. The bagging film is sealed around the edge of the lay-up by sealant tape, and the film is drawn down onto the composite laminate by pulling a vacuum under the bag. Vacuum bags must be heat resistant, flexible, nonvolatile, and resistant to tearing. Several high-temperature polymer films are used, including Kapton® (up to 316°C), nylon (180°C), and PVA (121°C). Silicon rubber bags are also used up to about 200°C, and they have the added advantage of being reusable.

14.2.2 Hand Lay-up

Even though the method has been replaced with automated techniques, the lay-up of preimpregnated material by hand is the oldest and most common fabrication method for advanced composite structures. Furthermore, the basic features of the method remain unchanged. A pictorial essay showing each step in the hand lay-up of a flat composite laminate is shown in Figures A.2–A.15. Each step must follow in successive fashion in order to obtain a high-quality composite laminate after final processing. A description of these steps follows.

Step 1, Figure A.2.   The surface of the tool is cleaned and a release agent is applied. If the surface is not clean, then the release agent will not function properly. The release agent can be in liquid form, or it may be a solid film. (In the photo-essay, to provide an indication of scale, a hand-held pointer or knife is included in the photographs.)

Step 2, not shown.   An optional sacrificial layer is laid up on the tool surface. This layer is usually a fiberglass fabric made with the same resin system as the composite laminate. The sacrificial layer protects the laminate from surface abrasion and surface irregularities during manufacturing.

Step 3, Figure A.3.   A peel ply is placed on top of the sacrificial layer. The peel ply will be removed after processing.

Step 4, Figure A.4.   The preimpregnated plies are cut according to design specifications. They can be cut by hand using shears or a steel blade knife. However, automated cutting machines have largely replaced hand cutting. The Gerber knife is a reciprocating-knife system originally developed for the textile industry. It is extremely fast and can cut up to 20 plies at one time. Lasers have been used for cutting, but they are expensive and have limitations on the number of plies that can be cut at one time. Water-jet cutters are also used extensively, and they can cut a large number of plies (> 40) at one time, but some moisture absorption occurs during the cutting operation. Ultrasonic cutters have been used as well.

Step 5, Figures A.5, A.6.   The first prepreg ply is oriented and placed upon the tool or mold. Subsequent plies are placed one upon another; a roller or other small hand tool is used to compact the plies and remove entrapped air that could later lead to voids or layer separations. It is important that the preimpregnated material have sufficient tack so that it sticks slightly to the peel ply and to the adjacent plies. Tackiness, a characteristic of preimpregnated material, quantifies the relative stickiness of the plies at room temperature. As the preimpregnated material ages, its tackiness is reduced. Eventually, the plies no longer stick together and they may have to be heated slightly to soften them during lay-up. Oils and dirt on the surface of the preimpregnated plies will contribute to reducing composite strength after processing. Technicians should wear gloves during lay-up so that oils and dirt from the hands do not contaminate the prepreg plies during lay-up. In some cases the hand lay-up procedure may be carried out in a clean room to reduce the risk of contamination of the prepreg plies.

Step 6, Figure A.7.   A flexible resin dam is anchored to the sacrificial layer approximately 3 mm from the edge of the laminate. The dam prevents resin flow out of the laminate, in the plane of the laminate. Flexible dams can be made from silicon rubber, cork, or release coated metal. (As no sacrificial layer is being used in the procedure here, the flexible dam is anchored to the peel ply.)

Step 7, Figure A.8.   Another peel ply is placed on top of the laminate to protect the laminate surface.

Step 8, not shown.   A sheet of porous release film is laid over the dam and the laminate. The porous release film will serve as a barrier to prevent bonding of the composite laminate to the secondary materials to follow.

Step 9, Figure A.9.   Next, bleeder plies are laid up over the release film, in this case the peel ply. The bleeder plies extend to the edge of the laminate. The number of bleeder plies to be used for a given laminate can be determined by using a resin flow process model or through empirical observation. As the number of bleeder plies increases, the final fiber volume fraction of the composite laminate increases. Eventually, a maximum number of bleeder plies is reached and no further increase in fiber volume fraction occurs.

Step 10, not shown.   Another porous release ply is next laid up over the bleeder plies extending past the flexible dam. This prevents excessive resin flow into the breather material while maintaining a vacuum pathway into the composite laminate.

Step 11, not shown.   Breather plies are placed over the entire lay-up. The breather plies will conduct the vacuum path into the laminate. It is critically important that sufficient breather material is used throughout the entire laminate. Creases and areas with shallow curvature are sometimes reinforced with additional layers of breather material to ensure that the breather plies do not collapse in these areas. Usually, two or three breather plies are sufficient.

Step 12, not shown. An edge bleeder is used to connect to the vacuum ports. An edge bleeder is nothing more than a strip of breather material folded along its length several times. It is placed so that it overlays the breather material surrounding the laminate and extends out to a convenient location for the placement of the vacuum port.

Step 13, Figure A.10. Caul plates are sometimes placed on top of the lay-up. The caul plate is a steel or aluminum plate that protects the surface from sharp temperature increases (it acts as a heat sink) and it gives a smooth nonwavy surface texture.

Step 14, Figure A.11. If a caul plate is used, then additional breather or bleeder plies are placed over the plate to protect the vacuum bag from puncture.

Step 15, Figure A.12. Sealant tape is placed around the entire periphery of the lay-up.

Step 16, Figure A.12. The vacuum bag is cut to size and placed over the lay-up.

Step 17, Figure A.13. The bag is sealed by pressing the bag over the sealant tape. It is critically important to ensure that the bag is adequately sealed before proceeding to the processing cycle. Many parts are scrapped because the vacuum fails during processing, causing excessive voids, inadequate resin flow, or incomplete consolidation.

Step 18, Figures A.14 and A.15. The vacuum port is installed through the bag and the contents are evacuated. The bag is now checked for leaks. If any are detected, they are repaired before processing. Usually a leak test calls for application of a vacuum to some specified level (cm of Hg), followed by a 30–60 minute hold. During the hold the bag is disconnected from the vacuum source and the pressure level within the bag is monitored. If the bag is sealed well and there are no leaks, then the vacuum level should not change for the 30–60 minutes. Some leaking generally occurs, so it is a question of having sufficient vacuum pump capacity to maintain the specified vacuum level. When the vacuum is satisfactory, the composite part is ready for processing. The specific processing steps depend on the particular composite material being used, and the operation of the autoclave depends on the specific make and model. General discussions of processing and autoclave features are presented in the sections to follow.

Obviously, there is a significant amount of skilled labor necessary for the hand lay-up of composite parts. Each step has a specific purpose and function. This type of fabrication is the most time-consuming, but it is also the most flexible and when combined with autoclave processing, it results in high-quality parts.

Automated equipment can be used to cut and place the preimpregnated material onto the tool surface. The economics of manufacturing dictate that a relatively large volume of parts must be made to make automated equipment cost-effective. Some of these automated methods will be discussed later.

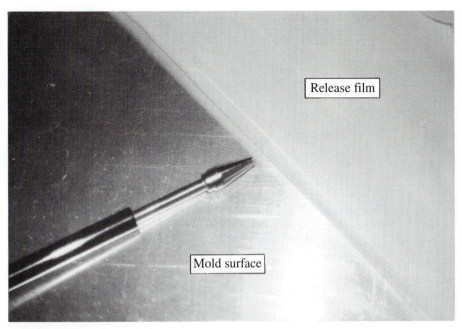

FIGURE A.2
Step 1 in the hand lay-up method: The mold is covered with a release film.

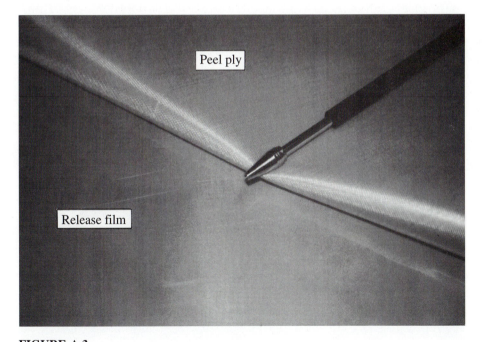

FIGURE A.3
Step 3 in the hand lay-up method: A peel ply is laid on top of release film. No sacrificial layer is used in the example lay-up.

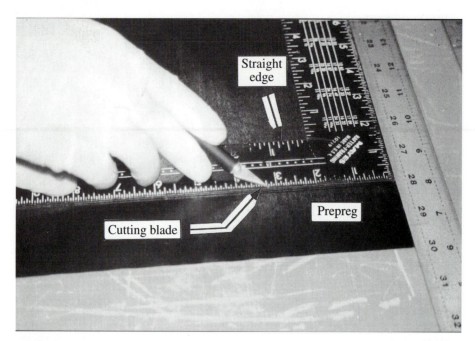

FIGURE A.4
Step 4 in the hand lay-up method: The prepreg plies are cut to design specifications.

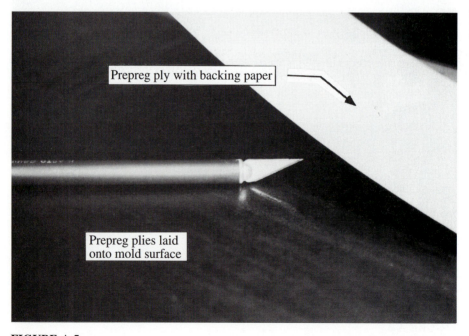

FIGURE A.5
Step 5 in the hand lay-up method: The prepreg plies are oriented and laid on the tool surface.

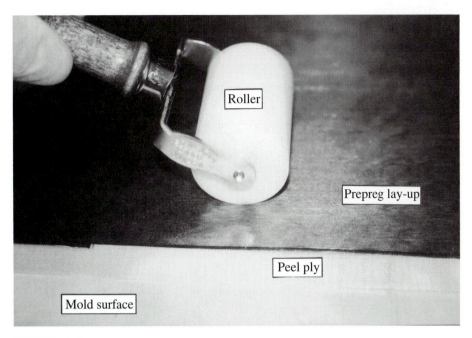

FIGURE A.6
Step 5 (continued) in the hand lay-up method: The prepreg plies are rolled out to remove wrinkles and air bubbles trapped during lay-up.

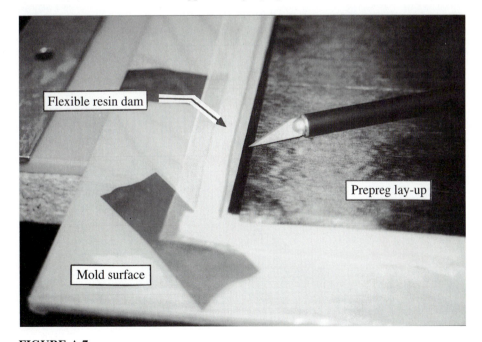

FIGURE A.7
Step 6 in the hand lay-up method: A flexible resin dam is placed around the edge of the
laminate. The dam prevents resin flow in the plane of the laminate.

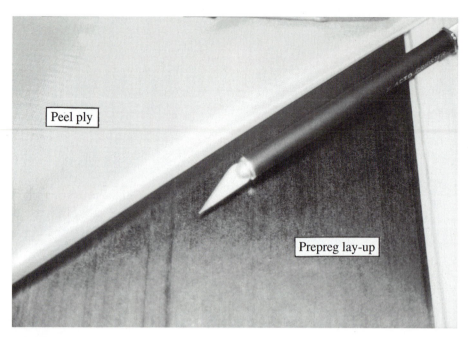

FIGURE A.8
Step 7 in the hand lay-up method: Another peel ply is placed on top of the laminate to protect the laminate surface.

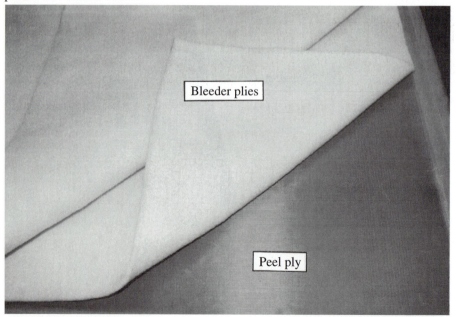

FIGURE A.9
Step 9 in the hand lay-up method: Bleeder plies are cut and placed on top of the lay-up to absorb excess resin. Note: No porous release film was used in the example lay-up. The peel ply serves as a release film in this case.

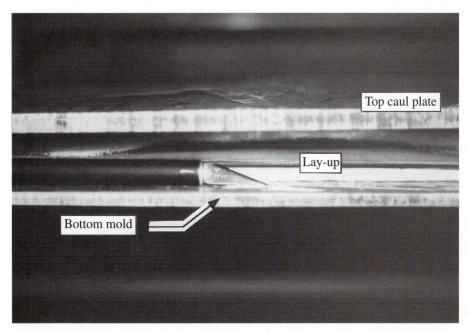

FIGURE A.10
Step 13 in the hand lay-up method: A caul plate is placed on top of the lay-up. Bleeder and breather plies can be seen directly underneath the caul plate.

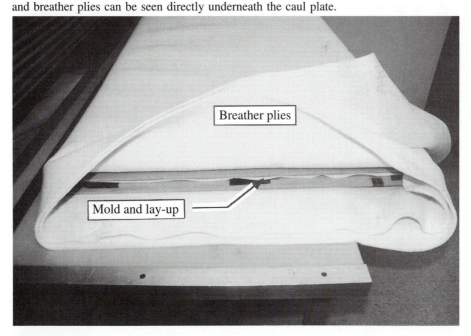

FIGURE A.11
Step 14 in the hand lay-up method: Additional breather plies are wrapped around the entire lay-up to protect the vacuum bag from puncture and to provide a vacuum pathway into the laminate.

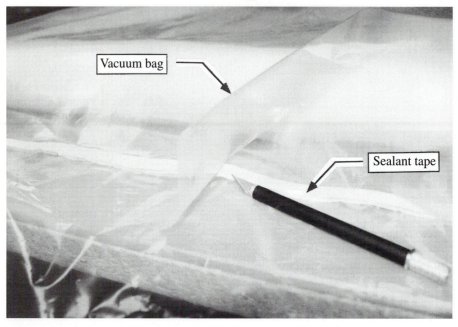

FIGURE A.12
Steps 15 and 16 in the hand lay-up method: Sealant tape is placed around the periphery of the lay-up, and a vacuum bag is cut to size to cover the lay-up.

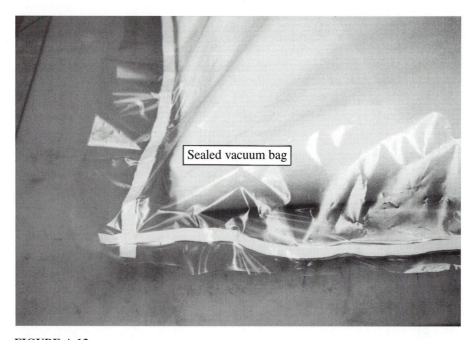

FIGURE A.13
Step 17 in the hand lay-up method: The bag is sealed by pressing the bag over the sealant tape.

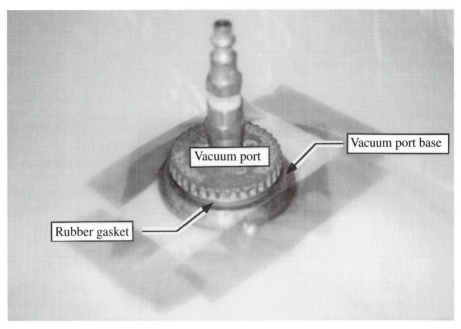

FIGURE A.14
Step 18 in the hand lay-up method: A vacuum port is installed through the vacuum bag.

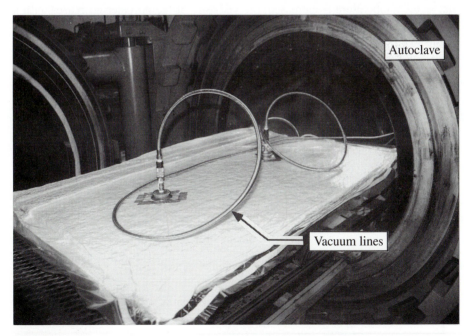

FIGURE A.15
Finished lay-up after vacuum has been applied ready for autoclave processing.

14.3
PROCESSING

14.3.1 Overall Considerations

Once the matrix and fibers are combined, and they have the desired structural shape, it is necessary to apply the proper temperature and pressure for specific periods of time to produce the fiber-reinforced structure. A judicious choice of temperature, pressure, and time produces composites that are fully cured, compacted, and of high quality. Slight deviations from the recommended processing conditions can result in unacceptable quality. The temperature cycle is usually referred to as the cure cycle, as it is the heating of the resin that initiates the cure reaction. The overall cycle, which includes pressurization and the temperature cycle, is referred to as the process cycle.

The typical cure cycle for thermosetting polymer matrix composites is a two-step cycle shown in Figure A.16. In such cycles the temperature of the material is increased from room temperature to some elevated temperature, and this temperature is held constant for the first dwell period. Afterwards, the temperature is increased again to a second temperature and held constant for the second dwell period. After the second dwell, the part is cooled to room temperature at a constant rate. Because there are two dwell periods, this type of cure cycle is referred to as a two-step cure cycle. The purpose of the first dwell is to allow gases (entrapped air, water vapor, or volatiles) to escape from the matrix material and to allow the matrix to flow, facilitating compaction of the part. Thus, the viscosity must be low during the first dwell. Typical viscosity versus temperature profiles of polymer matrices show that as the temperature is increased, the viscosity of the polymer decreases until a minimum viscosity is reached. As the temperature is increased further, the polymer begins to cure rapidly and the viscosity increases dramatically. Thus, the first dwell temperature must be chosen judiciously to allow the viscosity of the resin to be low, while keeping the cure to a minimum. Isothermal viscosity versus time profiles of the resins involved are useful in determining pot life, namely, the maximum length of time at a specific temperature for the resin to remain fluidlike. The first dwell time must be less than the pot life of the polymer at the dwell temperature.

The purpose of the second dwell is to allow cross-linking of the resin to take place. Here the strength and related mechanical properties of the composite are

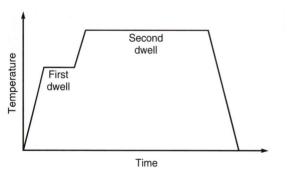

FIGURE A.16
Typical two-step cure cycle

developed. What is important to realize is that the cross-linking, or curing, process gives off heat (i.e., it is exothermic). Thus, temperatures can increase during cure even with no heat being added. However, since curing is accelerated by supplying heat, care must be taken not to overheat the composite by a combination of the exothermic nature of cure and the heat added to speed up the process. To characterize the exothermic cross-linking reaction of a thermosetting polymer matrix, a thermal cure monitor technique such as isothermal differential scanning calorimetry (DSC) is commonly used. Figure A.17 shows a typical isothermal DSC trace for a thermosetting polymer. The resin releases energy as the exothermic cross-linking reaction proceeds. Eventually, the DSC trace approaches a flat line as the cross-linking reaction nears completion. If the applied temperature, T, is increased, the reaction rate increases and the time to complete the reaction decreases. If the applied temperature is decreased, the reaction rate decreases and the time to complete the reaction increases.

Several competing priorities take place in the choice of the second dwell temperature. First, a low temperature is desirable to ease manufacturing and to reduce thermally induced stresses at the micromechanics level that are a result of the mismatch in the coefficients of thermal expansion between the fiber and matrix, and at the layer level that are a result of the mismatch in coefficients of expansion between layers with different fiber orientations. Second, the processing time should be as short as possible for economic considerations. Because low temperatures require longer dwell times, these two concerns must be compromised. Third, however, the temperatures due to the exothermic nature of curing must be kept in check. Often a vacuum is applied to the part during processing, typically during the first dwell to help facilitate removal of entrapped gases. Vacuum is discontinued after the viscosity of the resin increases significantly, and pressure is then applied to consolidate the laminate and to ensure fiber-matrix interaction. Pressure is removed either after significant cross-linking, or after completion of the process cycle. Thus, we see that

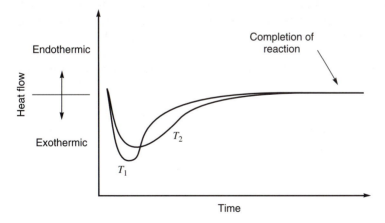

FIGURE A.17
Typical isothermal DSC relations for a polymer resin: Applied temperature T_1 > applied temperature T_2

the second dwell temperature is one of the most critical parameters in the process cycle. Its choice is largely material dependent. A certain minimum temperature must be reached before the cross-linking reaction begins.

Demands for increased performance have led to the development of several high-temperature resins (e.g., polyimides and bismaleimides). These high-temperature resins retain good mechanical properties at elevated temperatures. Processing these resins requires higher temperatures than conventional epoxy-matrix composites, and the higher temperatures lead to higher residual stresses. In some cases, processing-induced residual stresses can be high enough to cause cracking within the matrix even before a load is applied. This microcracking of the matrix can expose the fibers to degradation by chemical attack, and strength is adversely affected as stresses are initially present.

Chemically, the reinforcing fibers are affected very little during the process cycle. The polymer matrix, on the other hand, will contract by as much as 6 percent due to chemical skrinkage during cross-linking. Stress relaxation during the second dwell period can reduce chemical shrinkage effects, and this relaxation behavior increases with higher temperatures. However, these higher temperatures will increase thermally induced residual stresses. Increased pressure during cooldown reduces thermal contraction, and thus thermally induced residual stresses are reduced. However, too much pressure could lead to damage of the fibers or matrix cracking. Thus, we can see that the optimization of processing parameters is a complicated and inter-related problem. Once a process cycle is chosen, the resulting mechanical properties of the composite must be evaluated. Significant degradation in strength and stiffness or other mechanical properties would not generally be acceptable.

For processing thermoplastic matrix composites, as no cross-linking process occurs, there is no need to maintain elevated temperatures for extended periods. However, the processing temperatures are generally much higher than for thermosets, and therefore thermally induced residual stresses are an issue.

The area of process modeling is an attempt to quantify the effects of processing on physical parameters such as degree of cure, temperature, fiber volume fraction, and residual stresses. These models can be used to search for optimal processing conditions for specific material systems and structural shapes. Process modeling is an important component of the analysis of composite materials.

What really distinguishes the various processing methods that are available is how pressure, vacuum, and temperature are applied. Autoclave curing is discussed in the next section; other methods are discussed in a later section.

14.3.2 Autoclave Curing

The best quality parts are cured using an autoclave. Autoclaves have been used extensively for processing high-performance composite materials in the civilian and military aerospace industries. An autoclave consists of a large cylindrical metal pressure vessel with end enclosures that is thermally insulated and heated. Most autoclaves have a forced-hot-gas circulation system as well. An autoclave is pressurized using air or an inert gas such as nitrogen. What distinguishes the autoclave from the curing oven and hot press, to be discussed later, is the ability to cure parts using large, hydrostaticlike pressure. A typical autoclave can pressurize up to 20 atm. The large majority of composite structures can be processed using autoclaves with 2–4 m internal diameter, although some extremely large aerospace structures require autoclaves over 20 m in diameter. The capital equipment costs and operating costs for large autoclaves make this type of processing very costly. However, the high quality and high performance of autoclaved parts makes them attractive for certain applications. A typical autoclave is shown in schematic form in Figure A.18. The primary component is the pressure vessel itself, which is cylindrical and contains embedded heaters and cooling coils. A door at one end allows access to the interior to load parts and perform periodic maintenance. Also, several ports may be installed through the autoclave wall for access to the interior. Some of these ports are dedicated to vacuum lines connected to the parts to be cured. Others are used for control functions such as thermocouples, dielectric sensors, and pressure sensors, all of which help monitor the curing of the material. The interior of the autoclave is heated by radiation from the vessel walls and convection of hot gases as they circulate through the vessel. A circulating fan forces the hot gases through a series of baffles within the autoclave in a circulation loop that runs the length of the autoclave. Typically, this fan is housed at one end of the autoclave and the interior gases are drawn from the central portion of the cylinder, through the baffles, and they return to the other end through a jacket that covers the interior wall. The autoclave applies a pressure to the outer surface of the composite part through pressurization of the interior gases. This pressure is then transferred through the tool plate(s), breather plies, bleeder plies, and other secondary materials to the laminate surface. From there the pressure is shared between the fiber and matrix during curing. The most important aspect is the matrix resin pressure during cure. If it is too low, then voids can grow in the resin or inadequate resin bleeding may occur. In general, composite structures which have been processed in an autoclave exhibit uniform thicknesses, good consolidation, and very low void content.

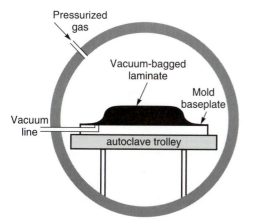

FIGURE A.18
Schematic of an autoclave

14.4
MANUFACTURING BY OTHER METHODS

Obviously, hand lay-up represents an extreme that cannot be scaled to high volume or large components; to address these issues, and because fibers are not always in the form of impregnated parallel tows in a layer, other fabrication techniques have been developed. In addition, it may be easier in some instances to bring the fibers and matrix together when forming the structural component. We begin this discussion of other fabrication methods by surveying other ways to make fibers available, either by themselves or with matrix material.

14.4.1 Fiber-Only Preforms

If the resin and fiber are to be combined during the fabrication of a composite structure, then they are both supplied as separate materials. Fibers when supplied as a separate material can come in many different forms. Most commonly they are continuous and grouped into bundles, as with tows. The bundles, or strands, are a collection of many hundreds or thousands of fibers twisted or bound together. These strands are wound onto a spool. The fibers usually have a binder that keeps them together and other coating agents to provide better handleability. A yarn is a twisted assemblage of fibers, usually less than 10,000. By contrast, a tow, for the most part, is untwisted. The simplest yarn is made from a single strand of fibers, and heavier yarns are obtained by twisting and plying several strands together. Typically, this consists of twisting two or more individual strands together, then twisting two or more twisted strands together. Yarns which are simply twisted will kink, corkscrew, and unravel because the twist is only in one direction. This problem is normally eliminated by countering the twist in the twisted yarns with the opposite twist when plying together the twisted yarns.

A woven fabric is a material with interlaced yarns, strands, or fibers. Typical fabrics are manufactured by interlacing warp (lengthwise) yarns or strands with fill (crosswise) yarns or strands on a conventional weaving loom. The weave of a fabric determines how the warp and fill yarns are interlaced. Popular weave patterns include plain, twill, crowfoot satin, long-shaft satin, leno, and unidirectional. The plain weave, shown in Figure A.19, is the oldest and most common textile weave. Each fill yarn is repetitively woven over one warp yarn and under the next. It is the most stable of the weave constructions and mechanically it behaves much like a cross-ply laminate. Twill weaves have one or more warp yarns passing over and under two, three, or more fill yarns in a regular pattern. Theses weaves drape better than the plain weave. In the crowfoot or long-shaft satin weaves, one warp yarn is woven over several successive fill yarns, then under one fill yarn. A weave pattern in which one warp end passes over four and under one fill yarn is called a five-harness satin weave and is shown in Figure A.20. Satin weaves are less open than other weaves and the strength is high in both the warp and fill directions. The unidirectional weave has a large number of yarns in the warp direction with fewer and generally smaller

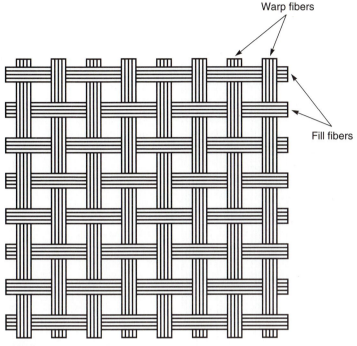

FIGURE A.19
Plain weave fiber preform

yarns in the fill direction, and the resulting product has much greater strength in warp direction. Several types of two-dimensional weaves are shown in Figure A.21. More recently, three-dimensional weaves, shown in Figure A.22, have been developed to provide better through-thickness strength in a composite structure. Three-dimensional weaves behave orthotropically because there are three principal material directions.

A nonwoven fabric is a sheet of parallel yarns or tows held together by an occasional transverse yarn or tow, or by a periodic crossbond with a binder.

Mats are blankets of chopped fibers or continuous fibers formed as a continuous flat sheet. The fibers are evenly and randomly distributed and are held together by a binder. The binder used must be able to dissolve in the liquid matrix once it infiltrates the mat. Low-solubility binders are used when the fabrication procedure is such that the matrix may wash out the fibers and create resin-rich regions during infiltration. The binder, due to its low solubility, remains intact until the infiltration process is complete, after which it dissolves in the matrix. A mat acts like a single layer of material with nearly the same properties in all inplane directions (i.e., isotropic behavior).

Continuous fibers can be chopped into short lengths, usually between 3 and 50 mm in length. These chopped strands can then be mixed with a liquid resin and

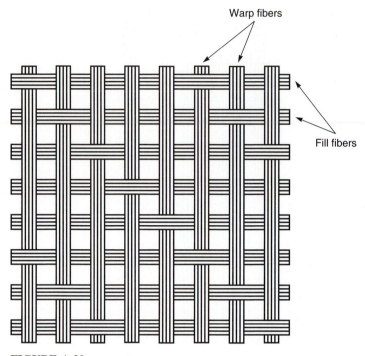

FIGURE A.20
Five-harness satin weave fiber preform

injected into a mold, sprayed onto a mold surface, or sprinkled onto a polymer sheet; these sheets then act like a single isotropic layer.

14.4.2 Other Combined Fiber-Matrix Preforms

As with tows, yarns, woven forms, nonwoven forms, and mats can be preimpregnated with resin before forming them into a structural shape. Combining the fiber and resin in a separate step makes the fabrication process simpler and results in composite structures with better quality. In particular, the proportion of resin to fiber is kept within very close tolerance and fiber orientation is more controlled and reproducible.

Sheet molding compound (SMC) is a type of preform used in the automotive industry. It consists of chopped glass fibers randomly distributed in a polyester sheet; these sheets are stacked one on top of the other and molded to shape with heat and pressure.

Some thermoplastic polymers, for example polyphenylene sulfide (PPS), can be drawn into fiber form. Once the polymer fibers are formed, they can be comingled with reinforcing fibers. The resulting strands of reinforcing fibers and polymer are

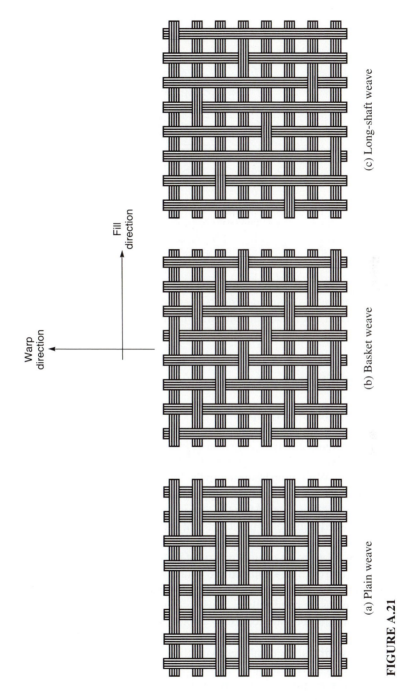

Warp direction

Fill direction

(a) Plain weave

(b) Basket weave

(c) Long-shaft weave

FIGURE A.21
Several two-dimensional weaves for fiber preforms

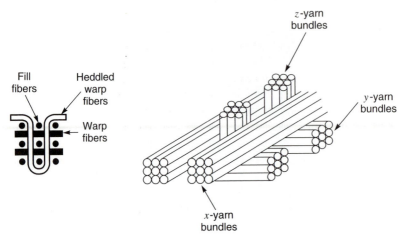

FIGURE A.22
Examples of three-dimensional weave patterns for fiber preforms

then wound onto a mandrel or pulled through a heated mold. During fabrication and processing the polymer fibers melt and infiltrate the reinforcement.

14.5
FORMING STRUCTURAL SHAPES

14.5.1 Wet Lay-up and Spray-up

Two other manual techniques have a long history in the composites industry, wet lay-up and spray lay-up. Both were developed for the fiberglass industry and they are still used very extensively. Wet lay-up is similar to the lay-up of preimpregnated material discussed previously. The only difference is that the reinforcement and matrix are now combined during fabrication. After the tool is properly prepared and a gel coat (a layer of unreinforced pigmented polyester that serves as the outer surface of the laminate when completed) is applied on the tool or mold, the reinforcement is put down in the form of chopped strand mat or woven materials. Once the reinforcement is in place, the resin and catalyst are mixed and poured onto the tool surface. To ensure that the resin infiltrates the reinforcement and that no air bubbles remain, it is rolled or brushed into place using hand rollers, brushes, or paddles. The resin mixture can also be sprayed onto the tool surface using a spray gun that automatically meters the appropriate mixture. This type of fabrication is called spray laminating. In some cases, chopped glass fibers are combined with the resin in the metering head and deposited onto the tool surface by spraying.

Wet lay-up and spray-up fabrication produces a glass-reinforced structure with 30–50 percent glass content. Higher glass concentrations are not easily fabricated using these techniques; however, the use of prepreg can give higher fiber volume fractions (50–75 percent).

14.5.2 Filament Winding

Filament winding is a fabrication technique developed in the late 1940s and early 1950s in which continuous tows of fibers are wound onto a rotating tool, or mandrel. The construction of the mandrel is a key step in the filament-winding process. The choice of material is critical, and of course, the mandrel must be removed from the inside of the finished composite structural form after processing. When complex enclosed shapes are manufactured by filament winding, it is sometimes difficult to remove the mandrel. Mandrels that can be disassembled in sections from inside the shape must be designed, or plaster or sand polyvinyl acetate (PVA) mandrels that can be dissolved with a solvent after processing can be used.

The filament-winding machine dispenses, or "pays out," the fiber tows while traversing along the mandrel axis of rotation. Some types of filament-winding machines and control systems allow very complex winding patterns to be generated. Filament winding is an automated fabrication method. Once the mandrel has been installed and the fiber or tow material loaded, an operator can start the machine and the fabrication proceeds automatically. The two types of filament winding are wet winding and dry winding. Wet winding refers to the use of a wet resin during winding. Fibers are passed through a resin bath before being wound onto the mandrel; the reinforcement and matrix are combined during fabrication. In dry winding, preimpregnated fibers, in tape or tow form, are used instead, and the preimpregnated material is wound directly onto the mandrel. Sometimes preheaters are used to soften the preimpregnated material before it is placed onto the mandrel surface. If thermoplastic composites are being wound, they are consolidated directly on the mandrel surface by using highly localized heat sources that soften the preimpregnated tape or tow right at the point of contact on the mandrel surface.

Surfaces of revolution are appropriate candidates for filament winding. These types of structures include piping, pressure vessels, tubing, rotor blades, or any spherical, conical, or geodesic shape. Other shapes, such as flat panels, can be made by winding onto a rectangular mandrel and then cutting the structure along the winding axis after fabrication.

Winding methods and patterns

The two basic types of winding patterns are polar and helical. The polar (or planar) winding pattern results when the mandrel does not rotate, but the head that dispenses the tows, commonly called the payout head, rotates about the longitudinal axis. To produce a given orientation of the tow along the axis, the payout head is inclined at the appropriate winding angle, and the mandrel moves longitudinally as the payout head rotates. This pattern is described as a single-circuit polar wrap and is shown in Figure A.23. As more tows are wound, each tow is placed adjacent to the next. A completed layer consists of many tows integrally woven at plus and minus the orientation angle covering the entire mandrel surface.

A helical pattern results when the mandrel rotates continuously while the fiber payout head traverses back and forth over the mandrel, as in Figure A.24. The carriage speed and mandrel rotation are controlled to generate the desired winding angle. A hoop winding can be made by advancing the payout head slowly along the

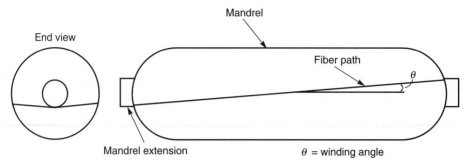

FIGURE A.23
Polar filament winding pattern

mandrel axis so that the fiber tows are wound transversely to the axis of rotation. As the longitudinal speed of the payout head is increased, the angle that the tows make with the axis of the mandrel rotation decreases. In general, a helical winding pattern does not deposit the next tow adjacent to the previous one. In fact, several circuits may be required before the full mandrel surface is covered. Because of tension in the tow as it is wound on the mandrel, the tows flatten out and the fibers tend to spread into bands. Individual winding patterns can be calculated from the mandrel geometry and the desired lay-up.

Winding machines

Filament winders are designed for either polar or helical winding. Polar winders are usually operated with the mandrel in the vertical position to eliminate deflections due to weight. A major advantage of the polar type is that machine control is much simpler. The rotation of the payout head is continuous and at a uniform speed. This eliminates inertial effects that occur in a helical winder when the traverse speed is changed abruptly or when the traverse direction is changed. However, polar winders are generally limited to dispensing only preimpregnated material.

Helical winders require at least two degrees of freedom: mandrel rotation and traverse of the payout head along the longitudinal axis. More complicated machines include motion of the payout head perpendicular to the mandrel axis, and rotation of the feed eye; both of these motions permit more accurate placement of the fiber tows around the ends. Figure A.25 shows a typical helical filament winder used in composites manufacturing.

Another component of the filament winder is the fiber tensioner. Fiber tensioning plays an important role in filament winding, as it helps to control the placement of the tow, the resin content of the structure, the width the tow flattens to, and the layer thickness. Improper tensioning of a filament-wound structure can result in unacceptable quality. Tensioning is accomplished through the use of guide eyes, brakes of the drum type, scissor bars, and, for wet winding, the viscous drag through the resin bath. Generally, the fiber tow unwinds from a spool of tow and passes through a series of rollers that are designed to impart a modest amount of tensioning, say, 5 to 50 Newtons/tow. Normally, the tensioning on dry tows is kept to a minimum to reduce abrasion. As the tows pass through the resin bath, the tension level is increased before the tow is placed onto the mandrel.

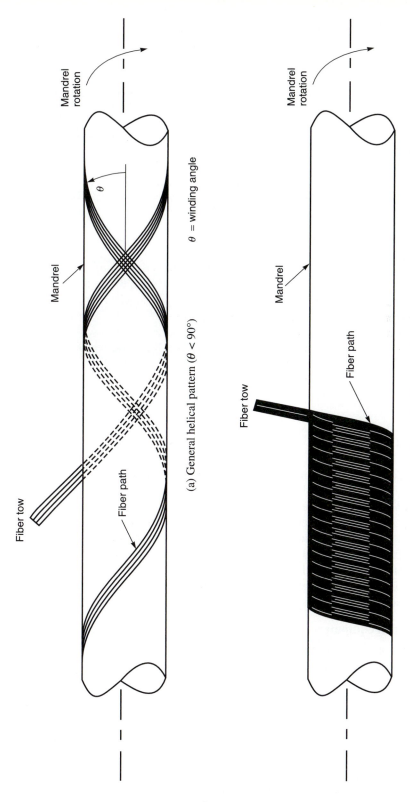

Fiber tow

Mandrel

θ

Mandrel
rotation

θ = winding angle

(a) General helical pattern (θ < 90°)

Fiber path

Fiber tow

Fiber path

Mandrel

Mandrel
rotation

(b) Hoop winding pattern (θ ≃ 90°)

FIGURE A.24
Helical filament winding patterns

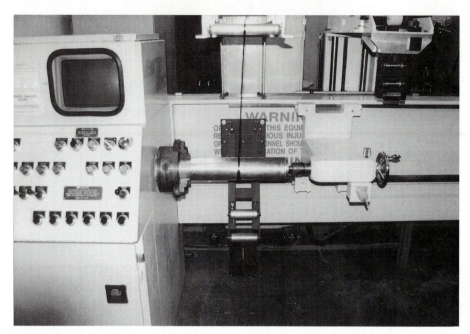

FIGURE A.25
En-Tec model 5K240-060-4 four-axis helical filament winder.

14.5.3 Pultrusion

During the early 1950s a continuous-processing technique called pultrusion was developed. The process involves the pulling of reinforcing fibers and resin matrix through a die or series of dies that shape and cure the material. Structures fabricated using pultrusion must possess a constant cross-sectional shape. Examples of pultruded structures include tubing, box beams or I-beams, C-channels, rectangular strips, or other cross-sectional shapes that are invariant with length. As with filament winding, both wet and dry pultrusion processes exist. In wet pultrusion, dry tows are pulled through a resin bath before being drawn into a die, whereas in dry pultrusion preimpregnated material is used as the raw material, often called feedstock.

Pultrusion equipment

The basic parts of the pultrusion process, shown schematically in Figure A.26, are the resin impregnator (for wet pultrusion), preheaters, shaping die, curing and forming die, pullers, and cut-off saw. Resin impregnation is usually by a resin bath through which the fiber tows are pulled, although prepreg tape may also be used. Guide rollers and tensioning rollers may be used to flatten and spread the tows and hence allow more complete infiltration of the resin.

During the forming process in pultrusion, it is very difficult to make dramatic changes in the feedstock shape. Gradual changes in shape, for example, from a flat plate to a V-shaped section, are much more efficient. Thus, there are often a series

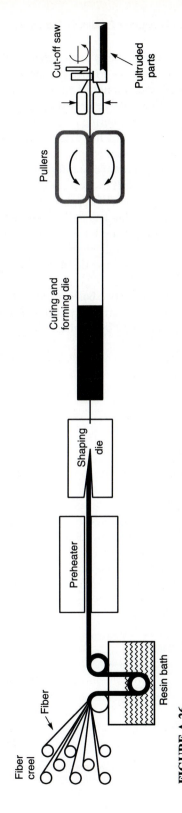

FIGURE A.26
Schematic of a pultrusion processing line

607

of shaping dies before the final curing die is approached. Interspersed throughout the forming line may be several preheaters. Preheating the feedstock material before it reaches the curing die reduces the amount of time (and line length) devoted to curing. Alternatively, the shaping dies may be heated to a temperature that facilitates forming, but does not promote excessive curing.

The curing and forming die is one of the most costly parts of the pultrusion process. It is usually a tool steel mold that is split along the longitudinal axis, and the interior is polished to extreme smoothness and then plated with chrome. The surface smoothness of the die is critical. If the resin sticks to a surface scratch while being pulled through the die, this feature will imprint on the material emerging from the die. The curing die may be quite long (up to 2 m) and it is usually heated to a temperature necessary to initiate the curing reaction of the resin. Sometimes oven heaters are used after the material emerges from the die to finish the curing reaction. Before the material emerges from the curing die it must be fully gelled so that it retains its structural shape.

As Figure A.27 shows, pullers are of two basic designs. The first consists of two rubber-cleated tread or chain drives that grip the pultruded shape and continuously pull it; the gripping pressure is controlled by mechanical or pneumatic control. The

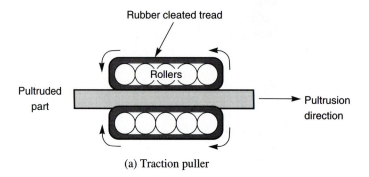

(a) Traction puller

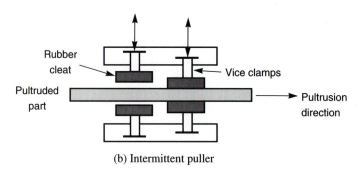

(b) Intermittent puller

FIGURE A.27
Types of pullers used in pultrusion

second design uses an intermittent vise clamp and rubber cleat mechanism that grips on the forward stroke and releases on the back stroke. Two of the intermittent clamps may be combined to give a continuous pulling action.

The cut-off saw is usually a wet or dry silicon carbide grit or diamond blade. While the pultrusion line is operating, the pultruded section is gripped and clamped to the saw table, which then travels with the pultruded structure. An appropriate length of structure is cut off and the saw returns to its original position to await for the next cut.

14.5.4 Resin Transfer Molding

Resin transfer molding (RTM) is a process in which a structural component consisting only of dry fibers, the fiber preform, is infiltrated with resin and cured in a tool. There are three basic steps: (1) making the fiber preform and loading it into the tool or mold, (2) infiltrating the resin into the fiber preform, and (3) curing the composite structure in the mold. One requirement of the fiber preform is that it retain its shape during resin infiltration. Thus, most preforms are stitched or bonded together with a binder. Dry fabrics can be laid up in a desired orientation and then stitched together. Some current development efforts are focused on braiding fiber preforms for RTM operations; the braided preform retains its shape through the interlocking of the reinforcing strands. Most fiber preforms are limited to 50–60 percent fiber volume fraction before infiltration, as higher fiber volume fractions cannot be infiltrated properly. The resin infiltration step is usually accomplished by pumping the resin into the mold under pressure. The resin viscosity must be low enough to allow full infiltration, and it must not cure too quickly so that sufficient fiber wet-out occurs. Both polyesters and epoxies have been used successfully in RTM operations. Curing is accomplished by heating the mold either before or after infiltration. If the mold is heated before infiltration, the resin must be slow curing so that it does not cross-link before the entire preform is infiltrated. Usually a vacuum is applied to the mold during infiltration and curing so that voids are expelled or reduced.

RTM equipment

Figure A.28 shows a typical set-up for an RTM process, with the mold as the most important component. It must be constructed so that resin is completely and uniformly distributed throughout the mold. The infiltration step must occur before cross-linking of the resin occurs and the fiber preform must remain stationary throughout the process. All of this should occur with a minimum resin pumping pressure. The mold should also be vented so that air is expelled from the mold as it fills with resin. If the mold is heated, then it should have uniform temperature control, though some molds are designed to be heated in an oven after resin infiltration. Resin pumping equipment is generally of simple design, as pumping pressures are low. Pressure-driven pistons that move through a cylindrical resin cavity are used to force resin through feedlines into the mold. In other cases, gas-pressurized diaphragms are used.

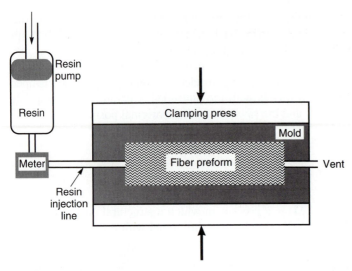

FIGURE A.28
Schematic of an RTM process

Sufficient clamping pressure must be available to keep the two halves of the mold together while the resin is pumped into it.

14.6
NONAUTOCLAVE CURING

While the use of an autoclave is the most desirable way to process a composite laminate, other methods of curing can, with care, lead to similar high-quality components. Two of these are described below.

14.6.1 Oven Curing

Ovens are the least-expensive processing equipment. However, unlike an autoclave, they have no capability to apply pressure during cure. Oven-cured parts are either vacuum bagged so that atmospheric pressure consolidates the part, or they are sealed in a mold with an expansive insert so that thermal expansion forces the composite component against the mold walls. A typical curing oven is a large, thermally insulated, forced-air-circulating metal oven with large access doors at one end. Curing ovens hundreds of cubic meters in volume are common. The most important aspect for curing ovens is temperature uniformity. If the volume of circulating air within the oven is small, then the response time to temperature fluctuations will be long. If there is insufficient baffling of the circulating air, then hot spots or zones will develop in the oven. Often oven-cured parts cured with vacuum-bagging techniques suffer from lack of uniformity in thickness, excessive voids, and lack of consolidation. However, the equipment is relatively inexpensive and for many low-performance applications the quality is sufficient.

14.6.2 Hot Pressing

Good quality parts can be obtained by hot pressing. A hot press is a mechanical or hydraulic press with heated platens. Processing using the hot press requires that the composite structure be enclosed between two matched-metal molds. The molds are placed between the platens and the platens are forced together to apply pressure to the mold. The platens are heated so that the mold temperature increases to a sufficient level to promote curing of the matrix. If the composite part is simply a flat plate, then the "mold" is simply two flat plates, one on the top of the laminate and one on the bottom.

Pressurization is by compression of the mold through the thickness; the pressurization forces the laminate to conform to the contour of the mold surface. The mold must be sealed to prevent uncontrolled resin loss and to promote hydrostatic pressurization of the resin. The pressurization of the resin is important to reduce voids created during cure or entrapped during the fabrication procedure. To mimic the autoclave, the composite part can be enclosed in a vacuum bag as part of the curing process. Alternatively, a special diaphragm mold can be used to apply a vacuum to a hot-pressed laminate during curing. Application of the vacuum helps to eliminate voids during cure.

The platens in the hot press must be uniformly heated. Typically, this is accomplished by having multiple heating sources embedded in the platens. More elaborate platens have multiple cooling channels bored throughout the platens so that the capability exists to both heat and cool specific regions of the platen. The transfer of thermal energy from the platens to the mold is largely by conduction, although some radiation and convection occurs. Platen and mold materials should be chosen to minimize the response time for thermal conduction. In general, platens are made from tool steel, and molds are either steel or aluminum. Molds are often insulated on the edges to increase temperature uniformity.

Hot-pressed composite structures show good consolidation and moderate void content. Some variation in thickness occurs due to lack of uniformity in resin flow and small misalignment of the matched-metal surfaces.

14.7
MANUFACTURING DEFECTS

No matter how carefully the manufacturing process is carried out, all composite structures exhibit processing defects. When a material is designated as high-quality, it has relatively few defects. Some processing defects have the potential to be quite detrimental to the mechanical performance of a composite structure. In other cases, they are more of a nuisance than a significant problem.

There are many different types of manufacturing defects: voids, delaminations, residual stress-induced cracking, resin starvation, resin-rich pockets, damaged fibers, fiber-matrix debonding, thermal decomposition, and undercured and overcured regions. The processing parameters have a profound impact on when and where these defects occur, and how they can be prevented.

Voids are extremely detrimental to the mechanical performance of composite materials. Their effect on structural performance has been extensively studied in the past. For a number of reasons voids can be created during processing. Air pockets can be trapped during the lay-up procedure, absorbed water in the resin can vaporize during the cure cycle, and gaseous by-products of the cure reaction may be released during curing. In resin transfer molding, voids can occur if the resin flow front does not uniformly infiltrate the reinforcement. The current methodology to control voids in processing is to apply sufficient pressure to prevent their growth during the cure cycle.

Delaminations or separated layers can arise during processing as a result of transport of gas bubbles, emitted from the resin as it is curing, to the interface between layers. If a sufficient number of voids collect at the interface, then a delamination is produced. In addition, dirt, grease, or other contaminants on the surface of the prepreg layers during lay-up may prevent layer bonding and consolidation. Most aerospace companies nondestructively evaluate each composite laminate for the presence of delaminations before releasing the part for service.

Residual stresses arise in composite materials due to the mismatch in thermal expansion between the constituents. As the material is cooled down from processing temperatures, this mismatch in thermal expansion leads to the build-up of residual stresses. If the mismatch is too great or if the processing temperature is too high, then residual stresses can lead to matrix cracking during cooldown.

If the consolidation has not been properly carried out, then either resin starvation or resin-rich pockets can result. Resin starvation is seen when the applied pressure is too high, causing too much resin to be squeezed out of the material. If too little pressure is applied or if resin flow from the material is not uniform, then resin-rich regions may be created. These regions are weak areas for the composite and they may ultimately be the initiation site for final failure.

Similarly, nonuniform curing can lead to weak regions of the material where the local degree of cure is reduced. This type of defect occurs when the temperature distribution during cure is nonuniform. In the extreme case, excessive temperatures may occur when the thickness of the article is large. The energy released by the resin during curing is trapped within the material and the local temperature can rise to very high levels, ultimately leading to thermal decomposition of the matrix.

Even when all reasonable precautions are taken, some types of defects are to be expected in composites processing. No process is perfect and a typical component made from a composite material will show some voids, resin-rich regions, delaminations, and residual stress-induced cracking. Figure A.29 shows such a part with several defect regions labeled. Understanding the effects of these kinds of defects on the mechanical performance of composite structures is the challenge that scientists and engineers must continually address.

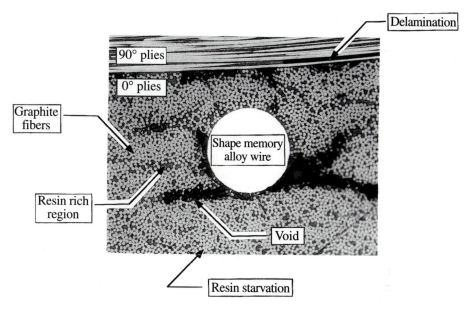

FIGURE A.29
Examples of manufacturing defects in a polymer matrix composite. The material is a graphite-epoxy composite laminate with an embedded shape memory alloy wire.

14.8
SUGGESTED READINGS

1. Lubin, G., ed. *Handbook of Composites*. New York: Van Nostrand Reinhold, 1982.
2. Schwartz, M. M., ed. *Composite Materials Handbook*. 2nd ed. New York: McGraw-Hill, 1992.
3. Strong, A. B. *Fundamentals of Composites Manufacturing: Materials, Methods, and Applications*. Dearborn, MI: Society of Manufacturing Engineers, 1989.

Companies that supply various forms of composite materials, and the specialty materials needed for fabrication, usually advertise in trade publications. The following is one such publication:

4. *SAMPE Journal*. Published by the Society for the Advancement of Materials and Process Engineering, 1161 Parkview Drive, Covina, CA 91724.

Index

F

M

T